Fungal infection of plants

Fungal infection of plants

SYMPOSIUM OF
THE BRITISH MYCOLOGICAL SOCIETY

EDITED BY
G. F. PEGG & PETER G. AYRES

The right of the
University of Cambridge
to print and sell
all manner of books
was granted by
Henry VIII in 1534.
The University has printed
and published continuously
since 1584.

CAMBRIDGE UNIVERSITY PRESS

CAMBRIDGE
NEW YORK   PORT CHESTER
MELBOURNE   SYDNEY

CAMBRIDGE UNIVERSITY PRESS
Cambridge, New York, Melbourne, Madrid, Cape Town,
Singapore, São Paulo, Delhi, Tokyo, Mexico City

Cambridge University Press
The Edinburgh Building, Cambridge CB2 8RU, UK

Published in the United States of America by Cambridge University Press, New York

www.cambridge.org
Information on this title: www.cambridge.org/9780521106283

First published 1987
Reprinted 1990
First paperback edition 2011

A catalogue record for this publication is available from the British Library

Library of Congress Cataloguing in Publication data

British Mycological Society. Symposium.
  Fungal infection of plants.
(British Mycological Society symposium series; 13)
  Revised papers from symposia held in 1982, 1983, and 1984.
  Includes index.
  1. Fungus diseases of plants-Congresses. I. Pegg, G.F. (George Frederick), 1930-. II.
Ayres, P.G. (Peter G.) III. Title. IV. Series. SB733.B75 1987   581.2'326   87-13780

ISBN 978-0-521-32457-1 Hardback
ISBN 978-0-521-10628-3 Paperback

Contents

Contributors

S. Al-Nahidh, *Department of Plant Sciences, University of Leeds, Leeds, LS2 9JT, UK*

Bat-Sheba Ben-Zvi, *Department of Microbiology, Faculty of Life Sciences, Tel-Aviv University, Ramat Aviv, Israel*

C.M. Brasier, *Forest Research Station, Alice Holt Lodge, Farnham, Surrey GU10 4LH, UK*

R.J.W. Byrde, *Long Ashton Research Station, Department of Agricultural Sciences, University of Bristol, Long Ashton, Bristol BS18 9AF, UK*

Y. Cohen, *Department of Life Sciences, Bar Ilan University, Ramat Gan 52100, Israel*

R. M. Cooper, *School of Biological Sciences, University of Bath, Claverton Down, Bath BA2 7AY, UK*

P.G.M. De Wit, *Department of Plant Pathology, Agricultural University, Binnenhaven 9, 6709 Wageningen, The Netherlands*

J.W. Deacon, *Microbiology Department, School of Agriculture, West Mains Road, Edinburgh EH9 3JG, UK*

Ralph A. Dean, *Department of Plant Pathology, University of Kentucky, Ky 40546, USA*

J.A. Duddridge, *Oxford Forestry Institute, Department of Plant Sciences, South Parks Road, Oxford OX1 3RB, UK*

Jonathan P. Duvick, *Department of Biotechnology Research, Pioneer Hi-Bred International Inc., Johnston, Iowa 50131-0085, USA*

Albert H. Ellingboe, *Department of Plant Pathology and Genetics, University of Wisconsin, Madison, Wisconsin 53706, USA*

A.R. Entwistle, *Institute of Horticultural Research, Wellesbourne, Warwick CV35 9EF, UK*

J.F. Farrar, *School of Plant Biology, University College of North Wales, Bangor, Gwynedd LL57 2UW, UK*

Aliza Finkler, *Department of Microbiology, Faculty of Life Sciences, Tel-Aviv University, Ramat-Aviv, Israel*

L.V. Fleming, *Microbiology Department, School of Agriculture, West Mains Road, Edinburgh EH9 3JG, UK*

F.M. Fox, *Microbiology Department, School of Agriculture, West Mains Road, Edinburgh EH9 3JG, UK*

J.L. Gay, *Department of Pure and Applied Biology, Imperial College of Science and Technology, University of London, Prince Consort Road, London SW7 2 BB, UK*

John A. Hargreaves, *Long Ashton Research Station, Department of Agricultural Sciences, University of Bristol, Bristol BS18 9AF, UK*

Peter Jeffries, *Biological Laboratories, University of Kent, Canterbury, Kent CT2 7NJ, UK*

J.P.R. Keon, *Long Ashton Research Station, Department of Agricultural Sciences, University of Bristol, Long Ashton, Bristol BS18 9AF, UK*

Ian Knights, *Schering Agrochemicals Limited, Chesterford Park Research Station, Saffron Walden, Essex CB10 1XL, UK*

Herman W. Knoche, *Department of Agricultural Biochemistry, University of Nebraska, Lincoln, Nebraska 68583-0718, USA*

Yigal Koltin, *Department of Microbiology, Faculty of Life Sciences, Tel-Aviv University, Ramat Aviv, Israel*

Joseph A. Kuć, Department of Plant Pathology, University of Kentucky, Ky 40546, USA

Paul J. Kuhn, *Shell Research Ltd., Sittingbourne Research Centre, Sittingbourne, Kent ME9 8AG, UK*

F.T. Last, *Institute of Terrestrial Ecology, Bush Estate, Penicuik, Midlothian EH26 0QB, UK*

D.H. Lewis, *Department of Botany, University of Sheffield, Sheffield S10 2TN, UK*

John Lucas, *Department of Botany, University of Nottingham, University Park, Nottingham NG7 2RD, UK*

P.A. Mason, *Institute of Terrestrial Ecology, Bush Estate, Penicuik, Midlothian EH26 0QB, UK*

A.G. Mitchell, *Forest Research Station, Alice Holt Lodge, Farnham, Surrey GU10 4LH, UK*

R.B. Pearce, *Oxford Forestry Institute, Department of Plant Sciences, University of Oxford, South Parks Road, Oxford OX1 3RB, UK*

J. Rotem, *Department of Plant Pathology, Agricultural Research Organization, The Volcani Center, Bet Dagan 50250, Israel*

F.E. Sanders, *Department of Plant Sciences, University of Leeds, Leeds LS2 9JT, UK*

Christopher Walker, *Forestry Commission, Northern Research Station, Roslin, Midlothian EH25 9SY, UK*

Joan F. Webber, *Forest Research Station, Alice Holt Lodge, Farnham, Surrey GU10 4LH, UK*

J. Wilson, *Institute of Terrestrial Ecology, Bush Estate, Penicuik, Midlothian EH26 0QB, UK*

A.M. Woods, *Department of Pure and Applied Biology, Imperial College of Science and Technology, University of London, Prince Consort Road, London SW7 2BB*

Preface

The infection of aerial and subterranean parts of plants by fungi is ubiquitous. It is the rule in plant biology rather than the exception. Disease by its definition is illustrated by the development of obvious symptoms and, perhaps of this reason and economic considerations, studies in plant microbiology over the last two-and-a-half centuries have been predominantly of pathological topics. By far the majority of plant infections however, the mycorrhizal associations, do not result in aerial symptoms and may be shown to confer benefits on plant species, often determining their ecological distribution. Since these two types of fungal infection were first recognised, research workers have confined their interests and activities to one or the other. This has led inevitably to the divergence of two distinct groups of plant microbiologist, each for the most part failing to keep abreast of developments in the other field. This isolation has been fostered by separate scientific meetings and specialist texts which have excluded advances on a particular topic, either mycorrhizal or pathological, but which nevertheless was common to both types of host infection.

The aim of this volume is to treat fungal infection (parasitism) as a theme common to pathogenic and symbiotic interactions. It is hoped to provide plant pathologists and mycorrhizal workers with up to date information about their own field of research and in the other area most closely related to it, in a form which stimulates comparison and encourages further experimental work. The chapters by Gay & Woods and by Farrar & Lewis develop this idea and underline the broad similarities that exist in the basic nutritional relationships between both pathogenic and mycorrhizal fungi and their hosts; together with the chapter by Jeffries, they show that a spectrum of interfacial and nutritional relationships exists between plants and infective fungi, where in each case structure is closely related to function. Chapters by de Wit, Duddridge and Sanders & Al Nahidh stress important differences that exist between the two groups of fungi where the phenomena of recognition, host specificity and response are concerned and highlight the different levels of progress that have been achieved in each field.

In the many different plant : fungus relationships referred to in the text, overall emphasis is placed on the fungal organism as it progresses from the establishment of infection through to the completion of its interaction with the host. Thus, the fungus is followed from spore germination (Chapter by Lucas & Knights), through the spread of mycelium, facilitated by the production of extracellular plant cell wall degrading enzymes (Chapter by Keon *et al.*), to the final, often neglected, stages of sporulation (Chapter by Cohen & Rotem) and resting body formation (Chapters by Entwistle and Walker). Detailed consideration, however, is also given to physiological and biochemical responses of the host (Chapters by Kuhn & Hargreaves and Pearce), and to the effects that phytotoxins liberated by the fungus may have on the host's ability to respond to infection (Chapter by Duvick & Knoche), since each of these significantly influences fungal development.

The effects that the condition of the host may have on fungal development are also explored, in physiological or ecological contexts, in chapters by Deacon, Mason *et al.* and Webber *et al.* There is a single question underlying many of these chapters; why is it that some infections lead to disease while others remain symptomless and beneficial to the host? This question, which is of great biological and economic importance, is far from being answered but we hope that the book will make a positive contribution to its solution. Chapters by Koltin, Ellingboe and Dean & Kuć examine the nature of virulence, avirulence and plant resistance mechanisms, and propose ways in which molecular biology and genetic engineering may lead to improved methods for controlling plant disease and promoting beneficial fungal associations.

The origins of this book lie in a series of meetings, on the theme of fungal infection, that were organised from 1982 onwards by the Pathogens and Mutualistic Interactions Special Interest Committee of the British Mycological Society. The three symposia, 'Establishment', 'Progress' and 'Outcome of Infection' covered the broad spectrum of plant parasitism, integrating mycorrhizal and pathogenic contributions under each heading. The success of the meetings prompted us to invite some of the original speakers to reconsider their subject in the greater depth which writing allows. These contributions were supplemented by several from international authorities selected to give balance to the book. We are grateful to all authors for preparing their manuscripts so promptly and for accepting so willingly those editorial changes that limitations of space made necessary. We would like to thank members of the B.M.S. for initiating and planning the original meetings. Finally we thank Mrs Pauline Donaldson for supervising the retyping of manuscripts on to computer disc. This method of book production was an experiment for the B.M.S. Symposium Series and one which could not have been successfully completed without her.

G.F. Pegg
University of Reading

P.G. Ayres
University of Lancaster

1

Specificity of active resistance mechanisms in plant-fungus interactions

P . J . G . M . DE WIT

Department of Plant Pathology, Agricultural University, Binnenhaven 9, 6709 PD Wageningen, The Netherlands

Introduction

In gene-for-gene systems involving biotrophic parasites, active defence mechanisms are usually induced by avirulent but initially not by virulent races of a parasite. The reason why active defence responses do not result from compatible race-cultivar combinations or are ineffective is of considerable interest. Heath (1982) has advanced the hypothesis that compatibility could be the result of a suppression of host resistance responses by a virulent parasite (induced susceptibility). However, if such induced susceptibility was an active plant response rather than a constitutive failure of recognition, metabolic inhibitors might be expected to turn a compatible combination into an incompatible one. This has usually not been found to occur. Conversely there are many reports of metabolic inhibitors turning an incompatible combination into a compatible one (Yoshikawa, Yamauchi & Masago, 1978; Keen, Holliday & Yoshikawa, 1982). This indicates that generally resistance needs to be actively induced, but it does not preclude the possibility that susceptibility results from a constitutive absence of the induction stimulus in the parasite.

In host-parasite interactions involving non-biotrophic parasites, the situation seems to be more complex; here induction and breakdown or tolerance of defence mechanisms seem to occur simultaneously and there is a delicate balance between inducing and overcoming or withstanding induced defence mechanisms which turns in favour of the parasite in cases of compatibility and in favour of the host in cases of incompatibility. A good example of this type of interaction is that between pea (*Pisum sativum*) and *Nectria haematococca* (Van Etten *et al.*, 1980; Tegtmeier & Van Etten, 1982; Denny & Van Etten, 1983*a,b*) where the pathogen causes the accumulation of the phytoalexin pisatin in a spreading lesion, but at the same time it tolerates or detoxifies the phytoalexin.

Other examples of non-biotrophic parasites interacting in a complex manner with their hosts are the vascular parasites such as *Verticillium* and *Fusarium* species (Pegg, 1985). Here the parasite usually enters the host through wounds and traverses the cortex and endodermis before the xylem, which appears to be a preferred habitat, can be colonised. In general a response of the host in this pre-xylem phase is limited or absent, independently of whether the host is susceptible or resistant. Resistance to vascular pathogens is expressed in the xylem region. It should be recognised however, that within these vascular parasites, physiologic specialisation – a well known phenomenon within biotrophic parasites – also occurs.

The emphasis within this chapter is on a limited number of fairly simple biological model systems. Concentration on model systems alone however, may produce a fragmentary picture of the nature of resistance. Extrapolation of findings from one system to another can be dangerous. The primary aim of this chapter is to examine mechanistic relationships between host and parasite characters that influence the extent to which the host is colonised given 'basic compatibility' (Heath, 1982). There will be no attempt to explain the basis for variation in severity of disease symptoms (e.g. as a result of toxin production). Neither will there be a consideration of the mechanisms that may be responsible for variation in how host growth or yield may be affected as a result of becoming diseased (tolerance). This chapter is concerned only with mechanisms which contribute to an understanding of differences in active resistance to biotrophic parasites between hosts of different genotype.

Genetic analyses of gene-for-gene systems support the contention that incompatibility as opposed to compatibility is the specific event (Ellingboe, 1981, 1982) since the former is uniquely conditioned by the interaction of resistance and avirulence alleles. This in turn implies that avirulent races of a parasite should specifically induce defence mechanisms in cultivars carrying resistance genes to these races. Instances of where resistance can be attributed solely to a single defence mechanism are very few. Several responses may occur simultaneously or consecutively. It is likely therefore that more than one biochemical mechanism may be responsible for the restricted growth of a pathogen. For example the hypersensitive response (HR) is often associated with the accumulation of phytoalexins (Bailey, 1982) or with lignification (Vance, Kirk &

Sherwood, 1980), although frequently the induction of only one response has been studied experimentally.

In a search for putative products of avirulence genes as inducers of defence responses, various molecules have been isolated from cell walls, and culture filtrates of different parasitic fungi. These molecules have come to be known as (specific) elicitors (Keen, 1975). A survey of elicitors and their possible biological functions will be given, especially those concerned with HR, phytoalexins and lignification.

Elicitors of defence responses
The hypersensitive response (HR)

HR is an ill-defined term usually describing host cell necrosis which is associated with little extensive parasite colonisation. HR is a symptom of host-parasite incompatibility and was first related to host resistance eighty two years ago (Ward, 1905). The common association between HR and race-specific resistance to biotrophic parasites led to the assumption that HR caused resistance because of the biotroph's dependence on living cells. This interpretation did not explain resistance to necrotrophs, where HR is nevertheless common. As a consequence, the concept of an association of HR with phytoalexin production as an anti-fungal reaction has become commonly accepted. However, elicitors of HR induction solely are discussed here.

Jones & Deverall (1978) and Deverall & Deakin (1985) found indirect evidence for the existence of an $Lr20$ gene-specific elicitor of HR produced by avirulent races of *Puccinia recondita* f.sp. *tritici*. This elicitor induced HR when it came into contact with mesophyll cells of resistant cultivars carrying the $Lr20$ gene. De Wit & Spikman (1982), De Wit, Hofman & Aarts (1984) and De Wit *et al.* (1985) found race-specific elicitors of chlorosis and necrosis in the intercellular fluids extracted from compatible combinations of *Cladosporium fulvum* and tomato. The chlorosis and necrosis induced by these specific elicitors reflected the HR which is typical for incompatible interactions between races of *C. fulvum* and tomato cultivars.

Phytoalexin accumulation

Phytoalexins are generally defined as low-molecular-weight antimicrobial compounds that are synthesised by and accumulate in plants after exposure to micro-organisms. HR and the accumulation of phytoalexins often coincide: this and their antimicrobial activity made phytoalexins obvious candidates for a role in plant resistance to microbial parasites. In a recent review, Keen (1982) states: 'There is substantial evidence from several gene-for-gene systems, that phytoalexins accumulate to high quantity after inoculation of plants with avirulent but not virulent parasite races' and further: 'The consistent association of high phytoalexin production only with incompatible host reactions in these systems strongly supports their role as mediators to disease resistance'. Various types of elicitors will induce the accumulation of phytoalexins.

Glucan elicitors of phytoalexins

Glucans have been studied extensively in the relationship between *Phytophthora megasperma* f.sp. *glycinea* and soybean (Albersheim & Valent, 1978; Darvill & Albersheim, 1984). The elicitors which have been isolated from culture filtrates and cell walls of the fungus are glucans consisting of terminal 3-,6- and 3,6-β-linked glucosyl residues. Partial acid hydrolysis has led to the isolation of the smallest cell wall component able to elicit the accumulation of glyceollin: hexa (β-D-glucopyranosyl)D-glucitol (Darvill & Albersheim, 1984; Sharp, McNeil & Albersheim, 1984). Eight of these heptamers have been described but only one isomer appeared to be active as an elicitor of glyceollin accumulation (Sharp *et al.*, 1984). The structure of the biologically active heptamer has now been confirmed by chemical synthesis (Ossowski *et al.*, 1984).

Keen, Yoshikawa & Wang (1983) isolated glucomannans enzymatically from the cell wall of *P. megasperma* f.sp. *glycinea* with a soybean 1,3-β-glucanase. These glucomannans are reportedly ten times more active as elicitors of glyceollin in soybean cotyledons than the glucan mentioned above. They also showed race-specificity in that they elicited more glyceollin in cultivars with which the isolate concerned was incompatible.

An elicitor from *Colletotrichum lindemuthianum* found in culture filtrates and cell walls was presumed to be a 3-β- and 4-β-linked glucan. This compound induced the accumulation of phaseollin, hydroxyphaseollin and other phytoalexins in bean (Anderson, 1980). Hadwiger & Beckman (1980) and Hadwiger, Beckman & Adams (1981) found that chitosan, a 1,4-β-linked glucosamine isolated from cell walls of *Fusarium solani*, induced the accumulation of pisatin in pea.

Protein and glycoprotein elicitors of phytoalexins

Cruickshank & Perrin (1968) isolated a protein from mycelium of *Monilinia fructicola*, monilicolin A, that induced the accumulation of phaseollin in bean. Potent glycoprotein inducers of phytoalexins have been isolated from a number of fungi. Keen & Legrand (1980) found glycoproteins from *P. megasperma* f.sp. *glycinea* on the mycelial wall surface which induced glyceollin in soybean cotyledons. However, the concentration of glycoproteins required was considerably higher than with the glucomannan and glucan elicitors from the same fungus. Stekoll & West (1978) similarly found a glycoprotein elicitor in culture filtrates of *Rhizopus stolonifer* that induced the accumulation of casbene in castor bean (*Ricinus communis*). It also showed endo-polygalacturonase activity (Lee & West, 1981). De Wit & Roseboom (1980) and De Wit & Kodde (1981) found peptidogalactoglucomannans in culture filtrates and cell walls of *C. fulvum* capable of inducing rishitin accumulation in tomato fruits. These same elicitors from *C. fulvum* also induced pisatin in pea pods and glyceollin in soybean cotyledons.

Unsaturated fatty acid elicitors

Bostock, Kuć & Laine (1981) found that arachidonic and eicosapentaenoic acids were effective inducers of sesquiterpenoid accumulation in potato tubers. These acids were released from the lipophylic materials of cell wall preparations of *Phytophthora infestans*. According to Bostock *et al.* (1981), the lipids alone were sufficient to induce phytoalexin accumulation, but Kurantz & Zacharius (1981) only found significant accumulation of sesquiterpenes when carbohydrate and lipid fractions from *P. infestans* were applied to potato tuber tissue in combination. More recently, Bostock, Laine & Kuć (1982) and Maniara, Laine & Kuć (1984) reported that the elicitor activity of these fatty acids could be greatly enhanced by addition of 1,3-β- and 3,6-β-glucans. The combination of carbohydrate fractions with other unsaturated C_{20} fatty acids also led to sesquiterpenoid accumulation (Preisig & Kuć, 1985). Arachidonic and eicosapentaenoic acid also induced capsidiol accumulation in pepper but not, however, rishitin in tomato fruits (Bloch, De Wit & Kuć, 1984). Apparently, species of the genus *Phytophthora* produce at least three types of elicitors: glucans, glycoproteins and unsaturated fatty acids. The first two induce glyceollin accumulation in soybean cotyledons, but not rishitin in potato tubers, while the latter induces rishitin accumulation in potato tubers, capsidiol in pepper fruits but not rishitin in tomato fruits.

Constitutive (host endogenous) elicitors

It is suggested that these elicitors, of host origin, are released (i) as a consequence of damage to the plant (Hargreaves & Bailey, 1978;

Darvill & Albersheim, 1984), for example during HR or the necrotrophic phase of parasitism or (ii) are released by pectic enzymes of pathogen or host origin during early pathogenesis. Pectic or other cell wall fragments released by pectolytic enzymes exhibit elicitor activities (Bruce & West, 1982; Darvill & Albersheim, 1984; Davis *et al.*, 1984). The endogenous elicitor from plant cell walls identified by Nothnagel *et al.* (1983) appeared to be a dodeca α-1,4-D galacturonide. Some cell wall derived fragments also show proteinase inhibitor inducing activity (Bishop *et al.*, 1984; Darvill & Albersheim, 1984; Walker-Simmons, Hadwiger & Ryan, 1983; Walker-Simmons *et al.*, 1984). Very small fragments could well be the proteinase inhibitor inducing factors themselves. Bailey (1982) has proposed a hypothesis to explain the mechanism of phytoalexin synthesis and accumulation. This is that cell death gives rise to endogenous elicitors which in turn induce phytoalexin accumulation. It takes account of the facts that: (i) phytoalexins are not induced in compatible interactions while the fungus is in its biotrophic phase; (ii) at the onset of the necrotrophic phase, phytoalexin accumulation has often been reported in susceptible cultivar-isolate combinations, and (iii) in incompatible interactions of the HR-type, phytoalexins accumulate immediately. The hypothesis also suggests that necrotrophic parasites inevitably induce phytoalexin accumulation but may have evolved a way of escaping their deleterious effects (Van Etten *et al.*, 1980; Tegtmeier & Van Etten, 1982; Denny & Van Etten, 1983*a,b*). It would be also compatible with the abiotic induction of phytoalexins (Hargreaves & Bailey, 1978). Alternatively, necrotrophic parasites might suppress the accumulation of phytoalexins by killing host cells so quickly that the host synthetic capability is impaired.

Lignification

Lignification has been associated with resistance in many host-parasite associations (Vance *et al.*, 1980). Histochemical studies show that lignification often occurs before penetration, as well as during colonisation, in either the epidermis or the internal cells of many plant organs. Both biotrophic and necrotrophic parasites may induce lignification (Bird & Ride, 1981; Beardmore, Ride & Granger, 1983).

There have been few studies of the eliciting of lignification; most have been conducted with living organisms (Vance *et al.*, 1980). The induction of lignin in wheat appears to be almost specifically a feature of filamentous fungi (Pearce & Ride, 1980). Chitin, chitosan and ethylene glycol induce lignification in wounded wheat leaves (Pearce & Ride, 1982); the first of these is of interest since it is a major constituent of the cell walls of many fungal parasites. As chitin itself is highly insoluble and may be unlikely to come in contact with plant cell membranes during the infection process, degradation products of chitin (N-acetylglucosamine, small chitin oligomers and glucosamine) have also been examined. However, all these substances appeared to be inactive as inducers of the lignification.

Additionally purified cell walls from *Botrytis cinerea* and *Agaricus bisporus*, which do not contain chitin, nevertheless elicited lignification. Thus it appears that cell wall components other than chitin can elicit the lignification response also. There are no reports of species or race-specific induction of lignification.

Elicitors and their specificity

HR and the accompanying accumulation of phytoalexins are frequently associated with incompatibility in gene-for-gene systems. For this reason a great deal of effort has been put into looking for molecules produced by avirulent races of a pathogen which specifically induce HR and the accumulation of phytoalexins: the so-called race-specific elicitors.

Most of the elicitors that have been isolated to date do not exhibit race-specificity; many do not even have species-specificity (De Wit & Roseboom, 1980; Bloch *et al.*, 1984). The relationships that have been most studied are those for which a reciprocal check arrangement is evident (Fig. 1.4). These are the associations between potato and *Phytophthora infestans,* soybean and *P. megasperma* f.sp. *glycinea,* and bean and *C. lindemuthianum.* The conclusions from these studies are contradictory and undoubtedly the procedures used to isolate elicitors are crucial. Elicitors obtained from cell walls of different fungal races using crude homogenisation procedures appear to be non-specific. These elicitors in addition seem to be rather diverse (unsaturated fatty acids from *P. infestans;* 1,3-β- and 1,6-β- branched glucans from *P. megasperma* f.sp. *glycinea*; and peptidogalactoglucomannans from *C. fulvum*).

When milder extraction methods have been used, some elicitors obtained have indeed appeared to be race-specific (Anderson, 1980; Keen & Legrand, 1980). Keen & Yoshikawa (1983) found that a 1,3-β-endoglucanase from soybean released potent race-specific elicitors from *P. megasperma* f.sp. *glycinea.* These appeared to be glucomannans and were different from the compounds previously reported for this fungus. De Wit & Spikman (1982) have obtained race-specific elicitors of chlorosis and necrosis from *in vivo* intercellular fluids extracted from compatible combinations of *C. fulvum* and tomato. These elicitors are probably of fungal origin (De Wit *et al.*, 1984) and are different from the non-specific glycoproteins isolated from the fungus grown *in vitro* (De Wit & Kodde, 1981). The elicitor of necrosis appears to be a small peptide (De Wit *et al.*, 1985) of which the amino acid sequence has recently been determined (unpublished results).

Suppressors and their specificity

In some studies, no evidence for the existence of specific elicitors has been obtained and a contrary hypothesis has emerged. It is suggested that specificity observed *in vivo* could be explained by the existence of specific suppressor molecules which eliminate or decrease the effect of non-specific elicitors in compatible interactions. In other words, non-specific elicitors and specific suppressors produced by virulent races only, would work in concert resulting in the observed *in vivo* specificity. Garas, Doke & Kuć (1979), Doke, Garas & Kuć (1980) and Doke & Tomiyama (1980) have isolated high molecular weight non-specific phytoalexin elicitors from mycelium of *P. infestans*. In addition, however, low-molecular-weight glucan molecules from homogenates and germination fluids of the same pathogen appeared to function as race-specific suppressors of HR and phytoalexin accumulation. These workers have proposed that race-specific suppressors confer gene-for-gene specificity. These suppressor molecules are branched 1,3-β-glucans of 17-23 glucose units. However, they did not suppress, but indeed enhanced the phytoalexin-inducing activity of arachidonic and eicosapentaenoic acids and even other C_{20} unsaturated fatty acids (Bostock *et al.*, 1982; Maniara *et al.*, 1984; Preisig & Kuć,1985).

Ziegler & Pontzen (1982) have found specific inhibition of glucan-elicited glyceollin accumulation in soybeans by an extracellular mannan-glycoprotein obtained from virulent races of *P. megasperma* f.sp. *glycinea*. The mannan-glycoprotein appeared to be an extracellular invertase. From these results, it was concluded that a susceptible host response (suppression of glyceollin accumulation) and not a resistance response (high glyceollin accumulation) is specifically induced by the fungus.

Receptors for elicitors of active resistance mechanisms

Elicitors of active defence responses in host plants have been studied more extensively than their receptors. Keen & Bruegger (1977) distinguished distinct determinative (recognition) and expressive phases in their model to explain gene-for-gene complementation. Recognition was envisaged as a static process while expression was thought of as dynamic. The contention is that recognition of an elicitor confers resistance by setting in motion a series of steps from, for example, *de novo* mRNA synthesis, through protein synthesis to the accumulation of phytoalexins (or any other active resistance response).

Recognition factors in plants are envisaged as constitutively produced surface molecules, as are elicitors from the parasite. Their occurrence at host and parasite cell surfaces would facilitate contact during the early stages of host-parasite relationship. The recognition factors in plants that have attracted most attention are lectins. Some lectins agglutinate bacteria and fungal spores. Their hapten specificity makes lectins plausible candidates as receptors of elicitors. Recognition events between two organisms through complementary surface-macromolecules including lectins have been the subject of intensive study (Dazzo & Truchet, 1983). The role of lectins in bacterial host-pathogen interactions especially has been studied extensively (Sequeira, 1978).

There are few indications that lectins could also play a role in relationships between fungal parasites and their hosts. Furuichi, Tomiyama & Doke (1980) suggested that potato lectin is involved in binding the cell wall surfaces of *P. infestans* to cell membranes. However, no race-specificity was found. Garas & Kuć (1981) found that potato lectin precipitated elicitors of terpenoid accumulation which had been extracted from the same fungus, while Nozue, Tomiyama & Doke (1980) found that the induction of HR in incompatible race-cultivar combinations could be inhibited by adding N,N'-diacetylchitobiose, a specific hapten for the potato lectin.

Kojima, Kawakita & Uritani (1982) reported that a lectin-like agglutination factor from sweet potato roots can agglutinate non-germinated spores of seven strains of *Ceratocystis fimbriata* including one parasitic on sweet potato. This factor, however, also showed agglutinating activity with germinated spores of five strains of the fungus parasitic on hosts other than sweet potato, while germinated spores of strains parasitic on sweet potato and almond were not agglutinated. The agglutination factor in this case was thought to be a high-molecular-weight polysaccharide. Yoshikawa, Keen & Wang (1983) have recently reported a receptor on soybean membranes for a nonspecific fungal elicitor (mycolaminarin). This receptor appears to be a protein or glycoprotein. However, binding of ^{14}C-mycolaminarin was irreversible and could not be displaced with an excess of unlabelled ligand. This finding suggests that the receptor was not likely to be a lectin.

There are other reports of lectins binding fungal elicitors, but often the lectins concerned are unrelated to those found in the host of the parasite in question. Hinch & Clarke (1980) reported an interesting inverse binding phenomenon: a fungal lectin interacting with a carbohydrate receptor from the host, but the significance of this is unclear. Little convincing experimental evidence has been accumulated to date to suggest a dominant role for lectins in the recognition of fungal parasites.

Conversion of elicitor-'signal' into metabolic events

In previous sections we have seen that there is preliminary evidence for existence of receptors on plant cell walls or membranes. What do we understand of the cellular machinery located between the primary recognition event and the eventual defence reponse?

In a number of cases, *de novo* synthesis of mRNA encoding enzymes involved in isoflavonoid-derived phytoalexin accumulation have been found after treatment with an avirulent race or elicitor (Lawton *et al.*, 1983; Ryder *et al.*, 1984; Schmelzer *et al.*, 1984; Cramer *et al.*, 1985), such as chalcone synthase (CHS), phenyl alanine ammonia lyase (PAL) and chalcone isomerase (CHI). This demonstration of specific early changes in gene expression after treating plant tissues with elicitors may lead to the elucidation of signal, response-coupling mechanisms in plant pathogen interactions. However, one should be aware that changes in mRNA synthesis are as yet no proof for the concomitant accumulation of phytoalexins. Ebel, Schmidt &Loyal (1984) observed that xanthan gum like a glucan elicitor stimulated *de novo* mRNA synthesis for CHS in soybean cells, but did not induce the subsequent accumulation of glyceollin; indicating that activation of enzyme activity as such does not always include accumulation of phytoalexin. Unfortunately, in most studies involving rapid switching of plant gene expression, nonspecific elicitors were used as inducers (Lawton *et al.*, 1983; Ebel *et al.*, 1984; Schmelzer *et al.*, 1984; Cramer *et al.*, 1985). It would be important to know therefore whether race-specific elicitors induce a similar type of gene expression.

Physiological models to explain gene-for-gene relationships

A widely held although speculative opinion is that basic compatibility between host and parasite evolved first and that the gene-for-gene relationship has become superimposed upon this (Ellingboe, 1981, 1982). A possible theory that plant parasites evolved from saprophytes raises the question of how many different attributes are required for an organism to be a successful parasite. Clearly, the evolutionary changes required for a saprophyte to become a parasite are greater than for a race of a parasitic species avirulent on a particular cultivar to become virulent to it. Most characteristics conferring parasitic ability are envisaged as positive functions enabling successful host colonisation. Some of these functions, for example, could be the production of toxins, cell-wall degrading enzymes, enzymes that degrade preformed antimicrobial compounds or phytoalexins, substances that mask recognition factors at parasite surfaces and substances that suppress the expression of HR (Heath, 1982; Keen, 1982). Undoubtedly, many genes must be involved in the successful development of a parasite in its host and the relationship with host gene functions could be complex.

Gabriel, Ellingboe & Rossman (1979) attempted to obtain temperature-sensitive parasite mutants of *Phyllosticta maydis* to probe for characters which were crucial for basic compatibility. A class of mutants was obtained which was temperature-sensitive *in vivo* but not *in vitro*. The conclusion was therefore that these mutations affected genes crucial to parasitism. If large numbers of this type of mutant could be obtained, it would be possible both to map the loci concerned and estimate how many genes are necessary for the establishment of basic compatibility. Perhaps more important than the numbers of genes would be knowledge of the relationship between genes controlling parasitic capability and host genes governing susceptibility. Is there a relationship analogous with the gene-for-gene system for race-specific resistance? Since almost nothing is known about the genetics of basic compatibility, only speculation is possible. Lack of knowledge is not surprising since such an investigation is equivalent to unravelling the course of the presumed evolution.

Given the existence of basic compatibility, there is likely to be a selection pressure on the plant towards resistance. In the species of agricultural crops, the plant breeder's role in the selection of resistant cultivars accelerates this process. Once resistant genotypes emerge either naturally or under man's influence, selection pressure is again imposed on the parasite to respond by overcoming the resistance. It has been suggested that this is how the gene-for-gene relationship comes into operation. Only after basic compatibility has been established can specificity at the race-cultivar level develop. Heath (1982) has pointed out that mechanisms controlling basic compatibility and race-specific resistance may operate simultaneously and this has a bearing on the interpretation of biochemical and physiological data as well as the design of models to explain gene-for-gene relationships. However, race-specific resistance has generally been studied with little regard to the processes responsible for basic compatibility. Many of the apparent contradictions between studies may be a reflection of this, and paying too little regard to basic compatibility processes may be one reason why mechanisms of race-specific resistance are still poorly understood.

Ellingboe (1981) has summarised the development of research on the physiology of host resistance since the 1950s. Much of the early work was conducted using two host lines, one susceptible and one resistant to a particular isolate of a parasite (Fig. 1.1). In most studies the different host lines may have differed for many characters other than resistance. Thus, it is inevitable that any difference between the two lines is absolutely correlated with resistance and susceptibility, and the same argument holds for investigation where a single cultivar or a line of a plant species is inoculated with virulent or avirulent isolates of a parasite (Fig. 1.2). Any difference between two isolates is absolutely correlated with virulence and avirulence. But establishing 'cause and effect' is almost impossible from this type of study. Rowell, Loegering & Powers (1963) influenced thinking by introducing the concept of the quadratic check as a genetical model for physiological studies of resistance to *Puccinia graminis* in wheat (Fig. 1.3). Examination of the quadratic check leads to the conclusion that three different genotype combinations $(A_1/r_1; a_1/r_1; a_1/R_1)$ all lead to a compatible phenotype while only one unique genotype combination (A_1/R_1) leads to incompatibility. This can be interpreted to mean that specificity is controlled by the interaction between the gene products of the dominant A and R alleles. In all other cases, it is envisaged that there is no recognition or interaction between gene products and as a consequence compatibility results. However, Martin & Ellingboe (1976) found small differences between the phenotypes of the three allele combinations leading to compatibility (see Chapter 19).

Loegering & Harmon (1969) introduced the use of near-isogenic lines, more or less identical to one another except for their resistance genes. This made it possible to study differences between resistant and susceptible lines that were more likely to be related to genes for resistance. The arrangement between isolates and cultivars employed in this study was called the (double) reciprocal check (Fig. 1.4). This arrangement can be interpreted to mean that A_1 interacts specifically with R_1 and not with R_2 and that A_2 interacts specifically with R_2 and not with R_1. There is a strict one gene-for-gene relationship. In addition, this arrangement implies that incompatibility is epistatic to compatibility.

Ellingboe (1982) and Keen (1982) have individually reviewed a number of models which have been proposed to explain the gene-for-gene relation-

Host lines

	A	B
Parasite	C	I

Fig. 1.1. The possible combinations between one parasite and two host lines (A and B) which vary in resistance. C = compatible or susceptible; I = incompatible or resistant.

Parasite isolates

	X	Y
Host	C	I

Fig. 1.2. The possible combinations between two parasite isolates (X and Y) which differ in parasitic capability, and one host line. C and I as in Fig. 1.1.

Genotypes of host lines

Genotypes of parasite isolates	R_1R_1	r_1r_1
A_1	I	C
a_1	C	C

Fig. 1.3. The 'quadratic check'. Possible combinations between two host lines and two haploid parasite isolates differing in resistance and parasitic capability respectively. R_1 and A_1 are alleles for resistance and avirulence respectively, and r_1 and a_1 are the opposite alleles for susceptibility and virulence. C and I as in Fig. 1.1.

Genotypes of host lines

Genotypes of parasite isolates	$R_1R_1r_2r_2$	$r_1r_1R_2R_2$
A_1a_2	I	C
a_1A_2	C	I

Fig. 1.4. The 'double reciprocal check'. Possible combinations between two host lines and two haploid parasite isolates differing in resistance and parasitic capability respectively. R_1 and R_2 are different alleles for resistance and r_1 and r_2 are the opposite alleles for susceptibility. A_1 and A_2 are different alleles for specific avirulence on R_1 and R_2 respectively; a_1 and a_2 are the opposite alleles for virulence. C and I as in Fig. 1.1.

ship on a mechanistic basis. Two of them are briefly elaborated and commented upon here.

The specific elicitor-specific receptor model

In this model it is suggested that the primary product of the dominant allele for avirulence (A_1) interacts with the primary product of the dominant allele for resistance (R_1), which results in a sequence of events culminating in locally induced resistance (HR, accumulation of phytoalexins, lignification etc.; Fig. 1.5). The locally induced resistance response would be initiated by a second messenger-like type of substance transferring information to the host genome. The nature of these second messengers is unknown, but endogenous elicitors mentioned earlier might have second messenger-like functions. The release of the avirulence gene product (specific elicitor) *in vivo* may be host-mediated. Plant enzymes like 1,3-β-glucanases and chitinases were found to release glucomannans with race-specific elicitor activity from cell walls of the fungus *P. megasperma* f.sp. *glycinea* (Keen *et al.*, 1983; Keen & Yoshikawa, 1983). Much of our thinking on resistance models has been based on studies on *Phytophthora megasperma* f.sp. *glycinea* infection of soybean and the assumption that inhibited growth of the pathogen was the result of active induction of the phytoalexin glyceollin. It is salutary to note therefore that Hahn, Bonhoff & Grisebach (1985) showed clear evidence for restricted hyphal growth in incompatible combinations *prior* to the induc-

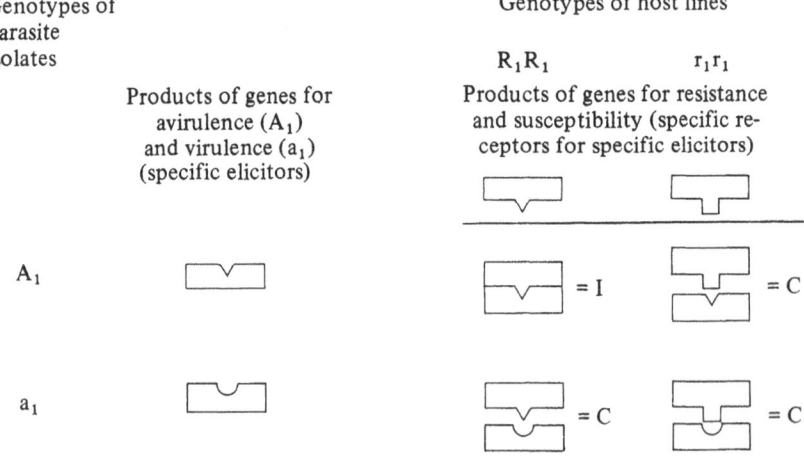

Fig. 1.5. A physiological model for expression of incompatibility after interactions between the product of the gene for resistance (R_j) and the product of the gene for avirulence (A_j). C and I as in Fig. 1.1.

tion of glyceollin. Also De Wit & Spikman (1982) found host-mediated release (or production) of race-specific elicitors of necrosis in intercellular fluids of compatible combinations between *C. fulvum* and tomato. The factor mediating the production of these elicitors in tomato is unknown but it is not dependent upon the R-gene carried by the cultivar in which the fungus is growing (De Wit *et al.*, 1984, 1985). However, race-specific elicitors might also be bound to the cell surface of the pathogen; thus physical contact between parasite and host is necessary for signal transfer.

Jones & Deverall (1978), Deverall & Deakin (1985), Anderson (1980) and Keen & Legrand (1980) also found evidence for the existence of race-specific elicitors of defence responses. The specific elicitor-specific receptor model does not conflict with the idea of the existence of basic compatibility since the gene-for-gene system is envisaged as superimposed upon it. The implication is that the general suppressors (Heath, 1981, 1982; Davidse & Boekeloo, 1984) are components of basic compatibility.

Ellingboe (1982) has argued that if responses such as phytoalexin accumulation are important in gene-for-gene controlled resistance, one gene-for-gene relationship would not be expected. His argument is that many enzymes and hence many genes are responsible for phytoalexin synthesis and that mutations at any of the loci should impair resistance. Similarly, since glycoprotein elicitors are also synthesised by many enzymes it is argued that their production is unlikely to relate to a single avirulence gene. Ellingboe (1982) (see this volume) is of the opinion that if this model were to operate, race-specific resistance and avirulence would both be polygenic traits. However, his arguments are hypothetical, as host mutants lacking enzymes or having less-efficient enzymes for synthesis of phytoalexins have not been studied and may be lethal; similarly with respect to the production of a glycoprotein elicitor the glycosyl transferase which provides the final structure to an elicitor could be the most important in terms of potential specificity, and this could be coded by the avirulent gene. Altogether the specific elicitor-specific receptor model is in agreement with genetical studies of gene-for-gene relationships (Flor, 1956).

The nonspecific elicitor and specific suppressor model

This model is based on the premise that nonspecific elicitors in concert with race-specific suppressors confer specificity in host-pathogen associations. It is envisaged that race-specific suppressors produced by virulence genes specifically bind to host receptors (products of genes for resistance). In this way, they either prevent the binding of nonspecific elicitors, thus causing the suppression of a resistance response, or alternatively prevent the expression of resistance in another way (Fig. 1.6). Doke *et al.* (1979, 1980), Doke & Tomiyama (1980) and Garas *et al.* (1979) found experimental evidence to support this hypothesis from studies with potato and *P. infestans*. Ziegler & Pontzen (1982) obtained data from studies with soybean and *P. megasperma* f.sp. *glycinea* which could also be interpreted on the basis of this hypothesis.

The suppressors identified by Doke *et al.* (1979, 1980) and Garas *et al.* (1979) are active in crude preparations of non-specific elicitors from *P. infestans* but unaccountably, they enhanced the activity of purified elicitors (Kurantz & Zacharius, 1981; Bostock *et al.*, 1982; Kurantz & Osman, 1983; Maniara *et al.*, 1984; Preisig & Kuć, 1985). Hence the role of race-specific suppressors is unclear.

Apart from the confusing experimental evidence, this model is not entirely in agreement with genetic studies of gene-for-gene relationships. In the model, compatibility is thought to be the specific event, yet the combinations A_1/r_1, a_1/R_1 and a_1/r_1 (Fig. 1.3) all lead to compatibility. As represented in Fig. 1.6 it can be seen that specific recognition between products of A_1 and r_1, and a_1 and R_1 could occur, but not between a_1 and r_1. It seems to be more feasible to consider suppressors to be one of the many characters needed by a parasite to establish basic compatibility.

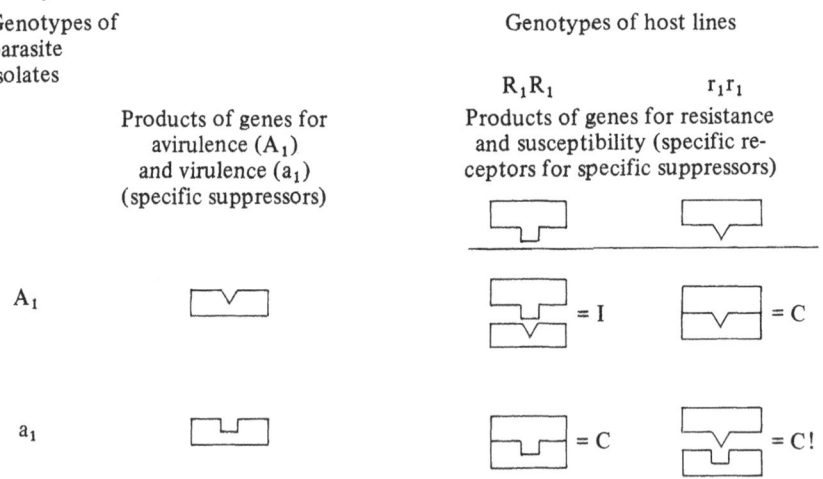

Fig. 1.6. A physiological model for expression of compatibility after interaction between products of genes for resistance (R_i) and susceptibility (r_i) and products of genes for avirulence (A_i) and virulence (a_i). Note that products of a_i and r_i do not interact yet a compatible association is observed in practice. C and I as in Fig. 1.1.

In this context, Davidse & Boekeloo (1984) have found evidence for the production of race-nonspecific suppressors by *P. infestans*. These suppressors inhibited the induction of necrosis in potato leaves caused by nonspecific elicitors from *P. infestans*. Heath (1981) also presented evidence for race-nonspecific suppressors and has suggested that they may act below the level of race-cultivar specificity. Bushnell & Rowell (1981) have proposed a model in which race-specific suppressors may have evolved from species-specific suppressors. This implies that one macromolecule possesses species as well as race-specific suppressor activity, and that basic compatibility in addition to race-specific compatibility could be determined by one macromolecule. According to this model, both basic compatibility and race-specific compatibility would be inherited as dominant characters. But while almost nothing is known about the genetics of basic compatibility, race-specific compatibility in host-parasite combinations involving diploid biotrophic fungi usually involves combinations of recessive alleles and not dominant alleles as would be expected from this model.

Constitutive models

One model advanced by Ellingboe (1982), and derived from genetic studies of the gene-for-gene relationship, suggests that the primary product of the dominant allele for avirulence (A_1) interacts directly with the primary product of the dominant gene for resistance (R_1) and that the interaction itself is responsible for incompatibility (Fig.1.5). It is suggested that a structural dimer is directly responsible for inhibiting parasite growth.

Lectins might contribute to inhibition of growth on the basis of this model, but there are no reports of lectins with race-specific binding properties, nor are there indications that cultivar or R-gene-specific lectins exist. Plant lectins can inhibit fungal growth (Mirelman *et al.*, 1975; Gibson *et al.*, 1982) but this is probably a non-specific phenomenon. A further objection that has been raised to this model is that compatibility can be induced with metabolic inhibitors, thus suggesting that resistance is not a constitutive character and that the expression of incompatibility requires active processes.

Another but different type of constitutive model has been described in relation to the action of host-selective toxins (Daly, 1984). These toxins have been shown to reproduce many if not all of the macroscopic, histological and biochemical symptoms associated with successful fungal colonisation. Genetic data support the contention that cultivar susceptibility to the fungus is due to toxin sensitivity and that cultivar resistance is a direct consequence of insensitivity to the toxin. There is thus circumstantial evidence for a passive mechanism of resistance based on the lack of a toxin receptor. This model of toxin action implies that the induction of host susceptibility and virulence are under control of dominant alleles in the host and the parasite, respectively, and this is exactly opposite to the situation normally encountered with gene-for-gene controlled host-parasite systems, where it is resistance that is considered to be the specific event (Ellingboe, 1982). However, as there is no reciprocal check arrangement in diseases involving host-selective toxins, such as *Helminthosporium* and *Alternaria* species, evidence for a gene-for-gene relationship is lacking.

Concluding remarks

The concentration on model systems produces a fragmentary picture. There is a temptation to extrapolate findings to other cases but the enormous diversity of reaction types makes this a dangerous practice, and valid generalities are hard to come by, even for closely related systems. Fortunately, studies on disease resistance mechanisms have moved on from an early preoccupation with the isolation and characterisation of phytoalexins, to molecular biology and molecular genetics. Effort is now being devoted to answering more important questions on the mechanisms that lead to the accumulation of phytoalexins and other defence responses.

Further molecular studies on products of genes for resistance and genes for avirulence will also tell us more about the biological functions of these genes. There is still the intriguing phenomenon of the abundance of genes for resistance and genes for virulence in the gene centres of crop plants and the opposite at the periphery: that is abundance of genes for 'susceptibility' and genes for avirulence (Vanderplank, 1978).

Vanderplank's conclusion is that genes for susceptibility are needed by the host plant for reasons unconnected with the presence or absence of the parasite; similarly genes for avirulence have a positive value to the parasite, a hidden value not apparent from the name. It is only under high selection pressure of the parasite that genes for resistance and genes for virulence have a more positive function relative to genes for 'susceptibility' and genes for avirulence.

Much progress has been made with bacterial plant pathogens. Staskawicz, Dahlbeck & Keen (1984) have cloned an avirulence gene of *Pseudomonas syringae* pv. *glycinea* which determines race-specific incompatibility. This avirulence gene was subcloned to obtain a 3 kb. fragment and experiments are under way to isolate the avirulence gene product and assess its phytoalexin-inducing activity. In the *C. fulvum-*

tomato interaction research has moved from another direction (De Wit *et al.*, 1985). A necrosis-inducing peptide, which has been suggested to be an avirulence gene product, has now been characterised. The amino acid sequence of this peptide has been determined and the search for the gene for avirulence from genomic libraries can begin with the help of a synthetic nucleotide probe. Alternatively purified specific elicitors will also permit a search for their receptors, the putative resistance gene products. It will be clear that in future recombinant DNA technology enables us to investigate a number of problems which were hitherto difficult or even impossible to study.

References

Albersheim, P. & Valent, B.S. (1978). Host-pathogen interactions in plants. Plants when exposed to oligosaccharides of fungal origin defend themselves by accumulating antibiotics. *Journal of Cell Biology*, **78**, 627–43.

Anderson, A.J. (1980). Differences in the biochemical composition and elicitor activity of extracellular components produced by three different races of fungal plant pathogen, *Colletotrichum lindemuthianum. Canadian Journal of Microbiology*, **26**, 1473–9.

Bailey, J.A. (1982). Mechanisms of phytoalexin accumulation. In: *Phytoalexins*, ed. J.A. Bailey & J.A. Mansfield, pp. 289–318. London: Blackie.

Beardmore, J., Ride, J.P. & Grainger, J.W. (1983). Cellular lignification as a factor in hypersensitive resistance of wheat to stem rust. *Physiological Plant Pathology*, **22**, 209–20.

Bird, P.M. & Ride, J.P. (1981). The resistance of wheat to *Septoria nodorum;* fungal development in relation to host lignification. *Physiological Plant Pathology*, **19**, 289–99.

Bishop, P.D., Pearce, G., Bryant, J.E. & Ryan, C.A. (1984). Isolation and characterization of the proteinase inhibitor-inducing factor from tomato leaves. Identity and activity of polygalacturonide and oligogalacturonide fragments. *Journal of Biological Chemistry*, **250**, 13172–7.

Bloch, C.B., De Wit, P.J.G.M. & Kuć, J.A. (1984). Elicitation of phytoalexins by arachidonic and eicosapentaenoic acids: a host survey. *Physiological Plant Pathology*, **25**, 199–208.

Bostock, R.M., Kuć, J.A. & Laine, R.A. (1981). Eicosapentaenoic and arachidonic acids from *Phytophthora infestans* elicit fungitoxic sesquiterpenes in the potato. *Science*, **212**, 67–9.

Bostock, R.M., Laine, R.A. & Kuć, J.A. (1982). Factors affecting the elicitation of sesquiterpenoid phytoalexin accumulation by eicosapentaenoic and arachidonic acids in potato. *Plant Physiology*, **70**, 1417–24.

Bruce, R.J. & West, C.A. (1982). Elicitation of casbene synthetase activity in castor bean *Ricinus communis*. The role of pectic fragments of the plant cell wall in elicitation by a fungal endo-polygalacturonase. *Plant Physiology*, **69**, 1181–8.

Bushnell, W.R. & Rowell, J.B. (1981). Suppressors of defence reactions: a model of roles of specificity. *Phytopathology*, **71**, 1012–14.

Cramer, C.L., Ryder, T.B., Bell, J.N. & Lamb, C.J. (1985). Rapid switching of plant gene expression induced by fungal elicitor. *Science*, **227**, 1240–3.

Cruicshank, I.A.M. & Perrin, D.R. (1968). The isolation and partial characterization of monilicolin A, a polypeptide with phaseollin-inducing activity from *Monilinia fructicola*. *Life Science*, **7**, 449–58.

Daly, J.M. (1984). The role of recognition in plant disease. *Annual Review of Phytopathology*, **22**, 273–307.

Darvill, A.G. & Albersheim, P. (1984). Phytoalexins and their elicitors. A. defence against microbial infection in plants. *Annual Review of Plant Physiology*, **35**, 243–76.

Davidse, L.C. & Boekeloo, M. (1984). Elicitation and suppression of necrosis in potato leaves by culture filtrate compounds of *Phytophthora infestans* (Mont.) de Bary. *Acta Botanica Neerlandica*, **33**, 234.

Davis, K.R., Lyon, G.D., Albersheim, P. & Darvill, A.G. (1984). Host- pathogen interactions. XXV. Endo-polygalacturonic acid lyase EC-4.2.2.2 from *Erwinia carotovora* elicits phytoalexin accumulation by releasing plant cell wall fragments. *Plant Physiology*, **74**, 52–60.

Dazzo, F.B. & Truchet, G.L. (1983). Interactions of lectins and their saccharide receptors in the *Rhizobium*-like symbiosis. *Journal of Membrane Biology*, **73**, 1–16.

Denny, T.P. & Van Etten, H.D. (1983*a*). Tolerance of *Nectria haematococca* MPVI to the phytoalexin pisatin in the absence of detoxification. *Journal of General Microbiology*, **129**, 2893–901.

Denny, T.P. & Van Etten, H.D. (1983*b*). Characterization of an inducible non-degradative tolerance of *Nectria haemotococca* MPPVI to phytoalexins. *Journal of General Microbiology*, **129**, 2903–13.

Deverall, B.J. & Deakin, A.L. (1985). Assessment of Lr20 gene-specificity of symptom elicitation by intercellular fluids from leaf rust-infected wheat leaves. *Physiological Plant Pathology*, **27**, 99–107.

De Wit, P.J.G.M., Hofman, J.E. & Aarts, J.M.M.J.G. (1984). Origin of specific elicitors of chlorosis and necrosis occurring in intercellular fluids of compatible interactions of *Cladosporium fulvum* (syn. *Fulvia fulva*) and tomato. *Physiological Plant Pathology*, **24**, 17–23.

De Wit, P.J.G.M., Hofman, J.E., Velthuis, G.C.M. & Kuć, J.A. (1985). Isolation and characterization of an elicitor of necrosis isolated from intercellular fluids of compatible interactions of *Cladosporium fulvum* (syn. *Fulvia fulva*) and tomato. *Plant Physiology*, **77**, 642–7.

De Wit, P.J.G.M. & Kodde, E. (1981). Further characterization and cultivar specificity of glycoprotein elicitors from culture filtrates and cell walls of *Cladosporium fulvum* (syn. *Fulvia fulva*). *Physiological Plant Pathology*, **18**, 297–314.

De Wit, P.J.G.M. & Roseboom, P.H.M. (1980). Isolation, partial characterization and specificity of glycoprotein elicitors from culture filtrates, mycelium and cell walls of *Cladosporium fulvum* (syn. *Fulvia fulva*). *Physiological Plant Pathology*, **16**, 391–408.

De Wit, P.J.G.M. & Spikman, G. (1982). Evidence for the occurrence of race and cultivar-specific elicitors of necrosis in intercellular fluids of compatible interactions of *Cladosporium fulvum* and tomato. *Physiological Plant Pathology*, **21**, 1-11.

Doke, N., Garas, N.A. & Kuć, J. (1979). Partial characterization and aspects of the mode of action of a hypersensitivity-inhibiting factor (HIF) isolated from *Phytophthora infestans*. *Physiological Plant Pathology*, **15**, 127–40.

Doke, N., Garas, N.A. & Kuć, J. (1980). Effect on host hypersensitivity of suppressors released during the germination of *Phytophthora infestans* cytospores. *Phytopathology*, **70**, 35–9.

Doke, N. & Tomiyama, K. (1980). Suppression of the hypersensitive response of potato tuber protoplasts to hyphal wall components by water-soluble glucans isolated from *Phytophthora infestans*. *Physiological Plant Pathology*, **16**, 177–86.

Ebel, J., Schmidt, W.E. & Loyal, R. (1984). Phytoalexin synthesis in soybean cells: elicitor induction of PAL and chalcone synthase mRNA's and correlation with phytoalexin accumulation. *Archives of Biochemistry and Biophysics*, **232**, 240–8.

Ellingboe, A.H. (1981). Changing concepts in host-pathogen genetics. *Annual Review of Phytopathology*, **19**, 125–43.

Ellingboe, A.H. (1982). Genetic aspects of active defence. In: *Active Defence Mechanisms in Plants*, ed. R.K.S. Wood, pp. 179–92. London: Plenum Press.

Flor, H.H. (1956). Complementary genic systems in flax and flax rust. *Advances in Genetics*, **8**, 29–54.

Furuichi, N., Tomiyama, K. & Doke, N. (1980). The role of potato lectin in binding of germ tubes of *Phytophthora infestans* to potato cell membrane. *Physiological Plant Pathology*, **16**, 249–56.

Gabriel, D.W., Ellingboe, A.H. & Rossman, E.C. (1979). Mutations affecting virulence in *Phyllosticta maydis*. *Canadian Journal of Botany*, **57**, 2639–43.

Garas, N.A., Doke, N. & Kuć, J. (1979). Suppression of the hypersensitive reaction in potato tubers by mycelial components from *Phytophthora infestans*. *Physiological Plant Pathology*, **15**, 117–26.

Garas, N.A. & Kuć, J. (1981). Potato lectin lyses zoospores from *Phytophthora infestans* and precipitates elicitors of terpenoid accumulation produced by the fungus. *Physiological Plant Pathology*, **18**, 227–37.

Gibson D.M., Stack, S., Knell, K & House, J. (1982). A comparison of soybean agglutinin in cultivars resistant and susceptible to *Phytophthora megasperma* var. *sojae* (race 1). *Plant Physiology*, **70**, 560–66.

Hadwiger, L.A. & Beckman, J.M. (1980). Chitosan as a component of pea-*Fusarium solani* interactions. *Plant Physiology*, **66**, 205–11.

Hadwiger, L.A., Beckman, J.M. & Adams, M.J. (1981). Localization of fungal compounds in the pea-*Fusarium* interaction detected immunochemically with anti-chitosan and anti-fungal cell wall antisera. *Plant Physiology*, **67**, 170–5.

Hahn, M.G., Darvill, A.G. & Albersheim, P. (1981). Host-pathogen interactions. XIX. The endogenous elicitor, a fragment of a plant cell wall polysaccharide that elicits phytoalexin accumulation in soybeans. *Plant Physiology*, **68**, 1161–9.

Hahn, M.G., Bonhoff, A. & Grisebach, H. (1985). Quantitative localisation of the phytoalexin glyceollin in relation to fungal hyphae in soybean roots infected with *Phytophthora megasperma* f.sp. *glycinea*. *Plant Physiology*, **77**, 591–601.

Hargreaves, J.A. & Bailey, J.A. (1978). Phytoalexin production by hypocotyls of *Phaseolus vulgaris* in response to constitutive metabolites released by damaged cells. *Physiological Plant Pathology*, **13**, 89–100.

Heath, M.C. (1981). A generalised concept of host-parasite specificity. *Phytopathology*, **71**, 1121–3.

Heath, M.C.(1982). Absence of active defence mechanisms in compatible host- pathogen interactions. In: *Active Defence Mechanisms in Plants,* ed. R.K.S. Wood, pp. 143–56. London: Plenum Press.

Hinch, J.M. & Clarke, A.E. (1980). Adhesion of fungal zoospores to root surfaces is mediated by carbohydrate determinants of the root slime. *Physiological Plant Pathology*, **16**, 303–7.

Jones, D.R. & Deverall, B.J. (1978). The use of leaf transplants to study the cause of hypersensitivity to leaf rust, *Puccinia recondita*,in wheat carrying the Lr20 gene. *Physiological Plant Pathology*, **12**, 311–19.

Keen, N.T. (1975). Specific elicitors of plant phytoalexin production: determinants of race specificity in pathogens. *Science*, **187**, 74–5.

Keen, N.T. (1982). Specific recognition in gene-for-gene host parasite systems. *Advances in Plant Pathology*, **1**, 35–82.

Keen, N.T. & Bruegger, B.B. (1977). Phytoalexins and chemicals that elicit their production in plants. In: *Host Plant Resistance to Pests,* ed. P. Hedin, pp. 1–26. American Chemical Society Symposium Series 62.

Keen, N.T., Holliday, M.J. & Yoshikawa, M. (1982). Effects of glyphosate on glyceollin production and the expression of resistance to *Phytophthora megasperma* f.sp. *glycinea* in soybean. *Phytopathology*, **72**, 1467–70.

Keen, N.T. & Legrand, M. (1980). Surface glycoproteins: evidence that they may function as the race-specific phytoalexin elicitors of *Phytophthora megasperma* f.sp. *glycinea*. *Physiological Plant Pathology*, **17**, 175–92.

Keen, N.T. & Yoshikawa, M. (1983). β-1,3-endoglucanase from soybean releases elicitor-active carbohydrates from fungal cell walls. *Plant Physiology*, **71**, 460–5.

Keen, N.T., Yoshikawa, M. & Wang, M.C. (1983). Phytoalexin elicitor activity of carbohydrates from *Phytophthora megasperma* f.sp. *glycinea* and other sources. *Plant Physiology*, **71**, 466–71.

Kojima, M., Kawakilta, K. & Uritani, I. (1982). Studies on a factor in sweet potato roots which agglutinates spores of *Ceratocystis fimbriata* black rot fungus. *Plant Physiology*, **69**, 474–8.

Kurantz, M.J. & Osman, S.F. (1983). Class distribution, fatty acid composition and elicitor activity of *Phytophthora infestans* mycelial lipids. *Physiological Plant Pathology*, **22**, 363–70.

Kurantz, M.J. & Zacharius, R.M. (1981). Hypersensitive response in potato tuber: elicitation by combination of non-eliciting components from *Phytophthora infestans*. *Physiological Plant Pathology*, **18**, 67–77.

Lawton, M.A., Dixon, R.A., Hahlbrock, K. & Lamb, C.J. (1983). Elicitor induction of messenger RNA activity. Rapid effects of elicitor on phenylalanine ammonia lyase EC-4.3.1.5 and chalcone synthase messenger activities in bean *Phaseolus vulgaris* cells. *European Journal of Biochemistry*, **130**, 131–40.

Lee, S.C. & West, C.A. (1981). Polygalacturonase from *Rhizopus stolonifer,* an elicitor of casbene synthetase activity in castorbean *Ricinus communis* seedlings. *Plant Physiology*, **67**, 633–9.

Loegering, W.Q. & Harmon, D.L. (1969). Wheat lines near-isogenic for reaction to *Puccinia graminis tritici*. *Phytopathology*, **59**, 456–9.

Maniara, G., Laine, R. & Kuć, J. (1984). Oligosaccharides from *Phytophthora infestans* enhance the elicitation of sesquiterpenoid stress metabolites by arachidonic acid in potato. *Physiological Plant Pathology*, **24**, 177–86.

Martin, T.J. & Ellingboe, A.H. (1976). Differences between compatible parasite/host genotypes involving the Pm4 locus of wheat and the corresponding genes in *Erysiphe graminis* f.sp. *tritici*. *Phytopathology*, **66**, 1435–8.

Mirelman, D., Galun, E., Sharon, N. & Lotan, R. (1975). Inhibition of fungal growth by wheat germ agglutinin. *Nature*, **156**, 414–16.

Nothnagel, E.A., McNeil, M., Albersheim, P. & Dell, A. (1983). Host- pathogen interactions. XXII. A galacturonic acid oligosaccharide from plant cell walls elicits phytoalexins. *Plant Physiology*, **71**, 916–26.

Nozue, M., Tomiyama, K. & Doke, N. (1980). Effect of *N, N'*-diacetyl-D-chitobiose, the potato lectin hapten and other sugars on hypersensitive reaction of potato tuber cells infected by incompatible and compatible races of *Phytophthora infestans*. *Physiological Plant Pathology*, **17**, 221–7.

Ossowski, P., Pilotti, A., Garegg, P.J. & Lindberg, B. (1984). Synthesis of a glucoheptaose and a glucooctaose that elicit phytoalexin accumulation in soybean. *Journal of Biological Chemistry*, **259**, 11337–40.

Pearce, R.B. & Ride, J.P. (1980). Specificity of induction of the lignification response in wounded wheat leaves. *Physiological Plant Pathology*, **16**, 197–204.

Pearce, R.B. & Ride, J.P. (1982). Chitin and related compounds as elicitors of the lignification in wounded wheat leaves. *Physiological Plant Pathology*, **20**, 119–23.

Pegg, G.F. (1985). Life in a black hole—the micro-environment of the vascular pathogen. *Transactions of the British Mycological Society*, **85**, 1–20.

Preisig, C.L. & Kuć, J.A. (1985). Arachidonic acid-related elicitors of the hypersensitive response in potato and enhancement of their activities by glucans from *Phytophthora infestans*. *Archives of Biochemistry and Biophysics*, **236**, 379–89.

Rowell, J.B, Loegering, W.Q. & Powers, H.R. (1963). Genetic model for physiologic studies of mechanisms governing development of infection type in wheat stem rust. *Phytopathology*, **53**, 932–7.

Ryder, T.B., Cramer, C.L., Bell, J.N., Robbins, M.P., Dixon, R.A. & Lamb, C.J. (1984). Elicitor rapidly induces chalcone synthase messenger RNA in *Phaseolus vulgaris* cells at the onset of the phytoalexin defence response. *Proceedings of the National Academy of Sciences USA*, **81**, 5724–8.

Sequeira, L. (1978). Lectins and their role in host-pathogen specificity. *Annual Review of Phytopathology*, **16**, 453–81.

Schmelzer, E., Boerner, H., Grisebach, H., Ebel, J. & Hahlbrock, K. (1984). Phytoalexin synthesis in soybean (*Glycine-max*). Similar time courses of messenger RNA induction in hypocotyls infected with a fungal pathogen and in cell cultures treated with fungal elicitor. *FEBS Letters*, **172**, 59–63.

Sharp, J.K., McNeil, M. & Albersheim, P. (1984). The primary structures of one elicitor-active and seven elicitor-inactive hexa (β-D-glucopyranosyl)-D-glucitols isolated from the mycelial walls of *Phytophthora megasperma* f.sp. *glycinea*. *Journal of Biological Chemistry*, **259**, 11321–6.

Staskawicz, B.J., Dahlbeck, D. & Keen, N.T. (1984). Cloned avirulence gene of *Pseudomonas syringae* pathovar *glycinea* determines race-specific incompatibility of *Glycine max*. *Proceedings of the National Academy of Sciences USA*, **81**, 6024–8.

Stekoll, M. & West, C.A. (1978). Purification and properties of an elicitor of castor bean phytoalexin from culture filtrates of the fungal *Rhizopus stolonifer*. *Plant Physiology*, **61**, 38–45.

Tegtmeier, K.J. & Van Etten, H.D. (1982). The role of pisatin tolerance and degradation in the virulence of *Nectria haematococca*. A genetic analysis. *Phytopathology*, **72**, 608–12.

Vance, T., Kirk, K. & Sherwood, R.T. (1980). Lignification as a mechanism of disease resistance. *Annual Review of Phytopathology*, **18**, 259–88.

Vanderplank, J.E. (1978). *Genetic and molecular basis of plant pathogenesis*, pp. 27–34. New York: Springer-Verlag.

Van Etten, H.D., Matthews, P.S., Tegtmeier, K.J., Dietert, M.F. & Stein, J.I. (1980). The association of pisatin tolerance and demethylation with virulence on pea in *Nectria haematococca*. *Physiological Plant Pathology*, **16**, 257–68.

Walker-Simmons, M., Hadwiger, L.A. & Ryan, C.A. (1983). Chitosan and pectic polysaccharides both induce the accumulation of the antifungal phytoalexin pisatin in pea pods and anti nutrient proteinase inhibitors in tomato leaves. *Biochemical and Biophysical Research Communications*, **110**, 194–9.

Walker-Simmons, M., Jin, D., West, C.A., Hadwiger, L.A. & Ryan, C.A. (1984). Comparison of proteinase inhibitor-inducing activities and phytoalexin elicitor activities of a pure fungal endopolygalacturonase pectic fragments and chitosans. *Plant Physiology*, **76**, 833–6.

Ward, H.M. (1905). Recent researches on the parasitism of fungi. *Annals of Botany*, **19**, 1–54.

Yoshikawa, M., Keen, N.T. & Wang, M.C. (1983). A receptor on soybean membranes for a fungal elicitor of phytoalexin accumulation. *Plant Physiology*, **73**, 497–506.
Yoshikawa, M., Yamauchi, K. & Masago, H. (1978). Glyceollin: its role in restricting fungal growth in resistant soybean hypocotyls infected with *Phytophthora megasperma* f.sp. *sojae. Physiological Plant Pathology*, **12**, 73–82.
Ziegler, E. & Pontzen, R. (1982). Specific inhibition of glucan-elicited glyceollin accumulation in soybeans by an extracellular mannan-glycoprotein of *Phytophthora megasperma* f.sp. *glycinea. Physiological Plant Pathology*, **20**, 321–31.

2
Specificity and recognition in ectomycorrhizal associations

J. A. DUDDRIDGE

Oxford Forestry Institute, Department of Plant Sciences, South Parks Road, Oxford OX1 3RB, UK

Introduction

Ectomycorrhizal associations, unlike interactions between plant and biotrophic fungal pathogens, show relatively low specificity (Gianinazzi-Pearson, 1984; Harley, 1985; Duddridge, 1986c). This is demonstrated by the large number of potential partners that exist for both host and fungus (Trappe, 1962, 1964; Kropp & Trappe, 1982; Danielson, 1984).

While ectomycorrhizal plants were once thought to make up only 3% of the Spermatophyta (Meyer, 1973), it is now realised that symbiosis occurs in a much wider range of plants (Harley & Smith, 1983). Many ectomycorrhizal hosts have been found also to form endomycorrhizas with vesicular-arbuscular fungal symbionts under certain ecological conditions (Vozzo & Hacskaylo, 1974; Warcup, 1980; Warcup & McGee, 1983; Iqbal, Yousaf & Younus, 1981). The importance is now recognised of the mycorrhizal status of the whole plant community in which a potential host is growing and can be illustrated by the situation found in some ferns. Among the ferns endomycorrhizas are usually predominant in natural communities but are displaced by ectomycorrhizas where ferns form the only understorey plants in pure stands of beech or pine (Cooper, 1976; Iqbal *et al.*, 1981). The converse situation may apply since there is also a report of endomycorrhizal infection in *Abies balsamea*, previously only recorded until now as an ectomycorrhizal host, when the associated ground cover in the plant community was of mostly endomycorrhizal species (Malloch & Malloch, 1981). Some ectomycorrhizal hosts appear to be more selective towards their fungal symbionts than others. Most ectomycorrhizal genera such as *Pinus, Fagus, Picea* (Trappe, 1962), *Betula* (Giltrap, 1979) and *Eucalyptus* (Chilvers, 1973; Malajczuk, Molina &

Trappe, 1982) will form associations with a wide range of symbionts, whereas *Alnus* is restricted to only a few (Molina, 1981).

Ectomycorrhizal fungi are not limited to one taxonomic group and can be found in all four of the main groups of the Eumycota. They are also able to form different types of mycorrhizal association with other hosts. For example many ectomycorrhizal fungi can form arbutoid mycorrhizas on hosts such as *Arbutus* and *Arctostaphylos* (Zak, 1976a,b; Molina & Trappe, 1982a;) and monotropoid mycorrhizas on achlorophyllous members of the Ericaceae such as *Monotropa* (Duddridge & Read, 1982b), *Pterospora* and *Sarcodes* (Robertson & Robertson, 1982). It has also been suggested that an ectomycorrhizal symbiont of birch may be the same endophyte found in the thalli of a British liverwort, *Cryptothallus mirabilis* (Pocock & Duckett, 1984).

Although most ectomycorrhizal fungi have a low degree of specificity, different symbionts may vary in their ability to increase the effectiveness of nutrient absorption by the host under different conditions (e.g. Theodorou & Bowen, 1970; Thomas & Jackson, 1983). These differences may even occur between different strains of the same fungus (Mason, 1975; Marx, 1979). Too much importance, however, must not be attached to the differences between fungal symbionts observed *in vitro*, because any host at any time simultaneously may have a number of different symbionts in the field (Trappe & Fogel, 1977).

Specificity in ectomycorrhizal associations

Not all ectomycorrhizal symbionts show a total lack of specificity. There are examples of specificity at the generic level, although very few species-specific interactions (e.g. *Suillus plorans* with *Pinus cembra*; Moser, 1978) have been reported. Molina & Trappe (1982b) have divided ectomycorrhizal fungi into three categories, (1) Broad host range/sporocarp-diverse, (2) Intermediate host range/sporocarp-specific and (3) Narrow host range/sporocarp-specific (see Table 2.1). Sporocarp-specific fungi are those which have a specific or limited sporocarp-host association as judged by field observations. This may tend to ignore those fungi which fruit infrequently. Kropp & Trappe (1982) suggested that host specificity of ectomycorrhizal fungi may have evolved mostly with pioneering host species that grow in pure stands, e.g. *Pinus*, and that understorey species such as hemlock *(Tsuga heterophylla)* (Kropp & Trappe, 1982) and *Arctostaphylos* and *Arbutus* spp. (Molina & Trappe, 1982a) are much less restrictive in their mycorrhizal associations because they have had to adapt to the 'mycorrhizal regime' of the overstorey.

Table 2.1 *Specificity of ectomycorrhizal fungi*

1 Broad host range/sporocarp-diverse	
Thelephora terrestris	Molina & Trappe, 1982*b*
Paxillus involutus	
Cenococcum graniforme	Trappe, 1964
Pisolithus tinctorius	Marx, 1977
Laccaria laccata	Trappe, 1962
Amanita muscaria	
2 Intermediate host range/sporocarp-specific	
Suillus lakei/Pseudotsuga menziesii	Molina & Trappe, 1982*b*
Rhizopogon vinicolor/Pinus spp.	
Suillus breviceps/Tsuga mertensiana, Pinus spp.	
3 Narrow host range/sporocarp-specific	
a) Conifer specific:	Molina & Trappe, 1982*b*
Suillus spp.	Trappe, 1962
Rhizopogon spp.	
Gomphidius spp.	
b) Genus specific:	
Rhizopogon/Pinus spp.	Molina & Trappe, 1982*b*
	Malajczuk *et. al.,* 1982
Alpova diplophloeus/Alnus spp.	Molina, 1981
Hydnangium carneum/Eucalyptus spp.	Malajczuk *et. al.,* 1982
Suillus grevillei/Larix spp.	Melin, 1922
Fuscoboletinus aeruginascens/Larix spp.	Molina & Trappe, 1982*b*
c) Species specific:	
Suillus plorans/Pinus cembra	Moser, 1978

Suillus grevillei, a fungus reputedly specific for *Larix* species (Melin, 1922), was selected for use in the present study so that a comparison could be made between incompatible and compatible interactions using a range of different ectomycorrhizal hosts. Ectomycorrhizas were aseptically synthesised in specially adapted plastic Petri dishes containing peat-vermiculite moistened with a nutrient solution (Duddridge, 1986*a*) (Fig.2.1a). Using these chambers, syntheses were set up between *Suillus grevillei* and a range of ectomycorrhizal hosts in the absence of any exogenous carbon. Table 2.2 shows the results of these syntheses. Although no Hartig net development occurred in *Allocasuarina inophloia* and *Pinus nigra*, the hyphae grew in the rhizosphere, frequently penetrating and killing the epidermal cells (fig. 2.1b). Apart from *Larix* species, mycorrhizas were only formed with *Pinus sylvestris* and *Pseudotsuga*

Table 2.2 *Interactions between* Suillus grevillei *and some ectomy-corrhizal hosts*

Mycorrhizal formation	
+	−
Larix decidua	*Alnus glutinosa*
Larix kaempferi	*Picea abies*
Pinus sylvestris	*Betula pubescens*
Pseudotsuga menziesii	*Pinus nigra* (r)
	Allocasuarina inophloia (r)

(r) = growth in rhizosphere

menziesii, and in the latter two cases ultrastructural analysis showed that the symbionts were probably not completely compatible. Short roots of *P. sylvestris* appeared to be superficially mycorrhizal (Fig. 2.1c) but a compact sheath was never formed and Hartig net development was irregular. Although mycorrhizal laterals of *P. menziesii* had well-developed sheaths and Hartig nets (Fig. 2.1d) there were abnormalities at the host-fungus inter-face which suggested a degree of incompatibility between the symbionts. The host cell wall adjacent to the fungus was thickened and contained electron-opaque deposits and host wall appositions were sometimes found (Fig. 2.1e). An electron-opaque layer was frequently observed between the fungus and host walls at points of contact (Fig. 2.1d) and in places there were signs of fungal lysis in the Hartig net. Large electron-opaque inclusions were often found in the fungus (Fig. 2.1d) and preliminary digestion with chlorine dioxide, a delignifying agent (Whistler, Bachrach & Bowman, 1948) suggested that they were of phenolic nature.

The lignification of host cell walls, the deposition of phenolic compounds (Molina, 1981; Molina & Trappe, 1982*b*; Malajczuk, Molina & Trappe, 1984), and papilla formation (Nylund, Kasimir & Strandberg-Arveby, 1982) in host cells have all been observed previously in interactions between incompatible ectomycorrhizal symbionts. There is some controversy as to whether there is an increase (Sylvia & Sinclair, 1983; Coleman & Anderson, 1985) or not (Foster & Marks, 1967; Duddridge, 1980; Piché, Fortin & Lafontaine, 1981; Duddridge & Read, 1984*a,b*) of phenolic compounds in mycorrhizal compared to non-mycorrhizal roots. Moreover there is the question, is the host-fungus interface functional in those ectomycorrhizal associations where the fungal symbiont is able to form an intercellular Hartig net in spite of the stimulation of the host's defence system? Current studies (Duddridge, Finlay & Read, 1987) on carbon transfer between compatible and incompatible ectomycorrhizal associations linked by a common fungal symbiont may go some way to answering this question.

The induction of phenolic compounds in the host appears to play an important role in ectomycorrhizal compatibility but as Molina & Trappe (1982b) previously pointed out, a more detailed study needs to be conducted to determine the type of compounds involved and the subcellular location of host enzymes such as laccase, peroxidase and polyphenoloxidase. Several plant species have high peroxidase levels normally present on their root surface and it has been suggested that mycorrhizal fungi may have to circumvent this (Anderson, 1985). Anti-microbial phenoxy radicals are produced when phenolics are oxidised by peroxidase in the presence of hydrogen peroxide. These may condense to form lignin polymers in the host cell wall as a defence mechanism to fungal penetration (Vance, Kirk & Sherwood, 1980; Hammerschmidt & Kuć, 1982). Ronald & Soderhall (1985) found that there were no significant differences in phenylalanine ammonia lyase (PAL) and peroxidase (PO) activities between mycorrhizal and non-mycorrhizal short roots in syntheses between *Pinus sylvestris* and *Laccaria laccata*. *L. laccata* did not appear either to suppress or increase PAL activity. Preliminary work (Duddridge, unpub.) has shown that PO can be localised at the host-fungus interface of incompatible (Fig. 2.1g) but not compatible (Fig. 2.1f) associations between ectomycorrhizal symbionts.

The capacity of an ectomycorrhizal host to form mycorrhizas with a much larger range of fungal symbionts in aseptic culture than in the field is probably due to the high levels of exogenous carbohydrate which are often included in the synthesis medium. This upsets the physiological balance which exists between the symbionts and may have a marked effect on mycorrhizal development (Giltrap, 1979; Malibari, 1979; Thomas & Jackson, 1979) and the ultrastructure of the host-fungus interface (Duddridge & Read, 1984c; Duddridge, 1986b). High levels of exogenous carbon can induce abnormal changes in the host cell wall of ectomycorrhizas formed between normally compatible symbionts (Fig. 2.2a), initiate intracellular penetration in some associations (Fig. 2.2b) or cause a host to form wall appositions in response to a fungal symbiont with which it would not normally form ectomycorrhizas (Fig. 2. 2c). These results demonstrate that the nutrient status of the synthesis medium is extremely important in determining the outcome of an interaction between a given pair of symbionts in ectomycorrhizal associations.

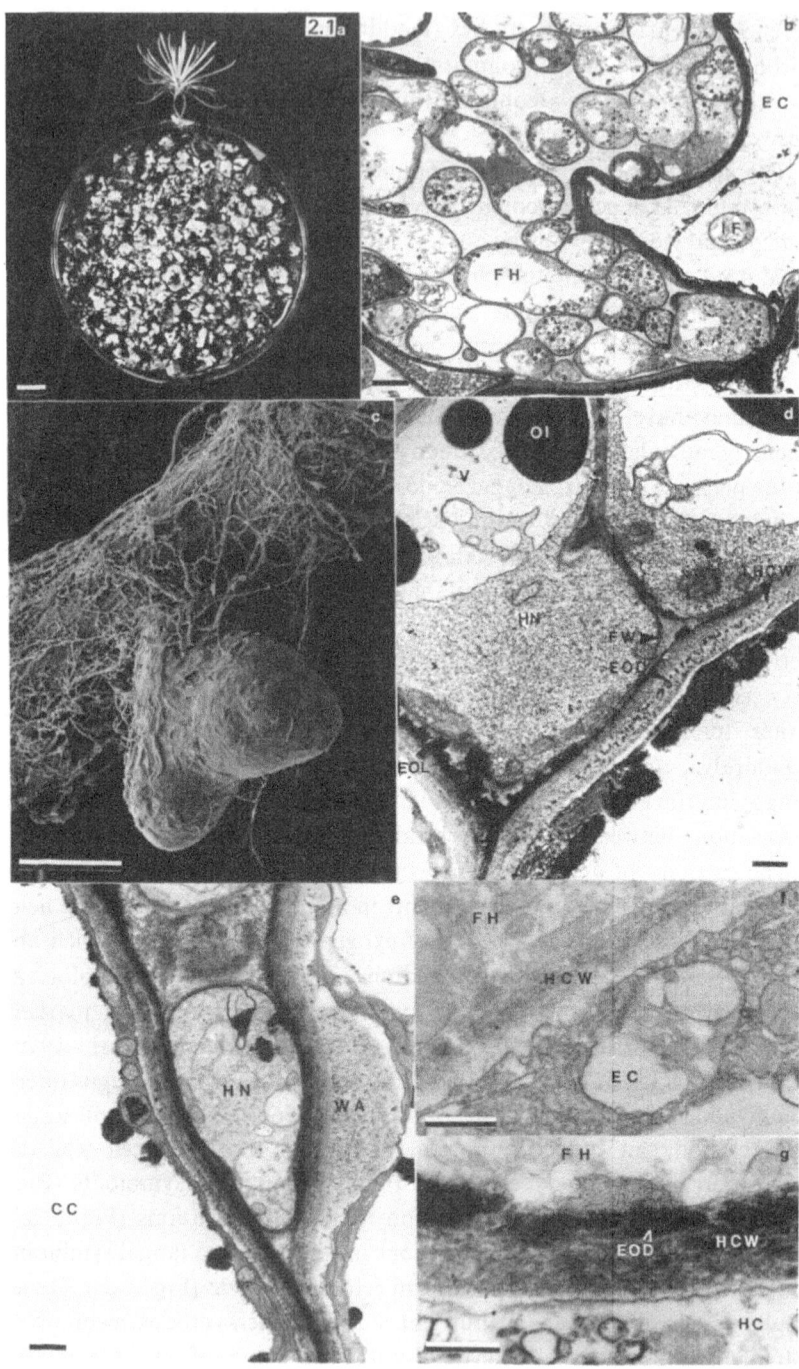

Tissue specificity

Mycorrhizal fungi as their name suggests almost exclusively infect root tissue. There are four categories of tissue which are generally believed to escape mycorrhizal infection: (1) actively photosynthetic tissue (2) vascular tissue (3) meristematic tissue and (4) suberised and secondary thickened tissue. The same fungal symbiont will form ectomycorrhizas of similar morphology on hosts as diverse as ferns and conifers, in spite of the large differences in the morphology and anatomy of their root tissues. Even the modification of root tissue by *Frankia* spp. to form non-leguminous root nodules of the *Alnus*-type (actinorhiza) does not prevent an ectomycorrhizal fungus recognising it and forming a sheath and Hartig net (Godbout & Fortin, 1983).

In ectomycorrhizal associations penetration by the fungal symbiont appears to occur between cells that have just reached or are almost to maturity and in which primary cell wall building is still occurring. Harley & Smith (1983) proposed that the fungus interferes with the cell wall building activity of the host, resulting in the formation of a zone of modified plant wall material at the host-fungus interface (Fig. 2. 2d). Nylund (1981) suggested that the cells in the so-called 'mycorrhizal infection zone'(Marks & Foster, 1973) were more susceptible to intercellular penetration because their high acid polysaccharide content made them more flexible. In ectomycorrhizal angiosperms (Clowes, 1951; Fontana,

---

Fig. 2.1. (a) Petri–dish synthesis chamber. (b) Hyphae (FH) of *Suillus grevillei* forming a loose weft round and penetrating into (IF) an epidermal cell (EC) of a short root of *Pinus nigra*. (c) A sparse covering of hyphae on a short root of *Pinus sylvestris,* 13 weeks after inoculation with *S. grevillei*. (d) and (e) T. S. of mycorrhizal short roots synthesised between *Pseudotsuga menziesii* and *S. grevillei*. (d) the host cell wall adjacent to the Hartig net (HN) is thickened (THCW) and contains electron–opaque deposits (EOD). Osmiophilic inclusions (OI), bounded by vacuoles(V), are present in the Hartig net and there is an electron–opaque layer (EOL) between the fungal (FW) and host cell walls. (e) A host wall apposition (WA) adjacent to the Hartig net (HN) which penetrates between the host cortical cells (CC). (Section stained with Thiéry's procedure for carbohydrate.)(f) and (g)Localisation of peroxidase at the host-fungus interface of ectomycorrhizal associations. (f) No localisation of peroxidase is observed in either the host cell wall (HCW) or epidermal cells (EC) adjacent to the fungus (FH), in a compatible interaction between *Larix kaempferi* and *S. grevillei*. (g) Electron-opaque deposits (EOD) in the cell wall (HCW) and cytoplasm (HC) of *P. menziesii* adjacent to hyphae of *S. grevillei* (FH) indicate the presence of peroxidase in an incompatible association. Scale bars: (a) 1.0 cm; (b) 2.0 μm; (c) 200 μm; (d)-(g) 0.5 μm.

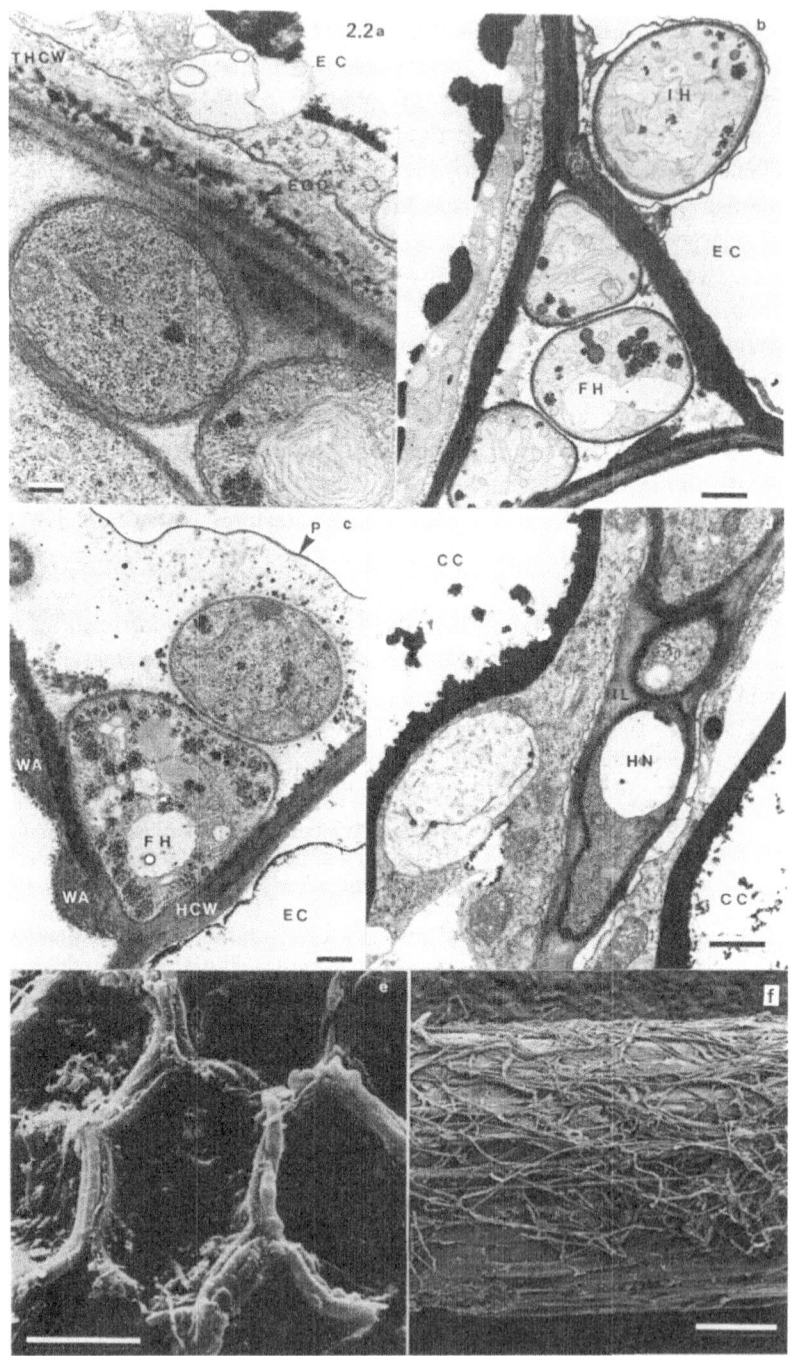

1962; Froidevaux, 1973; Malajczuk *et al.*, 1984) the Hartig net is usually limited to the epidermis, in contrast to gymnosperms where it penetrates several layers of cortical cells, although exceptions do occur (Debaud, Pepin & Bruchet, 1981; Fusconi, 1983).

Recognition in ectomycorrhizal associations

In interactions between fungal pathogens and plants, recognition appears to be for resistance; host receptors recognising markers on the fungal pathogens which alert the defence system (Callow, 1983; de Wit, Ch.1). In associations between plant roots and rhizobia, however, recognition appears to be for compatibility (see Stacey, Paau & Brill, 1980 for refs). How do plant roots therefore distinguish ectomycorrhizal fungi from the large population of saprophytic and parasitic fungi present in the rhizosphere? There are two possibilities; they may be either recognised as compatible symbionts or they are able to evade recognition as pathogens because of an inability to alert, or an ability to repress or tolerate the host defence system. Whether recognition is for compatibility or incompatibility the period immediately preceding and during contact is most important in the establishment of a successful symbiosis.

(1) Arrival of the fungal symbiont in the rhizosphere
The growing root provides an environment that is attractive to saprophytic, parasitic and mycorrhizal fungi alike, due to the presence of a high concentration of nutrients, in the form of exudates and sloughed-off cells (Starkey, 1959; Harley, 1969). Barber & Martin (1976) showed

---

Fig. 2.2. (a)-(c) Changes at the host-fungus interface of ectomycorrhizas induced by high levels of exogenous carbon in the medium. (a) A thickened epidermal cell wall (THCW) of *Larix kaempferi* with electron-opaque deposits (EOD), adjacent to hyphae of *Suillus grevillei* (FH). (b) Intracellular (IH) and intercellular (FH) hyphae of *S. grevillei* in a short root of *Pinus sylvestris*. (c) Wall appositions (WA) produced in the epidermal cell wall (HCW) of *Betula pubescens* at points of contact with hyphae (FH) of *S. grevillei*. The hyphae appear to be surrounded by a host 'pellicle' (P) on the root surface and adjacent epidermal cells (EC) are dead. (d) T.S. of a *Larix* mycorrhiza showing the Hartig net (HN), 'involving layer' (IL) and adjacent host cortical cells (CC). (e) Scanning electron micrograph (SEM) of the surface of a mycorrhiza of *Pyrola rotundifolia* showing the initial growth of the fungal symbiont along the cell junctions. (f) SEM of the root surface of *Allium porrum* showing the prolific growth of *S. grevillei* in the presence of exogenous carbon in the growth medium. Scale bars: (a) 0.2 μm; (b) 1.0 μm; (c) 0.5 μm; (d) 1.0 μm; (e) 10.0 μm; (f) 100.0 μm.

that the presence of non-pathogenic micro-organisms in the rhizosphere can increase the leakiness of cells. The growth of many root (Garrett, 1956) and shoot (Murray & Maxwell, 1975) infecting fungi is concentrated on cell junctions, areas of high exudation (Bowen & Rovira, 1976). This phenomenon is not so noticeable during the development of ectomycorrhizas but is illustrated well by the early stages of sheath formation in the arbutoid mycorrhizas of *Pyrola rotundifolia* (Duddridge, 1980) (Fig. 2e). The attraction of ectomycorrhizal fungi to plant roots appears to be non-specific. Melin (1959, 1963) found that an 'M' factor in plant exudates from both host and non-host roots stimulated the growth of ectomycorrhizal fungi. Other workers have shown that ectomycorrhizal fungi will grow in the rhizosphere of both incompatible (see previous section) (Duddridge, 1986*a*) and non-hosts (Filer & Toole, 1966; Theodorou & Bowen, 1971; Duddridge, 1986*c*) although in most cases not prolifically. Theodorou and Bowen (1971) found that the colonisation of non-host roots was greater than that of inert glass fibres, again suggesting the non-specific stimulation of the fungi by the roots. The only evidence of specific stimulation of ectomycorrhizal fungi by host roots is for basidiospores of *Thelephora terrestris* and a *Hebeloma* sp. (Fries & Birraux, 1980; Birraux & Fries, 1981). Many spores of pathogenic fungi can germinate equally well on host and non-host roots (Callow, 1983) and some respond positively and non-specifically to a wide range of compounds commonly contained in root exudates (Khew & Zentmeyer, 1973). However the capacity of fungal spores to germinate in the presence of non-host roots, or their exudates is not necesarily the same as the ability to infect them.

Current studies have shown that *S.grevillei* grows, albeit poorly, in the rhizosphere of non-hosts such as *Trifolium repens, Allium porrum* and *Rhododendron ponticum.* The presence of an exogenous carbon source in the synthesis medium dramatically increases the colonisation of the root surface (Fig. 2.2f).

Growth of the fungal symbiont on, and its attachment/adhesion to, the root surface
Attraction to, and growth in, the rhizosphere is followed by the adhesion of the ectomycorrhizal fungus to the host root. Surface contact between both plant pathogens (Wynn, 1976; Staples & Macko, 1980; Soulie, Vian & Guillot-Salomon, 1985) and mycorrhizal fungi (Molina & Trappe, 1982*b*; Malajczuk *et al.*, 1984; Duddridge, 1986*c*) and non-hosts are not as 'tight' as interactions between compatible symbionts. In ectomycor-

rhizas a compact sheath is never formed on non-hosts, although attachment of individual hyphae to the root surface occurs (Fig. 2.3b).

An electron-opaque fibrillar matrix has been implicated in the attachment of both pathogenic (Edwards & Allen, 1970; McKeen, 1974; Murray & Maxwell, 1975; Evans, Stempen & Stewart, 1981) and mycorrhizal (Piché *et al.*, 1983; Duddridge, 1986*a,b*) fungi. In ectomycorrhizal associations this matrix stains positively for PAS-sensitive carbohydrates (Piché *et al.*, 1983; Duddridge, 1986*a*) and Piché *et al.*(1983) suggest that it is only produced around fungal hyphae close to the root surface. In interactions where the physiological balance between the ectomycorrhizal symbionts has been upset abnormal cytological changes occur in both the host and fungal walls at the point of attachment. The host wall often swells and the 'cuticle' is broken. The fungal wall loses its integrity and becomes closely associated with the fibrils of the adjacent disrupted host wall (Fig. 2.3a).

Although dissolution of the 'cuticle' on the root surface has been previously observed during contact with ectomycorrhizal fungi (Duddridge & Read, 1984*a*; Duddridge, 1986*b*) the ultrastructural localisation of lipase and esterase at the host-fungus interface has proved unsuccessful (Duddridge unpub.).

Attachment of the fungal associate to the root surface appears to be non-specific in ectomycorrhizal association, since it occurs on non-host (Fig. 2.3b) as well as on host species (Duddridge, 1986*c*). In another group of mycorrhizas, the ericoid association, a fibrillar sheath formed round the endophyte, was originally thought to be involved in recognition because it was only produced in the presence of the host (Bonfante-Fasolo & Gianinazzi-Pearson, 1982). However later work (Bonfante-Fasolo, Gianinazzi-Pearson & Martinengo, 1984) showed that this fibrillar matrix was also produced in the presence of non-host roots and was therefore, probably a non-specific attachment mechanism.

Some avirulent and incompatible strains of bacterial pathogens are immobilised at the host cell wall by material of host origin, eventually becoming enveloped by a host pellicle, resulting in a hypersensitive host response (Bodgers, 1972; Goodman, Huang & White, 1976; Sequeira, Gaard & Zoeten, 1977; Fett & Jones, 1984). A similar phenomenon has been observed in interactions between some incompatible ectomycorrhizal symbionts (Duddridge, 1986*b*). The hyphae of *S.grevillei* were surrounded by a host pellicle on the root surface of *Betula pubescens* and adjacent epidermal cells were killed (Fig. 2.2c). Chemicals which initiate a hyper-

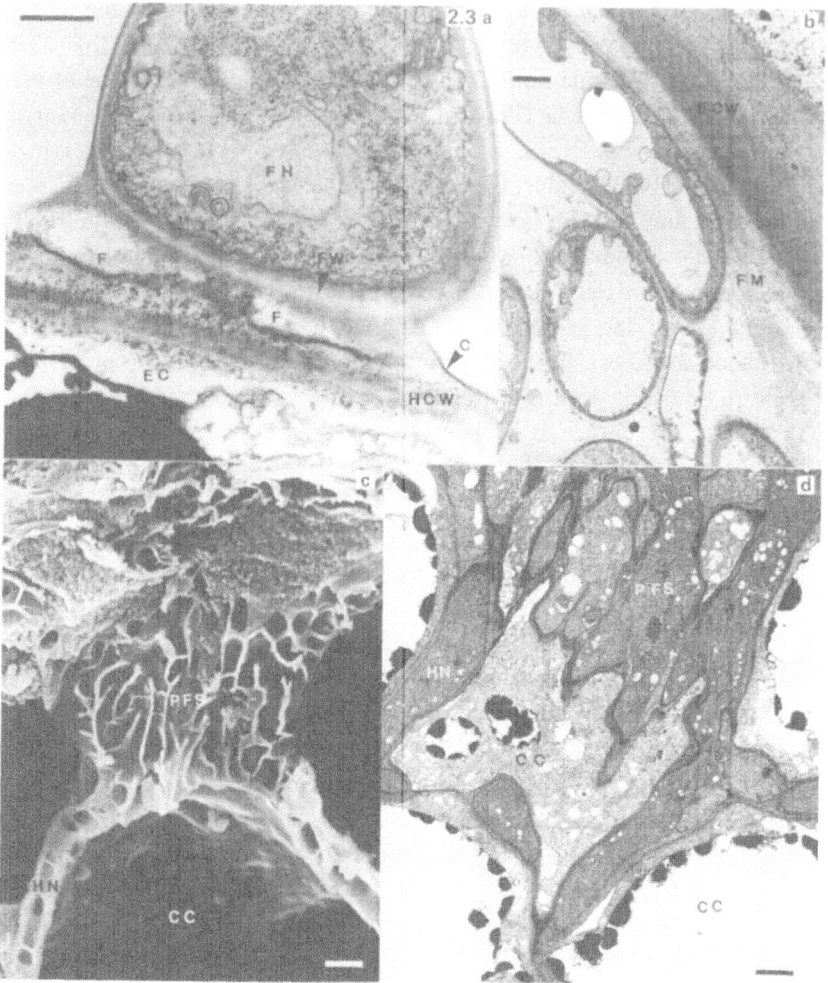

Fig. 2.3. (a) A hypha (FH) of *S. grevillei* attached to the cuticle (C) of the epidermal cell (EC) wall (HCW) of *L. kaempferi*. The fungal wall (FW) has lost its integrity and become closely associated with the fibrils (F) of the adjacent disrupted host wall. (b) Hyphae of *S. grevillei* are attached to the epidermal cell wall (ECW) of *Rhododendron ponticum*, a non-host, by fibrillar material (FM). An SEM (c) and a transmission electron micrograph (TEM) (d) of a surface of the obliquely cut outer cortical cell region showing that the Hartig net (HN) is made up of a pseudoparenchymatous fungal sheet (PFS) covering the surface of the cortical cells (CC). Scale bars: (a) 5.0 μm; (b) 0.5 μm; (c) 4.5 μm; (d) 2.0 μm.

sensitive host response are called elicitors (Albersheim & Anderson-Prouty, 1975). These have recently been isolated from some ectomycorrhizal fungi (Anderson, 1985) and the activity demonstrated on both incompatible and non-hosts (Coleman & Anderson, 1985). The latter authors suggest that fungal elicitors may be involved in the specificity of ectomycorrhizal fungi, triggering a host defence response only in incompatible interactions. In contrast in normal compatible interactions these elicitors are either masked by the production of fungal suppressors or are not produced (see de Wit, Chapter 1).

Formation of an intercellular network – the Hartig net
Attachment of ectomycorrhizal fungi to plant roots, although non-specific in itself appears to be essential before any molecular signals can take place between the fungus and plant, resulting in either a hypersensitive host response or the development of an intercellular fungal network.

The formation of an intercellular network appears to be a specific phenomenon. A regular Hartig net rarely occurs on incompatible hosts and has never been observed on non-hosts. This suggests that only a fungus which is either fully compatible or which can tolerate the host's defence system can enter into this intimate relationship. Previous ultrastructural studies (Nylund & Unestam, 1982; Duddridge & Read, 1984a,b) have revealed that intercellular penetration results in: (1) Change of mode of growth of the fungus; (2) Alterations in the cell walls of both host and fungus at the interface.

The mode of fungal growth in taxonomically diverse fungi changes from a hyphal form to the formation of a pseudoparenchymatous fungal tissue between the host cells (Figs. 2.3c, d). Tissue formation in fungi can be initiated by a wide range of stimuli (Watkinson, 1979). Nylund & Unestam (1982) considered that it was the result of substances produced by the host that subsequently resulted in tissue formation first in the Hartig net and then in the sheath. Some of the results obtained in a previous study (Duddridge & Read, 1984b) also show that tissue formation in the sheath does not occur until after the commencement of Hartig net formation. However Read & Armstrong (1972) showed that a pseudoparenchymatous fungal sheath was formed by an ectomycorrhizal fungus round artificial silicon 'roots', in oxygen-limiting, nutrient-rich conditions, only when the silicon tubing was exposed to the atmosphere. These results do not appear to support the existence of a recognition factor that is required for sheath formation, but as has been shown in an earlier section (page 28) ectomycorrhizal fungi will form a rudimentary sheath on non-hosts when an exogenous carbon source is present. It is therefore likely that recognition occurs after adhesion to the root surface but before the formation of a complete Hartig net.

At the interface between the plant cortical cells and the Hartig net, the host cell wall and to a lesser extent the fungal wall eventually lose their integrity and form an electron-opaque matrix called the 'involving' layer (Scannerini, 1968) in which the Hartig net is embedded (Duddridge & Read, 1984a,b) (Figs. 2.2d, 2.3d). The 'involving' layer appears to be made up of an area of modified plant material with a high pectin content (Duddridge, 1980), similar to that found round many intracellular fungal symbionts (Strullu & Gourret, 1975; Gil & Gay, 1977; Dexheimer, Gianinazzi & Gianinazzi-Pearson, 1979; Duddridge & Read, 1982a). The fungus may possibly interfere with the process by which the plant cell controls wall formation, thus altering the ratio of pectic to cellulosic material (Duddridge, 1980). No change in the cytoplasm and organelles of the cells adjacent to the Hartig net is observable in response to infection (Duddridge & Read, 1984a,b), as would be found in plant cells infected by either biotrophic fungal pathogens (Ehrlich & Ehrlich, 1966; Littlefield & Bracker, 1972) or intracellular mycorrhizal symbionts (Dörr & Kollmann, 1969; Strullu, 1976; Duddridge, 1980).

Morphological changes which occur in the plant cell wall during Hartig net formation represent the first indication of the host's response to infection at the molecular level. Harley & Smith (1983) suggested that proteins in the fungal membrane co-polymerise with host-wall building enzymes, inhibiting wall formation and allowing the fungus to obtain cell wall precursors. Carbohydrate-binding proteins called lectins are implicated in recognition between partners in plant symbioses such as the *Rhizobium*/legume association (see Stacey *et al.,* 1980 for refs.), lichens (Lockhart, Rowell & Stewart, 1978) plant pathogen interactions (Hinch & Clarke, 1980; Callow, 1983) and also in mycoparasitism (Elad, Barack & Chet, 1983). Hinch & Clarke (1980) found that the adhesion of zoospores of *Phytophthora infestans* was mediated by L–fucose determinants in the root surface slime and fucose receptors on the zoospore surface. Inspite of the increasing quantity of research on lectins, their role in recognition is still uncertain.

Conclusion

Although a few ectomycorrhizal associations show distinct patterns of specificity, they are generally non-specific interactions. The flexibility of both host and fungus which allows them to form different types of mycorrhizas with different partners makes difficult the search for specific molecules/substances involved in recognition. We do not understand the mechanism which allows an ectomycorrhizal fungus to form an intracellular association on an arbutoid host while remaining intercellular on an ectomycorrhizal host. Similary it is not known how some ectomycorrhizal hosts are able to form endomycorrhizas under ecological conditions. We are still ignorant of the exact stage of development epidermal and cortical cells have to reach before Hartig net formation will occur or why they are susceptible at this time. Harley & Smith (1983) suggested that they should be in the final stage of maturation in which primary wall building is still taking place but this has never been proved. Smith & Walker (1981) found with vesicular-arbuscular mycorrhizas that the root tip was ten times more likely to become infected than the rest of the root. Hepper (1985) also observed that the specific pattern of infection in the root systems of leek and clover was related to the age of the root cells. Having established which cells are susceptible the next step will be to determine the ultrastructural and enzymic changes that occur in the symbiont cell walls and membranes both before and after contact with the host. Current studies are investigating some of these aspects but many basic questions remain to be resolved.

References

Albersheim, P. & Anderson-Prouty, A. J. (1975). Carbohydrates, proteins, cell surfaces and the biochemistry of pathogenesis. *Annual Review of Plant Physiology,* **26,** 31–52.

Anderson, A. (1985). The problems of living with a plant-root: factors involved in root colonisation. In *Proceedings of the 6th North American Conference on Mycorrhizae,* ed. R. Molina, pp. 175–8. Corvallis: Oregon State University Press.

Barber, D. A. & Martin, J. K. (1976). The release of organic substances by cereal roots into soil. *New Phytologist,* **76,** 69–80.

Birraux, D. & Fries, N. (1981). Germination of *Thelephora terrestris* basidiospores. *Canadian Journal of Botany,* **59,** 2062–4.

Bodgers, R. J. (1972). On the interaction of *Agrobacterium tumefaciens* with cells of *Kalanchoe daigremontiana.* In: *Proceedings of the 3rd International Conference on Plant Pathogenic Bacteria,* ed. H. P. Maas Geesteranus, pp. 239–50. The Netherlands; Wageningen.

Bonfante-Fasolo, P. & Gianinazzi-Pearson, V. (1982). Ultrastructural aspects of endomycorrhiza in the Ericaceae. III. Morphology of the dissociated symbionts and modifications occurring during their reassociation. *New Phytologist,* **91,** 691–704.

Bonfante-Fasolo, P. Gianinazzi-Pearson, V. & Martinengo, L. (1984). Ultrastructural aspects of endomycorrhiza in the Ericaceae. IV. Comparison of infection by *Pezizella ericae* in host and non-host plants. *New Phytologist,* **98,** 329–33.

Bowen, G. D. & Rovira, A. D. (1976). Microbial colonisation of roots. *Annual Review of Phytopathology,* **14,** 121–44.

Callow, J. A. (1983). Cellular interactions and molecular recognition between higher plants and fungal pathogens. In *Cellular Interactions*, ed. H. F. Linskens & J. Heslop-Harrison, pp. 212–37. *Encyclopaedia of Plant Physiology* New Series vol. 17. New York: Springer-Verlag.

Chilvers, G. A. (1973). Host range of some eucalypt mycorrhizal fungi. *Australian Journal of Botany*, **21**, 103–11.

Clowes, F. A. L. (1951). The structure of mycorrhizal roots of *Fagus sylvatica*. *New Phytologist*, **50**, 1–16.

Coleman, M. E. & Anderson, A. J. (1985). The role of elicitors in ectomycorrhizal formation. In: *Proceedings of the 6th North American Conference on Mycorrhizas*, ed. R. Molina, pp. 361-2. Corvallis: Oregon State University Press.

Cooper, K. M. (1976). A field survey of mycorrhizas in New Zealand ferns. *New Zealand Journal of Botany*, **14**, 169–81.

Danielson, R. M. (1984). Ectomycorrhizal association in jack pine stands in northeastern Alberta. *Canadian Journal of Botany*, **62**, 932–9.

Debaud, J. C., Pepin, R. & Bruchet, G. (1981). Ultrastructure des ectomycorrhizes synthétiques à *Hebeloma alpinum* et *Hebeloma marginatalum* de *Dryas octopetala*. *Canadian Journal of Botany*, **59**, 2160–6. de Wit, (Chapter 1 this volume).

Dexheimer, J., Gianinazzi, S. & Gianinazzi–Pearson, V. (1979). Ultrastructural cytochemistry of the host-fungus interface in the endomycorrhizal association *Glomus mosseae/Allium cepa*. *Zeitschrift für Pflanzenphysiologie*, **86**, 189–201.

Dïr, I. & Kollman, R. (1969). Fine structure of mycorrhiza in *Neottia nidus-avis* (L) L.C. Rich (Orchidaceae). *Planta*, **89**, 373-5.

Duddridge, J. A. (1980). A comparative ultrastructural analysis of a range of mycorrhizal associations. *Ph.D. Thesis, University of Sheffield*.

Duddridge, J. A. (1986*a*). The development and ultrastructure of ectomycorrhizas. III. Compatible and incompatible interactions between *Suillus grevillei* (Klotzsch) Sing. and a number of ectomycorrhizal hosts *in vitro*, in the absence of exogenous carbohydrate. *New Phytologist*, **103**, 457-64.

Duddridge, J. A. (1986*b*). The development and ultrastructure of ectomycorrhizas. IV. Compatible and incompatible interactions between *Suillus grevillei* (Klotzsch) Sing. and a number of ectomycorrhizal hosts *in vitro*, in the presence of exogenous carbohydrate. *New Phytologist*, **103**, 465-71.

Duddridge, J. A. (1986*c*). Specificity and recognition in mycorrhizal associations. In *Physiological and Genetical Aspects of Mycorrhizae. Proceedings of the 1st European Symposium on Mycorrhizae. Dijon 1986*, ed. V. Gianinazzi-Pearson & S. Gianinazzi, pp. 145-58. Paris: INRA.

Duddridge, J. A. & Read, D. J. (1982*a*). An ultrastructural analysis of the development of mycorrhizas in *Rhododendron ponticum*. *Canadian Journal of Botany*, **60**, 2345–56.

Duddridge, J. A. & Read, D. J. (1982*b*). An ultrastructural analysis of the development of mycorrhizas in *Monotropa hypopitys* L. *New Phytologist*, **92**, 203–14.

Duddridge, J. A. & Read, D. J. (1984*a*). The development and ultrastructure of ectomycorrhizas. I. Ectomycorrhizal development on pine in the field. *New Phytologist*, **96**, 565–73.

Duddridge, J. A. & Read, D. J. (1984*b*). The development and ultrastructure of ectomycorrhizas. II. Ectomycorrhizal development on pine *in vitro*. *New Phytologist*, **96**, 575–82.

Duddridge, J. A. & Read, D. J. (1984*c*). Modification of the host-fungus interface in mycorrhizas synthesised between *Suillus bovinus* (Fr). O. Kuntz. and *Pinus sylvestris* L. *New Phytologist*, **96**, 583–8.

Duddridge, J. A., Finlay, R. D. & Read, D. J. (1987). The structure and function of vegetative mycelium of ectomycorrhizal roots. III. Ultrastructural and autoradiographic analysis of carbon distribution through intact mycelium systems. *New Phytologist* (in press).

Edwards, H. H. & Allen, P. J. (1970). A fine structure study of the primary infection process during infection of barley by *Erysiphe graminis* f.sp. *hordei*. *Phytopathology*, **60**, 1504-9.

Ehrlich, M. A. & Ehrlich, H. G. (1966). Ultrastructure of the hyphae and haustoria of *Phytophthora infestans* and hyphae of *P. parasitica*. *Canadian Journal of Botany*, **44**, 1495-1503.

Elad, Y., Barack, R. & Chet, I. (1983). Possible role of lectins in mycoparasitism. *Journal of Bacteriology*, **154**, 1431-5.

Evans, R. C., Stempen, H. & Stewart, S. J. (1981). Development of hyphal sheaths in *Bipolaris maydis* race T. *Canadian Journal of Botany*, **59**, 453-9.

Fett, W. F. & Jones, S. B. (1984). Stress metabolite accumulation, bacterial growth and bacterial immobilisation during host and non-host responses of soybean to bacteria. *Physiological Plant Pathology*, **25**, 277-96.

Filer, T. H. & Toole, E. R. (1966). Sweet gum mycorrhizae and some associated fungi. *Forest Science*, **12**, 432-7.

Fontana, A. (1962). Richerche sulle micorrhize del genere *Salix*. *Allionia*, **8**, 67-85.

Foster, R. C. & Marks, G. C. (1967). Observations on the mycorrhiza of forest trees. II. The rhizosphere of *Pinus radiata* D. Don. *Australian Journal of Biological Science*, **20**, 915-26.

Fries, N. & Birraux, D. (1980). Spore germination in *Hebeloma* stimulated by living plant roots. *Experientia*, **36**, 1056-7.

Froidevaux, L. (1973). The ectomycorrhizal association *Alnus rubra* + *Lactarius obscuratus*. *Canadian Journal of Forest Research*, **3**, 601-3.

Fusconi, A. (1983). The development of the fungal sheath on *Cistus incanus* short roots. *Canadian Journal of Botany*, **61**, 2546-53.

Garrett, S. D. (1956). *The Biology of Root-infecting Fungi*. Cambridge: Cambridge University Press.

Gianinazzi-Pearson, V. (1984). Host-fungus specificity, recognition and compatibility in mycorrhizae. In *Genes involved in Microbe-Plant Interactions. Advances in Plant Gene research, Basic Knowledge and Application*, volume 1. pp. 225-53. New York: Springer-Verlag.

Gil, F.& Gay, J. L. (1977). Ultrastructural and physiological properties o the host interfacial components of haustoria of *Erysiphe pisi in vivo* and *in vitro*. *Physiological Plant Pathology*. 10, 1-12.

Giltrap, N. J. (1979). Experimental studies on the establishment and stability of ectomycorrhizas. Ph.D. Thesis, University of Sheffield.

Godbout, C. & Fortin, J. A. (1983). Morphological features of synthesised ectomycorrhizae of *Alnus crispa* and *A. rugosa*. *New Phytologist*, **94**, 249-62.

Goodman, R. N., Huang, P-Y. & White, J. A. (1976). Ultrastructural evidence for immobilisation of an incompatible bacterium *Pseudomonas pisi* in tobacco leaf tissue. *Phytopathology*, **66**, 754-64.

Hammerschmidt, R. & Kuć, J. (1982). Lignification as a mechanism for induced systemic resistance in cucumber. *Physiological Plant Pathology*, **20**, 61-71.

Harley, J. L. (1969). *The Biology of Mycorrhiza*. 2nd edition. London: Leonard Hill.

Harley, J. L. (1985). Specificity and penetration of tissues by mycorrhizal fungi. *Proceedings of the Indian Academy of Science* (Plant Science), **94**, 99-109.

Harley, J. L. & Smith, S. E. (1983). *Mycorrhizal Symbiosis.* London, New York: Academic Press.

Hepper, C. M. (1985). Influence of age of roots on the pattern of vesicular–arbusclar mycorrhizal infection in leek and clover. *New Phytologist,* **101**, 685–93.

Hinch, J. M. & Clarke, A. E. (1980). Adhesion of fungal zoospores to root surfaces is mediated by carbohydrate determinants of root slime. Binding of *Phtophthora cinnamomi* zoospores to *Zea mays. Physiological Plant Pathology,* **16**, 303-7.

Iqbal, S. H., Yousaf, M. & Younus, M. (1981). A field survey of mycorrhizal associations in ferns of Pakistan. *New Phytologist,* **87**, 69–79.

Khew, K. L. & Zentmeyer, G. A. (1973). Chemotactic responses of zoospores of five species of *Phytophthora. Phytopathology,* **63**, 1511–17.

Kropp, B. P. & Trappe, J. M. (1982). Ectomycorrhizal fungi of *Tsuga heterophylla. Mycologia,* **74**, 479–88.

Littlefield, L. J. & Bracker, C. E. (1972). Ultrastructural specialisation at the host-pathogen interface in rust-infected flax. *Protoplasma,* **74**, 271–305.

Lockhart, C. M., Rowell, P. & Stewart, W. D. P. (1978). Phytohaemagglutinin from TM nitrogen fixing lichens *Peltigera canina* and *P. polydactyla. FEMS Microbiological Letters,* **3**, 127-30.

Malajczuk, N., Molina, R. & Trappe, J. M. (1982). Ectomycorrhiza formation in *Eucalyptus.* I. Pure culture synthesis, host specificity and mycorrhizal compatibility with *Pinus radiata. New Phytologist,* **91**, 467–82.

Malajczuk, N., Molina, R. & Trappe, J. M. (1984). Ectomycorrhiza formation in *Eucalyptus.* II. The ultrastructure of compatible and incompatible mycorrhizal fungi and associated roots. *New Phytologist,* **96**, 43–53.

Malibari, A. A. (1979). Biology of ectomycorrhizas with special reference to their possible role in plant water relations. *Ph.D. Thesis, University of Sheffield.*

Malloch, D. & Malloch, B. (1981). The mycorrhizal status of boreal plants: species from northeastern Ontario. *Canadian Journal of Botany,* **59**, 2167–72.

Marks, G. C. & Foster, R. C. (1973). Structure, morphogenesis and ultrastructure of ectomycorrhizas. In *Ectomycorrhizae: their ecology and physiology,* ed. G. C. Marks & T. T. Kozlowski, pp. 1–41. New York: Academic Press.

Marx, D. H. (1977). Tree host range and world distribution of the ectomycorrhizal fungus *Pisolithus tinctorius. Canadian Journal of Microbiology,* **23**, 217–23.

Marx, D. H. (1979). Synthesis of *Pisolithus* mycorrhizae on White Oak seedlings in fumigated nursery soil. *Forest Service Research Note, USDA,* 3E. 280.

Mason, P. (1975). The genetics of mycorrhizal associations between *Amanita muscaria* and *Betula verrucosa.* In *The Development and Function of Roots,* ed. J. G. Torrey & D. T. Clarkson, pp. 567–74. New York & London: Academic Press.

McKeen, W. E. (1974). Mode of penetration of epidermal cell walls of *Vicia faba* by *Botrytis cinerea. Phytopathology,* **64**, 461–7.

Melin, E. (1922). Untersuchungen uber die *Larix* mykorrhiza. I. Synthese der mykorrhiza in reinculture. *Svensk botanisk tidskrift,* **16**, 165-96.

Melin, E. (1959). Mykorrhiza. In: *Handbuch der Pflanzenphysiologie XI,* ed. W. Ruland, pp. 605–38. Berlin: Springer–Verlag.

Melin, E. (1963). Some effects of forest tree roots on mycorrhizal basidiomycetes. In: *Symbiotic Associations,* ed. B. Mosse & P. S. Nutman, pp. 124–45. Cambridge: Cambridge University Press.

Meyer, F. H. (1973). Distribution of ectomycorrhizae in natural and man–made forests. In *Ectomycorrhizae: their ecology and physiology,* ed. G. C. Marks & T. T. Kozlowski, pp. 79–105. New York & London: Academic Press.

Molina, R. (1981). Ectomycorrhizal specificity in the genus *Alnus. Canadian Journal of Botany,* **59**, 325–34.

Molina, R. & Trappe, J. M. (1982a). Lack of mycorrhizal specificity by the ericaceous hosts *Arbutus menziesii* and *Arctostaphylos uva-ursi*. *New Phytologist*, **90**, 495-509.

Molina, R. & Trappe, J. M. (1982b). Patterns of ectomycorrhizal specificity and potential among Pacific Northwest conifers and fungi. *Forest Science*, **28**, 423-58.

Moser, M. (1978). Rohrlinge and Blatterpilze. 4 Auflage. Band 11b *der Kleinen Kryptogamenflora*, ed. H. Gams. Stuttgart: Gustav Fischer Verlag.

Murray, G. M. & Maxwell, D. P. (1975). Penetration of *Zea mays* by *Helminthosporium carbonum*. *Canadian Journal of Botany*, **53**, 2872-83.

Nylund, J-E. (1981). The formation of ectomycorrhiza in conifers: structural and physiological studies with special reference to the mycobiont *Piloderma croceum* Erikss & Hjortst. *Ph.D. Thesis, University of Uppsala*.

Nylund, J-E., Kasimir, A. & Strandberg-Arveby, S. (1982). Cell wall penetration and papilla formation in senescent cortical cells during ectomycorrhiza synthesis *in vitro*. *Physiological Plant Pathology*, **21**, 71-3.

Nylund, J-E. & Unestam, T. (1982). Structure and physiology of ectomycorrhiza. I. The process of mycorrhiza formation in Norway spruce *in vitro*. *New Phytologist*, **91**, 63-79.

Piché, Y., Fortin, J. A. & Lafontaine, J. G. (1981). Cytoplasmic phenols and polysaccharides in ectomycorrhizal and non-mycorrhizal short roots of pine. *New Phytologist*, **88**, 695-703.

Piché, Y., Peterson, R. L., Howarth, M. J. & Fortin, J. A. (1983). A structural study of the interaction between the ectomycorrhizal fungus *Pisolithus tinctorius* and *Pinus strobus* roots. *Canadian Journal of Botany*, **61**, 1185-93.

Pocock, K. & Duckett, J. G. (1984). A comparative ultrastructural analysis of the fungal endophytes in *Cryptothallus mirabilis* Malm. and other British thalloid hepatics. *Journal of Biology*, **13**, 227-33.

Read, D. J. & Armstrong, W. (1972). A relationship between oxygen transport and the formation of the ectotrophic mycorrhizal sheath in conifer seedlings. *New Phytologist*, **71**, 49-53.

Robertson, D. C. & Robertson, J. A. (1982). Ultrastructure of *Pterospora andromedea*. Nuttall and *Sarcodes sanguinea* Torrey mycorrhizas. *New Phytologist*, **92**, 539-51.

Ronald, P. & Soderhall, K. (1985). Phenyl alanine ammonia lyase and peroxidase activity in mycorrhizal and non-mycorrhizal short roots of Scots pine. *Pinus sylvestris* L. *New Phytologist*, **101**, 487-94.

Scannerini, S. (1968). Sull'ultrastructuttura delle ectomicorrize. II. Ultrastruttura di una micorriza di ascomicete: *Tuber albidum* x *Pinus strobus* L. *Allionia*, **14**, 77-95.

Sequeira, L., Gaard, G. & Zoeten, G. A. (1977). Attachment of bacteria to host cell walls: its relation to mechanisms of induced resistance. *Physiological Plant Pathology*, **10**, 43-50.

Smith, S. E. & Walker, N. A. (1981). A quantitative study of mycorrhizal infection in *Trifolium:* separate determination of rates of infection and mycelial growth. *New Phytologist*, **89**, 225-40.

Soulie, M-C., Vian, B. & Guillot-Salomon, T. (1985). Interactions hôte-parasite lors de l'infection par *Cercosporella herpotrichoides*, agent du pietin-verse: morphologie du parasite et ultrastructure des parois d'hôtes sensibles et resistants. *Canadian Journal of Botany*, **63**, 851-8.

Stacey, G., Paau, A. S. & Brill, W. J. (1980). Host recognition in the *Rhizobium*-soybean symbiosis. *Plant Physiology*, **66**, 609-14.

Staples, R. & Macko, V. (1980). Formation of infection structures as recognition response in fungi. *Experimental Mycology*, **4**, 2–16.

Starkey, R. L. (1959). Interrelations between microorganisms and plant roots in the rhizosphere. *Bacteriology Review*, **22**, 154–72.

Strullu, D. G. (1976). Recherches de biologie et de microbiologie forestières. *Ph.D. Thesis, University of Rennes, France.*

Strullu, D. G. & Gourret, J. P. (1975). Ultrastructure et évolution du champignon symbiotique des racines de *Dactylorchis maculata* (L). Verm. *Journal de Microscopie*, **20**, 285–94.

Sylvia, D. M. & Sinclair, W. A. (1983). Phenolic compounds and resistance to fungal pathogens induced in primary roots of Douglas Fir seedlings by the ectomycorrhizal fungus *Laccaria laccata*. *Phytopathology*, **73**, 390–7.

Theodorou, C. & Bowen, G. D. (1970). Effects of non-host plants on the growth of mycorrhizal fungi of radiata pine. *Australian Forestry*, **35**, 17–22.

Theodorou, G. & Bowen, G. D. (1971). Mycorrhizal responses of radiata pine in experiments with different fungi. *Australian Forestry*, **34**, 183-91.

Thomas, G. W. & Jackson, R. M. (1979). Sheathing mycorrhizas of nursery-grown *Picea sitchensis*. *Transactions of the British Mycological Society*, **73**, 117–25.

Thomas, G. W. & Jackson, R. M. (1983). Growth responses of sitka spruce seedlings to mycorrhizal inoculation. *New Phytologist*, **95**, 223–9.

Trappe, J. M. (1962). Fungus associates of ectotrophic mycorrhizas. *Botanical Review*, **28**, 538–606.

Trappe, J. M. (1964). Mycorrhizal hosts and distribution of *Cenococcum graniforme*. *Lloydia*. **27**, 100-6.

Trappe, J. M. & Fogel, R. D. (1977). Ecosystematic functions of mycorrhizae. *Coniferous Forest Biome*, Contribution 85.

Vance, C. P., Kirk, T. K. & Sherwood, R. T. (1980). Lignification as a mechanism of disease resistance. *Annual Review of Phytopathology*, **18**, 259–88.

Vozzo, J. A. & Hacskaylo, E. (1974). Endo- and ectomycorrhizal associations. *Bulletin of the Torrey Botanical Club*, **101**, 182–6.

Warcup, J. H. (1980). Ectomycorrhizal associations of some Australian indigenous plants. *New Phytologist*, **85**, 531–5.

Warcup, J. H. & McGee, P. A. (1983). The mycorrhizal associations of some Australian Asteraceae. *New Phytologist*, **95**, 667–72.

Watkinson, S. C. (1979). Growth of rhizomorphs, mycelial strands, coremia and sclerotia. In *Fungal Walls and Hyphal Growth*, ed. J. H. Burnett & A. P. J. Trinci, pp. 93–113. Cambridge: Cambridge University Press.

Whistler, R. L., Bachrach, J. & Bowman, D. R. (1948). Preparation and properties of corn cob holocellulose. *Archives of Biochemistry*, **19**, 25–33.

Wynn, W. K. (1976). Appressorium formation over stomates by the bean rust fungus: response to a surface contact stimulus. *Phytopathology*, **66**, 136–46.

Zak, B. (1976a). Pure culture synthesis of bearberry mycorrhizae. *Canadian Journal of Botany*, **12**, 1297–1305.

Zak, B. (1976b). Pure culture synthesis of Pacific madrone ectendomycorrhizae. *Mycologia*, **68**, 362–9.

3

Spores on leaves: endogenous and exogenous control of development

JOHN LUCAS[1] and IAN KNIGHTS[2]

[1]*Department of Botany, University of Nottingham, University Park, Nottingham NG7 2RD, UK.* [2]*Schering Agrochemicals Ltd., Chesterford Park Research Station, Saffron Walden, Essex CB10 IXL, UK*

The interaction between an aerially dispersed parasitic fungus and a potential plant host begins when a spore impacts or settles on the leaf surface. What follows depends upon conditions prevailing in the phylloplane environment and also the capacity of the spore to enter a complex morphogenetic programme in response to appropriate environmental signals. For obligately biotrophic fungi this prepenetration phase of development is critical because the establishment of contact with nutrient resources within host cells is a prerequisite for further growth. These fungi possess adaptations ensuring that germination and subsequent developmental events are regulated by endogenous and exogenous controls which together optimise their chances of successfully colonising the host.

Fungal propagules are extremely diverse in form and function, ranging from thick-walled resistant resting spores to short-lived motile zoospores. A comprehensive survey of the many different spore types and their modes of germination and host penetration is clearly beyond the scope of the present review and has been documented elsewhere (Weber & Hess, 1976). This review is concerned with a single spore type, the aerially dispersed uredospore of rust fungi (Uredinales) and concentrates on a very limited number of species from the genera *Puccinia* and *Uromyces*.

The rust fungi are ecologically obligate parasites that have undergone close co-evolution with their host plants (Peterson, 1974). These fungi have complex life cyles involving as many as five different spore stages; the uredospore is the repeating stage in the cycle and its production underlies the exponential growth of the pathogen population. It has an essential role in the progress of disease epidemics.

The rust uredospore is an appropriate model for study by developmental biologists because germination on the leaf surface is followed by a finite period of germ tube growth and the differentiation of specialised infection structure. A number of steps in the penetration process can be distinguished (Fig. 3.1.) – spore hydration and swelling, term tube emergence, germ tube adherence to the leaf surface, extension and orientation of the germ tube, appressorium formation, infection peg development, and production of a vesicle in the substomatal cavity. This phase of development is fuelled entirely by endogenous energy reserves. Further growth inside the host requires the formation of functional haustoria and the subsequent uptake of host nutrients.

Factors inducing the differentiation of infection structures from germinating rust uredospores have received much attention and are described in detail in several recent articles (Littlefield & Heath, 1979; Macko, 1981; Staples & Macko, 1980, 1984). This review will focus on the early stages of germination, in particular germ tube emergence, and discuss evidence that this step also is regulated by a very sensitive system of induction or inhibition which ensures survival of the developing germling.

Factors controlling uredospore germination

Uredospores are comparatively large propagules containing reserves of lipid and carbohydrate (Daly, Knoche & Wiese, 1967), which

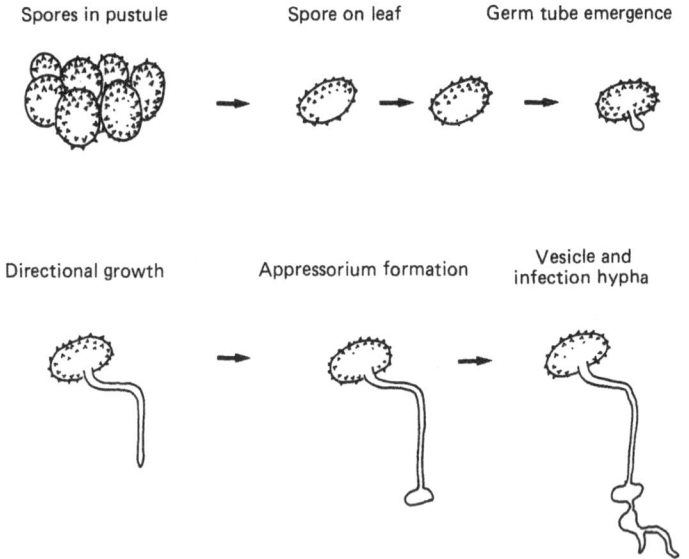

Fig. 3.1. Diagrammatic sequence showing uredospore germination and the development of infection structures.

Table 3.1. *Factors known to influence germination of rust uredospores*

Major	Other
Endogenous	**Endogenous**
Self-inhibitors	Germination stimulants
Exogenous	**Exogenous**
Temperature	Microbial antagonists
Light	Host factors
{ Relative humidity	{ Cultivar
{ Surface moisture	{ Leaf age
	{ Germination stimulants
	{ Nutrients/inhibitors

provide ample energy and materials for the synthesis of new cell wall and membranes (Staples & Macko, 1984) and all the components necessary for the synthesis of proteins (Yaniv & Staples, 1974). These spores are therefore not dependent on an exogenous supply of nutrients or mineral ions and can germinate in distilled, deionised water, although experiments with exogenously applied ^{14}C-labelled sugars and amino acids have shown that germinating uredospores take up extremely low amounts of these nutrients (Parker & Blakeman, 1984).

Many factors influence the germination of uredospores; in Table 3.1 these are grouped as either endogenous (for instance metabolic factors), or exogenous (environmental factors). It should be stressed that the further subdivision into factors of major or minor importance may simply reflect our incomplete knowledge of the latter. Of the many exogenous factors affecting rust spores, least is known about the significance of microbial antagonists, and compounds originating from the host plant itself.

The effects of the phylloplane microflora on uredospore germination are not clear. It is questionable whether bacterial populations ever reach levels sufficient to affect germination, except perhaps on lower leaves splashed by soil particles. *In vitro* studies (Mishra & Tewari, 1976; Reinecke, 1981) have shown however, that many leaf saprophytes are potentially antagonistic. Doherty & Preece (1978) found that a high proportion of leek rust, *Puccinia allii,* uredospores carried bacteria, principally *Bacillus cereus.* Pretreating leaves with high concentrations of this bacterium dramatically reduced rust infection, but low concentrations had no effect. Several of the phylloplane bacteria which inhibit germination or adversely affect germ tube growth in *P. striiformis* also appear to be *Bacillus* species (C. McLaughlin & J. G. Manners, unpublished). Overall, there is a need for more information relating to the behaviour of uredospores on mature, field-grown leaves populated by a diverse microflora.

Leaf exudates might be of importance in signalling a favourable environment or a potential host plant (Macko, 1981), but while several germination stimulants of host origin have been detected (French, 1985) there is little convincing evidence for a significant role for such compounds in host recognition by rust fungi. In normal circumstances uredospore germination does not appear to be greatly influenced by external biotic factors, and it seems that the physical environment is therefore of more significance.

Initiation of germination on the surface of the host requires free water and a favourable temperature (Sharp *et al.*, 1958; Burrage, 1970). The optimum temperature depends upon the species concerned; for instance *Puccinia striiformis* germinates best around 10°C (Manners, 1981), while the optimum for *P. graminis* is around 22°C. However, even if viable uredospores are incubated at a favourable temperature and relative humidity, germination will not necessarily commence. This may be due either to dormancy imposed by endogenous self-inhibitors, or to the influence of a further environmental factor, light.

Germination self-inhibitors

Regulatory molecules involved in the control of germination have been identified in a number of fungi and slime moulds (Trione, 1981). When uredospores are floated on water germination is dependent on spore density, with inhibition at high densities. Allen (1955) was the first to show that this phenomenon is due to compounds released from within the spores. Several germination self-inhibitors from rust fungi have now been isolated and characterised (Macko, 1981). All are low molecular weight compounds related to cinnamic acid; the self-inhibitor from black stem rust, *Puccinia graminis,* has been identified as methyl-*cis*-ferulate (MF), while uredospores of several other species, including yellow stripe rust, *P. striiformis,* peanut rust, *P. arachidis,* and bean rust *Uromyces phaseoli,* contain methyl-*cis*-3,4-dimethoxycinnamate (MDC). These compounds are volatile and highly potent, showing biological activity at concentrations as low as a few pg cm^{-3} (Macko, 1981).

Germ tube emergence commences with digestion of a distinct region of the uredospore wall known as the germ pore. On the basis of an ultrastructural study, Hess *et al.,* (1975) proposed that methyl-*cis*-ferulate acts by inhibiting enzymic dissolution of the germ pore plug. The significance of this observation will be considered later.

Photoinhibition of uredospore germination

An early study by Dillon-Weston (1931) showed that sunlight is inhibitory to uredospore germination. Sharp *et al.,* (1958) noted an inverse relationship between germination of *P. graminis* f.sp. *tritici* (Pgt) uredospores and light intensity. Similar photosensitivity of germination has now been reported in a wide range of rust species, including coffee rust, *Hemileia vastatrix* (Nutman & Roberts, 1963), as well as many of the major cereal rusts. Detailed analysis of this photoresponse, however, has been conducted almost entirely with Pgt.

The main features of photodormancy in Pgt uredospores can be summarised as follows: continuous irradiation with high intensity white light inhibits germination on both artificial substrates such as agar, and on leaves (Knights & Lucas, 1980). The threshold level of irradiation for complete photoinhibition of germination appears to be lower on leaf surfaces than on agar. If removed from light, spores recover to germinate normally; furthermore, partial recovery takes place eventually even under continuous irradiation (Fig. 3.2). Light therefore delays germination rather than imposes an indefinite dormancy.

Calpouzos & Chang (1971) determined a preliminary action spectrum for this photoresponse and showed that blue and far–red light were most inhibitory. Subsequent studies (Lucas, Kendrick & Givan, 1975) suggested that a red-far red photoreversible system similar to phytochrome might be involved. It is of interest that photosensitivity in the blue region of the spectrum is similar to numerous other fungal photoresponses in which carotenoid or flavoprotein photoreceptors have been implicated (Tan, 1978).

Photosensitivity is affected by the physiological state of the spore and in particular the degree of hydration. Work with several cereal rust species indicates that fully hydrated spores are more photosensitive than dehydrated spores (Givan & Bromfield, 1964*a* 1964*b*; Zadoks & Groenewegen, 1967; Chang, Calpouzos & Wilcoxson, 1973). During hydration, the fresh weight of spores approximately doubles and their volumes increase by one third (Strobel, 1965; Knights & Lucas, 1980). At the same time photosensitivity gradually increases (Fig. 3.3). The degree of photosensitivity of spores is not simply related to their water content, since re-drying hydrated spores does not alter their response to

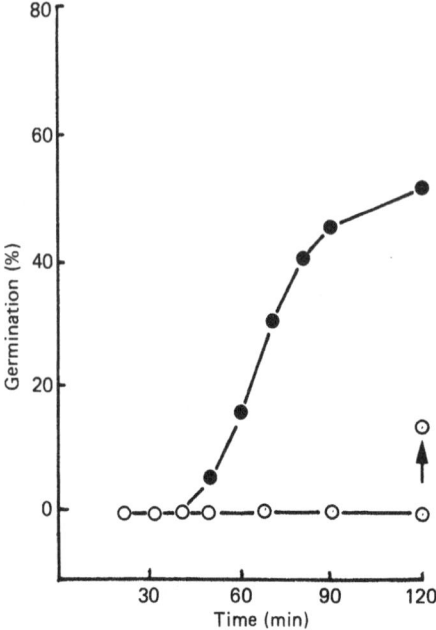

Fig. 3.2. Time course of uredospore germination in dark (●) and in light (O) on excised wheat leaves. Arrow indicates germination after 24h continuous light. (From Knights & Lucas, 1980.)

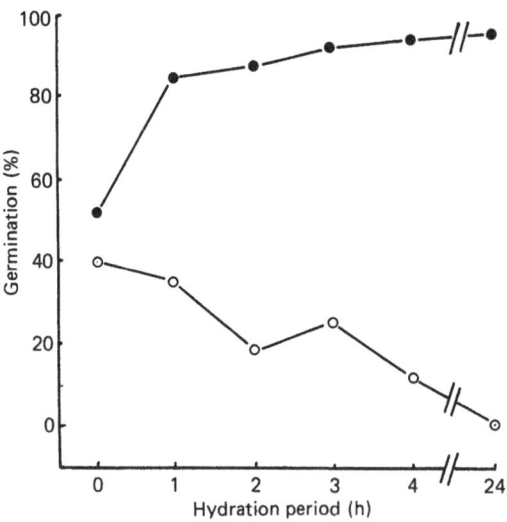

Fig. 3.3. Effect of length of hydration period on subsequent germination of *P. graminis* f.sp. *tritici* uredospores on tap water agar in the dark (●) and in the light (O). (10,000 lux, approx. 4.5×10^4 erg cm^{-2} s^{-1})

light. Rather, the evidence suggests that during the hydration process physiological changes take place within spores that enhance their photosensitivity. Interestingly, if spores are irradiated throughout the hydration period, they can be shown in subsequent germination tests to have substantially lost their photosensitivity (Fig. 3.4). Another treatment reducing photosensitivity is washing or soaking spores (Chang, Wilcoxson & Calpouzos, 1974). Knights (1981) found a marked negative correlation between the length of the washing treatment and subsequent photoinhibition. This suggests that a water-soluble photochemically active substance may be present in hydrated spores which inhibits their germination in the light.

A model of photoinhibition

On the basis of these observations a tentative model of the photoinhibition process can be proposed (Fig. 3.5). The key features of the model are as follows:

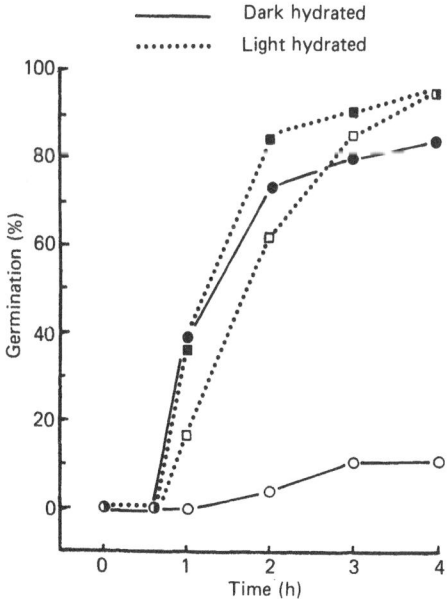

Fig. 3.4. Progress of germination of Pgt uredospores hydrated for 8 h in the dark, or under low-intensity white light (4500 lux, approx. 2.7×10^4 erg cm^{-2} sec^{-1}), then sown on tap water agar in the dark (closed symbols) or in high intensity (10000 lux, approx. 4.5×10^4 erg cm^{-1}) white light (open symbols).

1. A finite amount of precursor A is present in dehydrated spores.
2. During hydration in the dark, A is slowly converted to compound B, so that after 8h all A is lost.
3. Compound B is not itself inhibitory, so that germination commences immediately spores are transferred to favourable conditions in the dark.
4. In the presence of light, B is rapidly photoconverted to an inhibitory compound C. The degree of conversion depends on the level of irradiation.
5. Compound C is continually lost through either degradation or evaporation, so that in the absence of further production of C, the amount of inhibitor slowly declines.
6. A critical threshold concentration of C is required to prevent germination.

This model is consistent with most of the currently available data. Because the amount of compound A is finite, amounts of B and ultimately C are also limited. During continuous irradiation, therefore, the level of C will eventually fall below the critical threshold and recovery will take place. At low irradiances the degree of photoconversion of B to C is insufficient to raise the concentration of C above the inhibitory threshold. Washing the spores prior to exposure to light presumably removes B. If spores are hydrated in the light, A is slowly converted to B, but the latter is simultaneously photoconverted to C, and continuously lost. Such spores will therefore germinate irrespective of light. The model also accommodates the observation by Bromfield (1964) that cold, dormant uredos-

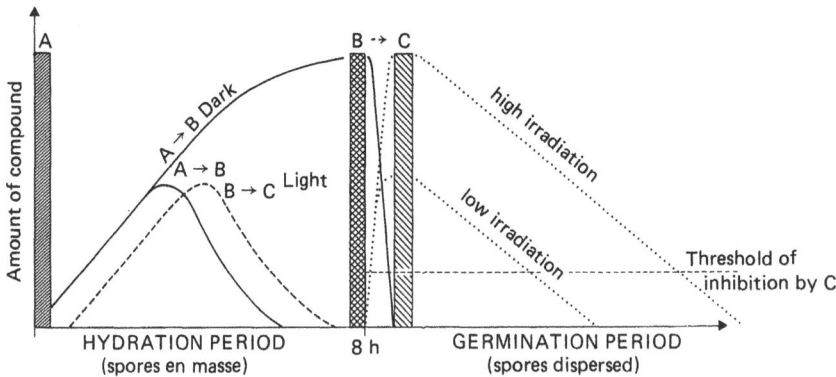

Fig. 3.5. Proposed model to explain the possible events leading to photoinhibition of uredospore germination. Broken arrow represents conversion by photoreceptor. See text for further explanation.

pores activated by heat shock are insensitive to light. As a hydration step is not involved in this procedure, the conversion of A to B will not take place, so inhibitor C will not be produced on exposure to light.

Confirmation or rejection of this model will depend upon identification of the postulated compounds, in particular B and C. The photoreceptor involved might be a separate molecule, or conceivably it might be compound B.

Mechanism of photoinhibition and links with self-inhibition

In thin section, the uredospore wall appears to consist of three layers (Strobel, 1965; Erlich & Erlich, 1969) with a dense innermost region adjacent to the plasma membrane (Knights, Davey & Lucas, 1982). This inner layer is missing at the four equatorial germ pores, the points at which a germ tube may emerge through the spore wall (Fig. 3.6). Germination might therefore be restricted to these sites through the specificity of enzymes attacking substrates only accessible in this region of the wall. Hess and co-workers (cited in Staples & Macko, 1984) have proposed that the germ pores are plugged with a mannoprotein material which is highly resistant to a range of hydrolases.

Time-course studies (Knights & Lucas, 1980) indicate that photoinhibition acts at a stage prior to digestion of the germ pore plug. Once the

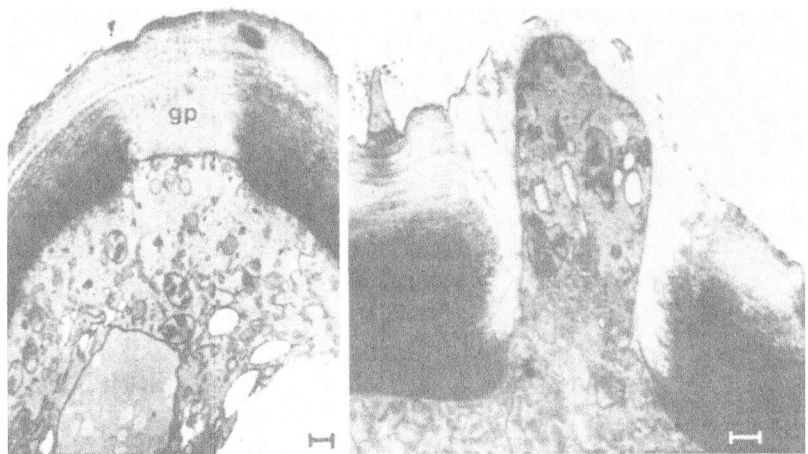

Fig. 3.6. Thin sections through dormant and germinating Pgt uredospores showing (left) the three-layered wall structure with the electron-dense inner layer absent in the germ pore (gp) region and (right) the developing germ tube breaking through the germ pore. Bar marker = 0.5 μm. (From Knights *et al.*, 1982)

spore wall is breached, light is no longer inhibitory and it has no effect upon the rate of germ tube growth. The similarities with the proposed site of action of the self-inhibitor (Hess *et al.*, 1975) are therefore striking. Ultrastructural evidence (Knights *et al.*, 1982) shows that most of the cytological events associated with germination also occur in photo-dormant spores but dissolution of the germ pore plug is apparently inhibited. Provided the spore wall has not been completely breached, either light or methyl-*cis*-ferulate can check the germination process. Elucidation of the enzymology of pore plug digestion is crucial to resolution of the mode of action of both photoinhibition and self-inhibition since current evidence is insufficient to distinguish whether germ pore dissolution or early growth of the germ tube is the process blocked. It has been argued (Staples & Macko, 1984) that the absence of pore plug digestion may be a consequence of inhibition at an alternative primary target.

Other similarities between photoinhibition and self-inhibition should be noted. Both processes can be alleviated by washing spores in water. In addition, uredospores of race 12b of Pgt, known to be insensitive to methyl-*cis*-ferulate (Macko *et al.*, 1972), are also relatively insensitive to light. While it is conceivable that both processes might be linked through a common pathway, it is unlikely that the self-inhibitor is directly involved in the photoresponse. Only the *cis* isomer of methyl ferulate shows inhibitory activity (Allen, 1972; Macko *et al.*, 1972), and during irradiation this undergoes conversion to a mixture containing the inactive *trans* isomer. Germination stimulants such as the alcohol nonanol, which relieve self-inhibition (Allen, 1972), have no effect on photoinhibition (Knights, 1981). Finally, the enhanced photosensitivity of hydrated uredospores does not appear to be correlated with changes in the content of self-inhibitor.

The ecological significance of spore dormancy

Almost all of the data on factors influencing uredospore germination have been acquired in the laboratory and one must therefore ask whether photoinhibition or self-inhibition has any proven ecological significance. Is there any evidence, for instance, that spore dormancy aids survival and subsequent host infection? The ecological advantage of self-inhibition is probably in precluding germination of spores en masse in the pustule and thus, preventing rapid simultaneous germination of all the spores (Staples & Macko, 1984). In contrast, photodormancy appears to be a device which synchronises germination with periods of darkness, thereby ensuring that germling development on the host surface takes place at night when the danger of desiccation is minimal. Spore trap data show that there is a marked diurnal periodicity in airborne rust uredospores, with the major peak around midday (Pady *et al.*, 1965).

Incubating spores within a wheat crop has shown that even on substrates highly conducive to germination, such as agar blocks, normal daylight is sufficient to delay germination until dusk (Knights & Lucas, 1981). Spores on leaves also remained dormant, and recovery from photoinhibition required dew formation as well as declining irradiation. The adaptive value of photoinhibition was demonstrated by breaking dormancy by preincubation on host plants in the dark and then transferring these plants to normal daytime environmental conditions. Very few pustules developed on these plants compared with others preincubated in the light (Knights & Lucas, 1981). The leaf surface environment during the day therefore appears to be hostile to the developing germling. Desiccation of the germ tube, or perhaps daylight itself may be lethal to germinated spores. Maddison & Manners (1972) showed that sunlight reduced the viability of uredospores of three different cereal rust species, with *P. striiformis* being the most sensitive. Lethal activity was mainly in the ultraviolet and near-ultraviolet regions of the spectrum, and irradiation reduced the ability to infect the host to a greater extent than germination itself (Maddison & Manners, 1973). More recent studies (Rotem, Wooding & Aylor, 1985) have concluded that solar radiation in general, and the UV portion in particular, is a major factor in the mortality of fungal spores, including rust uredospores. These findings are probably more relevant to the dispersal phase, especially to long distance dispersal in the upper atmosphere, than to survival on host surfaces. The concentration of spores falls off steeply from a point source (Roelfs & Martell, 1984), so most travel only a short distance. Nevertheless the results illustrate the potentially lethal effects of daylight on fungi, and suggest that further experiments should be carried out to determine the sensitivity of germinated spores to environmental factors, including light.

While the prepenetration development of several cereal rusts is favoured by periods of darkness (Kochman & Brown, 1976), the penetration process itself may be enhanced by light. Emge (1958) showed that light and temperatures around 27°C stimulated the formation of vesicles from Pgt germ tubes on artificial membranes. Politowski & Browning (1975) found that light near the end of the dew period enhanced penetration and pustule formation by Pgt and *P. graminis avenae,* but not by *P. coronata.* Appressorium formation from germinating spores usually takes from 4 to 8 h. Thus, at least with certain cereal rusts, increasing irradiation at dawn may favour the later stages of the infection process.

At present the most conspicuous gap in our knowledge concerns the physiological state of spores produced under natural conditions. There is little information, for instance, about the state of hydration of uredospores released from pustules in the field. This in turn will affect spore photosensitivity and possibly survival on the leaf surface. Environmental factors have major effects on the quality of rust inoculum (Wiese & Ravenscroft, 1979; Hollier, 1985), which will influence the viability and dormancy of spores and hence the proportion of the spore population capable of initiating lesions. Such effects are likely to be important in determining the rate of disease increase in rust epidemics.

Postscript – Responses of fungal plant pathogen spores to environmental stimuli

Given favourable conditions, rust uredospores germinate very quickly, sometimes in less than one hour. Yet even this brief phase of development appears to be regulated by a finely tuned system of environmental sensing. Once the germ tube has emerged, the spore is irreversibly committed to an intricate morphogenetic programme, controlled by further stimuli. These include a contact tropism in response to host surface topography (Dickinson, 1949) enabling correct orientation of the germ tube, followed by perception of stomatal guard cells, which triggers appressorium formation (Wynn, 1976; Hoch *et al.*, 1987). In Pgt the chemical environment around stomata may play a part in inducing differentiation of infection structures (Grambow & Grambow, 1978). Infection structure formation is accompanied by fundamental changes in the cell biology of germlings, including DNA synthesis, nuclear division (Staples, App & Ricci, 1975), and synthesis of novel proteins (Huang & Staples, 1981; Epstein *et al.*, 1985). This complex differentiation requires an extended period of favourable conditions to permit penetration of the host. Uredospore dormancy helps to ensure that germination is initiated only when these conditions are likely to prevail.

We have concentrated on a single type of fungal spore, but evidence suggests that even within this spore type there is significant variation in germination behaviour between species. This is to be expected as coevolution with different hosts will have resulted in quite different environmental optima for infection processes. The rust fungi as a whole are an excellent group in which to compare such environmental responses. Greater differences should be apparent with other types of spores serving a contrasting ecological function. Resting spores, involved in long-term survival of a pathogen in the absence of a host, may respond quite differently to certain combinations of environmental factors. For instance, soil-borne teliospores of the smut fungi *Tilletia indica* and *T. controversa* are stimulated by brief exposure to low irradiance (Smilanick, Hoffman & Royer, 1985). Dormancy in other propagules may be broken by chemical signals released from the host plant (French, 1985). The mechanisms involved in recognition of such environmental stimuli by fungi, and induction of the molecular changes underlying morphogenesis, pose fascinating questions for the developmental biologist.

References

Allen, P.J. (1955). The role of a self-inhibitor in the germination of rust uredospores. *Phytopathology*, **45**, 259–66.

Allen, P.J. (1972). Specificity of the *cis*-isomers of inhibitors of uredospore germination in the rust fungi. *Proceedings of the National Academy of Science, USA*, **69**, 3497–500.

Bromfield, K.R. (1964). Cold-induced dormancy and its reversal in uredospores of *Puccinia graminis tritici*. *Phytopathology*, **54**, 68–74.

Burrage, S.W. (1970). Environmental factors influencing the infection of wheat by *Puccinia graminis*. *Annals of Applied Biology*, **66**, 429–40.

Calpouzos, L. & Chang, H.S. (1971). Fungus spore germination inhibited by blue and far red radiation. *Plant Physiology*, **47**, 727–30.

Chang, H.S., Calpouzos, L. & Wilcoxson, R.D. (1973). Germination of hydrated uredospores of *Puccinia recondita* inhibited by light. *Canadian Journal of Botany*, **51**, 2459–62.

Chang, H.S., Wilcoxson, R.D. & Calpouzos, L. (1974). Inhibition of uredospore germination by light is partially relieved by soaking spores in water. *Phytopathology*, **64**, 158.

Daly, J.M., Knoche, H.W. & Wiese, M.V. (1967). Carbohydrate and lipid metabolism during germination of uredospores of *Puccinia graminis tritici*. *Plant Physiology*, **42**, 1633–642.

Dickinson, S. (1949). Studies in the physiology of obligate parasitism. I. The stimuli determining the direction of growth of the germ-tubes of rust and mildew spores. *Annals of Botany (N.S.)*, **13**, 89–104.

Dillon-Weston, W.A.R. (1931). Effect of light on the urediniospores of Black Stem Rust of Wheat, *Puccinia graminis tritici*. *Nature (London)*, **128**, 67–8.

Doherty, M.A. & Preece, T.F. (1978). *Bacillus cereus* prevents germination of uredospores of *Puccinia allii* and the development of rust disease of leek, *Allium porrum*, in controlled environments. *Physiological Plant Pathology*, **12**, 123-32.

Emge, R.G. (1958). The influence of light and temperature on the formation of infection-type structures of *Puccinia graminis* var. *tritici* on artificial substrates. *Phytopathology*, **48**, 649–52.

Epstein, L., Laccetti, L., Staples, R.C., Hoch, H.C. & Hoose, W.A. (1985). Extracellular proteins associated with induction of differentiation in bean rust uredospore germlings. *Phytopathology*, **75**, 1073–6.

Ehrlich, M.A. & Ehrlich, H.G. (1969). Uredospore development in *Puccinia graminis*. *Canadian Journal of Botany*, **47**, 2061–4.

French, R.C. (1985). The bioregulatory action of flavor compounds on fungal spores and other propagules. *Annual Review of Phytopathology*, **23**, 173–99.

Givan, C.V. & Bromfield, K.R. (1964a). Light inhibition of uredospore germination in *Puccinia recondita*. *Phytopathology*, **54**, 116–17.

Givan, C.V. & Bromfield, K.R. (1964b). Light inhibition of uredospore germination in *Puccinia graminis* var.*tritici*. *Phytopathology*, **54**, 382–4.

Grambow, H.J. & Grambow, G.E. (1978). The involvement of epicuticular and cell wall phenols of the host plant on the *in vitro* development of *Puccinia graminis* f.sp. *tritici*. *Zeitschrift für Pflansenphysiologie*, **90**, 1–9.

Hess, S.L., Allen, P.J., Nelson, D & Lester, H. (1975). Mode of action of methyl-*cis*-ferulate – the self-inhibitor of stem rust uredospore germination. *Physiological Plant Pathology*, **5**, 107–12.

Hoch, H.C., Staples, R.C., Whitehead, B., Comeau, J. & Wolf, E.D. (1987). Signaling for growth orientation and cell differentiation by surface topography in *Uromyces*. *Science*, **235**, 937–9.

Hollier, C.A. (1985). Effects of temperature and relative humidity on germinability and infectivity of *Puccinia polysora* uredospores. *Plant Disease*, **69**, 937–9.

Huang, B.F. & Staples, R.C. (1981). Synthesis of proteins during differentiation of the bean rust fungus. *Experimental Mycology*, **6**, 7–14.

Knights, I.K. (1981). *The photocontrol of rust uredospore germination*. PhD Thesis. *University of Nottingham, UK*.

Knights, I.K., Davey, M.R. & Lucas, J.A. (1982). The ultrastructure of dormant, germinating, and photo-inhibited uredospores of the rust fungus *Puccinia graminis* f.sp. *tritici*. *Protoplasma*, **113**, 57–68.

Knights, I.K. & Lucas, J.A. (1980). Photosensitivity of *Puccinia graminis* f.sp. *tritici* urediniospores *in vitro* and on the leaf surface. *Transactions of the British Mycological Society*, **74**, 543–9.

Knights, I.K. & Lucas, J.A. (1981). Photocontrol of *Puccinia graminis* f.sp. *tritici* urediniospore germination in the field. *Transactions of the British Mycological Society*, **77**, 519–27.

Kochman, J.K. & Brown, J.F. (1976). Effect of temperature, light and host on prepenetration development of *Puccinia graminis avenae* and *Puccinia coronata avenae*. *Annals of Applied Biology*, **82**, 241–9.

Littlefield, L.J. & Heath, M.C. (1979). *Ultrastructure of rust fungi*. New York: Academic Press.

Lucas, J.A., Kendrick, R.E. & Givan, C.V. (1975). Photocontrol of fungal spore germination. *Plant Physiology*, **56**, 847–9.

Macko, V. (1981). Inhibitors and stimulants of spore germination and infection structure formation in fungi. In: *The fungal spore: morphogenetic controls*, ed. G. Turian & H. R. Hohl, pp. 565–84. New York: Academic Press.

Macko, V., Staples, R.C., Renwick, J.A.A. & Pirone, J. (1972). Germination self-inhibitors of rust urediospores. *Physiological Plant Pathology*, **2**, 347–55.

Maddison, A.C. & Manners, J.G. (1972). Sunlight and viability of cereal rust uredospores. *Transactions of the British Mycological Society*, **59**, 429–43.

Maddison, A.C. & Manners, J.G. (1973). Lethal effects of artificial ultraviolet radiation on cereal rust uredospores. *Transactions of the British Mycological Society*, **60**, 471–94.

Manners, J.G. (1981). Biology of rusts on leaf surfaces. In: *Microbial Ecology of the Phylloplane*, ed. J. P. Blakeman, pp. 103–14. London: Academic Press.

Mishra, R.R. & Tewari, R.F. (1976). Studies on biological control of *Puccinia graminis tritici*. In: *Microbiology of aerial plant surfaces*, ed. C. H. Dickinson & T. F. Preece, pp. 559–67. London: Academic Press.

Nutman, F.J. & Roberts, F.M. (1963). Studies on the biology of *Hemileia vastatrix*, Berk and Br. *Transactions of the British Mycological Society*, **46**, 27–48.

Pady, S.M., Kramer, C.L., Pathak, V.K., Morgan, F.L. & Bhatti, M.A. (1965). Periodicity in airborne cereal rust urediospores. *Phytopathology*, **55**, 132–4.

Parker, A. & Blakeman, J.P. (1984). Nutritional factors affecting the behaviour of *Uromyces viciae-fabae* uredospores on broad bean leaves. *Plant Pathology*, **33**, 71-80.

Peterson, R.H. (1974). The rust fungus life cycle. *Botanical Reviews*, **40**, 453–513.

Politowski, K. & Browning, J.A. (1975). Effect of temperature, light and dew duration on relative numbers of infection structures of *Puccinia coronata avenae*. *Phytopathology*, **65**, 1400–4.

Rienecke, P. (1981). Antagonism and biological control on aerial surfaces of the Gramineae. In: *Microbial Ecology of the Phylloplane*, ed. J. P Blakeman, pp. 383–95. London: Academic Press.

Roelfs, A.P. & Martell, L.B. (1984). Uredospore dispersal from a point source within a wheat canopy. *Phytopathology*, **74**, 1262–7.

Rotem, J., Wooding, B. & Aylor, D.E. (1985). The role of solar radiation, especially ultraviolet light, in the mortality of fungal spores. *Phytopathology*, **75**, 510–4.

Sharp, E.L., Schmitt, C.G., Staley, J.M. & Kingsolver, C.H. (1958). Some critical factors involved in establishment of *Puccinia graminis* var. *tritici*. *Phytopathology*, **48**, 469-74.

Smilanick, J.L., Hoffman, J.A. & Royer, M.H. (1985). Effect of temperature, pH, light and desiccation on teliospores germination of *Tilletia indica*. *Phytophathlogy*, **75**, 1428–31.

Staples, R.C., App, A.A. & Ricci, P. (1975). DNA synthesis and nuclear division during formation of infection structures by bean rust urediospore germlings. *Archives für Microbiologie*, **104**, 123–7.

Staples, R.C. & Macko, V. (1980). Formation of infection structures as a recognition response in fungi. *Experimental Mycology*, **4**, 2–16.

Staples, R.C. & Macko, V. (1984). Germination of urediospores and differentiation of infection structures. In: *The Cereal Rusts* vol. 1, ed. W. R. Bushnell & A. P. Roelfs, pp. 255–89. New York: Academic Press.

Strobel, G.A. (1965). Biochemical and cytological processes associated with hydration of uredospores of *Puccinia striiformis*. *Phytopathology*, **55**, 1219–22.

Tan, K.K. (1978). Light-induced fungal development. In: *The Filamentous Fungi*, vol.3, ed. J. E. Smith & D. R. Berry, pp. 334–57. London: Edward Arnold.

Trione, E.J. (1981). Natural regulators of fungal development. In: *Plant Disease Control: Resistance and Susceptibility*, ed. R. C. Staples & G. H. Toenniessen, pp. 85–102. John Wiley: New York.

Weber, D. & Hess, W.M. (1976). *The Fungal Spore*. New York: John Wiley.

Wiese, M.V. & Ravenscroft, A.V. (1979). Environmental effects on inoculum quality of dormant rust urediospores. *Phytopathology*, **69**, 1106–08.

Wynn, W.K. (1976). Appressorium formation over stomates by the bean rust fungus: response to a surface contact stimulus. *Phytopathology*, **66**, 136–46.

Yaniv, Z. & Staples, R.C. (1974). Ribosomal activity in urediospores germinated on membranes. *Phytopathology*, **64**, 1111–14.

Zadoks, J.C. & Groenewegen, L.J.M. (1967). On light-sensitivity in germinating uredospores of wheat brown rust. *Netherlands Journal of Plant Pathology*, **73**, 83–102.

4

Pathways for the exchange of materials in mycoparasitic and plant-fungal interactions

PETER JEFFRIES

Biological Laboratory, University of Kent, Canterbury, Kent, CT2 7NJ, UK.

Fungi are chemoheterotrophic and the establishment of a successful symbiosis (in the broad sense of De Bary, 1887) depends, in part, on the exchange of materials between the symbionts. Cell-to-cell contact is usually necessary to facilitate this process and the fungi have evolved a range of structures or relationships whereby nutrient transfer is accomplished. This chapter is a descriptive review of several of these avenues of exchange between fungi and other organisms, and is not a review of the mechanisms of transfer. Furthermore, fungal-animal interactions will not be considered as they are outside the scope of this book, and infection of higher plant tissues by biotrophic pathogens will be discussed only briefly since the subject is discussed in detail in Chapter 5. Biotrophic relationships occurring during mycoparasitism will, however, be included here because a range of unique host-parasite interfaces can be found. Brief consideration will also be given to certain mutualistic relationships, such as lichens or vesicular-arbuscular mycorrhizas (VAM) in which structural similarities can be found with the haustoria of biotrophic plant pathogens. The intercellular relationships discussed have been divided into six main categories (see Table 4.1). Each will be discussed using specific examples of fungal interactions.

An excellent and comprehensive review of the structure of host-parasite interfaces in plant disease was given by Bracker & Littlefield (1973) and this information will not be repeated except to emphasise two striking generalities. Firstly, during the establishment, progress and outcome of a single host-parasite association, a succession of structurally different interfaces occurs. These are dependent on both parasite-mediated and host-mediated responses, but regardless of the developmental sequence there is always a stable condition during which the cells of the symbionts

Table 4.1. *Avenues for exchange of nutrients between fungi and other organisms*

Method of exchange	Examples
via direct cytoplasmic contact	some biotrophic mycoparasites
via the plasma membrane of protoplasts within host cells	intracellular chytrids
via haustoria or their equivalents	lichens, VA mycorrhizas, biotrophic parasites
by direct movement into or out of inter-cellular hyphae within necrotic or non-functional tissues	necrotrophic parasites
by direct movement into or out of inter-cellular hyphae within live tissues	ectomycorrhizas
by leakage across the walls of unmodified hyphae and cells in close contact with one another	colonisers of the plant surface

are juxtaposed and nutrient interchange can occur. This is the 'mature' symbiotic interface. Secondly, it is apparent over a broad taxonomic range of symbioses that a uniformity of interface types exists, and common features are often shared by relationships between fungi and widely different hosts.

The initial encounter between symbionts almost always involves nutrient transfer. For example, the stimulation of germination of spores or resting bodies of a parasite, and chemotaxis or chemotropism, are critical events in pathogenesis and must be considered in host-parasite nutritional relationships (Hancock & Huisman, 1981). Once contact between symbionts is achieved, the mature interface for exchange of materials is established and the outcome of infection determined.

Direct cytoplasmic contact

Burgeff (1924) described the interaction between *Chaetocladium* and several other fungi from the Mucorales. Mycoparasitism involved the contact of the hypha of the host by a hyphal tip of the parasite. A septum formed within the hypha of *Chaetocladium* near the contact zone to form a multinucleate 'cupping-cell'. Ultimately the interfacial wall between cupping-cell and the host appeared to be broken down, giving rise to a pore through which exchange of cytoplasm and nuclei occurred. In much more detailed ultrastructural studies, this unique form of host-fungus interface has been well characterised in three examples of biotrophic mycoparasitism; Hoch (1977*a,b*, 1978) has described the

contact zone between the mycoparasitic fungi *Gonatobotrys simplex,*
Stephanoma phaeospora and *Hansfordia parasitica* (= *Calcariosporium*
parasiticum), on the respective hosts *Alternaria tenuis, Fusarium* sp., and
Physalospora obtusa. These fungi infect their hosts by forming specialised
short globose protuberances or 'buffer cells' (Barnett, 1963) which contact
a hypha of the host but do not penetrate it. Cytoplasmic continuity
between partners is then achieved through the development of several
fine channels (24-31nm in diameter), resembling the plasmodesmata of
higher plants, through the closely adpressed hyphal walls *Gonatobotrys*
simplex, Stephanoma phaeospora) (Fig. 4.1) or by the dissolution of the
walls in a central area to form a large pore (0.1 to 1.0µm in diameter)
between the two fungi (*Hansfordia parasitica*) (Fig. 4.2). The size of the
portal between host and parasite thus varies. In *H. parasitica* the pore
is large enough for exchange of organelles, and nuclei of the parasite
have been observed to move into hyphae of the host, although the septal
pore of the contact cell in *Hansfordia* is always occluded after development of the mature host-parasite interface, and it may be that transport
of organelles does not occur further than this region (Hoch, 1977*b*). A
similar phenomenon has been reported during the alga-alga interaction
between *Polysiphonia* and its red algal parasite *Choreocolax.* Nuclei of
the latter are delimited within specialised 'conjunctor' cells and are then
transferred to the host via secondary pit connections (Goff & Coleman,
1984). In *Gonatobotrys* and *Stephanoma,* however, the narrowness of the
interconnecting channels prevents interspecific organelle exchange, but
obviously permits cytosolic movements and the diffusion of soluble

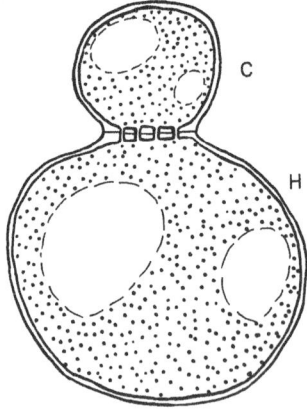

Fig. 4.1. Interface between the contact cell (C) of the biotroph *Gonatobotrys*
simplex and a hypha of the host *Alternaria tenuis* (H).

macromolecules. Virus-like particles, 40-47nm in diameter, were present in all nuclei of *Stephanoma* observed, but were never seen in the cytoplasm of the mycoparasite or in any part of the cells of *Fusarium*. Plasmodesmata have been observed previously in only one other type of host-parasite relationship (Hoch, 1977a). Species of *Cuscuta* have plasmodesmatal relationships with various plant hosts through which their protoplasts are possibly connected (Bennett, 1944; Dorr, 1969, 1972). The use of this plant as an agent of plant virus transmission is good circumstantial evidence for direct cytoplasmic connections with its hosts. Other channel-like connections have been reported to occur between the haustorial cells of *Puccinia graminis* f.sp. *tritici* and the cytoplasm of its host (Ehrlich & Ehrlich, 1963, 1971). Although micrographic evidence was presented, these observations have not been repeated for any other biotrophic plant pathogen, including several further studies of *Puccinia*. The phenomenon of cytoplasmic compatibility between unrelated fungi is unusual, but among related fungi is commonly demonstrated by anastomoses in Ascomycotina and Deuteromycotina, dikaryotisation in Basidiomycotina, or sexual contacts during the sexual process of the Mastigomycotina and Zygomycotina.

Totally intracellular relationships

Relationships which involve the complete protoplast of the symbiont entering the cell of the host occur in several examples of chytridiaceous fungi as well as in plasmodia of the Plasmodiophorales. Aist & Williams (1971) and Bracker & Littlefield (1973) described the host-

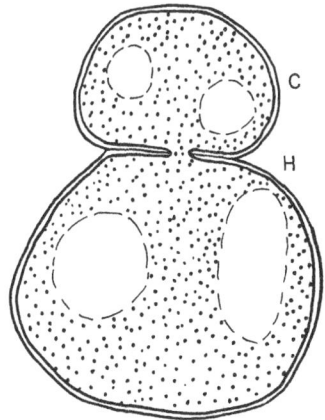

Fig. 4.2. Interface between the contact cell (C) of *Hansfordia parasitica* and a hypha (H) of the host.

pathogen interface in *Plasmodiophora brassicae* and concluded that in the mature interface materials must be exchanged via a compound double membrane consisting of parts derived from the plasmalemma of both the host and the parasite (interface type IT4 of Bracker & Littlefield, 1973). This means that the protoplasts of plant and fungus are in direct contact and the membranes of each organism must function together to ensure the survival of the pathogen. The outer membrane of the envelope is apparently not attached to the host plasmalemma and its functional equivalence is unknown.

The interface between endophytic chytrids and their partners can differ from that of *Plasmodiophora* in that the chytrid protoplast enters the cytoplasm of the host in the absence of a surrounding envelope derived from the host (i.e., interface IT1 of Bracker & Littlefield). In chytrids there are examples of both plant parasitic (e.g. the plasmodium of *Olpidium brassicae*) and mycoparasitic (e.g. *Rozella/Allomyces*) associations in which this occurs. The zoospores of these fungi are chemotropic towards their potential hosts, thus presumably some materials can be transferred to them directly in the rhizosphere. On contact, the zoospores bind strongly to the cell surface of the host and penetration by the protoplast occurs rapidly. In mycoparasitic species two distinct patterns of intracellular infection occur. The first is typified by species of *Rozella* (Held, 1980, 1981). When *Rozella* infects *Allomyces* the invasive protoplast fills a portion of the host hypha that becomes separated from non-infected regions by a septum. Thus, the hyphal wall of *Allomyces* now surrounds its own cytoplasm as well as that of *Rozella,* which retains its plasmalemma. At first, the presence of the mycoparasite appears to cause little harm to the host but later the host cytoplasm degenerates as the cytoplasm of the parasite differentiates into another generation of zoospores (Fig.4.3). In contrast, in most other endobiotic chytrids, such as *Olpidiopsis varians*, the invasive cell does not stimulate septum formation, thus several mycoparasitic individuals may share common cytoplasm. The protoplasts of these latter fungi apparently develop a wall–like layer at an early stage of their parasitic phase, thus the pathway of nutrient exchange has this additional component when compared to that in *Rozella* (Held, 1981).

Haustorial relationships

Many biotrophic mycoparasites form haustoria during their infection cycle. The development and mature structure of these organs are very similar to those exhibited by fungal pathogens of higher plants.

An appressorium is formed once a hypha of the mycoparasite contacts a susceptible region of the mycelium of the host. Hyphae of many biotrophic mycoparasites exhibit directed growth towards actively growing regions of host mycelia. Such hyphae do not respond to the proximity of non-host hyphae. On an agar medium this phenomenon may be noted at distances of several hundred μm, even against a background of a complex nutrient source such as malt extract. Growth of the mycoparasite *Dispira* is greatly stimulated by factors diffusing from adjacent hosts (Jeffries, 1985). There is thus an external pathway for the exchange of nutrients which operates prior to host-parasite contact and the development of the mature interface between them.

Once contact with the host is achieved, hyphae of haustorial mycoparasites form swollen infection structures, simple appressoria, often after a period of ectotropic growth along the host hypha (Jeffries & Young, 1976). Host penetration is achieved by growth of an infection peg through the wall of the host, and in a compatible interaction there is little evidence of extensive papilla formation at the penetration site (Jeffries, 1985). A neckband is present in the haustorial apparatus of *Piptocephalis unispora* (Jeffries & Young, 1976) and this presumably acts, in a similar way to those found in other haustorial pathogens, by sealing the apoplastic route for transport of materials (see Chapter 5). Once penetration is achieved,

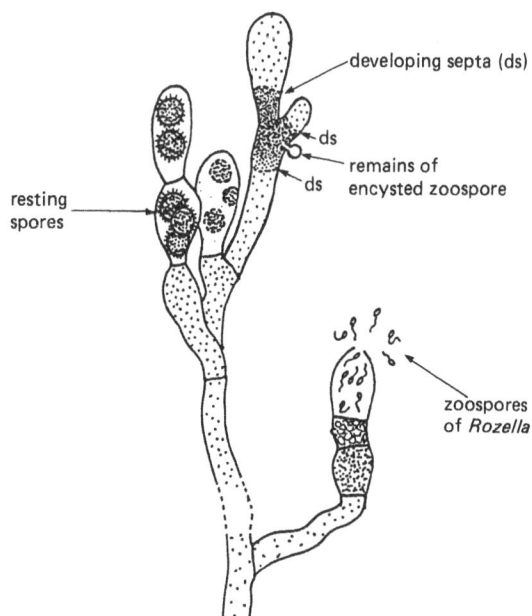

Fig. 4.3. A hypha of *Allomyces* parasitised by the chytrid *Rozella*.

the apex of the infection hypha develops into the body of the haustorium, and the host-parasite interface begins to mature. This region is the avenue through which materials are exchanged between host and parasite and it is here that the major differences between the infection cycle of the various haustorial mycoparasites are found (Fig. 4.4). At first, lobes of the haustorium of *Piptocephalis* (Jeffries & Young, 1976) are in contact with the extrahaustorial membrane (derived in a conventional way from the invaginated plasmalemma of the host) but as the haustorium matures, an amorphous matrix zone develops, presumably analogous to the extrahaustorial matrix of plant pathogens. In *Piptocephalis,* the haustorial lobes retain a wall which morphologically resembles that of the vegetative hyphae. Any material to be transferred between host and parasite via the haustorium must thus traverse the extrahaustorial membrane, extrahaustorial matrix, parasite cell wall and parasite plasmalemma (Fig. 4.4a). In contrast, the host–parasite interface in species from the Dimargaritales is less complex. Although the neck region of the haustorial complex is elaborate, and includes a penetration jacket (Jeffries & Young, 1981) and a septal complex, the wall of the invading hypha is substantially modified around the haustorial lobe. Host and parasite plasmalemmas are thus separated only by a narrow amorphous interfacial zone which is derived from the wall of the mycoparasite in the region at the base of haustorial neck (Fig. 4.4b). As a final example, a related mycoparasite, *Syncephalis,* has a different interfacial zone (Jeffries, 1985). Initial stages of parasitism resemble those of haustorial mycoparasites, but once inside the host cell the invading hypha does not develop into a haustorium but continues to grow and branch within the cytoplasm of the host. The plasmalemma of the host is pushed in around the hyphae of *Syncephalis* but an extensive interfacial zone does not develop prior to necrosis of the surrounding cytoplasm and disintegration of the invaginated plasmalemma. Material may thus be transferred between partners through the intact hyphal wall and plasmalemma of the parasite, via the plasmalemma of the host, or directly from the disorganised cytoplasm (Fig. 4.4c). Thus, although each of these three examples of mycoparasitic fungi develops infection structures that are similar in gross morphology, the ultrastructure of the host–parasite interface is substantially different, and contrasting pathways of nutrient exchange are involved.

Another example of a biotrophic interaction which involves the formation of a haustorium-like structure is found in certain mycorrhizas. These plant-fungus relationships also usually involve the formation of an extensive intercellular hyphal network by the fungal partner in the infected

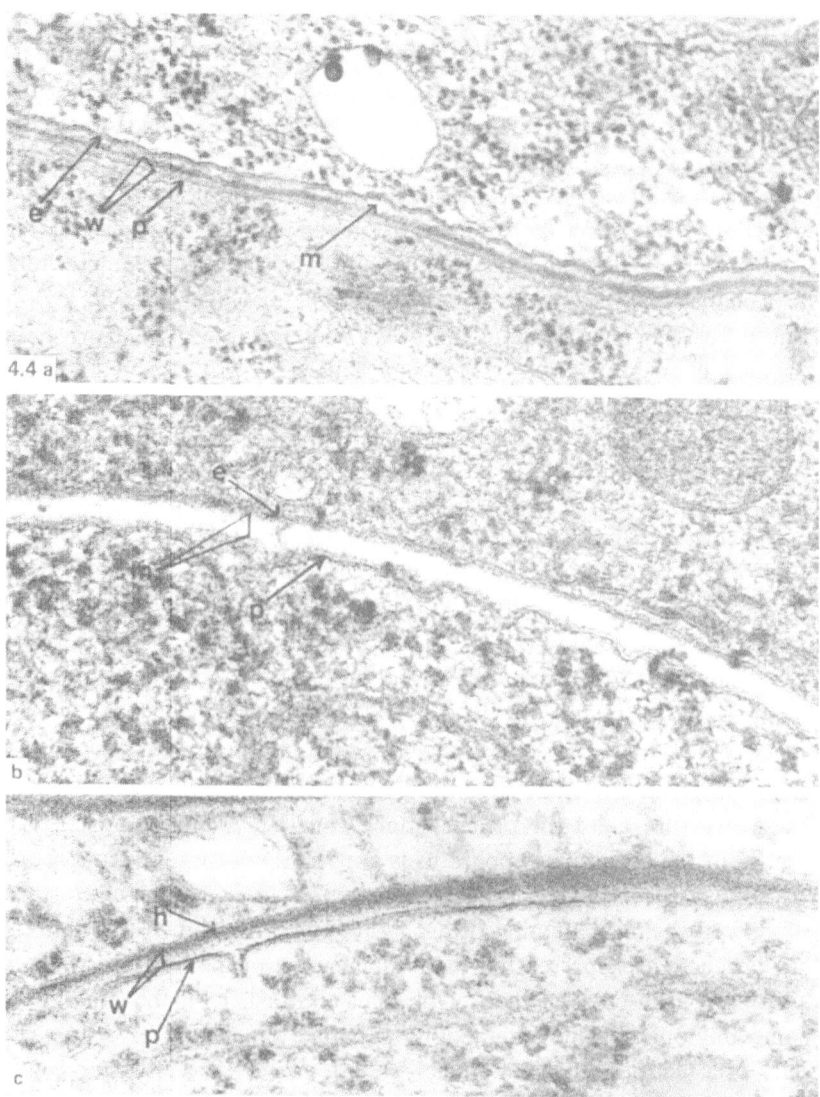

Fig. 4.4. Mature interfacial regions formed during interactions between various mycoparasitic fungi and the reference host *Cokeromyces recurvatus*. In all cases, the mycoparasite is the lower protoplast. (a) Interface around a haustorium of *Piptocephalis unispora* showing extrahaustorial membrane (e), extrahaustorial matrix (m), bilayered fungal wall (w) and plasmalemma of the parasite (p). (b) Interface around a haustorium of *Dimargaris cristalligena* showing extrahaustorial membrane (e), interstitial matrix (m) and plasmalemma of the parasite (p). × 100,000. (c) Interface around an intracellular hypha of *Syncephalis nodosa* showing plasmalemma of the host (h), bilayered fungal wall (w) and the plasmalemma of the parasite (p). ×90,000.

root cortex. There are few data available concerning the relative importance of these two avenues of exchange between partners. It is often
assumed, for example, that the arbuscules of VAM fungi act as the
primary avenue for nutrient uptake by the fungus, and it is often the
case that arbuscule formation coincides with the most active periods of
growth of the host roots. Nevertheless, there are often long periods when
root growth is slow and the fungus is present only as an intercellular
mycelium, as is the case for VAM infections of winter-sown cereals in
South-East England (Dodd & Jeffries, 1986). The intercellular mycelium
may act as the route of uptake for material necessary for the maintenance
energy for the fungus, while arbuscular uptake may be necessary as a
supplement when active growth and spread of the mycelium are occurring.

The fine structure of host-symbiont interfaces in a range of mycorrhizal
interactions has been discussed in detail elsewhere (e.g. Cooke, 1977;
Duddridge, 1985, and Chapter 2). Scannerini & Bonfante-Fasolo (1983)
considered that all interfaces between partners may have a role in nutrient
transfer, so an apoplastic route may reasonably be expected through the
host wall-fungal wall interface, with symplastic movements of materials
once they have entered the cytoplasm of the respective partners. Much
work has focused on the arbuscules of VAMs and close structural similarities have been noted with the haustoria of phytopathogens such as
Erysiphe graminis. There are, however, important differences between
these structures, and the plasmalemma which surrounds the interstitial
matrix of the arbuscular apparatus is apparently not modified morphologically as it is in phytopathogenic interactions (see Chapter 2). Furthermore, cytochemical staining (Marx *et al.*, 1982) has shown that marked
ATPase activity is present on the plasmalemma of the host cell where it
surrounds arbuscules of the fungus. This contrasts with the situation in
some phytopathogenic interactions (Chapter 2) and in mycoparasitic
haustoria (Jeffries, 1984) in which ATPase activity is depressed around
the extrahaustorial membrane. It is also of note that arbuscules do not
possess a neckband, found in the haustorial apparatus of several parasitic
fungi (e.g. Jeffries & Young, 1976; Manners & Gay, 1983; Woods & Gay,
1983), and it is thus feasible that an arbuscular fungus may obtain its
nutrients entirely by an apoplastic route. In some endophytic mycorrhizas,
'lysis' of intracellular structures of arbuscules occurs a few days after
their formation and it has been suggested that this represents a means
of nutrient transfer to the root (Lewis, 1973; Kinden & Brown, 1975).
There is now some evidence that this may not be the case (Cox &
Tinker, 1976; Smith, 1980; Scannerini & Bonfante–Fasolo, 1983) and

active mechanisms of transport are more important.

Associations between fungi and algae are common and widespread, although details of the pathways for exchange of materials between partners have not been well characterised for most examples. Not all such associations are lichens and there are examples in which the alga is the bulk tissue and the fungus grows in close association as intercellular hyphae. The relationship between *Mycosphaerella ascophylli* and the seaweed *Ascophyllum nodosum* is an example and the plant is never found without its fungal partner. Nutrient exchange presumably occurs directly via the intercellular hypha (see next section). Cooke (1977) has summarised the ultrastructural information available concerning interfaces in lichens and recognised three main types which represent a progression in complexity culminating in a haustorium-like structure. The commonest type involves simple wall-to-wall contact with very close attachment between the hypha and algal cell; there may be fusion of the two walls, those of both partners becoming thinner over the area of contact (for references see Cooke, 1977). In the second type, the hypha grows into the wall of the algal cell, which does not rupture but proliferates and becomes deeply invaginated. Within the invagination there may be fusion between the two walls, and ultrastructural changes in cell wall density and morphology. In the final type, the hypha enters the algal cell by penetrating the cell wall and pushing in the plasmalemma in an analogous way to a haustorium. The walls of the penetration hyphae are thinner than those of the external mycelium, but there is apparently no equivalent to the extrahaustorial matrix, the hyphal wall contacting the host plasmalemma directly.

Much more work needs to be carried out on the interrelationships between lichen partners before the significance of these interfaces in nutrient exchange is known. There is still considerable doubt as to whether such interfaces have a primary role in transfer of materials between partners. Chambers, Morris & Smith (1976) noted that in most intercellular contacts of lichens (type 1 above) carbohydrate efflux probably proceeds by means of facilitated diffusion. Cooke (1977) has also drawn attention to the fact that certain other structures found in some regions of fungal-algal contact are more likely to be important as avenues of nutrient exchange. In *Cornicularia normoerica* penetrating hyphae are absent, but the algal cells possess channels that originate from the chloroplast and then extend through the cell wall, flaring out at its periphery. These channels, which contain small tubules, are relatively infrequent (see Cooke, 1977).

Necrotrophic intercellular relationships

The characteristics of symbioses involving necrotrophic fungi vary widely depending on the particular host-parasite combination. Most comprise two phases, however, beginning with the penetration phase in which the pathogen is in contact with living cells of the host, followed by an extended phase characterised by the lack of direct contact between hyphae and living host cells, during which the parasite spreads through tissues killed as the infection front extends. The most significant movement of nutrients between hosts and parasites will occur during this post-penetration stage of pathogenesis. There is very little information to connect specific interfacial components with intercellular transport processes (Bracker & Littlefield, 1973) and the most intimate and direct avenue of transfer would be cell wall-to-cell wall contact of the respective partners. The secretion of diffusible enzymes, toxins, or other materials, however, is not limited to this one avenue of cell-to-cell transfer and such materials will diffuse along cell walls, the middle lamella, and through disorganised tissues so that drastic changes can be effected in cells far removed from the pathogen. In *Sclerotinia* infections, for example, the penetration of the cuticle and formation of a subcuticular infection vesicle involve a complex series of cytological events (Tariq, 1984; Tariq & Jeffries, 1984, 1985) but once the subcuticular vesicle is formed, colonisation of host tissues occurs rapidly by the intercellular growth of relatively undifferentiated hyphae. Cellular necrosis occurs in advance of the infection front and is evidenced initially by the appearance of proteinaceous deposits in the chloroplasts of host cells (Tariq & Jeffries, 1985). There is thus good evidence that material is transferred from pathogen to host prior to direct contact of hypha and host cell. There are many other such examples of necrotrophic interactions where, although morphologically distinct interfaces may exist, the functional boundaries of such interfaces are extremely difficult to identify (Bracker & Littlefield, 1973). For such parasites growing intercellularly without specialised absorptive structures, the only means of nutrient uptake is by intercepting solutes diffusing in hydrated host cell walls (Hancock & Huisman, 1981). Some pathogens may absorb nutrients direct from the apoplast in the immediate vicinity, whilst others may stimulate the diversion of host nutrients over some distance to the site of infection. Keon, Byrde & Cooper (Chapter 7) discuss secretion of material from hyphae of necrotrophic fungi into the adjacent cell walls of the host plant and, although they deal mainly with fungal pectinases, many of the principles apply to other materials being transferred between symbionts.

There are examples of necrotrophic mycoparasites in which entry and destruction of host tissues are accomplished by invasive hyphae of the mycoparasite (e.g. Dennis & Webster, 1971; Hoch & Fuller, 1977; Tzean & Estey, 1978; Tsuneda & Skoropad, 1980; Tu & Vaartaja, 1981; Kuter, 1984). Appressed growth and hyphal coiling are also frequently reported in such situations, although transfer of nutrients has not been demonstrated. *Pythium oligandrum*, however, has been shown to obtain carbon, organic nitrogen, thiamine (or pyrimidine), and sterols from other fungi (Deacon & Henry, 1978; Foley & Deacon, 1986), and yet there is no evidence that hyphae of its hosts are penetrated. Such an ability to obtain required nutrients from other fungi, presumably by external association, is clearly important in the physiological development of the mycoparasite.

Biotrophic intercellular relationships

In ectomycorrhizas, and in the intercellular hyphae of other kinds of mycorrhizas, the mature host-symbiont interface may consist of a relatively unmodified fungal wall and plant cell wall with the middle lamella disorganised to accommodate the presence of the fungal hypha. In other cases modifications in the interfacial region are more extreme and the host cell wall may become incorporated in an 'involving layer' or 'contact layer' which may be thin, thick or variable in width relative to the unmodified structure (see also Chapter 2). Damage to compatible host cells is minimal, however, and the intercellular hyphae have access to normal apoplastic host nutrient transport processes. In most mycorrhizas there is an extensive interface through which partners can communicate. The major component in ectomycorrhizal associations is the Hartig net, embedded in host cell wall material with a high pectin content (Scannerini, 1968; Duddridge, 1985; Nylund, 1980), but it must be noted that there is a separate, second interface between host cells and the inner sheath hyphae (Duddridge, 1985). Furthermore, there have been two different forms of interface described between the Hartig net and the host cortical cells of *Pinus* spp. (Duddridge & Read, 1984*a,b*). One is relatively simple, and involves direct contact of the juxtaposed cell walls, whilst the other is more complex and the two walls become merged in an electron–dense matrix. This matrix has been observed in a variety of ectomycorrhizal interactions (e.g. Atkinson, 1976; Strullu & Gerrault, 1977; Strullu, 1979; Debaud, Pepin & Bruchet, 1981) and is equivalent to the 'involving' layer of Scannerini (1968). Some earlier studies of ectomycorrhizal associations suggested that host cells involved contained no cytoplasm or were disorganised or dead (Foster & Marks, 1966;

Scannerini, 1968; Hofsten, 1969; Strullu & Gourret, 1973; Strullu & Gerrault, 1977). More recent studies (Atkinson, 1976; Duddridge, 1985; Duddridge & Read, 1984*a,b*) have shown that cortical cells enmeshed within the Hartig net in the developing region of the ectomycorrhiza are viable, containing active cytoplasm, confirming that translocation of substances between the symbionts occurs across a live interface. In the case of ectomycorrhizas formed between *Piloderma bicolor* and *Picea abies,* Nylund (1980) reported that pits with plasmodesmata maintained symplastic continuity between the epidermal cells within the Hartig net and those in the underlying cortex. Similar connections were observed by Warmbrodt & Eschrich (1985*a,b*) in *Pinus sylvestris* and it was suggested that both apoplastic and symplastic transport between cells of the vascular cylinder and host cells within the Hartig net was feasible in certain circumstances in which endodermal cells remained viable and available for transport across the periclinal wall (Warmbrodt & Eschrich, 1985*b*). This continuity is crucial with respect to direct nutrient exchange via a symplastic route, and Smith (1980) has suggested that transfer of phosphate to the host probably occurs through the fungus symplast in the sheath of ectotrophic mycorrhizas, and thence apoplastically across the walls of cells within the Hartig net and abutting cells of the root cortex. Piché *et al.* (1983) were unable to find similar plasmodesmatal connections in ectomycorrhizas of *Pinus strobus.*

In some monotropoid mycorrhizas, the fungus penetrates cortical cells forming specialised fungal pegs (Duddridge & Read, 1982; Robertson & Robertson, 1982; Duddridge, 1985). The invaginated host wall surrounds the fungal peg in a manner similar to that described earlier for certain lichen symbioses, and gives rise to cell wall extensions within the plant cells. However, the wall of the fungal peg in monotropoid interactions is highly modified and many ingrowths and protuberances have been observed. Such structures appear similar to transfer cells frequently observed in tissues of a wide variety of plants in regions where rapid and intensive short–distance solute transport has been observed or assumed (Pate & Gunning, 1972; Gunning, 1977; Duddridge, 1985). Similar modifications have also been noted in ectomycorrhizal infections of *Pisonia grandis* (Ashford & Allaway, 1982; Allaway, Carpenter & Ashford, 1985) and it has been estimated that the increase in interfacial surface area as a result of such modifications is of the order of 1.3-fold. A further avenue for nutrient exchange is provided during the later stages of fungal peg development in monotropoid interactions. The tip of the peg apparently breaks down, allowing material from it to enter a membranous sac delimited by the plasmalemma of the epidermal cell (Robertson & Robertson, 1982).

Direct transfer via unmodified hyphae

Many fungi which are associated with the surfaces of other living organisms are likely to be able to take up exudates from those surfaces directly. Thus rhizoplane and phylloplane fungi are well placed to utilise materials diffusing from the plant cell surface. For parasitic organisms, the flow of nutrients from the host is thought to be a major factor in determining the success or failure of infection in many diseases caused by soil- and air-borne pathogens (Schroth & Hildebrand, 1964; Blakeman, 1971). Much of the external surface of many plants is covered with relatively impermeable barriers such as the cuticle and it may be that few nutrients will be transferred via this route. Damage or degradation of these barriers, however, will increase the relative importance of this avenue of nutrient escape. Fokkema (1981) points out that most nutrients probably arrive at the plant surface via leaching and a wide variety of nutrients have been detected in surface fluids, albeit in trace amounts. There is circumstantial evidence that these nutrients exude largely from the apoplast (Hancock & Huisman, 1981). On roots, for example, the greatest growth of epiphytic bacteria occurs over the anticlinal walls where diffusion from the apoplast might be expected to be greatest (Van Vuurde, Kruyswyk & Schippers, 1979). Similarly several plant pathogens, such as *Sclerotinia sclerotiorum* (Tariq & Jeffries, 1984), form infection hyphae or appressoria over the junctions of epidermal cell walls. In some cases external sources of nutrients, such as insect honeydew or pollen deposits, may be more important than direct uptake of materials derived from within the plant. Also important is the effect of guttation fluids on the growth of plant surface fungi, and Frossard (1981) has discussed this aspect in relation to the phylloplane. At a secondary level, there will also be an exchange of materials between the different organisms which colonise the plant surface if their walls are in direct contact. There has been little critical work to assess such avenues of exchange, partly because these routes do not necessitate the differentiation of special structures. Hoch & Provvidenti (1979) reported that the interaction of *Sphaerotheca fuliginea* and *Tilletiopsis* sp. controlled the development of powdery mildew disease on detached cucumber leaves. Ultrastructural study of the relationship revealed that it was primarily a leaf surface phenomenon and not one involving direct cell penetration. The two fungi were closely associated and cells of *S. fuliginea* in contact with *Tilletiopsis* were usually necrotic. It was suggested that some substance was transferred to the host fungus, presumably across the closely adpressed hyphal walls, which killed the adjacent cells. *Tilletiopsis* then subsists on nutrients subsequently released from the dead cells of the host. This type of relationship is not unique and many similar cases of plant surface antagonism have been reported (e.g. Preece & Dickinson, 1971; Dickinson & Preece, 1976; Blakeman, 1981). In the case of necrotrophic mycoparasitism of *Ceratocystis fimbriata* by *Hirschioporus pargamenus* (Traquair & McKeen, 1978), hyphal penetration can occur, but host cell necrosis usually involves transfer of an unidentified toxin excreted by the parasite, for which contact is not essential.

This last example highlights the difficulties in deciding where the boundaries of host-parasite interactions lie, but it is evident from most of this discussion that a wide variety of avenues is utilised for the exchange of materials between fungi and their various partners from cell to cell. There are other examples of fungal structures that are adapted for transport of materials over much larger distances, for example rhizomorphs and mycelial strands. Such structures involve a host cell-to-fungal cell interface similar to those already discussed here, and excellent reviews describing the transport of materials from plant to plant via such fungal connections are available elsewhere (e.g. Read, Francis & Finlay, 1985).

References

Aist, J.R. & Williams, P.H. (1971). The cytology and kinetics of cabbage root hair penetration by *Plasmodiophora brassicae*. *Canadian Journal of Botany*, **49**, 2023–34.

Allaway, W.G., Carpenter, J.L. & Ashford, A.E. (1985). Amplification of inter-symbiont surface by root epidermal transfer cells in the *Pisonia* mycorrhiza. *Protoplasma*, **128**, 227–31.

Ashford, A.E. & Allaway, W.G. (1982). A sheathing mycorrhiza on *Pisonia grandis* R. Br. (*Nyctaginaceae*) with the development of transfer cells rather than a Hartig net. *New Phytologist*, **90**, 511–19.

Atkinson, M.A. (1976). *The Fine Structure of Mycorrhizas. D. Phil. Thesis, Oxford University, UK.*

Barnett, H.L. (1963). The nature of mycoparasitism by fungi. *Annual Review of Microbiology* **17**, 1–14.

Bennett, C.W. (1944). Studies of dodder transmission of plant viruses. *Phytopathology*, **34**, 905–32.

Blakeman, J.P. (1971). The chemical environment of the leaf surface in relation to the growth of pathogenic fungi. In: *Ecology of Leaf Surface Microorganisms*, ed. T. F. Preece & C. H. Dickinson, pp. 225–68. London: Academic Press.

Blakeman, J.P. (1981). *Microbial Ecology of the Phylloplane.* London: Academic Press.

Bracker, C.E. & Littlefield, L.J. (1973). Structural concepts of host–pathogen interfaces. In: *Fungal Pathogenicity and the Plant's Response*, ed. R. J. W. Byrde & C. V. Cutting, pp. 159–317. London: Academic Press.

Burgeff, H. (1924). Untersuchungen uber sexualität und parasitismus bei Mucorineen. I. *Botanische Abhandlungen*, **4**, 1–135.

Chambers, S., Morris, M. & Smith, D.C. (1976). Lichen physiology XV. The effect of digitonin and other treatments on biotrophic transport of glucose from alga to fungus in *Peltigera polydactyla*. *New Phytologist*, **76**, 485–500.

Cooke, R.C. (1976). *The Biology of Symbiotic Fungi*. Wiley: Chichester.

Cox, G.C. & Tinker, P.B. (1976). Translocation and transfer of nutrients in vesicular-arbuscular mycorrhizas I. The arbuscule and phosphorus transfer: a quantitative ultrastructural study. *New Phytologist*, **77**, 371–8.

Deacon, J.W. & Henry, C.M. (1978). Mycoparasitism by *Pythium oligandrum* and *P. acanthicum*. *Soil Biology and Biochemistry*, **10**, 409–15.

DeBaud, J.C., Pepin, R. & Bruchet, G. (1981). Ultrastructure des ectomycorrhizes synthettiques a *Hebeloma alpinum* et *H. marginatalum* de *Dryas octopetala*. *Canadian Journal of Botany*, **59**, 2160–6.

DeBary, A. (1887). *Comparative Morphology and Biology of the Fungi Mycetozoa and Bacteria*. Oxford: Oxford University Press.

Dennis, C. & Webster, J. (1971). Antagonistic properties of species-groups of *Trichoderma* III. Hyphal interaction. *Transactions of the British Mycological Society*, **57**, 363–9.

Dickinson, C.H. & Preece, T.F. (1976). *Microbiology of Aerial Plant Surfaces*. Academic Press: London.

Dodd, J.C. & Jeffries, P. (1986). Early development of vesicular-arbuscular mycorrhizas in autumn-sown cereals. *Soil Biology and Biochemistry*, **18**, 149–54.

Dorr, I. (1969). Feinstruktur intrazellular wachsender *Cuscuta* -hyphen. *Protoplasma*, **67**, 123–37.

Dorr, I. (1972). Der anschluss der Cuscuta-hyphen an die siebrohen ihrer wirtspflanzen. *Protoplasma*, **75**, 167–84.

Duddridge, J.A. (1985). A comparative ultrastructural analysis of the host-fungus interface in mycorrhizal and parasitic associations. In: *Developmental Biology of Higher Fungi*, ed. D. Moore, L. A. Casselton, D. A. Wood & J. C. Frankland, pp. 141–73. Cambridge University Press: Cambridge.

Duddridge, J.A. & Read, D.J. (1982). An ultrastructural analysis of the development of mycorrhizas in *Monotropa hypopitys* L. *New Phytologist*, **92**, 203–14.

Duddridge, J.A. & Read, D.J. (1984a). The development and ultrastructure of ectomycorrhizas I. Ectomycorrhizal development on pine in the field. *New Phytologist*, **96**, 565–73.

Duddridge, J.A. & Read, D.J. (1984b). The development and ultrastructure of ectomycorrhizas II. Ectomycorrhizal development on pine *in vitro*. *New Phytologist*, **96**, 575–82.

Ehrlich, M.A. & Ehrlich, H.A. (1971). Fine structure of host–parasite interfaces in mycoparasitism. *Annual Review of Phytopathology*, **9**, 155–84.

Fokkema, N.J. (1981). Fungal leaf saprophytes, beneficial or detrimental? In: *Microbial Ecology of the Phylloplane*, ed. J. P. Blakeman, pp. 433–54. London: Academic Press.

Foley, M.F. & Deacon, J.W. (1986). Physiological differences between mycoparasitic and plant–pathogenic *Pythium* spp. *Transactions of the British Mycological Society*, **86**, 225–31.

Foster, R.C. & Marks, G.C. (1966). Observations on the mycorrhiza of forest trees I. The fine structure of the mycorrhizas of *Pinus radiata*. D. Don. *Australian Journal of Biological Sciences*, **19**, 1027–38.

Frossard, R. (1981). Effect of guttation fluids on growth of micro–organisms on leaves. In: *Microbial Ecology of the Phylloplane*, ed. J. P. Blakeman, pp. 213–26. London: Academic Press.

Goff, L.J. & Coleman, A.W. (1984). Transfer of nuclei from a parasite to its host. *Proceedings of the National Academy of Science*, **81**, 5420–4.

Gunning, B.E.S. (1977). Transfer cells and their roles in transport of solutes in cells. *Science Progress, Oxford*, **64**, 539–68.

Hancock, J.G. & Huisman, O.C. (1981). Nutrient movement in host–pathogen systems. *Annual Review of Phytopathology*, **19**, 309–31.

Held, A.A. (1980). Development of *Rozella* in *Allomyces*: a single zoospore produces numerous zoosporangia and resistant sporangia. *Canadian Journal of Botany*, **58**, 959–79.

Held, A.A. (1981). *Rozella* and *Rozellopsis*: naked endoparasitic fungi which dress-up as their hosts. *The Botanical Review*, **47**, 451–515.

Hoch, H.C. (1977a). Mycoparasitic relationships: *Gonatobotrys simplex* parasitic on *Alternaria tenuis*. *Phytopathology*, **67**, 309–14.

Hoch, H.C. (1977b). Mycoparasitic relationships III. Parasitism of *Physalospora obtusa* by *Calcariosporium parasiticum*. *Canadian Journal of Botany*, **55**, 198–207.

Hoch, H.C. (1978). Mycoparasitic relationships IV. *Stephanoma phaeospora* parasitic on a species of *Fusarium*. *Mycologia*, **70**, 370–9.

Hoch, H.C. & Fuller, M.S. (1977). Mycoparasitic relationships I. Morphological features of interactions between *Pythium acanthicum* and several fungal hosts. *Archives of Microbiology*, **111**, 207–24.

Hoch, H.C. & Provvidenti, R. (1979). Mycoparasitic relationships: cytology of the *Sphaerotheca fuliginea-Tilletiopsis* sp. interaction. *Phytopathology*, **69**, 359–62.

Hofsten, A. von (1969). The ultrastructure of mycorrhiza I. Ectotrophic and ectendotrophic mycorrhiza of *Pinus sylvestris*. *Svensk Botanisk Tidskrift*, **63**, 455–63.

Jeffries, P. (1984). Outcome of mycoparasitism. *Bulletin of the British Mycological Society*, **18**, Supplement 2, 6.

Jeffries, P. (1985). Mycoparasitism within the Zygomycetes. *Botanical Journal of the Linnean Society*, **91**, 135–50.

Jeffries, P. & Young, T.W.K. (1976). Ultrastructure of infection of *Cokeromyces recurvatus* by *Piptocephalis unispora*. *Archives of Microbiology*, **109**, 277–88.

Jeffries, P. & Young, T.W.K. (1981). Ultrastructure of the haustorial apparatus of *Dimargaris cristalligena*. *Annals of Botany*, **47**, 107–19.

Kinden, D.A. & Brown, M.F. (1976). Electron microscopy of vesicular- arbuscular mycorrhizae of yellow poplar IV. Host-endophyte interactions during arbuscular deterioration. *Canadian Journal of Microbiology*, **22**, 64–75.

Kuter, G.A. (1984). Hyphal interactions between *Rhizoctonia solani* and some *Verticillium* species. *Mycologia*, **76**, 936–40.

Lewis, D.H. (1973). Concepts in fungal nutrition and the origin of biotrophy. *Biological Reviews*, **48**, 261–78.

Manners, J.M. & Gay, J.L. (1983). The host-parasite interface and nutrient transfer in biotrophic parasitism. In: *Biochemical Plant Pathology*, ed. J. A. Callow, pp. 163–95, Chichester: Wiley–Interscience.

Marx, C., Dexheimer, J., Gianinazzi-Pearson, V. & Gianinazzi, S. (1982). Enzymatic studies on the metabolism of vesicular-arbuscular mycorrhizas IV. Ultracytoenzymological evidence (ATPase) for active transfer processes in the host-arbuscule interface. *New Phytologist*, **90**, 37–43.

Nylund, J-E. (1980). Symplastic continuity during Hartig net formation in Norway spruce ectomycorrhizae. *New Phytologist*, **86**, 373–8.

Pate, J.S. & Gunning, B.E.S. (1972). Transfer cells. *Annual Review of Plant Physiology*, **23**, 173–96.

Piché, Y., Peterson, R.L., Howarth, M.J. & Fortin, J-A. (1983). A structural study of the interaction between the ectomycorrhizal fungus *Pisolithus tinctorius* and *Pinus strobus* roots. *Canadian Journal of Botany*, **61**, 1185–93.

Preece, T.F. & Dickinson, C.H., ed. (1971). *Ecology of Leaf Surface Microorganisms.* Academic Press: London.

Read, D.J., Francis, R. & Finlay, R.D. (1985). Mycorrhizal mycelia and nutrient cycling in plant communities. In: *Ecological Interactions in Soil*, ed. A. H. Fitter, pp. 193–217. Blackwell Scientific Publications: Oxford.

Robertson, D.C. & Robertson, J.A. (1982). Ultrastructure of *Pterospora andromedea* Nuttall and *Sarcodes sanguinea* Torrey mycorrhizas. *New Phytologist*, **92**, 539–51.

Scannerini, S. (1968). Sull ultrastruttura delle ectomicorrize II. Ultrastruttura di una micorriza di ascomicete: *Tuber albidum X Pinus strobus* L. *Allonia*, **14**, 77–95.

Scannerini, S. & Bonfante–Fasolo, P. (1983). Comparative ultrastructural analysis of mycorrhizal associations. (Proceedings of the 5th North American conference on Mycorrhizas). *Canadian Journal of Botany*, **61**, 917–43.

Schroth, M.N. & Hildebrand, D.C. (1964). Influence of plant exudates on root-infecting fungi. *Annual Review of Phytopathology*, **2**, 101–32.

Smith, S.E. (1980). Mycorrhizas of autotrophic higher plants. *Biological Reviews*, **55**, 475–510.

Strullu, D.G. (1979). The ultrastructure and spatial representation of the fungal mantle of ectomycorrhizae. *Canadian Journal of Botany*, **57**, 2319–24.

Strullu, D.G. & Gerrault, A. (1977). Étude des ectomycorrhizas à Basidiomycetes et à Ascomycetes du *Betula pubescens* (Ehrh.) en microscopie électronique. *Comptes Rendus Hebdomadaire des Séances de l'Academie des Sciences*, Paris, Série D, **284**, 2243-4.

Strullu, D.G. & Gourret, J.P. (1973).Étude des mycorrhizes ectotrophes de *Pinus brutia* Ten. en microscopie électronique à balayage et à transmission. *Comptes Rendus Hebdomadaire des Séances de l'Academie des Sciences*, **Paris,** Série D, 277, 1757–60.

Tariq, V.-N. (1984). *Sclerotinia* Diseases of Plants; Characterization of Isolates and Host-Parasite Interactions. *Ph.D. thesis, University of Kent.*

Tariq, V.-N. & Jeffries, P. (1984). Appressorium formation by *Sclerotinia sclerotiorium:* scanning electron microscopy. *Transactions of the British Mycological Society,* **82**, 645–51.

Tariq, V.-N. & Jeffries, P. (1985). Changes occurring in chloroplasts of *Phaseolus* following infection by *Sclerotinia:* a cytochemical study. *Journal of Cell Science,***75**, 195–-205.

Traquair, J.A. & McKeen, W.E. (1978). Necrotrophic mycoparasitism of *Ceratocystis fimbriata* by *Hirschioporus pargamenus* (Polyporaceae). *Canadian Journal of Microbiology*, **24**, 869–74.

Tsuneda, A. & Skoropad, W.P. (1980). Interactions between *Nectria inventa*, a destructive mycoparasite, and fourteen fungi associated with rapeseed. *Transactions of the British Mycological Society*, **74**, 501–7.

Tu, J.C. & Vaartaja, O. (1981). The effect of the hyperparasite (*Gliocladium virens*) on *Rhizoctonia solani* and *Rhizoctonia* root rot of beans. *Canadian Journal of Botany*, **59**, 22–7.

Tzean, S.S. & Estey, R.H. (1978). *Schizophyllum commune* as a destructive mycoparasite. *Canadian Journal of Microbiology*, **24**, 780–4.

Van Vuurde, J.W.L., Kruyswyk, C.J. & Schippers, B. (1979). Bacterial colonization of wheat roots in a root-soil model system. In: *Soil-borne Plant Pathogens*, ed. B. Schippers & W. Gams, pp. 229–34. London: Academic Press.

Warmbrodt, R.D. & Eschrich, W. (1985a). Studies on the mycorrhizas of *Pinus sylvestris*

L. as produced *in vitro* with the Basidiomycete *Suillus variegatus* (Sw. ex Fr.) O. Kuntze. I. Ultrastructure of the mycorrhizal rootlets. *New Phytologist*, **100**, 215–23.

Warmbrodt, R.D. & Eschrich, W. (1985*b*). Studies on the mycorrhizas of *Pinus sylvestris* L.produced *in vitro* with the Basidiomycete *Suillus variegatus* (Sw. ex Fr.) O. Kuntze. II. Ultrastructural aspects of the endodermis and vascular cylinder of the mycorrhizal rootlets. *New Phytologist*, **100**, 403–18.

Woods, A.M. & Gay, J.L. (1983). Evidence for a neckband delimiting structural and physiological regions of the host plasma membrane associated with haustoria of *Albugo candida. Physiological Plant Pathology*, **23**, 73–88.

5

Induced modifications in the plasma membranes of infected cells

J. L. GAY and A. M. WOODS

Department of Pure and Applied Biology, Imperial College of Science and Technology, University of London, Prince Consort Road, London, SW7 2BB, UK.

Plant pathologists are familiar with the modifications which fungal products cause in host cell membranes resulting in the rapid net efflux of cellular contents. For example, the toxin produced by *Helminthosporium victoriae* initiates ion efflux and consequent cell death. In another instance, fusicoccin, formed by *Fusicoccum amygdali* stimulates the extrusion of protons so that the solute balance of stomatal guard cells is severely disturbed and eventually the water relations of the whole plant are affected.

Plasma membranes of plant cells are also modified in infections in which haustoria are produced. Structural and physiological changes in them have been recorded in many infections and it is generally assumed that solutes pass across this interface. In one instance it has been proved that an abnormally high solute efflux occurs through the plant plasma membrane (Manners & Gay, 1983). This type of modification is unique because it is not associated with the immediate death of the plant cells and only part of the plasma membrane, namely the region adjacent to the haustorium, is affected. However, no molecular species or factor has been implicated in the modification associated with any haustorial infection (Chard & Gay, 1986) and the purpose of this chapter is to focus attention on the nature of the modifications and the variety of plants in association with diverse fungi in which they occur.

Evidence from isolated haustoria

The first indication of such phenomena was obtained by Hirata & Kojima (1962) who dissected individual haustoria of *Sphaerotheca fuliginea* from the epidermal cells of cucumber leaves. They showed that each haustorium was contained in a tough sac, but without corroborative electron microscopy it was impossible to equate the sac with an invagination of the plasma membrane of the host. Dekhuijzen (1966) extended this work devising a method for the mass isolation of haustoria of *S. fuliginea* and subsequently Dekhuijzen & van der Scheer (1967, 1969) examined sections of isolated haustoria by electron microscopy. They showed that each haustorium was contained in a membrane but, because the membrane was attached to the haustorium, deduced that it was of fungal origin. In retrospect it seems likely that this deduction precluded the proper recognition which the technical advance deserved, because it also precluded exploitation of the technique by other investigators who at the time believed it was the host plasma membrane that was invaginated when a haustorium penetrated a host cell.

Isolated haustoria from pea powdery mildew (*Erysiphe pisi*) were subsequently re-examined by Gay & Gil (1975) and Gil & Gay (1977). Dekhuijzen's method was substantially modified and parallel observations were made on haustoria *in vivo*. Electron microscopy then showed that the membrane enclosing such haustoria was a continuation of the host plasma membrane; *in vitro* studies showed that it was the external surface without any other host structures adhering to it. Because the isolated structures comprised both fungal and host components, they were named haustorial complexes.

Systematic light microscopy provided valuable information on the properties of the region of host plasma membrane surrounding each haustorium (the extrahaustorial membrane). Isolated complexes swelled and shrank in response to solutions of differing osmotic concentration which were applied to them (Fig. 5.1). This resulted from the change in the volume of material between the haustorial wall and the extrahaustorial membrane and the most likely explanation was that the membrane was selectively permeable and that it was attached to the haustorial neck by an impermeable junction. Subsequently, this was confirmed by immersing haustorial complexes in uranyl acetate and, after an interval, precipitating the heavy metal with phosphate. In electron micrographs, precipitates were absent from the interior except where the extrahaustorial membrane was broken (Gay & Manners, 1987).

The junction of the extrahaustorial membrane with the haustorial neck was examined by electron microscopy both in infected cells and in isolated complexes, and was seen to be intact even when the extrahaustorial membrane was greatly extended. From this finding it was inferred that the pathway of solutes transported into or out of haustoria is restricted to the extrahaustorial membrane and the infected epidermal cells, and that the haustorial surface has no connection, direct or indirect, with the apoplast of the leaf.

The extrahaustorial membrane proved to be abnormal in being thicker (230 nm) than the rest of the plasma membrane which lined the host cell wall. Furthermore, it was not disrupted in the detergents Tween-80, Triton X-100, Teepol (Fig.5.1b) or deoxycholate, although such treatment increased the rate at which the extrahaustorial membrane distended when isolated complexes were mounted in water. This observation indicated that extraction of membrane components increased its permeability. The extrahaustorial membrane proved to be extraordinarily stable, and at first was broken only by mechanical shock or 1N NaOH. However, the abundance of carbohydrate, shown in it by electron microscopy following periodate oxidation and silver coupling, suggested that cell wall degrading enzymes might affect it, and subsequently this proved to be correct. After the enzymic treatment, the extrahaustorial membrane was thinner, distended rapidly when complexes were placed in water (Fig. 5.1c) and then ruptured like the plasma membrane of a normal plant protoplast.

Recent work in which a variety of specific fluorescent probes, including coupled lectins, were applied to haustorial complexes isolated from pea epidermis has shown that components of the extrahaustorial membrane include β1-4 linked polysaccharides with small amounts of α-glucose, α-mannose, galactose and N-acetylglucosamine (Chard & Gay, 1984). The last was restricted to the haustorial face of the membrane. Protein, arginine, amino and sulphydryl groups and calcium were also detected in the membrane.

ATPase distribution

In parallel with the ultrastructural investigations on haustorial complexes, Manners & Gay (1978, 1982) examined the transfer to the fungus of radioactive solutes from infected plants photosynthesising in an atmosphere containing $^{14}CO_2$. Soluble and insoluble ^{14}C-labelled compounds were found in the haustorial complexes and it was deduced that sucrose and perhaps glycerol were the mostly likely translocates. Sucrose is the major currency in the plant's economy and it is therefore significant that when the rate of translocation from host to fungus was calculated for the extrahaustorial membranes involved, the rate of sucrose transport proved to be within the range characteristic of active transport of solute into plant and fungal cells (Manners, 1979). This was particularly surprising because the rate represented an efflux from the infected plant cells. The finding prompted ultrastructural investigations to localise the

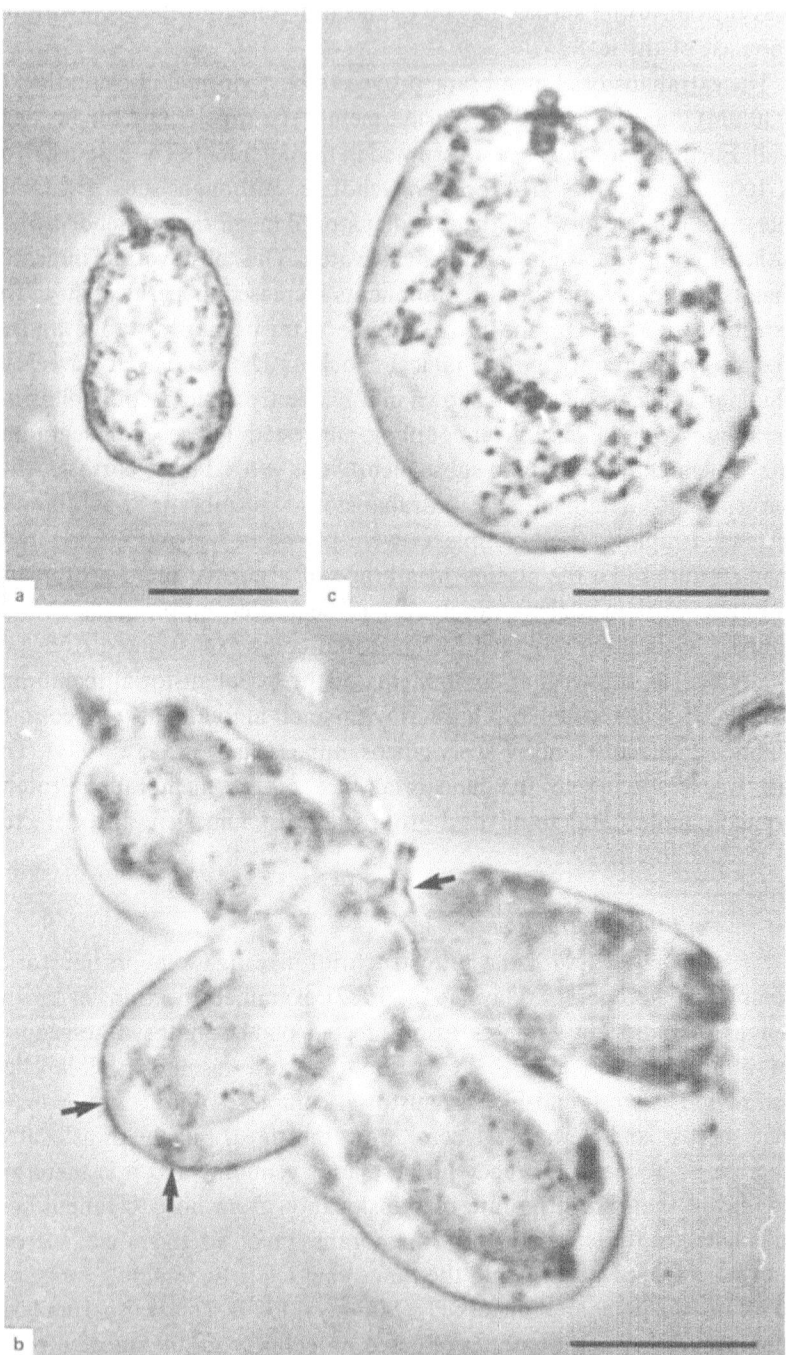

sites of ATPase activity (Spencer-Phillips & Gay, 1981) and it was discovered that infected cells had normal activity where the plasma membrane lined the host wall but none in the invaginated (extra-haustorial) region. Controls in which β-glycerophosphate replaced ATP as substrate showed a lack of enzymic activity throughout the whole of the plasma membrane, confirming the specificity of the activity with ATP and the inert nature of the extrahaustorial region.

The results led to the hypotheses outlined in Fig. 5.2a, b. It is proposed that the neckband seal is instrumental in coupling the transport systems of host and fungus and that the host's transport system is polarised by the formation of two domains in its plasma membrane. The host cell has no metabolic control to oppose solute efflux from its haustorial aspect and therefore an influx through its normal surface also increases the potential for efflux from its haustorial aspect.

The general applicability of the transport hypothesis and the pathogen's dependence on host pumps

In subsequent research, a variety of infections, both haustorial and intracellular, have all shown the same distribution of ATPase in the plant, and therefore the same polarisation of the infected cells. In French bean infected with the dikaryon of *Uromyces appendiculatus* (Spencer-Phillips & Gay, 1981), *Cardamine hirsuta* with *Albugo candida* (Woods & Gay, 1983), barley with *Erysiphe graminis* f.sp. *hordei* (Woods, 1985) and cabbage with *Plasmodiophora brassicae* (Gay, unpublished) the only significant difference was that the haustorial (or plasmodial) plasma membrane was devoid of enzymic activity, whereas in *E. pisi* ATPase and phosphatase were present. Experiments with hyphae of *Neurospora crassa* indicate that the negative results were not due to an inability to localise and identify ATPase in the plasma membrane. Therefore it is deduced that, in the absence of an energetic system at the fungal plasma membrane, the course of solute entry from the host depends on the

Fig. 5.1. Haustorial complexes isolated from *Pisum sativum* infected with *Erysiphe pisi*. (a) In an isolation medium containing sucrose the extrahaustorial membrane is indistinct as it is adpressed to the haustorial surface. (b) In detergent (5% Teepol) for 3 days. The extrahaustorial membrane is distended over an expanded extrahaustorial matrix. Arrows mark attachments of the extrahaustorial membrane to the haustorial neck (neckbands) and to haustorial lobes. (c) After treatment with pectinase and cellulase and transfer to water. The extrahaustorial membrane is distended even further and is at the point of rupture. The haustorium also shows signs of degradation. Scale bars = 10 μm. From Gil (1976).

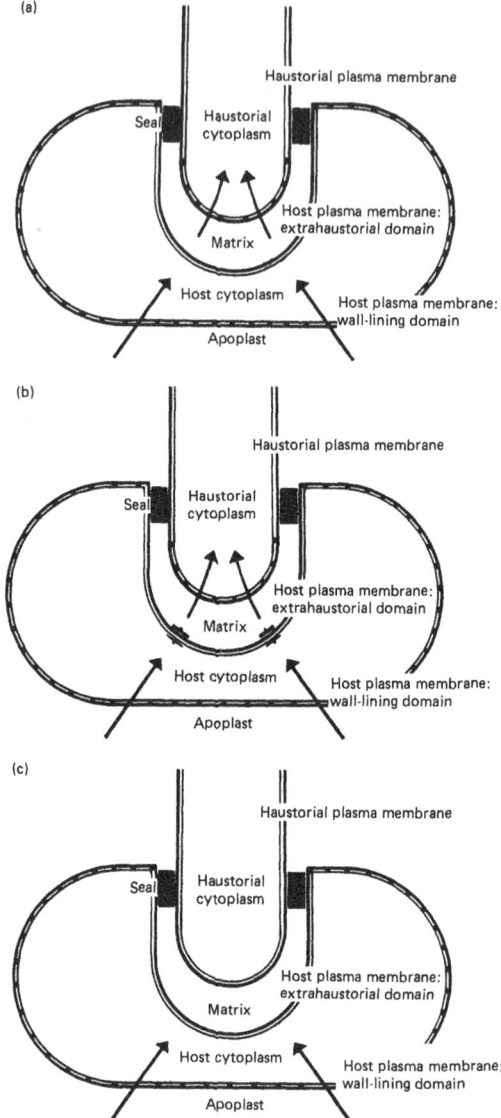

Fig. 5.2. Diagrams to illustrate hypotheses of solute transport into fungal haustoria. The plasma membranes of haustoria and infected cells are shown and regions with ATPase activity are dotted. (a,b) *Erysiphe pisi* infecting epidermal cells. The arrows indicate ATPase-activated influxes through the wall-lining domain of the host and the haustorial plasma membranes. It is proposed that, because the matrix is maintained as a closed compartment by the seal between host and haustorial plasma membranes, efflux through the invaginated domain of the host plasma membrane is effected by either a high concentration difference of the transported species (a) or the interaction of complementary ionic

distribution of host ATPase in the plasma membrane and a neckband whose constraints on diffusion effectively couple each haustorium to the transport system of its host cell (Fig. 5.2c). A structural entity constituting a neckband has been described in many rusts including *U. appendiculatus* (Hardwick, Greenwood & Wood, 1971) and Heath (1976) showed that it was a barrier to the apoplastic diffusion of solutes. In *Albugo candida*, although a neckband had not previously been recognised, evidence for it was provided both by the specific staining of sections examined by electron microscopy and by the close apposition of the plasma membrane to part of the haustorial neck, which resembles exactly that in other neckbands and apoplastic seals (Woods & Gay, 1983). Plasmolysis of infected cells detached the plasma membrane only from the cell wall and not from the haustorium (Coffey, 1983). In *E. graminis*, neckbands have been described from light, (Manners & Gay, 1977) and electron (Gay, unpublished) microscopy. In *P. brassicae*, transport coupling follows directly from the inclusion of the fungal protoplast within that of the host cell. In these instances, the findings strongly suggest that the parasitic mode is due, at least in part, to the parasite's dependence on host pumps.

The extreme thickness of extrahaustorial membranes is a feature peculiar to the Erysiphales on dicotyledonous hosts. However the distribution of ATPase in domains of the plasma membranes of infected cells is paralleled by the results of all freeze-fracture investigations at present available. These have been carried out in *Melampsora lini* (Littlefield & Bracker, 1972), *Erysiphe pisi* (Gil, 1976) and *Plasmodiophora brassicae* (Aist, 1974). In each instance the region lining the host cell wall was indistinguishable from the normal. The extrahaustorial region in *M. lini* exhibited few intramembrane particles, whereas in the other species particles were absent, although in *P. brassicae* envelopes they became evident as the plasmodia aged. Clearly the correlation with the distribution of ATPase may be significant but it has also been suggested that its absence may relate to the lack of an adjacent cell wall (Littlefield & Bracker,

Fig. 5.2. *(cont.)*

products of electrogenic pumps at the ATPase-active membranes (b). The seal also maintains the segregation of the two domains of the host plasma membrane. (From Spencer-Phillips & Gay (1981). Reproduced by permission of the trustees of the New Phytologist.) (c) In infections where there is no ATPase activity at the fungal plasma membrane, transport depends on concentration gradients generated by ATPase activity at the uninvaginated host plasma membrane, solute synthesis in the host, solute conversion in the fungus and the seal coupling the host and fungal cells. (From Woods & Gay, 1983).

1972). Sterols also were virtually absent from plasma membranes invaginated around rust haustoria (Harder & Mendgen, 1982). This was deduced from freeze-fracture preparations of material in which sterols had been tagged with filipin which was subsequently visualised by electron microscopy.

Variations on the theme

Recent research on two other haustorial infections further indicates that differentiation of part of the host plasma membrane is a common phenomenon but certain peculiarities characterise each example. Investigation of *Tussilago farfara* infected by the monokaryon of *Puccinia poarum* showed that the haustoria were filamentous, lacked any structure corresponding to a neckband and were enclosed to varying extents by material indistinguishable from the host cell wall (Al-Khesraji & Lösel, 1981; Woods & Gay, 1987). This material completely enclosed some haustoria but in others only a variable proximal region was so covered, the distal part being covered by other types of ensheathing materials. An extrahaustorial membrane enclosed all regions but only the part over the distal ensheathing materials lacked ATPase activity (Woods & Gay, 1987). Thus the population of infected cells had plasma membranes in which the distribution of ATPase varied continuously from instances which superficially resembled that in haustorial infections described above to those where ATPase was present throughout. A further difference lay in the gradual transition from activity to inactivity which occurred over a distance of 1.5μm (Fig. 5.2) instead of the abrupt change at a neckband which characterised all the examples (except *P. brassicae*) described above. This indicates that in *Puccinia poarum* the barrier maintaining the distinctness of the two plasma membrane domains is incomplete and this led Woods & Gay (1987) to propose that an altered composition caused a decrease in the fluidity of the plasma membranes of these infected cells. This hypothesis is strengthened by evidence that the transition region is mobile: neither Al-Khesraji & Lösel (1981) nor Woods & Gay (1987) studied the progress of haustorium development but, from the position of haustoria within colonies and the degree of their vacuolation, it was deduced that haustoria became gradually enveloped in cell wall. Woods & Gay (1987) proposed that the constituents of the normal plasma membrane invaded the abnormal extrahaustorial domain formed initially. Thus although the same general hypothesis for the differentiation of the plasma membranes of infected cells is applicable, the relationship is more variable and probably less permanent than where a neckband is developed. Furthermore in the absence of a neckband, the lack of a seal to the apoplast around each haustorium would render the haustorium less efficient. Woods & Gay (1987) proposed that the frequent infection of vascular tissue, where nutrients are abundant, probably compensates for the inefficiency. (Fig. 5.3)

A preliminary investigation of haustoria in lettuce infected by *Bremia lactucae* has shown another variation of this theme (Woods, 1985). Membrane differentiation was recognised by a procedure which distinguishes plasma membranes from other cell membranes and which was first applied to haustorial infections by Littlefield & Bracker (1972). (Sections prepared for electron microscopy are treated with periodic acid and subsequently with a mixture of chromic acid and phosphotungstic acids.) Staining is thought to indicate the presence of glycoproteins (Van der Woude, 1973). Where it has been applied to those membranes whose ATPase activity has also been investigated, the results have been exactly correlated. In infected lettuce, some cells conformed to the familiar pattern with two domains, but in others the whole plasma membrane was abnormal, although in some of the latter haustoria could not be seen. In infected cells with two domains of plasma membrane, neckbands were absent as in *P. poarum* but it was noted that the extrahaustorial region was smooth and adpressed to the haustorial wall, and was not undulate as around the haustoria of most other species unless fixed by freeze substitution (Mendgen, 1983). This suggests that the whole extrahaustorial membrane is tightly bound to the haustorial surface; the same proposal has been made for *Pseudoperonospora humuli* by Pares & Greenwood (1979). If this interpretation is correct, tight binding may be an alternative to the neckband as a means of preventing the lateral diffusion of the constituents of these plasma membranes and therefore a mechanism for maintaining two domains.

A final example illustrates that it should not be assumed that all invading intracellular structures are enclosed to some extent by a modified host plasma membrane. In infections of ovaries by *Claviceps purpurea*, development is entirely intercellular (Shaw & Mantle, 1980), with infrequent entry to dead xylem elements (Luttrell, 1980). However, *C. fusiformis*, infecting millet, occasionally penetrates living cells of the vascular parenchyma, but specific staining has not provided any evidence of abnormality in the structure of the invaginated plasma membrane despite its extensive proliferation (L. Faux & J. L. Gay, unpublished).

It is to be expected that in some instances aberrations and malfunction of the extrahaustorial membrane may result in haustorium inefficiency.

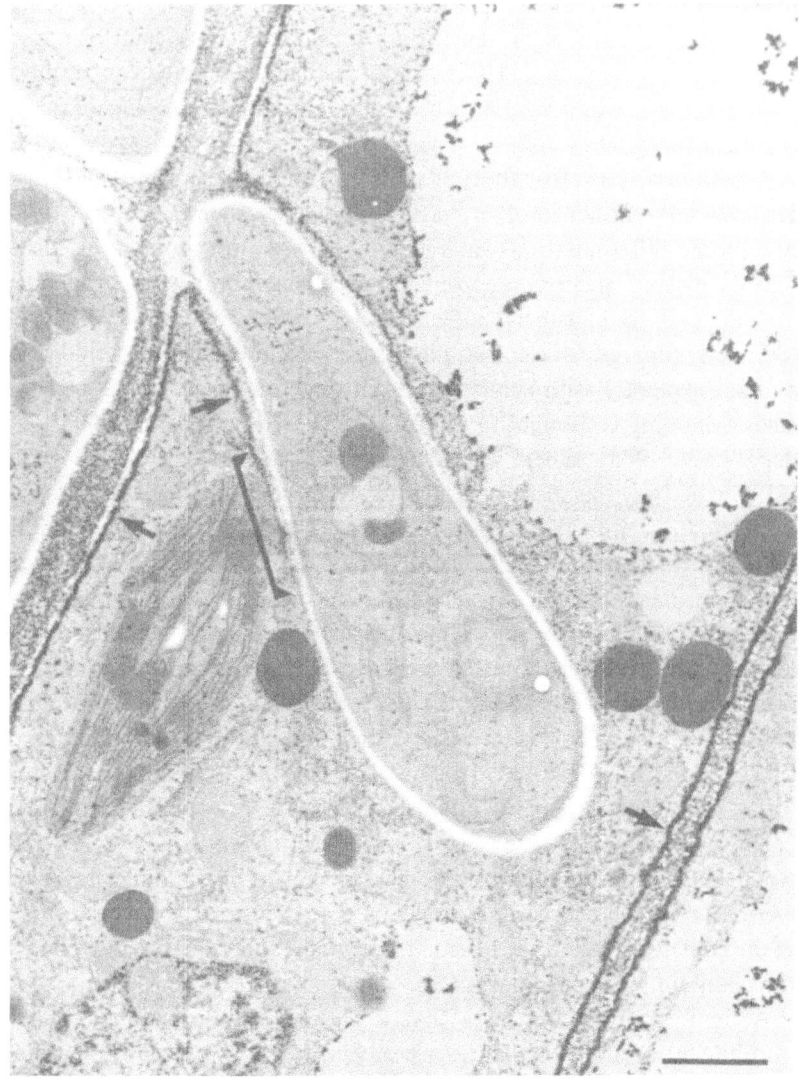

Fig. 5.3. Longitudinal section of haustorium of *P. poarum* infecting xylem parenchyma cell of *T. farfara* prepared from material incubated with ATP and Mg^{2+} at pH 7.2 to localise ATPase activity. The reaction product (arrowed) is localised to the host plasma membrane adjacent to the host cell wall and to host wall-like material which encloses the haustorium. Precipitates are absent from the fungal plasma membrane and the host plasma membrane over the distal part of the haustorium. The gradual transition (bracketed) from activity to inactivity occurs over a distance of approximately 1.3 μm. A fine lead precipitate, also observed

Manners & Gay (1980) and Spencer-Phillips & Gay (1980), found wide variation in the incorporation of radioactive products of photosynthesis into haustoria of a cultivar of pea susceptible to powdery mildew. This was not correlated with haustorial age (Manners & Gay, 1980) but with abnormalities in the structure of the extrahaustorial membrane some of which resembled damage by dehydration. Furthermore, in resistant cultivars permitting only slow colony growth, there was a marked delay in the onset of staining of extrahaustorial membranes by a reagent thought to be specific for anion transport sites (Chard & Gay, 1984). However, a correlation between slow colony growth and the distribution or intensity of ATPase activity or specific E. M. stains has not been found (M. A. Stumpf & J. Gay, unpublished).

There is wide circumstantial evidence that the extrahaustorial membrane develops by proliferation of the Golgi or endoplasmic reticulum (Woods & Gay, 1983) and it seems inevitable that the haustorium is the source of a metabolite which initiates its development. The effect is extremely localised and other parts of the plasma membrane, either existing or newly formed, which proliferate in relation to the development of new wall material in papillae and encasements, are not affected. Where neckbands occur, the initiator is likely to be concentrated between the haustorial surface and the extrahaustorial membrane.

In conclusion, now that it has been established that cell infection is normally associated with membrane differentiation, investigation should progress to the mechanisms initiating the differentiation.

Note added in proof. Since this Chapter was written, experiments with fluorescein used as an indicator of cytoplasmic pH and conducted over periods of minutes, have provided strong support for the ATPase deficiency coupled transport hypothesis. See Gay, J. L., Salzberg, A. & Woods, A. M., *New Phytologist* (in press).

Fig. 5.3. (cont.)

in controls without substrate, is present in the host and fungal cytoplasms, particularly within the nuclear envelope and endoplasmic reticulum. Scale bar = 1µm. (Reproduced from Woods & Gay, (1987) by permission of Academic Press Inc. (London) Ltd.)

References

Aist, J.R. (1974). A freeze etch study of *Plasmodiophora*-infected and non-infected cabbage root hairs. *Canadian Journal of Botany*, **52**, 1441-9.

Al-Khesraji, T.O. & Lösel, D.M. (1981). The fine structure of haustoria, intracellular hyphae and intercellular hyphae of *Puccinia poarum*. *Physiological Plant Pathology*, **19**, 301-311.

Chard, J.M. & Gay, J.L. (1984). Characterisation of the parasitic interface between *Erysiphe pisi* and *Pisum sativum* using fluorescent probes. *Physiological Plant Pathology*, **25**, 259-76.

Chard, J.M. & Gay, J.L. (1986). Cytochemical investigations of phospholipase and lipase activity at the parasitic interface between *Erysiphe pisi* and *Pisum sativum*. *Physiological and Molecular Plant Pathology*, **28**, 33-9.

Coffey, M.D. (1983). Cytochemical specialisation at the haustorial interface of a biotrophic fungal parasite, *Albugo candida*. *Canadian Journal of Botany*, **61**, 2004-14.

Dekhuijzen, H.M. (1966). The isolation of haustoria from cucumber leaves infected with powdery mildew. *Netherlands Journal of Plant Pathology*, **72**, 1-11.

Dekhuijzen, H.M. & van der Scheer, C. (1967). Electron microscopic observations on haustoria isolated from cucumber leaves infected with powdery mildew. *Netherlands Journal of Plant Pathology*, **73**, 121-5.

Dekhuijzen, H.M. & van der Scheer, C. (1969). The ultrastructure of powdery mildew, *Sphaerotheca fuliginea*, isolated from cucumber leaves. *Netherlands Journal of Plant Pathology*, **75**, 169-77.

Gay, J.L. & Gil, F. (1975). The fungus-host interface. *Proceedings 1st Intersectional Congress of the International Association of Microbiological Societies*, vol. 2 pp. 125-130. Tokyo: Science Council of Japan.

Gay, J.L. & Manners, J.M. (1987). Permeability of the parasitic interface in powdery mildews. *Physiological and Molecular Plant Pathology*, **30**, 389-99.

Gil, F. (1976). Ultrastructural and physiological properties of haustoria of powdery mildews and their host interfaces. *Ph.D. Thesis, University of London*.

Gil, F. & Gay, J.L. (1977). Ultrastructural and physiological properties of the host interfacial components of haustoria of *Erysiphe pisi in vivo* and *in vitro*. *Physiological Plant Pathology*, **10**, 1-12.

Harder, D.E. & Mendgen, K. (1982). Filipin-sterol complexes in bean rust and oat crown rust fungal/plant interactions; freeze-etch electron microscopy. *Protoplasma*, **112**, 46-54.

Hardwick, N.V., Greenwood, A.D. & Wood, R.K.S. (1971). The fine structure of the haustorium of *Uromyces appendiculatus* in *Phaseolus vulgaris*. *Canadian Journal of Botany*, **49**, 383-90.

Heath, M.C. (1976). Ultrastructural and functional similarity of the haustorial neckband of rust fungi and the Casparian strip of vascular plants. *Canadian Journal of Botany*, **54**, 2484-9.

Hirata, K. & Kojima, M. (1962). On the structure and the sack of the haustorium of some powdery mildews, with some consideration on the significance of the sack. *Transactions of the Mycological Society of Japan*, **3**, 43-6.

Littlefield, L.J. & Bracker, C.E. (1972). Ultrastructural specialisation at the host-pathogen interface of rust-infected flax. *Protoplasma*, **74**, 271-305.

Luttrell, E.S. (1980). Host-parasite relationships and development in the ergot sclerotium in *Claviceps purpurea*. *Canadian Journal of Botany*, **58**, 942-58.

Manners, J.M. (1979). Physiology of fungal haustoria (Erysiphales). *Ph.D. Thesis, University of London*.

Manners, J.M. & Gay, J.L. (1977). The morphology of haustorial complexes isolated from apple, barley, beet and vine infected with powdery mildews. *Physiological Plant Pathology*, **11**, 261–6.

Manners, J.M. & Gay, J.L. (1978). Uptake of ^{14}C photosynthates from *Pisum sativum* by haustoria of *Erysiphe pisi*. *Physiological Plant Pathology*, **12**, 199–209.

Manners, J.M. & Gay, J.L. (1980). Autoradiography of haustoria of *Erysiphe pisi*. *Journal of General Microbiology*, **116**, 529–33.

Manners, J.M. & Gay, J.L. (1982). Transport, translocation and metabolism of ^{14}C photosynthates at the host-parasite interface of *Pisum sativum* and *Erysiphe pisi*. *New Phytologist*, **91**, 221–44.

Manners, J.M. & Gay, J.L. (1983). The host-parasite interface and nutrient transfer in biotrophic parasitism. In: *Biochemical Plant Pathology*, ed. J. A. Callow, pp. 163–95. Chichester: John Wiley.

Mendgen, K. (1983). The interface between rust haustoria and their host cells. *Abstracts 3rd International Mycological Congress*, Tokyo, p. 185.

Pares, R.D. & Greenwood, A.D. (1979). The perihaustorial membrane in hop cells infected by *Pseudoperonospora humuli*. *New Phytologist*, **83**, 473–7.

Shaw, B.I. & Mantle, P.G. (1980). Host infection by *Claviceps purpurea*. *Transactions of the British Mycological Society*, **75**, 77–90.

Spencer-Phillips, P.T.N. & Gay, J.L. (1980). Electron microscope autoradiography of ^{14}C photosynthate at the haustorium-host interface in powdery mildew of *Pisum sativum*. *Protoplasma*, **103**, 131–54.

Spencer-Phillips, P.T.N. & Gay, J.L. (1981). Domains of ATPase in plasma membranes and transport through infected plant cells. *New Phytologist*, **89**, 393–400.

Van der Woude, W.J. (1973). Significance of the specific staining of plant plasma membranes by treatment with chromic acid-phosphotungstic acid. *Plant Physiology*, **51**, 15 (supplement).

Woods, A.M. (1985). Ultrastructural and cytochemical studies of higher plant-fungal interfaces, with special reference to biotrophy. *Ph.D. Thesis, University of London.*

Woods, A.M. & Gay, J.L. (1983). Evidence for a neckband delimiting structural and physiological regions of the host plasma membrane associated with haustoria of *Albugo candida*. *Physiological Plant Pathology*, **23**, 73–88.

Woods, A.M. & Gay, J.L. (1987). The interface between haustoria of *Puccinia poarum* (monokaryon) and *Tussilago farfara*. *Physiological and Molecular Plant Pathology*, **30**, 167–85.

6

Nutrient relations in biotrophic infections

J. F. FARRAR[1] and D. H. LEWIS[2]

[1]*School of Plant Biology, University College of North Wales, Bangor, Gwynedd LL57 2UW, UK and [2]Department of Botany, University of Sheffield, Sheffield S10 2TN, UK.*

Leaves and roots are rich sources of nutrients, both organic and inorganic, for invading fungi. In addition to the stores of many compounds at high concentrations within specific organelles and cells, a large traffic of many molecular and ionic species is constantly in progress, bathing hyphae in a continuously replenished solution of nutrients. In the early stages of infection the nutrient fluxes characteristic of an uninfected plant may be sufficient to sustain fungal growth; later, as the mass of fungal material is greater and the proportion of uninvaded host tissue correspondingly smaller, fungal growth may be possible only if an alteration of these nutrient fluxes is accomplished or access to particular stores of nutrient within the host achieved. Since the many different types of host cell have quite different physiologies, the expanding mycelium will experience a most heterogeneous environment. Further heterogeneity will be a consequence of infection as, in addition to cells functioning normally, the host will contain both cells directly invaded by the fungus and cells that are not invaded but which are modified in some way by its presence. Likely sources of nutrients for a fungus may thus be quite different in lightly infected leaves, with many host cells to support each mycelium, and in densely infected leaves with few; sources may also differ between resistant and susceptible reactions. Insufficient attention has previously been paid to this heterogeneity and to the localisation within infected plants of the physiological changes that have been reported to follow infection.

This chapter concentrates on effects of biotrophic pathogens on leaves but, at the end of each section, brief comparative references are made to those mutualistic mycorrhizas [ecto-, vesicular-arbuscular (VA) and ericoid (Lewis, 1986, 1987)] in which interchange of organic and inorganic metabolites, and water occurs.

Symptom spread, colony development and hyphal growth of biotrophic fungi in leaves and roots

The development of a mycelium from a germinating spore on the leaf surface has been described frequently and for many of the most commonly studied pathogens. Such studies are rarely of direct help in understanding the physiology of plant-pathogen interactions, since they are not sufficiently quantitative and are often conducted on material that is not genetically uniform. Furthermore, there is, to date, no means of assessing accurately the mass of fungal material within an infected leaf, much less the distribution between the two symbionts of any given compound common to fungus and host.

Careful work by Bushnell (1970) has shown that, in compatible infections of wheat with stem rust, branched hyphae define a region of a colony that has higher respiration rates, nitrogen content, and starch deposition than healthy tissue, but unbranched runner hyphae can extend about 0.5 mm beyond this region; incompatible infections lack these runner hyphae. By contrast, starch and increased respiration rates were detectable in powdery mildewed leaves up to 300 μm beyond the spread of mycelium.

In leaves of barley (*Hordeum distichum*) infected with brown rust (*Puccinia hordei*), there was a close relationship between the spread of visible symptoms and of mycelium; the area of starch deposition in host cells lagged slightly behind (J.D. Scholes & J.F. Farrar, unpublished). Interestingly, if the infected leaf was viewed by reflected light in the conventional way, both at flecking and at sporulation, pustules appeared lighter than the surrounding, uninfected tissue. This appearance, which has led to suggestions that there may be chlorophyll loss within pustules and regreening (synthesis of chlorophyll) prior to green island formation, is belied by viewing the leaves with transmitted light. Even 5d after inoculation, the pustules appeared darker green than the areas between them and seemed to be sites of chlorophyll retention. This was confirmed both by measuring the absorption of narrow beams of red light shone through the leaf, and by assay of chlorophylls extracted from carefully excised pustules and the areas between them. Loss of chlorophyll was greater from between than from within pustules (Fig. 6.1). Although this may indicate retention of chlorophylls by cells destined to be within green islands, there must be a population of cells which were invaded by hyphae only 6-9d after infection and which were previously between pustules. These may show net loss and resynthesis of chlorophylls.

By contrast, in bluebell (*Hyacinthoides non-scripta*) infected with *Uromyces muscari*, green islands extended 3-4 mm beyond the mycelium, whereas pre-sporulation pustules were congruent with mycelial spread. The chlorophyll concentration within green islands was much lower than in control tissue; it was lower still within the green islands where sporulation occurred (Scholes, 1985). Different again was the infection of *Phaseolus vulgaris* by *Uromyces phaseoli*: here, increased chlorophyll concentrations were found at flecking and at the green island stage, implying net chlorophyll synthesis consequent on infection (Sziraki *et al.*, 1984). The apparent confusion in the literature over whether green islands form by retention of chlorophyll or regreening probably reflects real differences between diseases. These would be made clear if rates of chlorophyll synthesis and degradation were measured separately throughout the infection process, both within and between pustules.

Mycelial growth within a leaf can be estimated most simply from the rate of radial expansion of colonies. Thus, hyphae of brown rust on barley elongated at 4.1 (J.D. Scholes & J.F. Farrar, unpubl.) and 2.6 µm h^{-1}

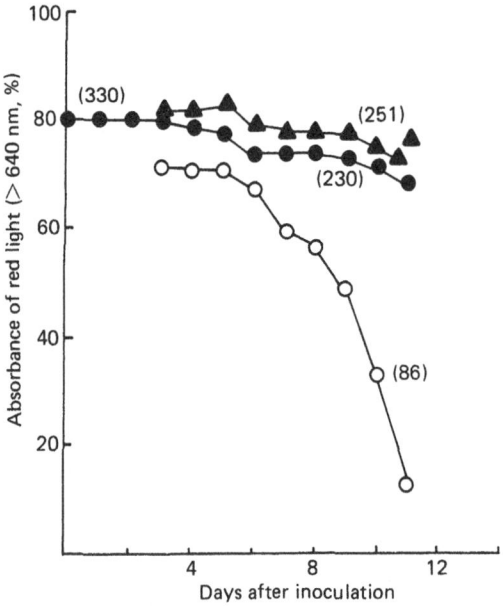

Fig. 6.1. *In vivo* assessment of chlorophyll density within different regions of a barley leaf infected with brown rust. A collimated beam of red light (40 µm, > 640 nm) was shone through leaves on the stage of a microscope fitted with a photomultiplier; allowance was made for absorption not due to extractable pigment. (●) control leaf; (▲) infected, non-sporulating area of pustule; (○) uninfected region between pustules. Each point is a mean of 100 determinations. Numbers in parentheses represent the total chlorophyll concentration (mg m^{-2}) at the green island stage, measured in extracts. (From Scholes, 1985.)

(Whipps, Roderick, Clifford & Lewis, 1985). Since hyphae usually develop tortuously around host cells, the true rate of extension will be greater and may reach 5-10 µm h^{-1} in rusts *in vivo* (Scott & Maclean, 1969). These rates are low compared with rates of germ tube extension from urediniospores (~ 75 µm h^{-1}) but are similar to those of axenically grown *Puccinia graminis* (1.3-8.3 µm h^{-1}) (Scott & Maclean, 1969); free living basidiomycetes have elongation rates of up to 280 µm h^{-1}, and ascomycetes and zygomycetes up to about 4 mm h^{-1}. After 5 d, a single colony of *P. hordei* on a barley leaf can have a total of 1 m of hyphae bearing more than 10^4 haustoria, an average of about 1.5 per host cell. Since all non-vascular cell types except guard cells are invaded, almost every host cell outside the mestome sheath is penetrated by at least one haustorium (Kneale & Farrar, 1985). To a first approximation, therefore, all host cells within a colony (which is well defined by visible symptoms, see above) are equally affected by the fungus and markedly different from those around the colony. The surface area of haustoria and hyphae will provide an upper limit to the amount of nutrient uptake that can be sustained by the fungus. In the barley/brown rust system, 15% of fungal surface area is accounted for by haustoria which clearly represent a large proportion of fungal biomass (Kneale & Farrar, 1985). In some biotrophic infections, vascular tissue is invaded: *Puccinia poarum* on coltsfoot invades both the bundle sheath and cells within the bundles (Al-Khesraji, Lösel & Gay, 1980). It would be very surprising if the invasion of specialised host cell types were not accompanied by considerable but currently unknown nutritional consequences for the fungus.

While there is no direct way of measuring fungal biomass within leaves, there is considerable information on the amounts of individual constituents, some of which are unique to the pathogen, present during the development of infection. As brown rust develops on barley, both mannan and chitin rise in amount after a 3d lag (Whipps, Clifford, Roderick & Lewis, 1980; Fig. 6.2a,b) with kinetics very similar to the spread of symptoms described above. Amounts of mannan and chitin correlated both with each other and with pustule numbers. A similar rise in glucosamine for rusted wheat leaves has been demonstrated (Mayama, Rehfeld & Daly, 1975). The fungal polyols, arabitol and mannitol, also rise (Fig. 6.2e; Mitchell, Fung & Lewis, 1978) in stem-rusted wheat; trehalose, a disaccharide often found in fungi, does not, although it accumulates in the oomycetous infections, for example *Albugo tragopogonis* on *Senecio squalidus* (Whipps, Haselwandter, McGee & Lewis, 1982). Oomycetes do not usually contain polyols although arabitol has been reported from

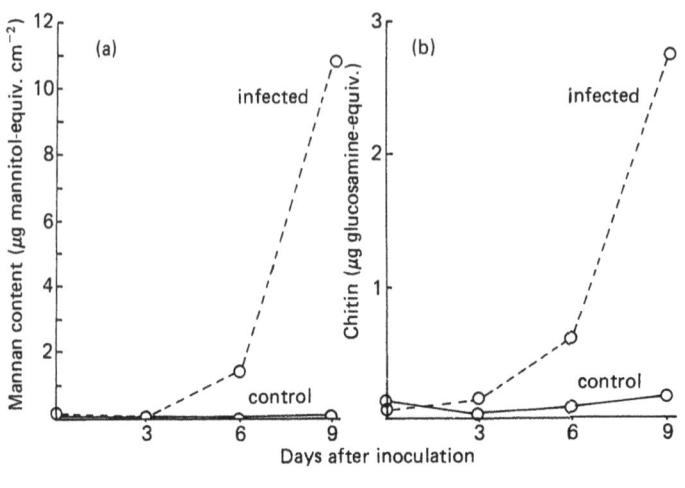

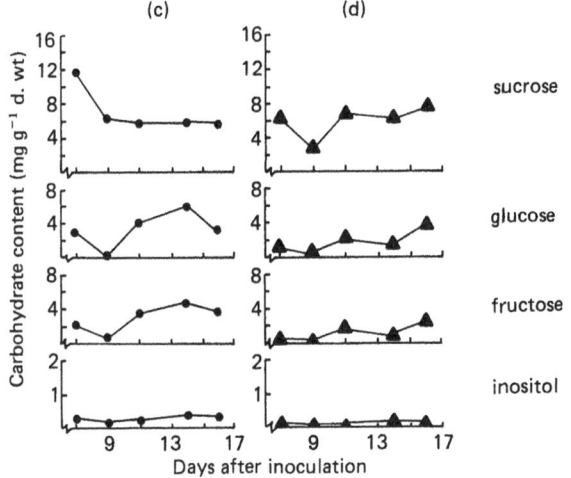

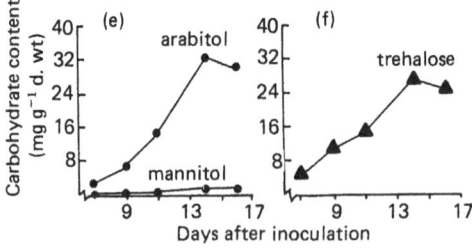

Phytophthora cinnamomi (Jennings, 1984). An extensive discussion of the use of chemical markers to estimate fungal growth is provided by Whipps *et al.* (1982). Here it should be noted that both symptom spread and accumulation of fungal metabolites show a lag phase of 2-3 d, a rapid rise to a linear rate of increase, and no or little net increase following sporulation. Whether a lag phase is present, or exponential growth occurs after the establishment of the fungus in the leaf, will need to be determined by more precise analyses. It would seem that fungal biomass will be rather low for about 3d after infection; the period of rapid increase in mass after this will be the first large drain on the reserves of the host.

The sugars, sucrose, glucose, fructose and inositol (all found in the host and all except sucrose in the fungus) show more complex changes following infection (Whipps & Lewis, 1981); some of the apparent complexity in the literature may be due to a failure to localise these sugars, as changes within and between pustules can be quite different (Fig. 6.2c,d; Mitchell *et al.* 1978).

The composition of host cell walls is often changed by pathogenic fungi (see chapter 7) but information pertinent to a discussion of fungal nutrition is not available. Extensive lignification of mesophyll cell walls would cause the apoplast to be a much less likely route for transport of water and nutrients to the fungus; other chemical changes in the walls, which might change the density of ionic exchange sites or the conductance to flow of liquid water, would also modify transport to the fungus.

In mutualistic symbioses between fungi and roots, the rate of spread of hyphae, whether they develop into a distinctive morphological entity, and the morphological effects they have on the roots themselves, varies greatly between ecto-, VA and ericoid mycorrhizas. They must be considered separately and generalisations for the three types are not possible.

Harley & Smith (1983) have reviewed the structure of ectomycorrhizal organs, their longevity and factors affecting the intensity of infection. They also consider the causal anatomy of the fungal sheath and the changes induced in roots by infection. Quantitative analyses of mycorrhizal development in both seedlings and adult trees are few but more

Fig. 6.2. Changes in amounts of carbohydrate during development of rust fungi in leaves. (a) Mannan and (b) chitin in barley (cv. Gold) infected with *Puccinia hordei*. (c) Sucrose, glucose, fructose and inositol within pustules and (d) in the region adjacent to pustules of stem rust on first leaves of wheat, and (e) arabitol, mannitol and (f) trehalose within pustules of this disease. (a,b from Whipps *et al.* (1980) and (c-f) from Mitchell, Fung & Lewis, 1978.)

studies can be expected now that mycorrhizal development can be followed in 'growth pouches' (Fortin, Piché & Lalonde, 1980). Relevant techniques have been compiled by Wilcox (1982) who makes important reservations about comparisons between laboratory and field studies (cf. Duddridge & Read, 1984a,b). Harley & Smith (1983) and Read (1983) have reviewed those few studies of development of ericoid mycorrhizas where intracellular penetration and development are important. For both ericoid and ectomycorrhizas, assays of chitin and mannan have been used to estimate fungal biomass. For the latter symbioses, mannan is not suitable for those of gymnosperms since the host plants themselves synthesise this polymer (Whipps *et al.*, 1982).

In contrast to the two other types of mycorrhizas, infection and spread by VA mycorrhizal fungi has been subjected to detailed mathematical analyses by a variety of approaches. The state of the art to 1983 was reviewed by Harley & Smith but there have been significant subsequent developments (e.g. Sanders & Sheikh, 1983; Buwalda, Stribley & Tinker, 1984; Walker & Smith, 1984; Amijee, Stribley & Tinker, 1986; Smith, Tester & Walker, 1986; Tester, Smith, Smith & Walker, 1986). These have concerned both the elaboration of the mathematical models themselves and the application of these models to data from experiments where the effects of biotic and abiotic variables on patterns of infection have been studied. Analyses of the formation of fungal entry points and rates of growth of infection units have been supplemented by estimates of the amounts of both internal and external hyphae (e.g. Sanders, Tinker, Black & Parmerley, 1977; Tisdall & Oades, 1979; Abbott & Robson, 1985), aspects particularly important in the acquisition and interchange of nutrients. In addition to these studies based on direct observation, assays of chitin and/or mannan have also been used to estimate fungal biomass (Hepper, 1977; Bethlenfalvay, Pacovsky & Brown, 1981; Whipps *et al.*, 1982).

Photosynthesis

Leaf-inhabiting pathogenic fungi affect photosynthesis in many different ways. An excellent way to appreciate this is to consider photosynthesis as a diffusion process, where F, the flux of CO_2, is driven by C, its concentration difference, across a conductance, K (Farquhar & Sharkey, 1982; Nobel, 1983; Sharkey, 1985). Thus, $F = KC$. This approach can be used to determine whether the conductance of stomata and cuticle, or of the mesophyll (the activity of the photosynthetic machinery of chloroplasts), or the concentration gradient driving entry of CO_2 into the leaf, is the prime mediator of the overall reduction in net photosynthetic rates so commonly seen. A further merit of this approach is that it emphasises the distinction, so crucial in diseased leaves, between net and gross photosynthesis.

The concentration of CO_2 in intracellular spaces within the leaf (C_i) determines the concentration gradient driving net photosynthesis, the net transfer of CO_2 from air to leaf. Rates of respiration are commonly greater in diseased than in healthy leaves, and the respiratory flux of CO_2 into intracellular air spaces raises C_i well above control values, as does reduced gross photosynthesis. An increase in C_i has been reported for several rust diseases (Owera, Farrar & Whitbread, 1981) but not for powdery mildew of beet (Gordon & Duniway, 1982a). Such an increase will itself decrease net photosynthesis. The conductance of stomata plus cuticle is the other determinant of net photosynthesis. Although, in general, biotrophic fungi reduce presporulation daytime transpiration (Fig. 6.3; Ayres, 1981a; Gordon & Duniway, 1982a), which is a measure of vapour-phase conductance to CO_2, both *Uromyces fabae* on *Phaseolus vulgaris* (Duniway & Durbin, 1971b) and brown rust on barley (C. Berryman & J.F. Farrar, unpublished) can cause increases. Following sporulation of rust fungi, the ruptured cuticle will have a high conductance to CO_2, which would itself increase net photosynthesis. The combined effect of raised C_i and altered diffusive conductance is, however, still a reduced rate of net photosynthesis.

Increased C_i gives a larger driving force for CO_2 transfer in the liquid phase of the leaf to the chloroplast, and potentially a higher chloroplastic CO_2 concentration, which would favour enhanced rates of gross photosynthesis (Colman & Espie, 1985). Indeed, for brown rust on barley (Owera *et al.*, 1981), and possibly for powdery mildew on barley (Last, 1963; Walters, 1985), an increase in gross photosynthesis per unit of green leaf area or of chlorophyll occurs; certainly reductions in gross-photosynthesis on a whole leaf basis are less than those in net-, for a range of foliar diseases. There is increased CO_2 cycling within a diseased leaf, with unexpectedly rapid gross photosynthesis and enhanced respiration, as summarised later, but a reduced net input of CO_2 from the environment. It is not yet completely clear how photorespiration contributes to CO_2 flux: most, but not all, workers find it to be decreased (Gordon & Duniway, 1982a; Bushnell, 1984) as would be expected if an increased C_i depresses the oxygenase function of ribulose 1,5-bisphosphate carboxylase (Rubisco).

The commonplace observation that concentrations of chlorophylls are

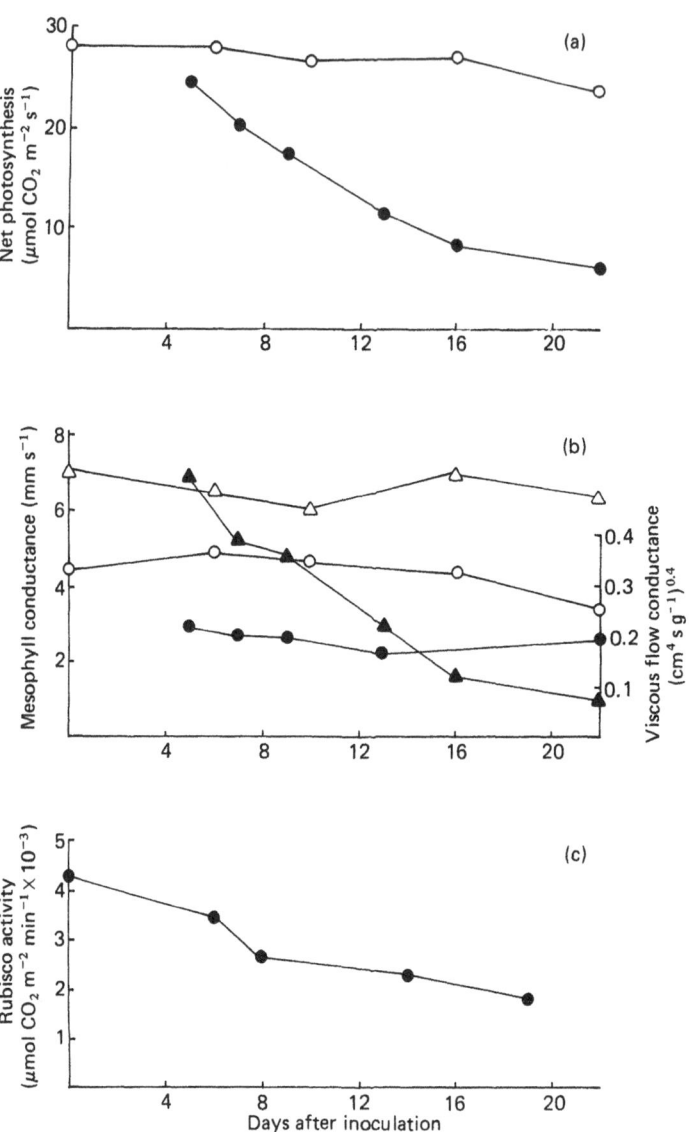

Fig. 6.3. Characteristics of photosynthesis in leaves of sugar-beet leaves infected with powdery mildew (*Erysiphe polygoni*). (a) Net photosynthesis, (b) conductances for CO_2 transfer (viscous flow conductance being largely a function of stomatal conductance and related to the rate of transpiration), and (c) activity of ribulose bisphosphate carboxylase (Rubisco) (day zero represents the value for an uninfected leaf). Open symbols, control; closed symbols, infected. (Redrawn from Gordon & Duniway, 1982*c*.)

lower in diseased leaves does not imply an effect on gross photosynthesis, since the distribution of chlorophylls across the diseased leaf may be far from uniform and, also, the amount of light-harvesting machinery may not limit carbon fixation. Very few diseases have been investigated in detail. In barley leaves infected with powdery mildew, and in mildewed beet, both the amount and activity of Rubisco were progressively reduced compared to controls (Gordon & Duniway, 1982c; Walters & Ayres, 1984) (Fig. 6.3c). In mildewed barley, chlorophyll concentration was also reduced along with the activities of two other enzymes of the Calvin cycle, phosphoglycerate kinase (also involved in glycolysis) and NADPH-dependent glyceraldehyde 3-phosphate dehydrogenase. Since, in mildewed pea reductions in photosynthesis (measured in 2% oxygen to depress photorespiration) occur sooner than reductions in stomatal aperture (Ayres, 1980), it is possible that a reduced amount of photosynthetic machinery is the primary cause of reduced photosynthesis. Gordon & Duniway (1982c) similarly ascribed the reduced photosynthesis of powdery mildew-infected sugar beet to changes within the mesophyll, including reductions in Rubisco.

Barley infected with brown rust shows reduced mesophyll conductance at sporulation, suggesting changes in the amount or activity of photosynthetic machinery. Since gross photosynthesis per unit chlorophyll is actually increased (Owera *et al.*, 1981), any such changes are likely to be complex and heterogeneous. When photosynthetic oxygen evolution per unit chlorophyll was measured (in 5% CO_2 to reduce photorespiration), considerable enhancement over controls occurred from 3 to 13 d after infection. As the concentration of chlorophylls fell over the same period, so too did oxygen evolution per unit area and apparent quantum yield (Scholes, 1985). It was argued earlier that this disease is characterised by loss of chlorophyll from between rather than within pustules. Thus, as expected, excised pustules showed rates of photosynthetic oxygen evolution considerably higher than regions excised from between pustules. Autoradiography of leaves briefly fed $^{14}CO_2$ confirmed that the bulk of CO_2 fixation was occurring within pustules, where nearly all cells are invaded by haustoria, rather than between them (Scholes, 1985). These changes could be accounted for by alterations either to the light reactions of photosynthesis or to carbon metabolism, but only the former have been investigated further.

In vivo chlorophyll fluorescence kinetics have been monitored by irradiating a previously darkened leaf with monochromatic light and measuring the kinetics of the resultant fluorescence at longer wavelengths for about two min. Precise interpretation of the resultant curves is complex and controversial (Papageorgiou, 1975; Lavorel & Etienne, 1977; Baker & Bradbury, 1981; Krause & Weis, 1984; Sivak & Walker, 1985) but, in much of the work with disease, they are of value if used comparatively. Infection of barley brown rust apparently preserves the fluorescence kinetics characteristic of juvenile control leaves (Fig. 6.4). Infiltration of leaves with the proton translocator, CCCP, prior to measurement of fluorescence showed that proton gradients were still maintained across the thylakoid lamellae of diseased leaves (Ahmad, Farrar & Whitbread, 1983, 1985). Qualitatively similar changes obtained from flecking to the green island stage and, more crucially, pustules and regions between pustules showed quite different kinetics. The variable fluorescence, fluorescence quenching and quantum yield of the primary photochemistry of photosystem II were all greater within than between pustules (Scholes, 1985) and the M peak, often considered to be related to the onset of carbon metabolism, was reduced in all parts of a diseased leaf relative to controls. Measurement of fluorescence at the wavelength

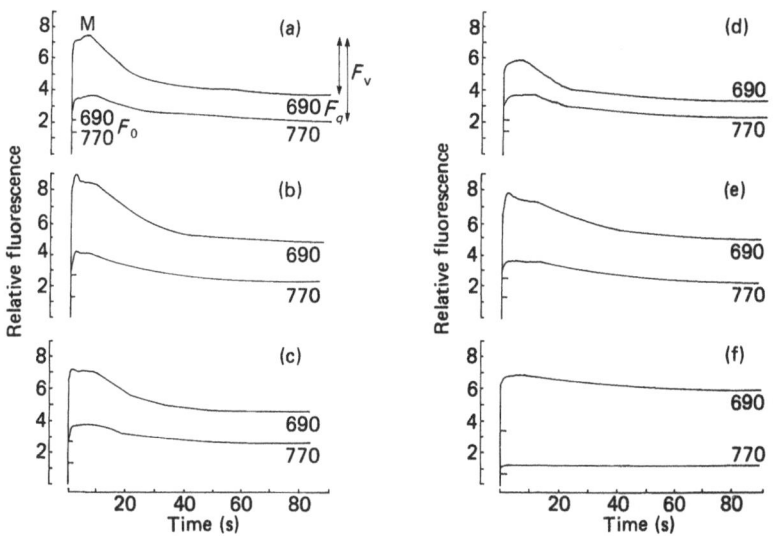

Fig. 6.4 Chlorophyll *a* fluorescence induction curves for barley leaves infected with brown rust: (a-c) at sporulation (d-f) the green island stage. Uninfected leaves (a,d), pustules (b,e) and regions between pustules (c,f) were irradiated at 632.8 nm, giving fluorescence curves at 690nm (mainly PS II) and 770 nm (not relevant to this paper). The position of F_o fluorescence seen within 0.5 ms of illumination, is indicated by a short horizontal line; that at 690 nm is the higher in every case; F_v, variable fluorescence; F_q, quenching of variable fluorescence. (From Scholes, 1985.)

characteristic of photosystem II may indicate that non-cyclic electron transport was enhanced within, but declined between, pustules (Scholes, 1985).

In contrast, within the large pustules of *Uromyces muscari* on bluebell leaves, there was a progressive decrease in chlorophylls, carotenoids, chloroplast volume, photosynthetic oxygen evolution per unit chlorophyll and fluorescence quenching, indicative of a decline in non-cyclic electron transport (Scholes, 1985; Scholes & Farrar, 1985). Beyond the visible boundaries of the lesion, photosynthesis appeared unaffected.

Clearly, these two rusts affected photosynthesis differently; most notably only brown rust caused increases in gross photosynthesis in host cells invaded by haustoria. It is difficult to interpret older experiments where chloroplasts isolated from whole diseased leaves were examined. Preferential reduction of non-cyclic photophosphorylation in chloroplasts from mildew-infected leaves of *Beta vulgaris* and *Vicia faba* has been reported (Buchanan, Hutcheson, Magyarsosy & Montalbini, 1981) but this result may have reflected the mean behaviour of a heterogeneous population of chloroplasts, as might the unaltered rate of the Hill reaction in chloroplasts isolated from rusted barley (Ahmad *et al.*, 1983).

Powdery mildew, which does not invade mesophyll cells and which can be confined to the adaxial leaf surface, has clear effects on the chlorophyll *a* fluorescence kinetics of barley leaves which are consistent with a reduction in non-cyclic photophosphorylation.

The mechanisms by which fungi within leaves affect photosynthesis remain uncertain. Whilst the production of specific compounds, by host or fungus, may explain some effects on stomata (Ayres, 1980), it is unlikely that any such compound is directly responsible for reduced non-cyclic photophosphorylation by inhibition of electron flow in the manner of DCMU (Buchanan *et al.*, 1981); a more subtle effect of some specific compound remains possible. The increased gross photosynthesis of fungally-invaded cells in barley leaves could simply be due to relaxation of normal feedback controls limiting the rate of host photosynthesis; the increased sink that the fungus provides for carbon compounds could produce such an effect, since the magnitude of the rise is within that produced by manipulating source/sink ratios in cereals (Rawson, Gifford & Bremner, 1976), as is that found in uninfected leaves on infected plants (Ayres, 1981*b*). Some of the decrease in photosynthesis could be due to altered water and solute relations within host cells. The reduced size of chloroplasts within pustules of rusted leaves of bluebell (Scholes & Farrar, 1985) and wheat (Whitney, Shaw & Naylor, 1962) is most easily explained as due to an increased concentration of solutes in the host cytosol (see later). The consequences of such changes for electron transport and carbon fixation are uncertain, although high concentrations of cations can change the rate of chloroplastic ferricyanide reduction and fluorescence parameters (Baker, 1978) and there is a negative relationship between leaf solute potential and variable fluorescence in excised barley leaves (J. F. Farrar, unpublished).

The conclusions of the last paragraph apply equally well to the effects mycorrhizal fungi have on photosynthesis of their hosts. The studies, which are relatively few compared with those concerning biotrophic leaf pathogens and which have been reviewed by Harley & Smith (1983) and Lewis (1986), have mostly concerned VA mycorrhizas. Here, infection generally increases rates of photosynthesis (although the method used for the calculation–by area or by weight–is crucially important) and diverts more carbon to roots. The most recent discussions of mechanisms (Harris, Pacovsky & Paul, 1985; Snellgrove, Stribley, Tinker & Lawlor, 1986; Dehne, 1986) emphasise many aspects already considered above for pathogens, or for mycorrhizas in general, by Harley & Smith (1983) and Lewis (1986), namely, source-sink relationships, starch mobilisation, concentration of P in leaves, and alterations to specific leaf area and leaf water relations.

Carbohydrates

Recent advances in our knowledge of carbohydrate storage and metabolism in vascular plants should lead to a more sophisticated view of the carbon nutrition of pathogenic fungi. For example, many vascular plants can be separated into groups according to whether the carbohydrate they store daily in leaves is starch (e.g. tomato, pepper, tobacco, soybean, beet) or sucrose and, perhaps, fructan (e.g. barley, ryegrass, wheat, maize) (Huber, 1981; Pollock, 1984, 1986; Farrar & Farrar, 1985a). Similarly, the magnitude of fluxes of carbohydrate and the localisation of spatially distinct carbohydrate pools have recently been described (Geiger, Ploeger, Fox & Fondy, 1983; Fondy & Geiger, 1985; Farrar & Farrar, 1985a,b, 1986). In addition to the nightly degradation of starch accumulated during the previous day, sucrose stored in the vacuole (which predominates in sucrose-strong species) is mobilised during darkness to maintain translocation. Sucrose in this pool, and in the cytosol, is also turned over during the light, as is vacuolar fructan (Farrar & Farrar, 1985b, 1986; Pollock, 1984, 1986). There is also considerable evidence for the existence of apoplastic pools of sucrose, which exchange with cytosolic pools via specific transport systems at the plasmalemma (Huber & Moreland, 1980). In common with many transport systems, these are biphasic with respect to sugar concentration, with a high-affinity energy-dependent system showing classical kinetics and a low-affinity system which is not energised and may show linear or hyperbolic kinetics (Maynard & Lucas, 1982; Reinhold & Kaplan, 1984). Both mesophyll cells and phloem tissue have these sugar transport systems (Van Bel & Koops, 1985).

These findings have numerous and profound implications for leaf-inhabiting fungi. Hyphae within the apoplast will be bathed in sugar solutions which, although dilute, will potentially be constantly replenished. Uptake of sugar by intercellular hyphae may be an important source of carbohydrate for the mycelium; there is certainly sufficient surface area of hyphae in barley leaves infected with brown rust (Farrar, 1984; Kneale & Farrar, 1985) and the flux of C (and P and K) into the apoplast is sufficient to support fungal growth (Ahmad *et al.*, 1985). However, the fungus will have to compete for these sugars with the transport systems of host plasmalemmae. Haustoria in the mesophyll of sucrose-storing species will be close to large stores of sucrose in host vacuoles; many haustoria may exploit the permeases of the host plasmalemma by control of conditions in the extrahaustorial matrix (see Chapter 5). The type of cell in which haustoria are found will determine not only the type and quantity of carbohydrate available but also the flux of those carbohydrates. The parenchymatous bundle sheath of temperate C3 Gramineae is probably physiologically distinct from the mesophyll around it since it has morphologically distinct chloroplasts (Carolin, Jacobs & Vesk, 1973) and a role in transport of assimilates which will result in a large flux of sucrose through it (Kuo, O'Brien, & Canny, 1974; Altus & Canny, 1985). In barley leaves, it is the tissue nearest to the vascular system that contains typical haustoria of brown rust; the epidermis contains smaller haustoria and may have a different sugar complement (Hwang, Ibenthal & Heitefuss, 1983; Kneale & Farrar, 1985).

These considerations emphasise the importance of localising events within tissues of an infected leaf, but the intracellular localisation of compartmentalised metabolites such as sucrose is also important.

The following simple calculation emphasises the necessity of measuring fluxes of sugars, rather than simply their quantity. A single colony of brown rust on a barley leaf absorbs about 6 μg of sugar daily (Kneale & Farrar, 1985); at a common infection density, this amounts to 150 mg carbohydrate m^{-2} leaf h^{-1}. The total carbohydrate available to the fungus is approximately 3000 mg m^{-2} leaf (Owera, Farrar & Whitbread, 1983) which is enough to supply the fungus for 20h. Replacement of these carbohydrates from photosynthesis or other host sources is obviously necessary to enable fungal growth to occur. A similar calculation for uptake of carbohydrate through haustoria of oat mildew shows that, 50h after infection, the photosynthetic capacity of a single mesophyll cell is substantially greater than the rate of uptake of carbohydrate by a mildew colony based in an adjoining epidermal cell. Thus, carbohydrate supply is unlikely to limit growth of mildew (T. Carver & J.F. Farrar, unpublished).

Sugars entering the fungus must do so via transport systems at the fungal plasmalemma. Surprisingly little is known about sugar transport into pathogenic fungi (Farrar, 1985*b*). Uptake of fructose, glucose and sucrose each showed biphasic kinetics in *Phytophthora palmivora* grown in pure culture (J. Sheard & J.F. Farrar, unpublished); these kinetics, and the presence of high activity of a particulate acid invertase, are consistent with sucrose being hydrolysed extracellularly by the fungus prior to transport of the hexoses, although some sucrose may be absorbed intact at high concentrations. There is no evidence for a high-affinity transport system for sucrose in any fungal species yet studied (Komor, 1982) but it would be of considerable interest to determine whether such a transport system can be demonstrated in plant pathogenic fungi, for which sucrose is usually the major carbon source.

Potential changes in carbohydrate metabolism are best defined by altered enzyme activities, and actual changes of contents and fluxes of sugars. These are now considered in turn. Acid invertase is the most-studied relevant enzyme in infected leaves and it is of central interest in view of its probable role in hydrolysis of sucrose prior to sugar uptake by fungi. Studies of invertase in such leaves face two major problems: it is extremely difficult to tell whether invertase is produced by host or by fungus and its function in mature, healthy leaves – where it is present in sufficient quantity to hydrolyse much of the sucrose produced in photosynthesis – is quite obscure. In the investigation of Greenland & Lewis (1981, 1983) for oats infected with *Puccinia coronata*, the activities of both soluble and particulate acid invertases increased following inoculation to about 7-8 times those of uninfected controls although total amounts of soluble protein were unchanged. The pH optima of soluble and particulate enzymes were slightly different and unaltered by infection. Most of the invertase from diseased leaves appeared as a novel peak separable by gel filtration; this invertase had a higher K_m and was more sensitive to the inhibitor, aniline, than the others. It is not clear if this enzyme originated from the host or the fungus. Similar observations made earlier on different host/pathogen combinations are equally difficult to interpret (Whipps & Lewis, 1981) although the increases in activity were confined to pustules (Long, Fung, McGee, Cooke & Lewis, 1975). Not all pathogenic fungi cause a stimulation of invertase activity: the powdery mildew *Microsphaera alphitoides* on oak does not (Hewitt & Ayres, 1976). Specifically fungal invertases have been recorded from *Phytophthora palmivora* (J. Sheard & J.F. Farrar, unpublished), *P. megasperma* (Ziegler & Pontzen, 1982), *Puccinia graminis* (Williams, Maclean & Scott, 1984), *Claviceps purpurea* (Dickerson, 1972) and *Ustilago maydis* (Callow, Long & Lithgow, 1980). These may, if produced in sufficient quantity in the infected plant, be concerned with hydrolysis of extracellular sucrose before uptake of hexose by the fungus. In spite of suggestions to the contrary (Manners & Gay, 1982), it currently seems most likely that extracellular hydrolysis of sucrose is more probable than fungal transport of intact sucrose. An α-glucosidase capable of hydrolysing sucrose occurs in mildewed leaves of pea (I. Donaldson, personal communication).

The result of changes in enzyme activities of the host, plus the presence of the fungus, is that carbohydrate metabolism is greatly modified in the infected leaf. For a discussion of changed concentrations of carbohydrates in many diseases, see Whipps & Lewis (1981). Here, the comparatively few attempts to assess carbohydrate fluxes will be considered. Disease-induced alterations in photosynthesis, detailed above, result in a reduced net flux of carbon into the leaf. The usual fate of carbon fixed by a mature leaf, i.e. translocation, would thus be expected to be reduced, especially because both fungal metabolism and host defence mechanisms will compete with the transport system for fixed carbon. Indeed, reduced translocation is a common feature of all diseases examined (Whipps & Lewis, 1981) and it may be due simply to reduced sucrose concentrations at source (Owera *et al.*, 1983). Fig. 6.5, a carbon budget for a brown-rust infected barley leaf, illustrates the quantitative relationship between net and gross photosynthesis, fungal growth and translocation. In addition to the large internal recycling of CO_2 by repeated respiration and photosynthesis in the diseased leaf, a smaller than normal proportion of net photosynthesis is subsequently available for translocation since both fungal growth and net synthesis of starch by the host reduce the amount of available assimilate. No net import of translocated carbon is seen in this disease (Farrar, 1984), although this may occur in others, especially of dicotyledonous hosts. It is not possible, in any disease, to be sure of the size of the gross flux of carbon into the fungus; this could be most easily measured for the largely superficial powdery mildews.

In a healthy leaf, considerable changes in carbohydrate status and flux take place during a diel cycle and it is important to note that these are greater than the overall changes induced by disease. Not only is starch or sucrose (depending upon species) stored during the day and degraded at night but rates of turnover of these compounds change as do rates of photosynthesis (Farrar & Farrar, 1985a,b; Fondy & Geiger, 1985). The ways in which disease affects these normal patterns of carbohydrate metabolism are quite unexplored. Two pertinent observations can be made. On the basis of profiles of ^{14}C-translocation, Owera *et al.* (1983) proposed that the fall in amount of sucrose in leaves of barley infected with brown rust was largely due to loss of sucrose from vacuoles. A more direct test is now needed of the suggestion that loss of vacuolar sucrose – and thus presumably a cessation of sucrose transport into and out of vacuoles – is characteristic of this, and possibly other, diseases.

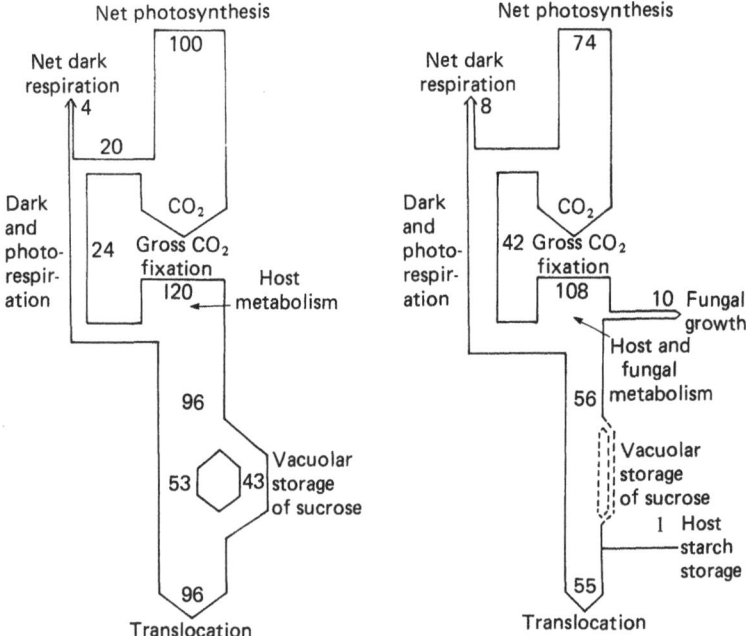

Fig. 6.5 Major carbon fluxes in leaves of barley infected with brown rust. Control (left) and infected (right). Arrow widths are proportional to fluxes, and the numbers are percentages of net photosynthesis in control leaves. Based on data from Farrar & Farrar (1985a, 1986), and Owera *et al.* (1981, 1983); see those papers for implicit assumptions.

Secondly, Holligan, Chen, McGee & Lewis (1974) and Whipps & Lewis (1981) have examined α-glucans and incorporation of ^{14}C into them in leaves of coltsfoot infected with *Puccinia poarum*. A relatively water-insoluble glucan–presumably starch–is lost during prolonged darkening of infected leaves and incorporates little ^{14}C from a pulse of $^{14}CO_2$ supplied prior to darkening; this glucan is localised outside pustules. Within pustules, a relatively water-soluble glucan–presumably fungal glycogen–shows very little loss in amount during prolonged darkening of the leaf but incorporates much ^{14}C. Investigations of diel changes will allow a more precise understanding of the alterations to starch metabolism of the host induced by infection. Certainly fluxes, rather than simply concentrations, of carbohydrates in diseased leaves must be considered, as must their localisation. The massive increases in fructans within pustules of *Puccinia poarum* on coltsfoot and the reduced sucrose but increased hexose content of these pustules (Holligan, Chen & Lewis, 1973) relative to surrounding, uninvaded, tissue show the importance of such localisation (see also Fig. 6.2).

Comparatively little is known of fungal carbohydrate metabolism in diseased tissues (Farrar, 1985*b*). Although hexitols such as mannitol may represent early products of fungal carbon metabolism, pentitols such as arabitol may represent carbohydrate storage in the more usual sense (Maclean, 1982; Farrar, 1985*b*). Turnover of mannitol in leaves of brown-rust infected barley proceeds at a high rate; by contrast, arabitol slowly accumulates ^{14}C of host origin (Owera *et al.*, 1983); hexitols–possibly mainly sorbitol–are readily labelled after feeding ^{14}C-glucose to cultured *Puccinia graminis* (Maclean, 1982; Manners, Maclean & Scott, 1982).

Harley & Smith (1983) and Lewis (1986) have reviewed the composition and metabolism of carbohydrates in the different mycorrhizal types and Lewis (1986) has indicated the futility of some studies on the carbon nutrition of cultured ecto-mycorrhizal fungi in relation to understanding the mycorrhizal condition. The activities of acid invertase in ecto- and VA mycorrhizas (Harley & Jennings, 1958; Lewis & Harley, 1965; Dehne, 1986) make it likely that sucrose, both apoplastic (ecto- and VA) and symplastic (VA), is hydrolysed by the fungi before absorption, as discussed above for pathogens. In the ecto- and ericoid mycorrhizas investigated, the same suite of storage carbohydrates as found in biotrophic ascomycetes and basidiomycetes of leaves, occur but the soluble carbohydrates of VA endophytes have yet to be identified and quantified. In these, it appears that lipids are quantitatively a more important sink for carbon from their hosts (Cooper & Lösel, 1978; Harley & Smith, 1983).

A start has now been made on the enzymology of carbohydrate metabolism in ecto- and VA mycorrhizal fungi (e.g. Dehne, 1986; Söderhäll, Jirgis & Ramstedt, 1986) but there is a long way to go. Although it is probable that the net flux of carbon in mature, well-illuminated plants with mutualistic mycorrhizas is always from plant to fungus, net movement of carbon from fungus to plant is a possibility, especially in shaded seedlings (Brownlee, Duddridge, Malibari & Read, 1983; Francis & Read, 1984; Finlay & Read, 1986; Finlay, Söderström & Read, 1986). Some 'reverse' flow of carbon will be a normal feature of mycorrhizas when nitrogen acquired from soil passes from fungus to plant as organic compounds (Lewis, 1976; Abuzinadah, Finlay & Read, 1986; Abuzinadah & Read, 1986*a,b*).

Respiration

Infected tissue has, in general, a higher rate of respiration than corresponding healthy tissues (Daly, 1976), placing a greater drain on reserves of respirable substrate. Respiration generates the energy that drives fungal growth, host defence, and the continual and normal maintenance processes of host and fungus.

Studies of the enzymic complement of diseased leaves and the results of feeding radiolabelled sugars agree that there is an increased contribution of the pentose phosphate pathway (PPP) relative to glycolysis (Daly, 1976), with the probable exception of downy mildew infections. This accords with the importance in diseased leaves of polyols and of phytoalexins (see Chapter 9), in part produced from intermediates of the PPP (Daly, 1976; Kosuge & Kimpel, 1981; Farrar, 1985*b*). In mildewed barley leaves, the activities of the first two enzymes of the PPP are increased whilst those of several glycolytic enzymes fall (Scott & Smillie, 1966). These changes only occur in susceptible interactions. In this host/pathogen system, malate metabolism in particular may be impaired as both malate dehydrogenase and NAD^+-dependent malic enzymes show reduced activities (Scott & Smillie, 1966; Walters & Ayres, 1984). It would be useful to show the subcellular localisation of these enzymes. Other glycolytic and Krebs cycle enzymes show unchanged activities following infection (Scott & Smillie, 1966). In rusted wheat, the first two enzymes of the PPP again rise in activity but the glycolytic enzymes, phosphoglucomutase and glucose phosphate isomerase, also show modest increases. The most dramatic increase recorded, however, was in hexokinase, the activity of which may regulate the rate of respiration (Lunderstadt, Heitefuss & Fuchs, 1962).

An overall increase in carbon flux through the respiratory network will produce increased amounts of NADPH but, although the detailed fate of this NADPH is unknown, most will presumably be oxidised by the mitochondrial respiratory chain. It is important to identify the engagement of the alternative pathway which is SHAM (salicylhydroxamic acid)-sensitive but cyanide-resistant and which is known from fungi as well as plants. Since it is non-phosphorylating (Fig. 6.6), it will contribute minimally to the cell's energy requirements. In non-thermogenic tissues, two functions have been suggested for the alternative pathway: it might be used to consume excess carbohydrates (Lambers, 1982) or it may enable a turnover of the Krebs cycle, providing substrates for amino-acid and other biosyntheses, under conditions of high energy charge (Moller & Palmer, 1984). The former is unlikely to be of major importance in leaves, which are carbohydrate-rich, but may have some alternative oxidase activity (Azcon-Bieto, Lambers & Day, 1983).

The latter has clear relevance to rapidly growing fungi, or to host cells involved in biosyntheses related to defence reactions. Even carbon skeletons withdrawn from the PPP may create a similar problem, since there is now evidence for an external NADPH dehydrogenase in plant mitochondria (Palmer, 1984). Infection of barley leaves with powdery mildew results in a rise in respiration, part of which is due to the alternative oxidase (F. Rayns & J.F. Farrar, unpublished). Both the increase in respiration and the activity of the alternative oxidase in part persist in leaves from which the superficial mycelium has been removed. *Ceratocystis ulmi* and *Ustilago maydis* also show some activity of the alternative

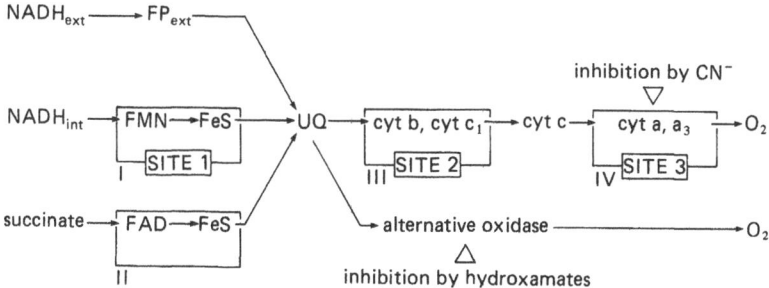

Fig. 6.6 A simplified scheme showing electron flow through the respiratory chain of plant mitochondria. The Roman numerals I-IV indicate presumed structural complexes, and 1-3 sites of proton translocation for subsequent ATP synthesis. FAD, flavin adenine dinucleotide; FMN, flavin mononucleotide; FP, flavoprotein; FeS, iron-sulphur proteins; UQ, ubiquinone, $_{ext}$, external; $_{int}$, internal.

oxidase (Sherald & Sisler, 1972). These results have several implications. Firstly, the lack of effect of uncouplers on the respiration of diseased tissue may be partly due to the reduction, in relative terms, of the role of the cytochrome pathway. Secondly, the increased CO_2 output from diseased tissue may not be matched by a proportionate increase in ATP production and, thus, available metabolic energy. Thirdly, since leaf-inhabiting fungi affect both non-photosynthetic and photosynthetic electron transport, it may be worth exploring mechanisms of fungal pathogenicity that can act at this level in both chloroplasts and mitochondria.

The localisation of respiratory increases is difficult but it is likely that a large proportion is due to the fungus itself. This would be expected since there are strong relationships between growth and respiration rates. Presumably, the respiratory efficiency of biotrophic fungi will be similar to that of related, free-living, fungi. In barley leaves infected with brown rust, the increase in respiration rate over uninfected controls (0.61 μmol CO_2 m^{-2} leaf h^{-1}; Owera *et al.*, 1981) is similar to the respiration rate of the fungus calculated from rates of sugar uptake and assuming that half of the absorbed sugar is subsequently respired (0.72 μmol CO_2 m^{-2} leaf^{-1} h^{-1}; Kneale & Farrar, 1985). These rates are close to the rate of turnover of mannitol (0.69 μmol cm^{-2} leaf^{-1} h^{-1}) in the same system. Additional support for the idea that rapidly growing fungal tissue is the source of much of the increased respiration in infected tissues comes from the observation that, in bluebell infected with *Uromyces muscari*, respiratory increases were confined to fungally invaded areas and only detectable before and during sporulation. At the green island stage, when fungal activity was reduced, respiration rates decreased below control values (Scholes & Farrar, 1985).

The control of respiration rate in infected leaves is probably largely by re-cycling of adenylates, much as in healthy plants (Farrar, 1985*a*). Sugar content of rusted wheat leaves does not correlate with respiration rates, implying that substrate supply is not limiting (Samborski & Shaw, 1956), although corroborative sugar-feeding experiments are needed to avoid problems owing to compartmentation. Exhaustive searches for endogenous uncouplers in leaves inhabited by biotrophic fungi have proved unsuccessful. In at least two diseases, *Pyrenophora teres* on barley and *Helminthosporium maydis* race T on maize, toxins produced by the fungus have been considered to stimulate host respiration. The toxin from the latter may have a mitochondrial site of action (see Chapter 8). In general, however, the demand for respiratory energy controls the respiration rate of diseased tissue via adenylates and pyridine nucleotides. As Harley & Smith (1983) point out, information on the respiration of mycorrhizal roots is scanty. The few manometric, cytochemical and enzymological studies available have been reviewed by them (see also Dehne, 1986; Söderhäll *et al.*, 1986) and, in general, the pattern is as expected from the discussion above. There appear to have been no studies of the pentose phosphate pathway but the existence of arabitol in at least some ecto- and ericoid mycorrhizas (Söderström, Finlay & Read, 1986) indicates its occurrence.

Cyanide-resistant, SHAM-sensitive, respiration occurs in both ecto- and VA mycorrhizas (Harley & ap Rees, 1959; Coleman & Harley, 1976; Antibus, Trappe & Linkins, 1980) but, as Harley & Smith (1983) note, its function and control is not yet understood. In line with the suggestions above concerning provision of carbon skeletons for synthesis of organic nitrogenous compounds, Lewis (1986) drew attention to the need to examine the effects of differing nitrogen sources on the activity of this pathway in mycorrhizas (see also Farrar,1985*a*).

Wedding & Harley (1976) proposed that fungal mannitol could control the activity of key glycolytic enzymes in the host, a suggestion denied by the evidence of Jirgis, Ramstedt & Söderhäll (1986). In view of recent work on the activity of pyrophosphate-phosphofructokinase and the control of glycolysis by concentrations of phosphate, adenylates and fructose 2,6-bisphosphate (see Farrar, 1985*a*), future work on the respiratory metabolism of both mycorrhizal systems and biotrophic pathogens of leaves should embrace the concepts that have emerged from these studies.

Water, ions and phosphate
Water

The water relations of a leaf are completely altered by fungal infection: typically, water potential is lowered as hydration of the leaf falls. Before sporulation, transpiration may rise or fall, depending on the disease. Where sporulation ruptures the epidermis, localised increases in transpiration will occur. Calculations of total flux of water have to be made with care, however, since large diel changes in the effect of disease are often seen (Ayres, 1980; Gordon & Duniway, 1982*b*). It is not clear from where in the healthy leaf, water evaporates resulting in high concentrations of solute; nor is it currently certain that most water crosses the mesophyll via cell walls, the apoplastic route. The problem is that a variable amount moves from cell to cell via the symplastic route. When the added complexities of localised lignification in the apoplast of diseased leaves and localised rupture of the cuticle are considered, it will be clear that the paths of water movement in diseased leaves are very difficult to describe.

The bulk concentration of numerous individual solutes changes during infection. Coupled with changes in leaf water status, it is possible that the total solute concentration is altered with profound consequences for (for example) the volume of organelles which are of necessity in osmotic equilibrium with the cytosol. As noted above, chloroplasts are smaller in some rust infections. It is difficult to estimate separately the solute potentials of fungus and host. An indirect approach adopted for brown rust of barley has been to measure the bulk solute potential and the apoplastic solute potential within pustules, to estimate the minimum solute potential of the fungus, and then to calculate the maximum possible solute of the host tissue within pustules (Table 6.1). Although this procedure is imperfect, the results seem fairly clear. The estimated solute concentration within the fungus is very high (solute potential low); thus, the solute concentration in the host symplast within pustules is less than the bulk solute concentration and is very close to values from control tissue. This would lead to the prediction that (for example) chloroplast size is virtually unchanged within pustules, quite in accord with the maintenance of chloroplast integrity and functioning within pustules. By contrast, solute concentration is greater in host cells between pustules, in just the area where photosynthetic function becomes impaired. It is of interest to consider the status of host cells just outside a small, developing pustule that subsequently become invaded by haustoria and eventually form green island tissue. It is possible that any small changes in solute concentration which occur develop only relatively late in infection.

There are few measurements of water potential and its components in diseased plants. It would be expected that water potential would fall in diseased leaves as their water content and status decline. This is precisely what occurs in leaves of *Phaseolus vulgaris* infected with *Uromyces phaseoli*, at a variety of water potentials in the rooting medium (Duniway & Durbin, 1971*b*) and in brown-rust infected barley (C. Berryman & J.F. Farrar, unpublished). Wheat infected with *Puccinia recondita* only showed reduced leaf water potentials when well-watered, however, and not when growing in dry soil (Cowan & van der Waal, 1975). Changes in turgor (pressure potential) would thus be expected as both water potentials and solute potentials change, especially since turgor is commonly the most sensitive component when plant water status is altered. Indeed, a large

Table 6.1. *The distribution of solutes within leaves of barley infected with brown rust (J.F. Farrar & C. Berryman, unpublished)*

	Volume $(cm^3 \, m^{-2}$ leaf)	Solute content $(mosmol \, m^{-2}$ leaf)	Solute concentration $(osmol \, m^{-3})$
Control			
Host symplast	178	99	582
Infected			
Host symplast			
(between pustules)	72	50	680
Host symplast			
(within pustules)	72	43	590
Fungus	17	17	1000
Apoplast	9	1	131

Figures for volume are derived from Kneale & Farrar (1985) and observations of the proportion of leaf infected. Solute concentrations of host symplast, both control and between pustules, are from direct psychrometric measurements; that for apoplast is derived from water potential isotherms. The solute concentration for the fungus is calculated assuming that all the arabitol and mannitol, and 150 mol m^{-3} K^+ with anion, is fungal. The solute in host symplast was then calculated by subtracting fungal solutes from the solute content of pustules, measured psychrometrically

fall in turgor was found in *Uromyces phaseoli*-infected bean leaves (Duniway & Durbin, 1971a). A fall in turgor, and some increase in solutes, occurs in uninfected leaves on plants of *Vicia faba* partially infected with *Uromyces viciae-fabae* (Tissera & Ayres, 1986).

Measurements averaged over whole leaves of barley infected with brown rust have indicated a lowering of turgor (pressure) potential in leaves at 100% relative water content (rwc) but no change in the initial relationship between loss of rwc and turgor (C. Berryman & J.F. Farrar, unpublished). Between pustules, turgor of epidermal cells, measured with a Zimmerman micro-pressure probe, is relatively unchanged; within pustules it falls progressively as disease develops (but never to zero). The extent of changes around a uredinium can be seen in Fig. 6.7. If it can be assumed that mesophyll cell turgor is similar to that of the overlying epidermis, in this disease, functional chloroplasts exist in cells of relatively unchanged solute potential but greatly lowered turgor. It is now believed that many processes in vascular plants are regulated by turgor (Zimmermann & Steudle, 1978) so changes in turgor may be the cause of several disease-mediated alterations of host physiology.

A reduced difference in solute concentration across the plasmalemma and some loss of turgor would occur if the concentration of solutes in the apoplast rose, with no change in properties of the plasmalemma or in cytosolic solute concentration. Indeed, apoplastic K^+ is raised in brown-rust infected barley leaves (Ahmad *et al.*, 1985) as are total apoplastic solutes (C. Berryman & J.F. Farrar, unpublished). Redirection of the transpiration stream into pustules would explain this. Reduced turgor could also arise as the result of reduced membrane integrity but it is argued below that integrity is maintained within pustules.

Whereas studies of the water relations of diseased leaves have concentrated on those of the host, a major interest concerning the, until recently, neglected investigations of mycorrhizal systems has been water flux from substrate to fungus, through fungal hyphae and from fungus to plant. As Hardie (1985) and Read (1986) have pointed out, interpretation of some studies which have compared mycorrhizal and non-mycorrhizal plants is complicated by effects attributable to changes in phosphate status of the two kinds of plant. There is now no doubt that hyphae (ecto- and VA) and mycelial strands (ecto-) are effective gatherers of water which becomes available to the plant (Duddridge, Malibari & Read, 1980;

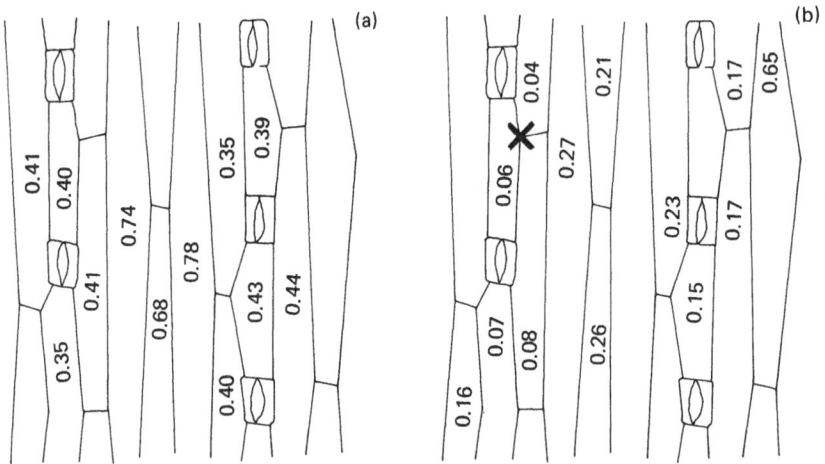

Fig. 6.7 Schematic maps of epidermal cell turgor (MPa) on the abaxial surface of barley first leaves: (a) control; (b) infected with brown rust 6d. Measurements were made with a Zimmerman micro-pressure probe; values were in each case compiled from measurements on 5 replicate leaves. The cross denotes the centre of a uredinium. C. Berryman & J.F. Farrar (unpublished).

Brownlee, Duddridge, Malibari & Read, 1983; Hardie, 1985, 1986). The changed water status of infected plants is then reflected in altered stomatal resistances and rates of photosynthesis (e.g. Allen, Smith, Moore & Christensen, 1981; Allen & Boosalis, 1983).

Ions

The concentrations of a variety of inorganic ions increase in diseased leaves and these ions accumulate at infection sites (Durbin, 1984). The overall increase in concentration has been ascribed to the balance between an increased or unaltered rate of transpiration, and thus import via the xylem, and greatly reduced export in the phloem (Ahmad, Owera, Farrar & Whitbread, 1982; Farrar, 1984). Concentrations of ions that are normally re-translocated from leaves, such as K^+, increase relatively more than Ca^{2+} and Mg^{2+} which are phloem-immobile and thus scarcely retranslocated. Some accumulation at infection sites may be the consequence of water flow in the apoplast being directed there, or to active accumulation by fungus or host within the pustule. One way to distinguish between these possibilities would be to use a radioactive isotope of a non-metabolised ion such as $^{86}Rb^+$, $^{36}Cl^-$ or $^{45}Ca^{2+}$. Ayres & Jones (1976) showed accumulation of root-fed $^{86}Rb^+$ at regions of barley leaves infected with *Rhynchosporium secalis* and the pattern of accumulation broadly followed that of transpiration. When $^{45}Ca^{2+}$ was fed to the transpiration stream of rusted or mildewed wheat and barley leaves, there was accumulation at infection sites, with that beneath uredinia detectable after 2-4 h (Shaw & Samborski, 1956). By contrast, when slices cut from barley leaves infected with brown rust were incubated on 1 mol m^{-3} $^{45}CaSO_4$ for 24 h, very little $^{45}Ca^{2+}$ could be found within pustules and most was in uninfected mesophyll (C. Berryman & J.F. Farrar, unpublished). Relatively little $^{45}Ca^{2+}$ was found in veins of control or diseased leaves. About 70% of this $^{45}Ca^{2+}$ could be exchanged from the tissue into unlabelled $CaSO_4$ over a 10-h period; there was no more loss from pustules than from regions between them. Within pustules, where relatively little $^{45}Ca^{2+}$ was found, the sporulating centre regularly contained more than the ring of infected tissue around it. The last results are consistent with the distribution of Ca^{2+} within healthy cells: it is maintained at low (*c.* 1 mmol m^{-3}) concentrations in the cytosol, most of the intracellular Ca^{2+} being within organelles, including the vacuole (Macklon, 1984). Thus, the relative exclusion of $^{45}Ca^{2+}$ from pustules may reflect the relatively high amount of Ca^{2+}-excluding cytosol, owing to the presence of mycelium, and perhaps lack of uptake by vacuoles within cells invaded by haustoria. The mycelial area of brown rust-infected barley leaves also excluded $^{36}Cl^-$ (although the central sporulating region accumulated some and the surrounding, uninfected, mesophyll more). Although strictly comparable experiments with feeding both to leaf segments and into the transpiration stream are needed, a tentative conclusion is that non-metabolised ions largely follow the transpiration stream in diseased leaves. The behaviour of metabolic compounds fed to leaves ($^{32}PO_4^{2-}$ and ^{14}C-glucose, for example) usually shows transpiration-independent accumulation in pustules.

Two other techniques have added to our understanding of nutrient relations within diseased leaves: electron microprobe analysis particularly of frozen or freeze-substituted material, and isotope efflux kinetics. Haustoria of *Puccinia coronata* in oat have been shown by electron microprobe analysis to contain iron and sulphur, in addition to polyphosphate (Harder & Chong, 1984). The neck ring deposited by the fungus around the site of haustorial penetration is rich in silicon (Chong & Harder, 1980): this deposit may be important in effecting a seal between the extrahaustorial matrix and host apoplast (see Chapter 5). Electron microprobe analysis of freeze-substituted barley leaves infected with brown rust has shown that rust spores are rich in Ca^{2+} but hyphae and host cells within pustules may have lower Ca^{2+} than uninfected tissue (Ahmad, Farrar & Whitbread, 1984), supporting the evidence from $^{45}Ca^{2+}$-feeding experiments discussed above. Use of this technique has also demonstrated Si in fungal material and high K^+ in spores. The dynamics of isotope movement can be monitored with isotope efflux analysis. The efflux of $^{86}Rb^+$, $^{45}Ca^{2+}$ and $^{36}Cl^-$ from discs of barley leaves infected with brown rust, preloaded with isotope, has been followed and in each case, the major result is that disease alters the pattern of efflux very little (Ahmad *et al.*, 1985; C. Berryman & J.F. Farrar, unpublished). There is certainly no indication of the massive and rapid efflux that would be expected if the integrity of host membranes were reduced by infection (see below). There is, however, relatively more K^+ in the apoplast in rusted barley leaves than in controls implying that a redistribution of ions may follow infection.

Fungal infection of leaves can also have effects on the long-distance movement of inorganic nutrients (see Walters, 1985) but, as seen above, different ions behave differently. For example, brown-rust-infected barley showed increased rates of uptake of Ca^{2+}, K^+ Mg^{2+} and NO_3^- but not of Pi per unit mass of root, with some redistribution of these nutrients in the intact plant. The nutrient content of xylem sap was little affected

(Ahmad *et al.*, 1982). By contrast, mildewed barley showed increased uptake of Pi and decreased uptake of NO_3^-; uptake of K^+ and Cl^- was also increased (Walters, 1985). Both this variability between diseases, and the lack of knowledge of the control of nutrient uptake in healthy plants, make speculation over the mechanism of these changes premature.

Qualitative and quantitative aspects of the absorption of ions of essential elements by mycorrhizas, their metabolism and translocation within the fungi, their transport from fungi to hosts and their consequent effects on growth and metabolism of hosts have been thoroughly investigated and are topics too large to be discussed here. They have been especially well-reviewed (Tinker, 1975*a,b*; Harley & Smith, 1983; Gianinazzi-Pearson & Gianinazzi, 1986).

Other important work has considered the role ecto-, VA and ericoid mycorrhizal fungi play in excluding toxic ions from their hosts (Gildon & Tinker, 1981; Bradley, Burt & Read, 1982; Read, 1983, 1986; Burt, Hashem, Shaw & Read, 1986) and in the transfer of metabolites *between hosts*, a topic which is generating some controversy but is potentially of great ecological importance (cf. Francis, Finlay & Read, 1986; Newman & Ritz, 1986).

Phosphate and biotrophic pathogens

Inorganic phosphate (Pi) has been invoked as a major controller of several areas of metabolism, including the partitioning of photosynthate between starch and triose phosphate within chloroplasts, export of triose phosphate from chloroplasts (Herold & Walker, 1979; Herold, 1984), and control of the production of the regulatory metabolite, fructose 2,6-bisphosphate (Preiss, 1984; Stitt, 1985). It is thus of considerable interest that P accumulates to a far greater extent in rusted barley leaves than K, Ca or Mg (Ahmad *et al.*, 1982) and has also been shown to accumulate in mildewed barley (Fric, 1978; Walters & Ayres, 1981), rusted wheat (Bennett & Scott, 1971) and rusted bean (Pozsar & Kiraly, 1966) leaves. The increased concentration of P in diseased leaves is due to greatly reduced export (Ahmad *et al.*, 1984). Much of the P is localised within pustules (Samborski & Shaw, 1956). Although some of the P accumulates as fungal polyphosphates, much remains soluble, probably as Pi (Bennett & Scott, 1971; Ahmad *et al.*, 1984; Harder & Chong, 1984). Most of the Pi in vascular plant cells is within the vacuole (Bieleski, 1973; Ahmad *et al.*, 1984; Woodrow, Ellis, Jellings & Foyer, 1984).

Three questions are especially relevant to the diseased state: what is the source of P for the fungus, what is the effect of infection on the concentration of cytosolic Pi and can cytosolic Pi concentration be regulated by exchange with the vacuolar Pi pool? Phosphorus budgets for leaves of barley infected with brown rust have shown that the rate of uptake of P by the fungus approximates to the rate of delivery of Pi to the leaf in the transpiration stream (Fig. 6.8). Moreover, ^{32}Pi entering the leaf via the xylem reached fungal spores with only a 25% fall in specific activity, suggesting that direct uptake from the transpiration stream accounts for the largest part of fungal acquisition of P (Ahmad *et al.*, 1984). There is no suggestion that net transfer of P occurs from host cells, either haustorially invaded or around a pustule. The second question is much harder to answer, yet the concentration of Pi in the cytosol is critical for metabolic control. Isotope efflux analysis for brown-rusted barley leaves implied a reduced Pi concentration in the cytosol (Ahmad *et al.*, 1985) but unfortunately this was a mean figure for the whole of an infected leaf, and the concentrations in the cytosol of invaded and non-invaded cells were not determined separately. The answer to the third question is that since the half-time for Pi in barley leaf vacuoles is about 10d, vacuolar Pi cannot readily and rapidly buffer cytosolic Pi (Ahmad *et al.*, 1984).

The consequences for host carbon metabolism and photosynthesis of the disease-induced changes in P status are thus uncertain, particularly because increased P contents of chloroplasts from diseased leaves have been reported (Ahmad *et al.*, 1983). One major symptom of reduced cytosolic Pi is increased synthesis of starch (Whipps & Lewis, 1981) and it would thus seem important to examine those cells showing increased deposition of starch. An alternative approach which has been attempted, would be to excise portions of diseased leaves and measure their rates of photosynthetic oxygen evolution in the presence of exogenously supplied Pi: neither pustules, nor regions between pustules, of brown rust on barley leaves showed any systematic response to treatment with 30 mol m^{-3} Pi (Scholes, 1985), implying that Pi-limitations cannot explain altered rates of photosynthesis in these diseased leaves.

Membrane permeability

An obvious and much canvassed means by which access to intracellular nutrients can be gained is by a reduction in the integrity of host membranes. While some diseases are undoubtedly characterised by massive increases in permeability of host membranes (Hanchey & Wheeler, 1979; Daly, 1981) and by changes in membrane potential (Tomiyama, Okamoto & Katou, 1983), it is by no means certain that

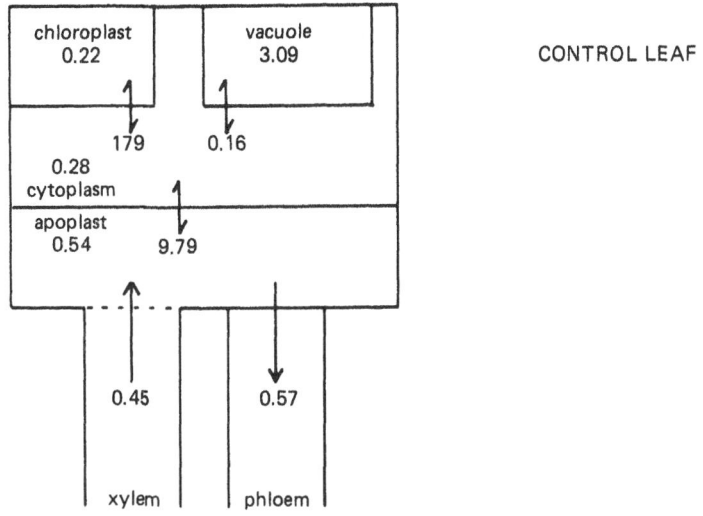

Fig. 6.8 Phosphorus budgets of (a) control and (b) brown-rusted barley leaf blades on the day of sporulation. Boxes represent discrete compartments; numbers within them are pool sizes (nmol P m^{-2} leaf). Only adjacent compartments can communicate directly; fluxes between them (nmol P m^{-2} leaf d^{-1}) are indicated by arrows. (From Ahmad *et al.*, 1984.)

this is a universal response to infection. Changes in specific membrane transport processes may, however, often occur (see Chapter 5, for example). Increased host membrane permeability will have potential disadvantages for a mycelium living and growing for a comparatively long period in its host. It will mean a fall in the concentration difference between apoplast and cytosol resulting in decreased turgor (see above); a fall in cytosolic solute concentration and thus a swelling of organelles such as chloroplasts and mitochondria. There will also be a loss of host cell function as the metabolites involved in and controlling them change greatly in concentration. Damage to the tonoplast will result in a fall of cytosolic pH, among other things. It is inconceivable that photosynthesis can proceed effectively in cells with leaky membranes, since not only will chloroplast volume change osmotically but the traffic of metabolites and ions across the chloroplast envelope, on which continued photosynthesis depends, will also be massively changed.

Several lines of evidence indicate that, in leaves infected with biotrophic fungi, membrane integrity is relatively unchanged (Farrar, 1984). The accumulation of many compounds at infection sites implies intact membranes. The high rates of photosynthesis found in cells invaded by fungal haustoria, mentioned above, are compatible only with a retained integrity of host plasma membranes. The use of ion efflux analysis enables a distinction to be made between total loss of ions from a tissue (which can increase simply because the apoplastic pool of that ion is greater than normal) and the ease with which ions move across membranes. The measure of this–the rate constant–is unaffected in brown rust-infected barley leaves for K^+ (Ahmad *et al.*, 1985), Cl^- and Ca^{2+} (C. Berryman & J.F. Farrar, unpublished) and Pi (Ahmad *et al.*, 1984). Autoradiographic evidence also supports the idea that membranes are intact. If membranes were damaged in localised regions of the leaf– for example, within pustules–more rapid washout of isotope from these regions should be detected autoradiographically from leaves pre-loaded with isotope. Brown-rusted barley leaves showed no such prefer-ential loss of either $^{45}Ca^{2+}$ or $^{36}Cl^-$ (C. Berryman & J.F. Farrar, unpub-lished). There is no reason, at least in this host/pathogen combination, to believe that the integrity of host membranes is impaired: the pathogen must obtain nutrients as they move through the leaf and not deplete host stores after damaging the membranes containing them. It is, however, unlikely that the changes in late infection in heavily infected leaves– loss of photosynthetic activity and pigments, for example–can occur without corresponding damage to membranes, if only because the supply of energy for maintenance of them will be reduced.

Biotrophic fungi which infect leaves are such powerful sinks for metabolites of the host that there is very limited transfer in the reverse direction. By contrast, massive exchange of metabolites is the hallmark of mutualistic mycorrhizas. It follows that, as there must be tight control of membrane permeability and transport systems (influx and efflux) at the plasmalemmae of the two symbionts in regions of contact, all comments made above about host membrane permeability in diseased leaves apply even more strongly to that of *both* partners in mycorrhizas. Smith & Smith (1986) make these points convincingly in a discussion of mycorrhizal exchange of metabolites in the context of a brief review of the general principles of membrane transport.

Conclusion

A key question is, will the changes induced in leaves and roots by biotrophic fungi result in increased growth of the fungi? There are several reasons for doubting whether this will occur. Firstly, the growth rate of fungi is more likely to be controlled endogenously rather than by a limiting flux of material from the host. Secondly, not only are the fungi apparently relatively slow growing but the proportion of gross photosynthesis which needs to be diverted to sustain their growth, at least in leaves, is not large. Further, the normal processes of the host, at least in leaves, result in the delivery of all nutrients so far studied to the apoplast at rates in excess of their uptake by fungi. Diversion of nutrients may be unnecessary; a nearly normal host leaf, photosynthesising behind functional membranes, may be the best guarantee of a fully supplied fungal mycelium, although the nitrogen requirements of pathogenic fungi are unknown and in need of detailed investigation.

Quantitative aspects of fluxes of carbohydrates and other nutrients into regions of roots from which they can be absorbed by mycorrhizal fungi and of fluxes of phosphate and other nutrients from the fungi still require more study. In ecto-mycorrhizas, the substantial fungal sheath and the massive production of sporophores, coupled with enhanced growth of the host, may well need some mechanism which partitions nutrients.

For mycorrhizal development, Lewis (1986) has proposed a speculative scheme for programmed responses which may be applicable to leaf-inhabiting fungi. This scheme demands that infection is the trigger for the synthesis of one or more second messengers (e.g. cyclic AMP, inositol tris-phosphate, polyamines) which affect metabolism and morphogenesis of both host and fungi via changes in cytosolic concentration of calcium and changes in sensitivity to growth regulators. The scheme receives some indirect support from the well-known general effects of calcium on membrane permeability (Epstein, 1972), the comments of Smith & Smith (1986) on the specific regulatory effects of calcium on ion channels in membranes and the observation that calmodulin is present in uredospores of bean rust (H.C. Hoch, personal communication). Furthermore, the effects of abscisic acid in plants, prokaryotes and animals are mediated through the control of calcium concentration (Huddart, Smith, Langton, Hetherington & Mansfield, 1986). Fungi have not yet been investigated and other growth regulators could well act similarly. The prospects are exciting.

References

Abbott, L.K. & Robson, A.D. (1985). Formation of external hyphae in soil by four species of vesicular-arbuscular mycorrhizal fungi. *New Phytologist*, **99**, 245–55.

Abuzinadah, R.A., Finlay, R.D. & Read, D.J. (1986). The role of proteins in the nitrogen nutrition of ectomycorrhizal plants. II. Utilisation of protein by mycorrhizal plants of *Pinus contorta*. *New Phytologist*, **103**, 495–506.

Abuzinadah, R.A. & Read, D.J. (1986*a*). The role of proteins in the nitrogen nutrition of ectomycorrhizal plants. I. Utilization of peptides and proteins by ectomycorrhizal fungi. *New Phytologist*, **103**, 481–93.

Abuzinadah, R.A. & Read, D.J. (1986*b*). The role of proteins in the nitrogen nutrition of ectomycorrhizal plants. III. Protein utilisation by *Betula, Picea* and *Pinus* mycorrhizal association with *Hebeloma crustuliniforme*. *New Phytologist*, **103**, 507–14.

Ahmad, I., Owera, S.A.P., Farrar, J.F. & Whitbread, R. (1982). The distribution of five major nutrients in barley plants infected with brown rust. *Physiological Plant Pathology*, **21**, 335–46.

Ahmad, I., Farrar, J.F. & Whitbread, R. (1983). Photosynthesis and chloroplast functioning in leaves of barley infected with brown rust. *Physiological Plant Pathology*, **23**, 411–19.

Ahmad, I., Farrar, J.F. & Whitbread, R. (1984). Fluxes of phosphorus in leaves of barley infected with brown rust. *New Phytologist*, **98**, 361–75.

Ahmad, I., Farrar, J.F. & Whitbread, R. (1985). Membrane integrity in leaves of barley infected by brown rust: an examination using tracer efflux and *in vivo* chlorophyll fluorescence. *New Phytologist*, **99**, 107–15.

Al-Khesraji, T.O., Lösel, D.M. & Gay, J.L. (1980). The infection of vascular tissue in leaves of *Tussilago farfara* L. by pycnial-aecial stages of *Puccinia poarum* Niels. *Physiological Plant Pathology*, **17**, 193–7.

Allen, M.F. & Boosalis, M.G. (1983). Effects of two species of VA mycorrhizal fungi on drought tolerance of winter wheat. *New Phytologist*, **93**, 67–76.

Allen, M.F., Smith, W.K., Moore, T.S. & Christensen, M. (1981). Comparative water relations and photosynthesis of mycorrhizal and non-mycorrhizal *Bouteloua gracilis* H.B.K. Lag ex Sted. *New Phytologist*, **88**, 683–93.

Amijee, F., Stribley, D.P. & Tinker, P.B. (1986). The development of endomycorrhizal root systems. VI. The relationship between development of infections, and intensity of infection in young leek roots. *New Phytologist*, **102**, 293–301.

Antibus, R.K., Trappe, J.M. & Linkins, A.E. (1980). Cyanide resistant respiration in *Salix nigra* endomycorrhizae. *Canadian Journal of Botany*, **58**, 14–20.

Altus, D.P. & Canny, M.J. (1985). Loading of assimilates in wheat leaves. II. The path from chloroplast to vein. *Plant, Cell & Environment*, **8**, 275–86.

Ayres, P.G. (1980). Responses of stomata to pathogenic microorganisms. In: *Stomatal Physiology*, ed. P.G. Jarvis & T.A. Mansfield, pp. 205–21. Cambridge University Press: Cambridge.

Ayres, P.G. (1981a). Effect of disease on plant water relations. In: *Effects of Disease on the Physiology of the Growing Plant*, ed. P.G. Ayres, pp. 131–48. Cambridge University Press: Cambridge.

Ayres, P.G. (1981b). Powdery mildew stimulates photosynthesis in uninfected leaves of pea plants. *Phytopathologische Zeitschrift*, **100**, 312–18.

Ayres, P.G. & Jones, P. (1976). Increased transpiration and the accumulation of root absorbed ^{86}Rb in barley leaves infected by *Rhynchosporium secalis* (leaf blotch). *Physiological Plant Pathology*, **7**, 49–58.

Azcon-Bieto, J., Lambers, H. & Day, D.A. (1983). Effect of photosynthesis and carbohydrate status on respiratory rates and the involvement of the alternative pathway in leaf respiration. *Plant Physiology*, **72**, 598–603.

Baker, N.R. (1978). Effects of high cation concentrations on photosystem II activities. *Plant Physiology*, **62**, 889–93.

Baker, N.R. & Bradbury, M. (1981). Possible applications of chlorophyll fluorescence techniques for studying photosynthesis *in vivo*. In: *Plants and the Daylight Spectrum*, ed. H. Smith, pp. 355–73. Academic Press: London.

Bennett, J. & Scott, K.J. (1971). Inorganic polyphosphates in the wheat stem rust fungus and in rust-infected wheat leaves. *Physiological Plant Pathology*, **1**, 185–98.

Bethlenfalvay, G.J., Packovsky, R.S. & Brown, M.S. (1981). Measurement of mycorrhizal infection in soybeans. *Soil Science Society of America Journal*, **45**, 871–75.

Bieleski, R.L. (1973). Phosphate pools, phosphate transport and phosphate availability. *Annual Review of Plant Physiology*, **24**, 225–52.

Bradley, R., Burt, A.J. & Read, D.J. (1982). The biology of mycorrhiza in the Ericaceae. VIII. The role of mycorrhizal infection in heavy metal resistance. *New Phytologist*, **91**, 197–209.

Brownlee, C., Duddridge, J.A., Malibari, A. & Read, D.J. (1983). The structure and function of mycelial systems of ectomycorrhizal roots with special reference to their role in forming inter-plant connections and providing pathways for assimilate and water transport. *Plant and Soil*, **71**, 433–43.

Buchanan, B.B., Hutcheson, S.W., Magyarosy, A.C. & Montalbini, P. (1981). Photosynthesis in healthy and diseased plants. In: *Effects of Disease on the Physiology of the Growing Plant*, ed. P.G. Ayres, pp. 13–28. Cambridge University Press: Cambridge.

Burt, A.J., Hashem, G., Shaw, G. & Read, D.J. (1986). Comparative analysis of metal tolerance in ericoid and ectomycorrhizal fungi. In: *Physiological and Genetical Aspects of Mycorrhizae*, ed. V. Gianinazzi-Pearson & S. Gianinazzi, pp. 683–7. INRA: Paris.

Bushnell, W.R. (1970). Patterns in growth, oxygen uptake, and nitrogen content of single colonies of wheat stem rust on wheat leaves. *Phytopathology*, **60**, 92–9.

Bushnell, W.R. (1984). Structural and physiological alterations in susceptible host tissue. In: *The Cereal Rusts*, vol. I., ed. W.R. Bushnell & A.P. Roelfs, pp. 477–508. Academic Press: Orlando.

Buwalda, J.G., Stribley, D.P. & Tinker, P.B. (1984). The development of endomycorrhizal root systems. V. The detailed pattern of development of infection and the control of infection level by host in young leek plants. *New Phytologist*, **96**, 411–27.

Callow, J.A., Long, D.E. & Lithgow, E.D. (1980). Multiple molecular forms of invertase in maize smut infections. *Physiological Plant Pathology*, 16, 93–107.

Carolin, R.C., Jacobs, S.W.L. & Vesk, M. (1973). The structure of cells of the mesophyll and parenchymatous bundle sheath of the Gramineae. *Botanical Journal of the Linnean Society*, 66, 259–75.

Chong, J. & Harder, D.E. (1980). Ultrastructure of haustorium development in *Puccinia coronata aveneae*. I. Cytochemistry and electron probe X-ray analysis of the haustorial neck ring. *Canadian Journal of Botany*, 58, 2496–505.

Coleman, B. & Espie, G.S. (1985). CO_2 uptake and transport in leaf mesophyll cells. *Plant Cell and Environment*, 8, 449–58.

Coleman, J.O.D. & Harley, J.L. (1976). Mitochondria of mycorrhizal roots of *Fagus sylvatica*. *New Phytologist*, 76, 317–30.

Cooper, K.M. & Lösel, D.M. (1978). Lipid physiology of vesicular-arbuscular mycorrhiza. I. Composition of lipids in roots of onion, clover and ryegrass infected with *Glomus mosseae*. *New Phytologist*, 80, 143–51.

Cowan, M.C. & van der Waal, A.F. (1975). An ecophysiological approach to crop losses, exemplified in the system wheat, leaf rust and glume blotch. IV. Water flow and leaf water potential of uninfected wheat plants and plants infected with *Puccinia recondita* f.sp *tritici*. *Netherlands Journal of Plant Pathology*, 81, 49–57.

Daly, J.M. (1976). The carbon balance of diseased plants: changes in respiration, photosynthesis and translocation. In: *Physiological Plant Pathology, Encyclopedia of Plant Physiology* 4, ed. R. Heitefuss & P.H. Williams, pp. 450–79. Springer: Berlin.

Daly, J.M. (1981). Mechanisms of action. In: *Toxins in Plant Disease*, ed. R.D. Durbin, pp. 331–94. Academic Press: New York.

Dehne, H.W. (1986). Influence of VA mycorrhizae on host plant physiology. In: *Physiological and Genetical Aspects of Mycorrhizae*, ed. V. Gianinazzi-Pearson & S. Gianinazzi, pp. 431–35. INRA: Paris.

Dickerson, A.G. (1972). Aβ-D-fructofuranosidase from *Claviceps purpurea*. *Biochemical Journal*, 129, 267–76.

Duddridge, J.A., Malibari, A. & Read, D.J. (1980). Structure and function of mycorrhizal rhizomorphs with special reference to their role in water transport. *Nature*, 287, 834–36.

Duddridge, J.A. & Read, D.J. (1984a). The development and ultrastructure of ectomycorrhizas. I. Ectomycorrhizal development in the field. *New Phytologist*, 96, 565–73.

Duddridge, J.A. & Read, D.J. (1984b). The development and ultrastructure of ectomycorrhizas. II. Ectomycorrhizal development of pine *in vitro*. *New Phytologist*, 96, 575–82.

Duniway, J.M. & Durbin, R.D. (1971a). Detrimental effects of rust infection on the water relations of bean. *Plant Physiology*, 48, 69–72.

Duniway, J.M. & Durbin, R.D. (1971b). Some effects of *Uromyces phaseoli* on the transpiration rate and stomatal response of bean leaves. *Phytopathology*, 61, 114–19.

Durbin, R.D. (1948). Effects of rust on plant development in relation to the translocation of inorganic and organic solutes. In: *The Cereal Rusts*, vol. I, ed. W.R. Bushnell & A.P. Roelfs, pp. 509–28. Academic Press: Orlando.

Epstein, E. (1972). *Mineral Nutrition of Plants: Principles and Perspectives*. Wiley: New York.

Farquhar, G.D. & Sharkey, T.D. (1982). Stomatal conductance and photosynthesis. *Annual Review of Plant Physiology*, 33, 317–45.

Farrar, J.F. (1984). Effects of pathogens on plant transport systems. In: *Plant Diseases: Infection, Damage and Loss*, ed. R.K.S. Wood & G.J. Jellis, pp. 87–104. Blackwell: Oxford.

Farrar, J.F. (1985a). The respiratory source of CO_2. *Plant, Cell and Environment*, **8**, 427-38.

Farrar, J.F. (1985b). Carbohydrate metabolism in biotrophic plant pathogens. *Microbiological Sciences*, **2**, 314–17.

Farrar, S.C. & Farrar, J.F. (1985a). Carbon fluxes in leaf blades of barley. *New Phytologist*, **100**, 271–83.

Farrar, S.C. & Farrar, J.F. (1985b). Fluxes of carbon compounds in leaves and roots of barley plants. In: *Regulation of Sources and Sinks in Crop Plants*. Monograph 12, ed. B. Jeffcoat, A.F. Hawkins & A.D. Stead, pp. 1–15. British Plant Growth Regulator Group: Bristol.

Farrar, S.C. & Farrar, J.F. (1986). Compartmentation and fluxes of sucrose in leaf blades of barley. *New Phytologist*, **103**, 645–57.

Finlay, R.D., Söderström, B. & Read, D.J. (1986). Factors influencing the flux of carbon through ectomycorrhizal mycelium forming inter plant connections. In: *Physiological and Genetical Aspects of Mycorrhizae*, ed. V. Gianinazzi-Pearson & S. Gianinazzi, pp. 301–6. INRA: Paris.

Finlay, R.D. & Read, D.J. (1986). The structure and function of the vegetative mycelium of ectomycorrhizal plants. I. Translocation of ^{14}C-labelled carbon between plants interconnected by a common mycelium. *New Phytologist*, **103**, 143–56.

Fondy, B.R. & Geiger, D.R. (1985). Diurnal changes in allocation of newly fixed carbon in exporting sugar beet leaves. *Plant Physiology*, **78**, 753–7.

Fortin, J.A., Piché, Y. & Lalonde, M. (1980). Technique for observation of early morphological changes during ectomycorrhiza formation. *Canadian Journal of Botany*, **58**, 361–65.

Francis, R., Finlay, R.D. & Read, D.J. (1986). Vesicular-arbuscular mycorrhiza in natural vegetation systems. IV. Transfer of nutrients in inter- and intra-specific combinations of host plants. *New Phytologist*, **102**, 103–11.

Francis, R. & Read, D.J. (1984). Direct transfer of carbon between plants connected by vesicular arbuscular mycorrhizal mycelium. *Nature*, **307**, 53–6.

Fric, F. (1978). Absorption and translocation of phosphate in barley plants infected with powdery mildew. *Phytopathologische Zeitschrift*, **91**, 23–32.

Geiger, D.R., Ploeger, B.J., Fox, T.C. & Fondy, B.P. (1983). Sources of sucrose translocated from illuminated sugar beet source leaves. *Plant Physiology*, **72**, 964–70.

Gianinazzi-Pearson, V. & Gianinazzi, S. (1986). The physiology of improved phosphate nutrition in mycorrhizal plants. In: *Physiological and Genetical Aspects of Mycorrhizae*, ed. V. Gianinazzi-Pearson & S. Gianinazzi, pp. 101–9. INRA: Paris.

Gildon, A. & Tinker, P.B. (1981). A heavy metal tolerant strain of a mycorrhizal fungus. *Transactions of the British Mycological Society*, **77**, 648–9.

Gordon, T.R. & Duniway, J.M. (1982a). Photosynthesis in powdery mildewed sugar beet leaves. *Phytopathology*, **72**, 718–23.

Gordon, T.R. & Duniway, J.M. (1982b). Stomatal behaviour and water relations in sugar beet leaves infected by *Erysiphe polygoni*. *Phytopathology*, **72**, 723–6.

Gordon, T.R. & Duniway, J.M. (1982c). Effects of powdery mildew infection on the efficiency of CO_2 fixation and light utilization by sugar beet leaves. *Plant Physiology*, **69**, 139–42.

Greenland, A.J. & Lewis, D.H. (1981). The acid invertases of the developing third leaf of oat. II. Changes in the activities of soluble isoenzymes. *New Phytologist*, **88**, 279–88.

Greenland, A.J. & Lewis, D.H. (1983). Changes in the activities and predominant molecular forms of acid invertase in oat leaves infected by crown rust. *Physiological Plant Pathology*, **22**, 293–312.

Hanchey, P. & Wheeler, H. (1979). The role of host cell membranes. In: *Recognition and Specificity in Plant Host-Parasite Interactions*, ed. J.M. Daly & I. Uritani, pp. 193–210. Japan Scientific Societies: Tokyo.

Harder, D.E. & Chong, J. (1984). Structure and physiology of haustoria. In: *The Cereal Rusts*, vol. I, ed. W.R. Bushrell & A.P. Roelfs, pp. 431–76. Academic Press, Orlando.

Hardie, K. (1985). The effect of removal of extraradical hyphae on water uptake by vesicular-arbuscular mycorrhizal plants. *New Phytologist*, **101**, 677–84.

Hardie, K. (1986). The role of extraradical hyphae in water uptake by vesicular-arbuscular mycorrhizal plants. In: *Physiological and Genetical Aspects of Mycorrhizae*, ed. V. Gianinazzi-Pearson & S. Gianinazzi, pp. 651–5. INRA: Paris.

Harley, J.L. & ap Rees, T. (1959). Cytochrome oxidase in mycorrhizal and non-mycorrhizal roots of *Fagus sylvatica*. *New Phytologist*, **58**, 364–86.

Harley, J.L. & Jennings, D.H. (1958). The effect of sugars on the respiratory responses of beech mycorrhiza to salts. *Proceedings of the Royal Society*, **B148**, 403–18.

Harley, J.L. & Smith, S.E. (1983). *Mycorrhizal Symbiosis*. Academic Press: London.

Harris, D., Packovsky, R.S. & Paul, E.A. (1985). Carbon economy of soybean - *Rhizobium-Glomus* associations. *New Phytologist*, **101**, 427–40.

Hepper, C.M. (1977). A colorimetric method for estimating vesicular-arbuscular mycorrhizal infection in roots. *Soil Biology and Biochemistry*, **9**, 15–18.

Herold, A. (1984). Biochemistry and physiology of synthesis of starch in leaves: autotrophic and heterotrophic chloroplasts. In: *Storage Carbohydrates in Vascular Plants*, ed. D.H. Lewis, pp. 181–204. Cambridge University Press: Cambridge.

Herold, A. & Walker, D.A. (1979). Transport across chloroplast envelopes - the role of phosphate. In: *Handbook of Transport Across Membranes*, ed. G. Giebrich, D.C. Tosteson & H.H. Ussing, pp. 411–39. Springer: Heidelberg.

Hewitt, H.G. & Ayres, P.G. (1976). Effect of infection by *Microsphaera alphitoides* (powdery mildew) on carbohydrate levels and translocation in seedlings of *Quercus robur*. *New Phytologist*, **77**, 379–90.

Holligan, P.M., Chen, C. & Lewis, D.H. (1973). Changes in the carbohydrate composition of leaves of *Tussilago farfara* during infection by *Puccinia poarum*. *New Phytologist*, **72**, 947–55.

Holligan, P.M., Chen, C., McGee, E.E. & Lewis, D.H. (1974). Carbohydrate metabolism in healthy and rusted leaves of coltsfoot. *New Phytologist*, **73**, 881–8.

Huber, S.C. (1981). Inter- and intra-specific variation in photosynthetic formation of starch and sucrose. *Zeitschrift für Pflanzenphysiologie*, **101**, 49–54.

Huber, S.C. & Moreland, D.E. (1980). Co-transport of potassium and sugars across the plasmalemma of mesophyll protoplasts. *Plant Physiology*, **67**, 163–9.

Huddart, H., Smith, R.J., Langton, P.D., Hetherington, A.M. & Mansfield, T.A. (1986). Is abscisic acid a universally active calcium agonist? *New Phytologist*, **104**, 161–73.

Hwang, B.K., Ibenthal, W.-D. & Heitefuss, R. (1983). Age, rate of growth, carbohydrate and amino-acid contents of spring barley plants in relation to their resistance to powdery mildew (*Erysiphe graminis* f.sp *hordei*). *Physiological Plant Pathology*, **22**, 1–14.

Jennings, D.H. (1984). Polyol metabolism in fungi. *Advances in Microbial Physiology*, **25**, 149–93.

Jirgis, R., Ramstedt, M. & Söderhäll, K. (1986). Mannitol does not inhibit glycolytic enzymes in roots of *Pinus sylvestris* and *Fagus orientalis*. *New Phytologist*, **102**, 285–91.

Kneale, J. & Farrar, J.F. (1985). The localisation and frequency of haustoria in colonies of brown rust on barley leaves. *New Phytologist*, **101**, 495–505.

Komor, E. (1982). Transport of sugar. In: *Plant Carbohydrates* I, *Encyclopedia of Plant Physiology*, 13A, ed. F.A. Loewus & W. Tanner, pp. 635–76. Springer: Berlin.

Kosuge, T. & Kimpel, J.A. (1981). Energy use and metabolic regulation in plant-pathogen interactions. In: *Effects of Disease on the Physiology of the Growing Plant*, ed. P.G. Ayres, pp. 29–45. Cambridge University Press: Cambridge.

Krause, G.M. & Weis, E. (1984). Chlorophyll fluorescence as a tool in plant physiology. II. Interpretation of fluorescence signals. *Photosynthesis Research*, **5**, 139–57.

Kuo, J., O'Brien, T.P. & Canny, M.J. (1974). Pit-field distribution, plasmodesmatal frequency, and assimilate flux in the mestome sheath of wheat leaves. *Planta*, **121**, 97–118.

Lambers, H. (1982). Cyanide-resistant respiration: a non-phosphorylating electron transport pathway acting as an energy overflow. *Physiologia Plantarum*, **55**, 478–85.

Last, F.T. (1963). Metabolism of barley leaves inoculated with *Erysiphe graminis*. *Annals of Botany*, **27**, 685–90.

Lavorel, J. & Etienne, A.L. (1977). *In vivo* chlorophyll fluorescence. In: *Primary Processes of Photosynthesis*, ed. J. Barber, pp. 203–68. Elsevier: Amsterdam.

Lewis, D.H. (1976). Interchange of metabolites in biotrophic symbioses between angiosperms and fungi. In: *Perspectives in Experimental Biology, vol. 2, Botany*, ed. N. Sunderland, pp. 207–19. Pergamon Press, Oxford.

Lewis, D.H. (1986). Inter-relationships between carbon nutrition and morphogenesis in mycorrhizas. In: *Physiological and Genetical Aspects of Mycorrhizae*, ed. V. Gianinazzi-Pearson & S. Gianinazzi, pp. 85–100. INRA: Paris.

Lewis, D.H. (1987). Evolutionary aspects of mutualistic associations between fungi and photosynthetic organisms. In: *Evolutionary Biology of the Fungi*, ed. A.D.M. Rayner, C. Brasier & D. Moore. Cambridge University Press: Cambridge.

Lewis, D.H. & Harley, J.L. (1965). Carbohydrate physiology of mycorrhizal roots of beech. II. Utilisation of exogenous sugars by uninfected and mycorrhizal roots. *New Phytologist*, **64**, 238–55.

Long, D.E., Fung, A.K., McGee, E.E., Cooke, R.C. & Lewis, D.H. (1975). The activity of invertase and its relevance to the accumulation of storage polysaccharides in leaves infected by biotrophic fungi. *New Phytologist*, **74**, 173–82.

Lösel, D.M. (1978). Lipid metabolism of leaves of *Poa pratensis* during infection by *Puccinia poarum*. *New Phytologist*, **73**, 1157–69.

Lunderstaadt, J., Heitefuss, R. & Fuchs, W.H. (1962). Aktivitat einiger enzyme des Kohlenhydratstoffwechsels aus Weizenkeimpflazen nach Infektion nict *Puccinia graminis tritici*. *Naturwissenschaften*, **49**, 403.

Maclean, D.J. (1982). Axenic culture and metabolism of rust fungi. In: *The Rust Fungi*, ed. K.J. Scott & A.K. Chakravorty, pp. 37–120. Academic Press: New York.

Macklon, A.E.S. (1984). Calcium fluxes at plasmalemma and tonoplast. *Plant, Cell and Environment*, **7**, 407–13.

Manners, J.M & Gay, J.L. (1982). Transport, translocation and metabolism of ^{14}C-photosynthates at the host: parasite interface of *Pisum sativum* and *Erysiphe pisi*. *New Phytologist*, **91**, 221–44.

Manners, J.M., Maclean, D.J. & Scott, K.J. (1982). Pathways of glucose assimilation in *Puccinia graminis*. *Journal of General Microbiology*, **128**, 2621–30.

Mayama, S., Rehfeld, D.W. & Daly, J.M. (1975). A comparison of the development of *Puccinia graminis tritici* in resistant and susceptible wheat based on glucosamine content. *Physiological Plant Pathology*, **7**, 243-57.

Maynard, J.W. & Lucas, W.J. (1982). A reanalysis of the two-component phloem loading system in *Beta vulgaris*. *Plant Physiology*, **69**, 734-9.

Mitchell, D.T., Fung, A.K. & Lewis, D.H. (1978). Changes in the ethanol-soluble carbohydrate composition and acid invertase in infected first leaf tissues susceptible to crown rust of oat and wheat stem rust. *New Phytologist*, **80**, 381–92.

Moller, I.M. & Palmer, J.M. (1984) Regulation of the tricarboxylic acid cycle and organic acid metabolism. In: *The Physiology and Biochemistry of Plant Respiration*, ed. J.M. Palmer, pp. 225–33. Cambridge University Press: Cambridge.

Newman, E.I. & Ritz, K. (1986). Evidence on the pathways of phosphorus transfer between vesicular-arbuscular mycorrhizal plants. *New Phytologist*, 104, 77–88.

Nobel, P.S. (1983). *Biophysical Plant Physiology and Ecology*. Freeman: San Francisco.

Owera, S.A.P., Farrar, J.F. & Whitbread, R. (1981). Growth and photosynthesis in barley infected with brown rust. *Physiological Plant Pathology*, 18, 79–90.

Owera, S.A.P., Farrar, J.F. & Whitbread, R. (1983). Translocation from leaves of barley infected with brown rust. *New Phytologist*, 94, 111–23.

Palmer, J.M. (1984). The operation and control of the respiratory chain. In: *The Physiology and Biochemistry of Plant Respiration*, ed. J.M. Palmer, pp. 123–40. Cambridge University Press: Cambridge.

Papageorgiou, G. (1975). Chlorophyll fluorescence: an intrinsic probe of photosynthesis. In: *Bioenergetics of Photosynthesis*, ed. Govindjee, pp. 319–71. Academic Press: London.

Pollock, C.J. (1984). Physiology and metabolism of sucrosyl-fructans. In: *Storage Carbohydrates in Vascular Plants*, ed. D.H. Lewis, pp. 97–114. Cambridge University Press: Cambridge.

Pollock, C.J. (1986). Fructans and the metabolism of sucrose in vascular plants. *New Phytologist*, 104, 1–24.

Pozsar, B.I. & Kiraly, Z. (1966). Phloem-transport in rust-infected plants and the cytokinin-directed long-distance movement of nutrients. *Phytopathologische Zeitschrift*, 56, 279–309.

Preiss, J. (1984). Starch, sucrose biosynthesis and partition of carbon in plants are regulated by orthophosphate and triose-phosphates. *Trends in Biochemical Science*, 9, 24–7.

Rawson, H.M., Gifford, R.M. & Bremner, P.M. (1976). Carbon dioxide exchange in relation to sink demand in wheat. *Planta*, 132, 19–23.

Read, D.J. (1983). The biology of mycorrhiza in the Ericales. *Canadian Journal of Botany*, 61, 985–1004.

Read, D.J. (1986). Non-nutritional effects of mycorrhizal infection. In: *Physiological and Genetical Aspects of Mycorrhizae*, ed. V. Gianinazzi-Pearson & S. Gianinazzi, pp. 169–76. INRA: Paris.

Reinhold, L. & Kaplan, A. (1984). Membrane transport of sugars and amino acids. *Annual Review of Plant Physiology*, 35, 45–83.

Samborski, D.J. & Shaw, M. (1956). The physiology of host-parasite relations. II. The effects of *Puccinia graminis tritici* Eriks. & Henn. on the respiration of the first leaf of resistant and susceptible varieties of wheat. *Canadian Journal of Botany*, 34, 601–19.

Sanders, F.E. (1986). Quantitative approaches to the analysis of the development of mycorrhizal root systems. In: *Physiological and Genetical Aspects of Mycorrhizae*, ed. V. Gianinazzi-Pearson & S. Gianinazzi, pp. 209–16. INRA: Paris.

Sanders, F.E. & Sheikh, N.A. (1983). The development of vesicular-arbuscular mycorrhizal infection in plant root systems. *Plant and Soil*, 71, 223-46.

Sanders, F.E., Tinker, P.B., Black, R.L.B. & Palmerley, S.M. (1977). The development of endomycorrhizal root systems. I. Spread of infection and growth-promoting effects with four species of vesicular-arbuscular endophytes. *New Phytologist*, 78, 257–268.

Scholes, J.D. (1985). The Effects of Biotrophic Pathogens on Photosynthesis, *Ph.D. Thesis, University of Wales*.

Scholes, J.D. & Farrar, J.F. (1985). Photosynthesis and chloroplast functioning within individual pustules of *Uromyces muscari* on bluebell leaves. *Physiological Plant Pathology*, 27, 387–400.

Scott, K.J. & Maclean, D.J. (1969). Culturing of rust fungi. *Annual Review of Phytopathology*, **7**, 123–46.

Scott, K.J. & Smillie, R.M. (1966). Metabolic regulation in diseased leaves. I. The respiratory rise in barley leaves infected with powdery mildew. *Plant Pathology*, **41**, 289–407.

Sharkey, T.D. (1985). Photosynthesis in intact leaves of C3 plants: physics, physiology and rate limitations. *Botanical Review*, **51**, 53–105.

Shaw, M. & Samborski, D.J. (1956). The physiology of host-parasite relations. I. The accumulation of radioactive substances at infections of facultative and obligate parasites including tobacco mosaic virus. *Canadian Journal of Botany*, **34**, 389–405.

Sherald, J.L. & Sisler, H.D. (1972). Selective inhibition of antimycin-A insensitive respiration in *Ustilago maydis* and *Ceratocystis ulmi*. *Plant and Cell Physiology*, **13**, 1039–52.

Sivak, M.N. & Walker, D.A. (1985). Chlorophyll *a* fluorescence: can it shed light on fundamental questions in photosynthetic carbon dioxide fixation? *Plant, Cell and Environment*, **8**, 439–48.

Smith, F.A. & Smith, S.E. (1986). Movement across membranes: physiology and biochemistry. In: *Physiological and Genetical Aspects of Mycorrhizae*, ed. V. Gianinazzi-Pearson & S. Gianinazzi, pp. 75–8. INRA: Paris.

Smith, S.E., Tester, M. & Walker, N.A. (1986). The development of mycorrhizal root systems in *Trifolium subterraneum* L.: growth of roots and the uniformity of spatial distribution of mycorrhizal infection units in young plants. *New Phytologist*, **103**, 117–31.

Snellgrove, R.C., Splittstoesser, W.E., Stribley, D.P. & Tinker, P.B. (1982). The distribution of carbon and the demand for the fungal symbiont in leek plants with vesicular-arbuscular mycorrhizas. *New Phytologist*, **92**, 75–87.

Snellgrove, R.C., Stribley, D.P., Tinker, P.B. & Lawlor, D.W. (1986). The effect of vesicular-arbuscular mycorrhizal infection on photosynthesis and carbon distribution in leek plants. In: *Physiological and Genetical Aspects of Mycorrhizae*, ed. V. Gianinazzi-Pearson & S. Gianinazzi, pp. 421–4. INRA: Paris.

Söderhäll, K., Jirgis, R. & Ramstedt, M. (1986). The mannitol cycle, a regulatory pathway in mycorrhizae? In: *Physiological and Genetical Aspects of Mycorrhizae*, ed. V. Gianinazzi-Pearson & S. Gianinazzi, pp. 297–9. INRA: Paris.

Söderström, B., Finlay, R.D. & Read, D.J. (1986). Qualitative analysis of carbohydrate contents of mycorrhizal mycelia after feeding of interconnected host plants with $^{14}CO_2$. In: *Physiological and Genetical Aspects of Mycorrhizae*, ed. V. Gianinazzi-Pearson & S. Gianinazzi, pp. 307–9. INRA: Paris.

Stitt, M. (1985). Control of photosynthetic sucrose synthesis by fructose 2,6-bisphosphate: comparative studies in C_3 and C_4 species. In: *Regulation of Sources and Sinks in Crop Plants*, ed. B. Jeffcoat, A.F. Hawkins & A.D. Stead, Monograph 12, pp. 35–50. British Plant Growth Regulator Group: Bristol.

Sziraki, I., Mustardy, L.A., Faludi-Daniel, A. & Kiraly, Z. (1984). Alterations in chloroplast ultrastructure and chlorophyll content in rust-infected pinto beans at different stages of disease development. *Phytopathology*, **74**, 77–84.

Tester, M., Smith, S.E., Smith, F.A. & Walker, N.A. (1986). Effects of photon irradiance on the growth of shoots and roots, on the rate of initiation of mycorrhizal infection and on growth of infection units in *Trifolium subterraneum* L. *New Phytologist*, **103**, 375–90.

Tinker, P.B. (1975*a*). Effects of vesicular-arbuscular mycorrhizas on higher plants. *Symposia of the Society for Experimental Biology*, **29**, 325–49.

Tinker, P.B. (1975*b*). Soil chemistry of phosphorus and mycorrhizal effects on plant growth. In: *Endomycorrhizas*, ed. F.E. Sanders, B. Mosse & P.B. Tinker, pp. 353–71. Academic Press: London.

Tisdall, J.M. & Oades, J.M. (1979). Stabilisation of soil aggregates by the root systems of ryegrass. *Australian Journal of Soil Research*, **17**, 429–41.

Tissera, P. & Ayres, P.G. (1986). Transpiration and the water relations of faba bean (*Vicia faba*) infected by rust (*Uromyces viciae-fabae*). *New Phytologist*, **102**, 385–96.

Tomiyama, K., Okamoto, H. & Katou, K. (1983). Effect of infection by *Phytophthora infestans* on the membrane potential of potato cells. *Physiological Plant Pathology*, **22**, 233–43.

Van Bel, A.J.E. & Koops, A.J. (1985). Uptake of ^{14}C sucrose in isolated minor-vein networks of *Commelina benghalensis* L. *Planta*, **164**, 362–9.

Walker, N.A. & Smith, S.E. (1984). The quantitative study of mycorrhizal infection. II. The relation of rate of infection and speed of fungal growth to propagule density, the mean length of infection unit and the limiting value of the fraction of root infected. *New Phytologist*, **96**, 55–69.

Walters, D.R. (1985). Shoot:root interrelationships: the effects of obligately biotrophic fungal pathogens. *Biological Reviews*, **60**, 47–70.

Walters, D.R. & Ayres, P.G. (1981). Phosphate uptake and transport by roots of powdery mildew infected barley. *Physiological Plant Pathology*, **18**, 195–205.

Walters, D.R. & Ayres, P.G. (1984). Ribulose bisphosphate carboxylase and enzymes of CO_2 assimilation in a compatible barley/powdery mildew combination. *Phytopathologische Zeitschrift*, **109**, 208-18.

Wedding, R.T. & Harley, J.L. (1976). Fungal polyol metabolites in the control of carbohydrate metabolism of mycorrhizal roots of beech. *New Phytologist*, **77**, 675–88.

Whipps, J.M., Clifford, B.C., Roderick, H.W. & Lewis, D.H. (1980). A comparison of development of *Puccinia hordei* Otth. on normal and slow rusting varieties of barley (*Hordeum vulgare* L.) using analyses of fungal chitin and mannan. *New Phytologist*, **85**, 191–9.

Whipps, J.M., Haselwandter, K., McGee, E.E.M. & Lewis, D.H. (1982). Use of biochemical markers to determine growth, development and biomass of fungi in infected tissues, with particular reference to antagonistic and mutualistic biotrophs. *Transactions of the British Mycological Society*, **79**, 385–400.

Whipps, J.M. & Lewis, D.H. (1981). Patterns of translocation, storage and interconversion of carbohydrates. In: *Effects of Disease on the Physiology of the Growing Plant*, ed. P.G. Ayres, pp. 47–83. Cambridge University Press: Cambridge.

Whipps, J.M., Roderick, H.W., Clifford, B.C. & Lewis, D.H. (1985). Relationship of fungal chemical markers to normal and benodanil-affected growth and development of *Puccinia hordei* in leaves of barley. *Transactions of the British Mycological Society*, **84**, 333–6.

Whitney, S.H., Shaw, M. & Naylor, J.M. (1962). The physiology of host-parasite relations. XII. A cytoplasmic study of the distribution of DNA and RNA in rust infected leaves. *Canadian Journal of Botany*, **40**, 1433–44.

Wilcox, H.E. (1982). Morphology and development of ecto- and ectendomycorrhizae. In: *Methods and Principles of Mycorrhizal Research*, ed. N.C. Schenck, pp. 103–13. The American Phytopathological Society: St. Paul, Minnesota.

Williams, A.M., Maclean, D.J. & Scott, K.J. (1984). Cellular localisation and properties of invertase in mycelium of *Puccinia graminis*. *New Phytologist*, **98**, 451–63.

Woodrow, I.E., Ellis, J.R., Jellings, A. & Foyer, C.H. (1984). Compartmentation and fluxes of inorganic phosphate in photosynthetic cells. *Planta*, **161**, 525–30.

Ziegler, E. & Pontzen, R. (1982). Specific inhibition of glucan-elicited glyceollin accumulations in soybeans by an extracellular mannan-glycoprotein of *Phytophthora megasperma* f.sp. *glycinea*. *Physiological Plant Pathology*, **20**, 321–31.

Zimmerman, U. & Steudle, E. (1978). Physical aspects of water relations of plant cells. *Advances in Botanical Research*, **6**, 45–117.

7

Some aspects of fungal enzymes that degrade plant cell walls

J. P. R. KEON, R. J. W. BYRDE[1] and R. M. COOPER[2]

[1]*Long Ashton Research Station, Department of Agricultural Sciences, University of Bristol, Long Ashton, Bristol BS18 9AF, UK.* and [2]*School of Biological Sciences, University of Bath, Claverton Down, Bath BA2 7AY, UK.*

Just 100 years ago de Bary (1886) first postulated that an extracellular enzyme, of undetermined nature, was involved in the infection of plant tissue by a parasitic fungus. He envisaged the enzyme's action resulting in the swelling and softening of the plant cell wall, and the protoplast becoming detached from it. He also believed that 'the destruction of the cell protoplasm is a direct, or more likely indirect, effect of the same enzyme'. So perceptive were these theories that their validity remains unchallenged (Byrde, 1982). Much scientific effort during the intervening century has been directed towards elucidating the biochemical and ultra-structural detail necessary for a full understanding of the role of cell wall degrading enzymes (CWDE).

In recent years, two advances beyond de Bary's conception have broad-ened the interest in CWDE and brought about recognition of their central role in plant pathology. Firstly, a purified CWDE has been shown to act as an 'elicitor' of phytoalexin biosynthesis (Lee & West, 1981a), probably by mediating, either directly or indirectly, the release of wall fragments that elicit this and other responses (McNeil *et al.*, 1984). Secondly, well characterised CWDE of plant pathogens represent valua-ble model systems for a genetic manipulation approach towards the understanding of host–pathogen interactions. So far these studies have been confined to bacteria, but the new potential for fungal transformation opens up many possibilities (see below).

Critical studies are still needed to show whether CWDE are patho-genicity determinants or whether they merely modify symptoms. Cooper (1983, 1984) listed up to nine criteria that should be fulfilled in a given host-pathogen system before a CWDE can be implicated in pathogenesis, adding that 'these ideals have been achieved in remarkably few cases'.

Limited space precludes a comprehensive treatment of all aspects of CWDE and we have selected three for special consideration: multiple forms; membrane leakage and cell death; and the role of CWDEs in biotrophy. Aspects not covered include the regulation of CWDE biosynthesis in fungi (Cooper, 1977, 1983), naturally occurring inhibitors, interaction with host defence systems (McNeil *et al.*, 1984), and the biochemistry of enzyme–substrate reactions (Rexova-Benkova & Markovic, 1976).

The plant cell wall

Essential to an understanding of CWDE is a knowledge of primary cell wall structure (Fig. 7.1), which is extremely complex (Cooper, 1983; McNeil *et al.*, 1984). Only a very brief summary is given here, derived from the reviews above.

Cellulose microfibrils give the wall much of its mechanical strength; they are formed of aggregates of linear β–1,4-linked glucan chains.

Xyloglucan, possibly hydrogen-bonded to the cellulose, consists of a backbone of β-1,4-linked D-glucosyl residues, with D-xylosyl sidechains. β-1,3- and β-1,4-glucans are present only in monocot cell walls.

The cellulose and xyloglucan are embedded in a matrix consisting largely of polymers of α-1,4-linked D-galacturonic acid, with or without rhamnose interspersed in the polyuronide chain. Two types of *rhamnogalacturonan* are recognised. The carboxyl groups of the uronic acid are often methoxylated.

Arabans, galactans and arabinogalactans are polymers of L-arabinofuranose and D-galactopyranose, and both, respectively; the main chains are α-1,5- and β-1,4-linked respectively. These neutral sugar polymers are closely associated with the rhamnogalacturonan.

Xylans are constituents of secondary walls of dicots, but in contrast form the main component of some monocot primary walls (see below); they comprise a β-1,4-backbone with linked arabinose and glucuronic acid side groups.

There is also a characteristic structural protein component ('extensin') rich in hydroxyproline residues, with α- and β-linked tetra-arabinosides attached.

The nature of the linkages between the different components is still not fully understood, but includes covalent bonds (e.g. rhamnogalacturonan-araban), hydrogen bonds (xyloglucan-cellulose) and ionic bonds (polygalacturonic acid-Ca^{2+}) (McNeil *et al.*, 1984).

A speculative model for general spatial relationships of the wall polymers is shown in Fig. 7.1.

Biochemical and physiological aspects of enzymes degrading acidic polymers

Nomenclature

The enzymes that degrade the important rhamnogalacturonan chains are classified on the basis of three criteria (Bateman & Basham, 1976):

(a) The preferred substrate – esterified or unesterified galacturonan – i.e. *pectin* or *pectate*.

(b) The nature of disruption of the glycosidic bond, hydrolytic (*polygalacturonase*) or lytic (*lyase*), where the elements of water are not added, but a double bond is formed between C_4 and C_5. 'Transeliminase' is a former term for lyase.

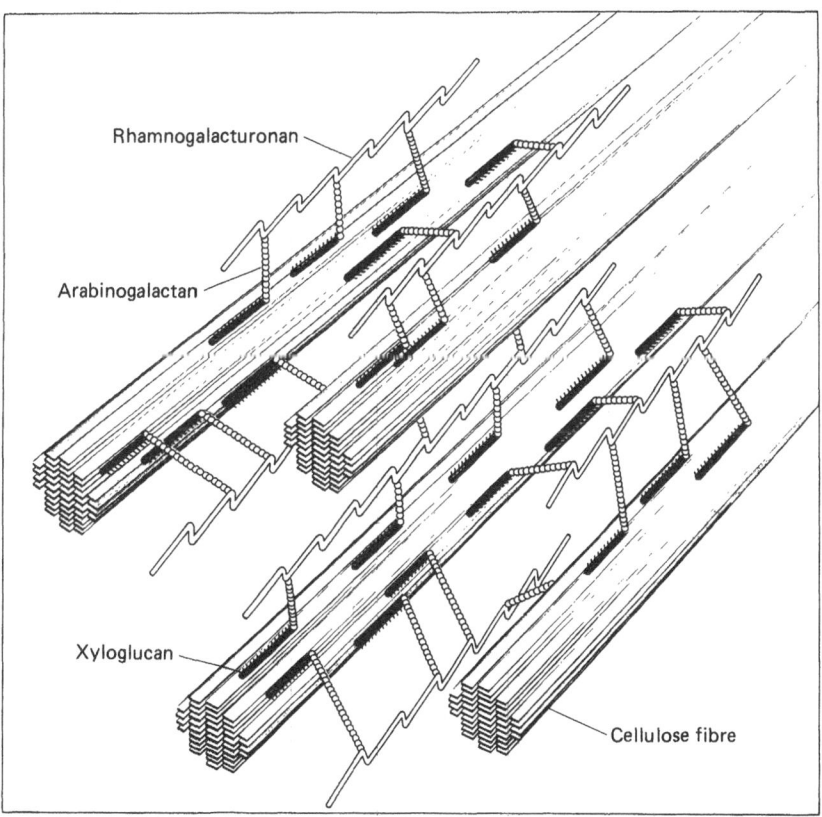

Fig. 7.1 A model of the primary cell walls of dicotyledons; protein molecules are not shown. (Reproduced with permission from Albersheim, P. (1975). The walls of growing plant cells. *Scientific American*, 232, 81-95. Copyright 1975 by Scientific American, Inc. All rights reserved.)

(c) Whether the attack on the chain is primarly random (*endo-*) or terminal (*exo-*): these categories are not as distinct as was once believed, and intermediate forms are recognised (Cooper, 1983).

Distinct from enzymes that cleave the galacturonan chain, *pectinesterase* de-esterifies individual methoxylated galacturonic acid units.

Because they can rapidly break down large chains to comparatively small fragments with relatively few cleavages, the endo-enzymes are those implicated in causing structural damage (Bateman & Basham, 1976).

Multiplicity of form: structure and function

Fungi frequently secrete several molecular forms of CWDE attacking the same substrate: these generally differ in isoelectric point (pI) and , often, in molecular weight. Such differences naturally give rise to differing electrophoretic characteristics. Multiplicity of enzyme form is widespread in living organisms, and Markert (1975) has drawn a colourful analogy with singers in a choir, each form with its own unique properties, yet contributing to the overall performance of the whole. The advantages of multiplicity of form in CWDE are well recognised, in that they confer flexibility to a potential pathogen (Hancock, 1976; Byrde, 1979).

Amongst fungal rhamnogalacturonan-degrading enzymes, molecular forms with a wide range of pI are known. At least one of the forms secreted by a number of fungal and bacterial plant pathogens is typically basic in nature, with a pI of 8.5 or above, and we have selected these for more detailed consideration. Examples are the endo-PGs secreted by *Verticillium albo-atrum* (Wang & Keen, 1970), three *Monilinia* spp. (Willetts *et al.,* 1977), several other fruit-rotting fungi (Fielding, 1981), *Colletotrichum lindemuthianum* (Wijesundera, Bailey & Byrde, 1984) and *Mucor* and *Rhizopus* spp. (Nguyen-The, Bertheau & Coleno, 1984); also the endopectate lyases secreted by *Hypomyces solani* f.sp. *cucurbitae* (Hancock, 1976) and by the bacteria *Erwinia atroseptica* (Quantick, Cervone & Wood, 1983), and *E. carotovora* (Tanabe *et al.,* 1984). In many instances these forms have been shown to be present in infected tissue as well as in culture filtrates. Fielding (1981) commented that, among the fruit-rotting fungi that he examined, the secretion of PGs with high pI was associated with some degree of host specialisation.

The relevance of the high pI value (which results in a positive charge at physiological pH values) is not fully understood, though it would obviously be likely to result in ionic binding to negatively-charged cell walls (Cooper, 1983) and to vascular elements (Pegg, 1985). As a further consequence, diffusion may also be limited (Pring, Byrde & Willetts, 1981), unless full solubilisation of the rhamnogalacturonan occurs.

A comparatively low molecular weight (*c.* 30 000 daltons); (see Table 7.1) is also well conserved among rhamnogalacturonan-degrading enzymes secreted by necrotrophic plant pathogens. Production of such an enzyme may relate to pore diameters in the plant cell wall which have been reported as 35-52 Å and 70 Å; the latter would allow permeation by globular proteins of 60 000 daltons or less (Cooper, 1983).

The question arises as to how these well conserved proteins have such high isoelectric points. Arginine and lysine are the two amino acids that are candidates for this function, with pIs of 10.76 and 9.74 respectively (Mahler & Cordes, 1969). Published information on the amino acid composition of 'basic' PGs is scanty. Table 7.1 lists the contents of strongly basic and of acidic amino acids for PGs of specialised facultative parasites (in descending order of pI) and for unspecialised necrotrophs and saprotrophs. Several conclusions can be drawn from the data: (a) there is more lysine than arginine in all PGs; (b) irrespective of pI, the lysine content is relatively constant (*c.* 20 mol/mol) in the PGs of specialised necrotrophs and significantly greater than in those of unspecialised necrotrophs and saprotrophs; (c) the acidic amino acids (aspartic and glutamic acids, pIs 2.87 and 3.22 respectively) are nominally more plentiful than the basic ones. However, in amino acid analysis, their amides, asparagine and glutamine, pIs 5.41 and 5.65 respectively, are recorded as the corresponding acids: extensive amidation, together with 'masking' of acidic amino acids in the interior, is perhaps responsible for the high pI of some forms.

The lysine residues are probably mainly on the surface of the tertiary structure; as this amino acid is the principal carrier of basic changes in these enzymes (Table 7.1), some of these residues may well be directly involved in their attachment to the acidic groups of the rhamnogalacturonan substrate. Thus Rexova-Benkova & Markovic (1976) postulated the presence of three to five attachment subsites for different PGs, and McClendon (1979*a*) seven, linearly arranged along the polyuronide chain. The unexpectedly greater lysine content in the PGs of more specialised necrotrophs cannot be fully explained, but suggests an increased need for such binding (cf. Cervone, Scala & Scala, 1978; Cooper, Wardman & Skelton, 1981). It also emphasises the relevance of lysine biosynthesis which, in the higher fungi, is by the unusual route of α-aminoadipic acid and saccharopine (Wade, Thomson & Miflin, 1980).

Table 7.1. *Content of strongly basic, and acidic, amino acids in fungal polygalacturonases*

Fungus	Enzyme	Mol. wt (kd)	pI	Arg[a]	Lys	Asp	Glu	Reference
Specialized facultative parasites								
Verticillium albo-atrum	PG	29	10.8[b]	4	20	41	13	Wang & Keen (1970)
Monilinia fructigena	PG4	33	9.7	7	21	31	23	M.M. Snape & R.J.W. Byrde (unpub.)
Stereum purpureum	PG	41	8.5	5	24	65	14	Miyairi, Okuno & Sawai (1985)
Fusarium oxysporum	PG1 + PG2	37	7.0	2	21	60	13	Strand, Corden & MacDonald
f. sp. *lycopersici*	PG2	37	7.0	2	22	62	13	(1976)
Rhizoctonia fragariae	PG1	36	6.8	3	18	38	17	Cervone *et al.* (1977, 1978)
	PG2	36	7.1	2	19	37	18	
Unspecialised necrotrophs and saprotrophs								
Galactomyces reesii	c	30	8.4	4	16	35	17	Sakai & Yoshitake (1984)
Rhizopus stolonifer	PG	32	8.0	3	14	59	15	Lee & West (1981b)
Trichosporon penicillatum	c	30	7.8	3	10	28	15	Sakai & Okushima (1982)
Botrytis cinerea	PG1	34	7.3	12	16	28	32	Marcus & Schejter (1983)
	PG2	56	7.6	8	12	40	64	
Kluyveromyces fragilis	PG F2	33	5.6	4	15	40	10	Sakai, Okushima & Yoshitake (1984)
Aspergillus niger	PG	35	5.1	10	11	38	24	Rexova-Benkova & Slezarik (1968)

(a) Arg = arginine; Asp = aspartic acid; Glu = glutamic acid; Lys = lysine.
(b) Theoretical value; Mussell & Strouse (1972) reported a pI of 9.7.
(c) Described as a protopectin-solubilising enzyme with PG activity.

Membrane leakage and cell death

It has long been apparent that some enzymes that degrade the pectic fraction of the cell wall are capable of macerating plant tissue and causing protoplast death (Tribe, 1955; Bateman, 1976). Studies with purified pectic enzymes show that this ability is restricted to endo-pectin or endo-pectate lyases, and endo-polygalacturonases (Basham & Bateman, 1975b; Hislop, Keon & Fielding, 1979). In contrast, purified enzymes that attack the neutral sugar fraction of the cell wall, e.g. arabinofuranosidases (Byrde & Fielding, 1968), arabanases (Baker *et al.*, 1979) and galactanases (Bauer, Bateman & Whalen, 1977) do not apparently macerate plant tissues when acting alone. Also initial attack by pectic enzymes is typically a prerequisite for cell wall breakdown by cellulases derived from pathogens (Bauer *et al.*, 1977). Phosphatidases and proteases that are capable of killing isolated protoplasts are not toxic to intact plant tissues (Stephens & Wood, 1975). Furthermore there have been no reports of exo-acting pectic enzymes capable of macertating tissues, although care must be taken in classifying the mode of action of pectic enzymes (Cooper, Rankin & Wood, 1978).

Recently the crucial role of pectic enzymes in tissue maceration has been elegantly demonstrated in experiments in which the non-phytopathogenic bacterium *E. coli* was transformed with genetic material coding for pectic enzymes from the phytopathogen *Erwinia chrysanthemi* (Keen *et al.*, 1984; Collmer *et al.*, 1985; Zink & Chatterjee, 1985); some endo-pectate lyase producing transformants were shown to be capable of macerating and killing plant tissues (Collmer *et al.*, 1985; Zink & Chatterjee, 1985). However, the very large inoculum required to initiate colonisation of tissues by these transformants, and by implication their restricted pathogenicity, suggests that the secretion of pectic enzymes is in itself insufficient to ensure plant parasitism.

Genetic engineering of plant pathogens that produce cell wall degrading enzymes is at present restricted to bacteria amenable to genetic manipulation, but transformation systems for use with filamentous fungi are rapidly being developed (Yoder *et al.*, 1986), which will make these exciting new techniques available for the study of fungal pathogens.

Experiments utilising genetic transformation and recombination (Collmer, 1986) have demonstrated that not all of the five isoenzymes of endo-pectate lyase produced by *Erwinia chrysanthemi* were required for pathogenesis; inactivation of two genes caused little alteration in the ability of the organism to macerate intact potato tubers. This suggests that some enzyme forms may be redundant or alternatively have roles limited to certain hosts or ecological niches. Some endo-pectic isozymes fail to macerate but these are exceptions to the general rule (Byrde & Fielding, 1962; Garibaldi & Bateman, 1971); whether this failure is a genuine reflection of lack of macerating ability or the result of enzyme inhibitors (Fielding, 1981) or of unfavourable physiological conditions (Cooper, 1983) is not clear at the present.

Maceration of tissues, typified by a separation of cells along the pectin-rich middle lamella (Bateman, 1976), is invariably accompanied by proto-plast injury which is manifested by a loss of control of cellular water relations, i.e. an efflux of ions and water (Hall & Wood, 1970), the inability of the protoplast to undergo plasmolysis and deplasmolysis (Brown, 1959) and loss of ability to retain vital dyes (Tribe, 1955).

A number of hypotheses have been advanced to account for injury of plant protoplasts by purified pectic enzyme preparations (Bateman, 1976). Those currently most plausible include:

1. A physical effect, whereby the random cleavage of the pectic fraction by pectic enzymes results in a dissolution of the middle lamella and a loosening of the primary cell wall structure such that the cell wall is unable to retain the protoplast which is fully turgid (Basham & Bateman, 1975*b*; Bateman, 1976).

2. A biochemical effect whereby pectic enzyme action releases some factor from the cell wall that is injurious to the protoplast. These factors fall into two categories: (a) architectural, in which elements of the cell wall structure, released by enzyme activity, interact with the protoplast, and (b) enzymic, in which immobilised enzymes within the cell wall are released after pectic enzyme degradation such that their products are toxic to the protoplast (Mussell, 1973).

Cell wall breakdown and tissue maceration closely parallel protoplast injury following exposure of tissue to pectic enzyme (Basham & Bateman, 1975*b*). These effects can only be separated by plasmolysing the tissue (Tribe, 1955; Basham & Bateman, 1975*b*). In the presence of plasmotica protoplasts retain vital dyes but maceration and cell wall breakdown proceed (Basham & Bateman, 1975*b*). However, protoplasts are rapidly killed when the tissue is deplasmolysed. These observations support the hypothesis that the turgid protoplast cannot be retained by a weakened cell wall. However, the lag period observed between application of enzyme to the tissue and the start of ion leakage, a symptom of injury, is very short (Basham & Bateman, 1975*a*; Hislop, Keon & Fielding, 1979), imply-ing that the cell wall degradation required to initiate injury is not exten-sive.

Fragments of plant cell wall polymers have been shown to possess a range of biological activities involved in the regulation of plant cell metabolism (McNeil *et al.*, 1984). Among these is a fragment that can cause the death of plant cells (Yamazaki *et al.*, 1983). This fragment, prepared by acid hydrolysis from cell walls, is at present undefined, but may derive from the cell wall pectic polysaccharide. The fragment inhibits the uptake of ^{14}C-leucine into cells, and causes the loss of vital dye staining, whilst some protection is conferred on the protoplast by plasmolysis. However, in contrast, it has earlier been demonstrated that the hydrolytic and lytic degradation products of the pectic fraction of cell wall digests are not toxic to cells in plant tissues (Fushtey, 1957; Basham & Bateman, 1975a). Thus at present the role of structural factors released from the cell wall during maceration remains unclear; so too is any involvement in cell killing of oxygen free radicals such as superoxide (Hislop *et al.*, 1979) and hydroxyl.

The plant cell wall has been considered to be part of the cell's lysozome system (Matile, 1975), enzymic degradation of which could release various enzymes with potentially injurious physiological activity, e.g. acid phosphatase, glucose oxidase, peroxidase and IAA oxidase (Mussell & Strand, 1977; Hislop *et al.*, 1979). However, to date, attempts to demonstrate a causal relationship between these enzymes, and their byproducts, and pectic enzyme induced protoplast injury, have been negative (Bateman & Basham, 1976; Hislop *et al.*, 1979).

Electron microscopy of *in vivo* host/pathogen interactions in which pectic enzymes are implicated clearly illustrates the profound effect of these enzymes on host tissues (Fig. 7.2). These include dissolution of the middle lamella and a swelling of the plant cell wall, finally leaving only a micro fibrillar skeleton after removal of the pectic matrix. Dramatic changes to the protoplast are also observed including psuedoplasmolysis, associated with a densely staining cytoplasm, and vesiculation and disorganisation of internal membranes and organelles (Fig. 7.2 a-c) (see review by Cooper, 1981). However, protoplast injury is not invariably co-existent with obvious cell-wall degradation (Kenning & Hanchey, 1980; O'Connell, Bailey & Richmond, 1985); both sets of authors observed dead protoplasts adjacent to apparently undegraded cell walls (Fig. 7.2d). However, the complexities of *in vivo* interactions make it impossible to draw conclusions as to the cause of injury of individual cells in infected tissues.

Studies using purified pectic enzyme on intact tissues are limited to that of endo-pectin lyase (PL) of *M. fructigena* on cultured apple cells

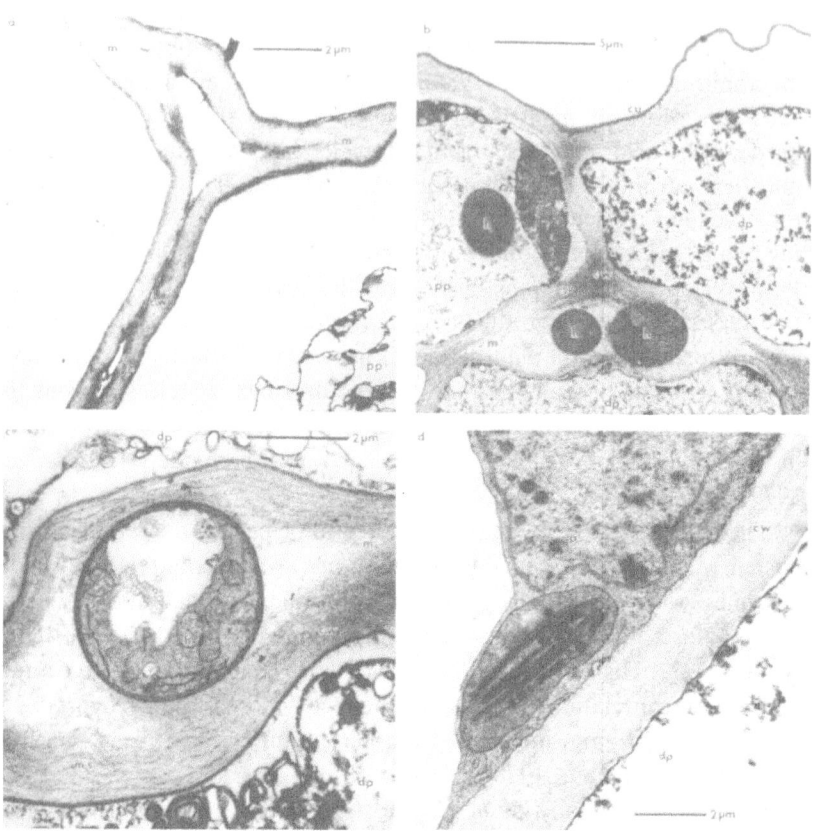

Fig. 7.2 Plant tissues invaded by pectic enzyme secreting fungi.

a,   Pear fruit invaded by *Monilinia fructigena* [courtesy of R. Pring, see also Pring *et al.*, 1981].

b,c,d, French bean invaded by *Colletotrichum lindemuthianum* [courtesy of Richard O'Connell, see also O'Connell *et al.*, 1985].

a,b,c, Note dead and pseudoplasmolysed protoplasts accompanying a swelling of the cell wall and considerable degradation of the middle lamella region. Considerable degradation leaves visible the cellulosic microfibrillar skeleton of the cell wall (cw).

d,   Shows cells at the edge of a lesion with a dead and a healthy protoplast adjacent to an apparently undegraded cell wall.

Abbreviations: cw, cell wall; dcw, degraded cell wall; p, apparently healthy protoplast; pp, pseudoplasmolysed protoplast; dp, dead protoplast; m, middle lamella; h, fungal hypha; mi, mitochondrion; pmb, paramural body; ver, vesiculated endoplasmic reticulum; ms, microfibrillar skeleton.

(Hislop *et al.*, 1979; Keon, 1985). A time course study (Keon, 1985) indicated that methylated pectin (the substrate for PL), as revealed by iron hydroxamate staining, was rapidly lost from the cell walls of cells treated with PL. This was concomitant with the loss of vital dye staining, indicating protoplast injury, and the earliest ultrastructural cytoplasmic changes detectable in the e.m. (Fig. 7.3a,b). However, conventional fixation and staining methods, and the histochemical staining of carbohydrates with silver proteinate, suggested that the overall integrity of the cell wall was intact even when protoplast injury was clearly apparent (Fig. 7.3 c,d). Thus PL-induced cellular injury apparently required neither the rupture and gross disorganisation of the plasmalemma, nor considerable degradation of the cell wall. However, no other factor was observed to act at an early stage of injury and the basis of killing is still unclear.

Cells with heavily degraded cell walls invariably contained grossly disorganised and pseudoplasmolysed protoplasts (Fig. 7.3c).

Status of CWDE in parasitism

Host penetration and necrotrophy

Although some necrotrophic fungal pathogens such as *Botrytis* spp., *R. solani* and *Sclerotium rolfsii*, cause wall breakdown even prior to or during penetration (Cooper, 1983), many pathogens are not truly necrotrophic at the time of, and soon after, initial infection of a healthy host. Instead, until better established, they use one of two strategies, presumably to avoid switching on with CWDE the cell's (or cells') defence mechanisms. One strategy is initially to live saprotrophically, either on a wound surface (e.g. *Monilinia fructigena* on apple) or in a quiescent infection (Swinburne, 1983), as in a lenticel on a fruit surface (e.g. *Pezicula malicorticis*) or in a pocket of dead tissue in the bark of a tree (Cooke & Rayner, 1984, pp. 123–5). The pathogen emerges from this quiescent infection when the host becomes senescent (fruit) or stressed (trees), e.g. by drought, and its defence mechanisms are weakened.

The term *hemibiotrophic* has been applied to a pathogen that initially invades a plant in a biotrophic mode, but later switches to a necrotrophic mode. Examples are *Phytophthora infestans* on potato and *Colletotrichum lindemuthianum* on French bean. In the biotrophic mode, CWDE secretion is greatly modified (see below); on reversion to necrotrophy there is extensive secretion of CWDE (Wijesundera *et al.*, 1984; O'Connell *et al*.,1985).

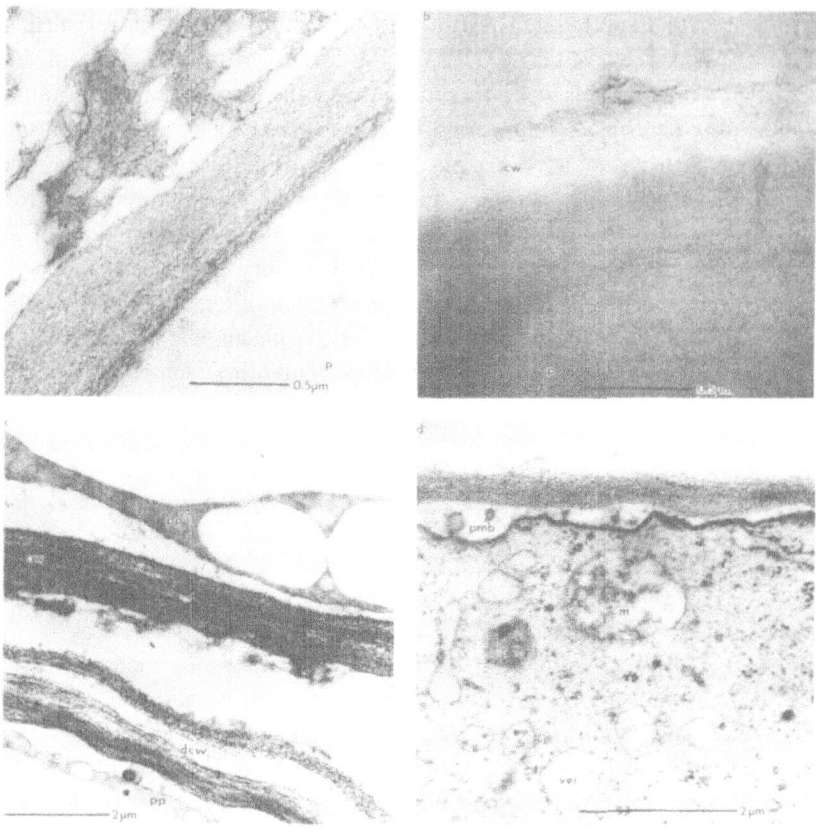

Fig. 7.3. The effect of a purified pectic enzyme on cultured apple cells (Keon, 1985). (Abbreviations as in Fig. 7.2).

a, b.   Untreated and enzyme-treated cells, respectively. Note removal of methylated pectin as revealed by histochemical staining. The protoplast was not well visualised by the method but parallel observations using conventional e.m. techniques and vital dye staining suggested that damage was occurring at this stage.

c,   Note pseudoplasmolysed protoplast adjacent to apparently little-degraded cell wall (top cell) and appearance of heavily degraded cell wall (bottom cell).

d,   Illustrates the degree of injury, e.g. vesiculation of e.r., disorganisation of organelles, and presence of paramural bodies, typically present before cell-wall degradation was visible by conventional e.m. preparation methods.

Hemibiotrophy

Hemibiotrophs include parasites which have a high potential to produce endo-PGs, such as *V. albo-atrum* (Cooper & Wood, 1975; Bishop & Cooper, 1983) and *C. lindemuthianum* (see above). Presumably they must differ from necrotrophs in terms of enzyme characteristics, regulation or secretion, but this remains to be clearly established. The initially highly restricted wall degradation could result from several causes which will be considered briefly here.

Localised breakdown of wall polymers may result from restriction by host walls of enzyme activity or movement. As already suggested, ionic binding or molecular sieving can have a profound influence on activity in walls. It is intriguing therefore that the endo-PG of *P. infestans* is exceptionally large (*ca.* 350,000 daltons) and virtually ineffective against wall polygalacturonide (Knee *et al.*, 1975). Alternatively, controlled degradation could arise from CWDE remaining bound to walls of the parasite, as with obligate biotrophs and some mycorrhizal fungi (see below).

Wall-bound PG inhibitors have now been isolated from many species and frequently proposed as factors involved in restriction of wall breakdown. However, their efficacy as inhibitors bound in native walls is questionable in the light of data from Cooper *et al.* (1981) and Turner & Hoffman (1985). Thus it has not been clearly established that these inhibitors are wall-bound rather than cytoplasmic proteins which adsorb to walls during extraction. Most of the 42 kilodalton PG inhibitor from pea leaflets occurs as a soluble intracellular protein (Turner & Hoffman, 1985). Moreover, walls from which proteins were removed or inactivated were no less inhibitory than those with active wall-bound proteins (Cooper *et al.*, 1981).

As already indicated, the mode of action of CWDE (terminal or random cleavage) has a marked influence on tissue destruction. The PGs of at least two hemibiotrophs, *V. albo-atrum* and *C. lindemuthianum*, degrade by 'multiple' attack (combined endo/exo action) (Cooper *et al.*, 1978), which can allow wall degradation without rapid maceration, as demonstrated with a PG of this type from *Aspergillus* (McClendon, 1979*b*).

The previous four mechanisms might contribute to biotrophy but do not explain the relatively sudden switch in mode of parasitism to necrotrophy. One tenable hypothesis is release of CWDE synthesis from catabolite repression (CR). Along with specific induction, CR is the main regulatory control over synthesis of most fungal CWDE (see Cooper,

1977, 1983). CR is exerted non-specifically by even low levels of sugars and amino acids. In this context depletion of soluble carbohydrates from invaded host cells may eventually allow pectic enzyme production. Areas infected by obligate biotrophs typically accumulate high levels of photosynthates, which has led to the suggestion by Lewis (1974) for evolution of biotrophy from facultative parasitism *via* repression of CWDE by host sugars. Unfortunately critical evidence is still wanting.

Most fungal parasites produce both PG and PL activities which presumably give flexibility under differing substrate, pH and ionic conditions. Coincident with their contrasting pH optima (see Cooper *et al.*, 1978), production is also pH dependent with hydrolases accumulating under acidic conditions and lyases predominating at pH $\geqslant 7$ (Cooper, 1977). This phenomenon might explain why the two activities in culture and *in vivo* are often mutually exclusive. Conceivably alteration in pH of infected tissue could switch production from a wall-bound non-destructive form to an extracellular 'toxic' enzyme. There is much evidence for substantial shifts in pH in either direction following infection, such as in stems rotted by *Fusarium solani* f.sp. *cucurbitae* or by *Sclerotinia sclerotiorum* (Hancock, 1968; Bateman, 1976). Recent preliminary observations with *C. lindemuthianum* suggest that in bean hypocotyls a change to alkaline conditions coincides with the appearance of extracellular PL and necrotrophy; the initial biotrophic phase may have been facilitated by wall-bound PG (J.A. Bailey, pers. comm.; Wijesundera *et al.*, 1984).

Mycorrhizal fungi

Mycorrhizal fungi face similar potential problems to biotrophic parasites as they must grow within walls (ecto-forms) or penetrate walls (endo-forms) but maintain host cell viability. Unfortunately little critical work on their CWDE has been attempted.

Many mycorrhizal species have a limited capacity to synthesise glycanases, and the range of mono- and especially of polysaccharides utilised may be very restricted (Lewis, 1974; Cooper, unpublished data). Thus a comparison of leaf litter-decomposing fungi with mycorrhizal species (including a *Boletus subtomentosus* strain from both groups) revealed utilisation of galacturonic acid and degradation of pectin only by the former group (Lindeberg & Lindeberg, 1977).

Nevertheless for some species strict control of CWDE must occur *in vivo*, because certain ectomycorrhizas can produce pectinases, hemicellulases and cellulases in culture (Lewis, 1974). Endo-PG production by strains of the orchid endophyte *Rhizoctonia solani* is similar to that of pathogenic strains (Perombelon & Hadley, 1964). If CR was responsible for maintaining biotrophy then reduction in root carbohydrates by shading plants might be expected to disrupt the interaction. However, this treatment did not affect the balance of several birch and pine mycorrhizal associations (Giltrap, 1979).

We have shown that mycelium of three out of six ectomycorrhizal fungi substantially reduced the viscosity of polygalacturonan, but in contrast to most necrotrophic parasites, only low extracellular activity was detected and that from only one of the isolates (R. Cooper & J. Fox, unpub.; Fig. 7.4). It would seem therefore that at least some mycorrhizal fungi produce *cell-bound* endo-PG which presumably affords another mechanism of achieving non-destructive localised wall breakdown.

Obligate biotrophy

Numerous ultrastructural studies on obligate biotrophs reveal that penetration of host cells is achieved by highly localised wall erosion, as around infection pegs from appressoria of lettuce downy mildew (Ingram, Sargent & Tommerup, 1976) or from haustorial mother cells of rusts (e.g. Pring, 1980). It could be predicted that fungi which must maintain prolonged viability of host cells would employ this subtle strategy.

It has long been considered that obligate biotrophs lack CWDE and penetrate walls mechanically; the difficulty of axenic culture of most of them seems to have thwarted attempts to clarify this point. However, our recent work on conidia of *Erysiphe graminis*, urediospores of *Uromyces* spp., sporangia of *Bremia lactucae*, and axenic culture (as sporidia) of *Ustilago maydis* has clearly shown the presence of CWDE (Cooper, 1983, 1984). In contrast to facultative parasites, activities are usually at much lower levels, largely cell–bound rather than extracellular (cf. *E. chrysanthemi* which secretes $> 98\%$ of its endo-PL: see Collmer *et al.* (1985)), and significantly they do not produce endo-PGs and lyases.

It is apparent that wall penetration is achieved by glycanases and glycosidases which degrade neutral wall polymers. The pathogens of dicots produce mainly arabanases, galactanases and glucanases, but the predominant enzymes of the maize smut pathogen – xylanase, β-xylosidase and α-arabinosidase – coincide with arabinoxylan, the principal matrix polymer of its host. *E. graminis* produces enzymes with a similar spectrum as shown in Fig. 7.5. At least some of the enzymes are capable of solubilising polymers from host walls, such as those extracted from

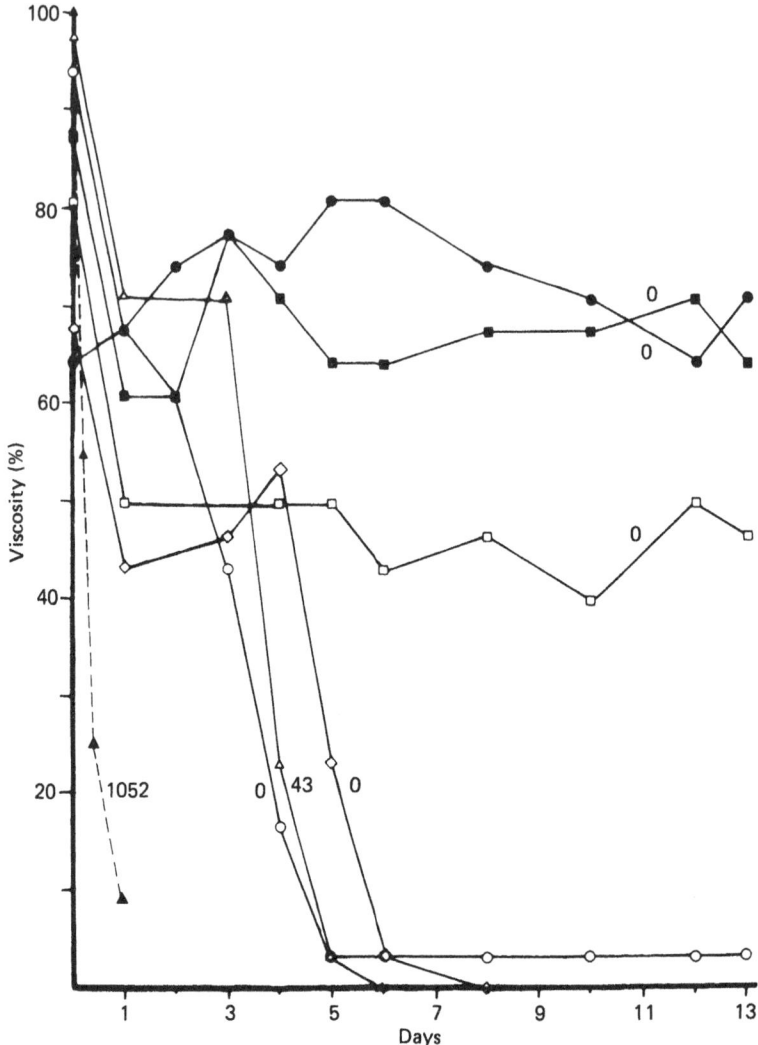

Fig. 7.4. Extracellular PG production and effect on polygalacturonan viscosity by mycelium from six mycorrhizal fungi. Mycelium (c. 200 mg d. wt) was first established by growth on glucose for 5 weeks, then washed and transferred to the polymer (pectate); the viscosity of the polymer was then determined at intervals. The numbers refer to extracellular endo PG levels (as Relative Viscosity Units; see Cooper & Wood, 1975) for each isolate at the end of the experiment. Viscosity of water is represented by the baseline. PG production and viscosity reduction by the necrotroph, *Botrytis fabae* (▲) is included for comparison. Mycorrhizal species were *Laccaria laccata* (△), *Paxillus involutus* (○), *Pisolithus tinctorius* (●), *Rhizopogon* sp. (□), *Rhizopogon roseolus* (isolate A) (■), *R. roseolus* (isolate B) (◇) (R.M. Cooper & J. Fox,

appressoria of *Uromyces viciae-fabae* and *E. graminis*. Nevertheless the extent of wall breakdown is generally very low, *ca* <5%. The monomeric analysis of oligomers released from barley walls by *E. graminis* CWDE are shown in Fig. 7.5 and reveal xylose as the main (92%) product.

Polysaccharidase activities are found in ungerminated spores, others increase or are produced only on germination or appressorium formation. Particular attention is being paid to CWDE associated with infection structures as many activities produced before this stage may relate to changes in *fungal* walls during germination, e.g. dissolution of glucans and glucomannans in uredospore walls (Trocha & Daly, 1974).

Proof of involvement in pathogenesis will be problematic. Activities are low and wall-bound and therefore difficult to detect. However, use of living discs cut from rust- and mildew-infected leaves and suspended directly into substrates has revealed up to five-fold increases of many glycosidases (see Cooper, 1984). Some hope is afforded for detection of low activities by immunocytochemistry using specific antibodies, as demonstrated *in vivo* with a fungal cutinase (Shaykh, Soliday & Kolattukudy, 1977) and a plant glucanase (Sexton *et al.*, 1980). Currently enzyme purification is impracticable with the small amounts obtained from infection structures, which we induce *en masse* on artificial membranes. However, insight into biotrophy may be provided by members of the Ustilaginales which are readily cultured on simple media and provide sufficient CWDE for characterisation and purification. Thus in infected maize the enhanced xylanase and α-arabinosidase derive from *U. maydis* as judged by the similarity of pIs *in vivo* and *in vitro* (I. Harding & R. Cooper, unpubl.).

'Neutral' glycanases and glycosidases in facultative parasitism

It is evident from our studies on obligate biotrophs that wall penetration can be achieved without degradation of polygalacturonan. Various evidence suggests that CWDE which attack neutral polysaccharides of the wall matrix are also employed by diverse facultative parasites.

Production *in vitro* of high levels of enzymes such as arabanases, galactanases and xylanases is common, and activities have often been detected

Fig. 7.4 *(cont.)*

unpubl.). In parallel experiments, viscosity of pectin was not affected and extracellular pectic enzymes were absent; all isolates reduced the viscosity of soluble cellulose (CMC) but no extracellular cellulases were detected.

in infected tissue from which the corresponding wall polymer has been depleted (Bateman, 1976). Possibly because none of these CWDE affects overall wall integrity (e.g. Bauer *et al.*, 1977; Cooper *et al.*, 1978) they have unfortunately been assigned a mere nutritional role.

Hydrolysis of a neutral polymer might be expected to be significant during invasion of a tissue which contains a high proportion of it in the wall matrix, e.g. β-1,4-galactan constitutes an unusually high amount (*c.* 45%) of potato tuber primary walls. Coincidentally endo-galactanases appear to function as key CWDE of *P. infestans* and *Phoma exigua* as revealed by rate and level of production in culture and in tubers, and

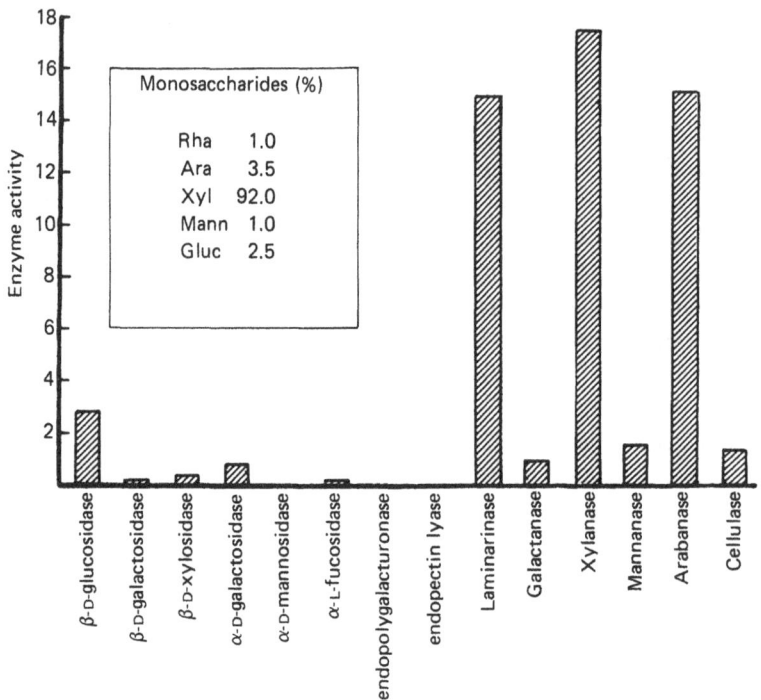

Fig. 7.5 Polysaccharidases in conidia and appressoria of *Erysiphe graminis* and effect on host walls. Disrupted conidia were extracted with 0.02 M phosphate buffer pH 6.0 to release 'soluble' activity, then 'bound' enzymes were obtained by successive extractions in 0.4 M NaCl, 0.01% Tween 80 and 0.02 M mercaptoethanol. Total activities are shown as nmol substrate monomer released ml^{-1} min $^{-1}$. Inset shows glc analysis (of alditol acetate following TFA hydrolysis) of monosaccharides derived from products solubilised from barley walls by CWDE extracted from appressoria. Rha (rhamnose), Ara (arabinose), Xyl (xylose), Mann (mannose), Gluc (glucose) (D. Longman & R.M. Cooper, unpubl.).

by depletion of host galactan (Jarvis, Threlfall & Friend, 1981; Keenan, 1984). It is evident from the apparent lack of structural effect on potato tuber tissue by a pure endo-galactanase from *Sclerotinia sclerotiorum* capable of releasing *c.* 56% of the wall matrix (Bauer *et al.*, 1977), that appraising the involvement of a CWDE should not be based simply on a crude determination of tissue strength.

A preoccupation with pectic enzymes is largely justified based on the bias of most investigations towards host-pathogen interactions involving dicots. However, in monocots (if Gramineae are representative), the predominant matrix polymer is a β-1,4-xylan with arabinose side chains (McNeil *et al.*, 1984). Non-cellulosic β-1,4- and β-1,3-glucans have also been reported in cereals. Recent evidence suggests that pathogens of cereals have developed a corresponding spectrum of CWDE. Thus α-arabinosidase and xylanase, and subsequently β-1,3-glucanase are the first of a sequence of CWDE produced *in vitro* on host walls by cereal parasites such as *Fusarium culmorum, Rhizoctonia cerealis* and *Pseudocercosporella herpotrichoides*. Pectic enzyme levels remain low and appear late, in sharp contrast to pathogens of dicots grown on host walls such as *V. albo-atrum* when PG is first produced, followed by 'hemicellulases' then cellulase (Cooper, 1977; Cooper *et al.*, 1987). Production of several CWDE of cereal parasites is constitutive which might suggest a continuing major role in their life cycles, possibly including saprotrophic periods (see Flannigan & Sellars, 1978). Xylanase and β-1,3-glucanase of *U. maydis* are under even looser control as synthesis is not subject to catabolite repression. As well as xylanase in maize smut galls (see above), activity has been detected in wheat roots infected with *Gaeumannomyces graminis* (J. Friend, pers. comm.; see also Cooper *et al., 1987*).

The disparity between walls of dicots and cereals is also clearly apparent from the relative efficacies of xylanase and PL on maize walls and bean walls (Baker *et al.*, 1977, 1980).

α-Arabinosidase production has also been correlated with pathogenicity of induced mutants of *M. fructigena* (Howell, 1975); and activity has been detected in infected fruits, together with the breakdown product arabinose (Byrde *et al.*, 1973). The function of such an exo-acting enzyme is not obvious, but glycosidases can act in concert with endoglycanases by removal of side groups which cause stearic hindrance. Thus arabinosidase could facilitate degradation of arabinoxylan and arabinogalactan chains, and analogously galactosidases enhance endomannanase activity (see Cooper, 1988).

Evidence relating to cellulases (endo- and exo-glucanases) generally reveals a lack of involvement in infection as might be predicted from the refractory nature of microfibrils in walls (Cooper, 1983). However, in view of the apparent role of xyloglucan in binding the amorphous phase to the fibrillar phase in plant cell wall structure (McNeil *et al.*, 1984), it is perhaps surprising that pathogens do not seem to exploit endoglucanases. Endoglucanase action reverses the hydrogen bonding of the polymer to cellulose and seems able to cause wall loosening as suggested by the appearance of enzyme activity in abscission zones of many plant species; in one instance abscission was prevented with antiserum to a specific glucanase isozyme (Sexton *et al.*, 1980).

A constitutive β-1,3-glucanase has been implicated in the infection of cereal florets by *Claviceps purpurea*. Its presence may partly reflect the presence of glucans in cereal walls but more importantly it seems to function in destruction of callose – a $1,3$-β-glucan formed as a wound polymer and often linked with resistance. During penetration of the stigma and style by *C. purpurea*, wound callose fails to accumulate and normal callose is absent from the adjacent phloem which therefore continues to supply nutrients to the developing ergot. The extracellular glucanase is present in infected tissue and in exuding 'honeydew' (Dickerson *et al.*, 1978).

Conclusions

Although not yet demonstrated unequivocally for any host-parasite combination, CWDE are probably involved in the majority of plant diseases. Their contribution can range from extensive tissue destruction caused by extracellular pectic enzymes of necrotrophs to highly localised wall changes effected by wall-bound glycanases and glycosidases of obligate biotrophs. Fascinating evidence is beginning to emerge as to how different fungi exploit variables such as enzyme size, charge, regulation, mode of action, and location for their own particular parasitic strategy. However, proof awaits analysis by classical and molecular genetics of defined mutants with altered synthesis or secretion of CWDE.

After 100 years of experimentation (de Bary, 1886) the mechanism of initial damage to host protoplasts by extracellular endo-pectic enzymes is still not fully understood. Also a paradox exists between the ability of these enzymes to suppress host resistance by cell killing and the potential to activate defence by release from walls of phytoalexin elicitors. Hopefully the newly available techniques of genetic engineering and e.m. localisation of proteins and carbohydrates using antibodies and lectins will enable rapid progress to be made in this field.

References

Baker, C.J., Whalen, C.H. & Bateman, D.F. (1977). Xylanase from *Trichoderma pseudokoningii*: purification, characterization, and effects on isolated plant cell walls. *Phytopathology*, **67**, 1250–8.

Baker, C.J. Whalen, C.H., Korman, R.Z. & Bateman, D.F. (1979). α-L-Arabinofuranosidase from *Sclerotinia sclerotiorum*: purification, characterization and effects on plant cell walls and tissue. *Phytopathology*, **69**, 789–93.

Baker, C.J., Aist, J.R. & Bateman, D.F. (1980). Ultrastructural and biochemical effects of endo-pectate lyase on cell walls from suspension cultures of bean and rice. *Canadian Journal of Botany*, **58**, 867—80.

Basham, D.F. & Bateman, H.G. (1975a). Killing of plant cells by pectic enzymes; the lack of direct injurious interaction between pectic enzymes or their soluble reaction products and plant cells. *Phytopathology*, **65**, 141–53.

Basham, H.G. & Bateman, D.F. (1975b). Relationship of cell death in plant tissue treated with a homogeneous endopectate lyase to cell wall degradation. *Physiological Plant Pathology*, **5**, 249–61.

Bateman, D.F. (1976). Plant cell wall hydrolysis by pathogens. In: *Biochemical Aspects of Plant-Parasite Relationships*, ed.J. Friend & D.R. Threlfall, pp. 79–113. Academic Press: London.

Bateman, D.F. & Basham, H.G. (1976). Degradation of plant cell walls and membranes by microbial enzymes. In: *Encyclopedia of Plant Physiology*, vol. 4, pp. 316–55, ed. R. Heitefuss & P.H. Williams. Springer-Verlag: New York.

Bauer, W.D., Bateman, D.F. & Whalen, C.H. (1977). Purification of an *endo*-β-1,4-galactanase produced by *Sclerotinia sclerotiorum*: effects on isolated plant cell walls and potato tissue. *Phytopathology*, **67**, 862–8.

Bishop, C.D. & Cooper, R.M. (1983). An ultrastructural study of root invasion in three vascular wilt diseases. *Physiological Plant Pathology*, **22**, 15–27.

Brown, W. (1915). Studies in the physiology of parasitism. I. The action of *Botrytis cinerea*. *Annals of Botany*, **29**, 313–48.

Byrde, R.J.W. (1979). Role of polysaccharide-degrading enzymes in microbial pathogenicity. In: *Microbial Polysaccharides and Polysaccharases*, ed. R.C.W. Berkeley, G.W. Gooday & D.C. Ellwood, pp. 417–36. Academic Press: London & New York.

Byrde, R.J.W. (1982). Fungal 'pectinases': from ribosome to plant cell wall. *Transactions of the British Mycological Society* **79**, 1–14.

Byrde, R.J.W. & Fielding, A.H. (1962). Resolution of endopolygalacturonase and a macerating factor in a fungal culture filtrate. *Nature*, **196**, 1227.

Byrde, R.J.W. & Fielding, A.H. (1968). Pectin methyl-*trans*-eliminase as the maceration factor of *Sclerotinia fructigena* and its significance in brown rot of apple. *Journal of General Microbiology*, **52**, 287–97.

Byrde, R.J.W., Fielding, A.H., Archer, S.A. & Davies, E. (1973). The role of extracellular enzymes in the rotting of fruit tissue by *Sclerotinia fructigena*. In: *Fungal Pathogenicity and the Plant's Response*, ed. R.J.W. Byrde & C.V. Cutting, pp.39–53. Academic Press, London & New York.

Cervone, F., Scala, A., Foresti, M., Cacace, M.G. & Noviello, C. (1977). Endopolygalacturonase from *Rhizoctonia fragariae*: purification and characterization of two isoenzymes. *Biochimica et Biophysica Acta*, **482**, 379–85.

Cervone, F., Scala, A. & Scala, F. (1978). Polygalacturonase from *Rhizoctonia fragariae*: further characterization of two isoenzymes and their action towards strawberry tissue. *Physiological Plant Pathology*, **12**, 19–26.

Collmer, A. (1986). The molecular biology of pectic enzyme production and bacterial soft

154 Enzymes that degrade plant cell walls

rot pathogenesis. In: *Biology of Plant Pathogen Interactions*, ed. J.A. Bailey, pp. 227–90. Plenum Press: New York.

Collmer, A., Schoedel, C., Roeder, D.L., Ried, J.L. & Rissler, J.F. (1986). Molecular cloning in *Escherichia coli* of *Erwinia chrysanthemi* genes encoding multiple forms of pectate lyase. *Journal of Bacteriology*, **161**, 913–20.

Cooke, R.C. & Rayner, A.D.M. (1984). *Ecology of Saprotrophic Fungi*. Longman: London & New York.

Cooper, R.M. (1977). Regulation of synthesis of cell wall–degrading enzymes of plant pathogens. In: *Cell Wall Biochemistry*, ed. B. Solheim & J. Raa, pp. 163–211. Universitetsforlaget: Tromso, Oslo & Bergen.

Cooper, R.M. (1981). Pathogen induced changes in host ultrastructure. In: *Plant Disease Control*, ed. R.C. Staples & G.H. Toenniessen, pp. 105–42. John Wiley & Sons: New York.

Cooper, R.M. (1983). The mechanisms and significance of enzymic degradation of host cell walls. In: *Biochemical Plant Pathology*, ed. J.A. Callow, pp. 101–37. Wiley: Chichester.

Cooper, R.M. (1984). The role of cell wall-degrading enzymes in infection and damage. In: *Plant Diseases: Infection, Damage and Loss*, ed. R.K.S. Wood & G.J. Jellis, pp. 13–27. Blackwell Scientific Publications: Oxford.

Cooper, R.M. (1988). Host cell wall degradation by microbial plant pathogens. *Biological Reviews* (In press).

Cooper, R.M., Longman, D., Campbell, A., Henry, M. & Lees, P.E. (1987). Enzymic adaptation of cereal pathogens to the monocot primary wall. *Physiological and Molecular Plant Pathology* (In press).

Cooper, R.M., Rankin, B. & Wood, R.K.S. (1978). Cell wall degrading enzymes of vascular wilt fungi. II. Properties and modes of action ofpolysaccharidases of *Verticillium albo-atrum* and *Fusarium oxysporum* f.sp. *lycopersici*. *Physiological Plant Pathology*, **5**, 135–6.

Cooper, R.M., Wardman, P.A. & Skelton, J.E.M. (1981). The influence of cell walls from host and non-host plants on the production and activity of polygalacturonide-degrading enzymes from fungal pathogens. *Physiological Plant Pathology*, **18**, 239–55.

Cooper, R.M. & Wood, R.K.S. (1975). Regulation of synthesis of cell wall-degrading enzymes by *Verticillium albo-atrum* and *Fusarium oxysporum* f.sp. *lycopersici*. *Physiological Plant Pathology*, **5**, 135–56.

De Bary, A. (1886). Ueber einige Sclerotinien und Sclerotien-Krankheiten. *Botanische Zeitung*, **44**, 376–480.

Dickerson, A.G., Mantle, P.G., Nisbet, L.J. & Shaw, B.I. (1978). A role for β-glucanases in the parasitism of cereals by *Claviceps purpurea*. *Physiological Plant Pathology*, **12**, 55–62.

Fielding, A.H. (1981). Natural inhibitors of fungal polygalacturonases in infected fruit tissues. *Journal of General Microbiology*, **123**, 377–81.

Flannigan, B. & Sellars, P.N. (1978). Production of xylolytic enzymes by *Aspergillus fumigatus*. *Transactions of the British Mycological Society*, **71**, 353–8.

Fushtey, S.G. (1957). Studies in the physiology of parasitism. XXIV. Further experiments on the killing of plant cells by fungal and bacterial extracts. *Annals of Botany, (N.S.)*, **21**, 273–86.

Garibaldi, A. & Bateman, D.F. (1971). Pectic enzymes produced by *Erwinia chrysanthemi* and their effects on plant tissue. *Physiological Plant Pathology*, **1**, 25–40.

Giltrap, N.J. (1979). Experimental studies on the establishment and stability of ectomycorrhizas. *Ph.D. Thesis, University of Sheffield.*

Hall, J.A. & Wood, R.K.S. (1970). Plant cells killed by soft rot parasites. *Nature*, **227**, 1266–7.

Hancock, J.G. (1968). Degradation of pectic substances during pathogenesis by *Fusarium solani* f.sp. *cucurbitae*. *Phytopathology*, **58**, 62–9.

Hancock, J.G. (1976). Multiple forms of endo-pectate lyase formed in culture and in infected squash hypocotyls by *Hypomyces solani* f.sp. *cucurbitae*. *Phytopathology*, **66**, 40–5.

Hislop, E.C., Keon, J.P.R. & Fielding, A.H. (1979). Effects of pectin lyase from *Monilinia fructigena* on viability, ultrastructure and localisation of acid phosphatase of cultured apple cells. *Physiological Plant Pathology*, **14**, 371–81.

Howell, H.E. (1975). Correlation of virulence with secretion *in vitro* of three wall-degrading enzymes in isolates of *Sclerotinia fructigena* obtained after mutagen treatment. *Journal of General Microbiology*, **90**, 32–40.

Ingram, D.S., Sargent, J.A. & Tommerup, I.C. (1976). Structural aspects of infection by biotrophic fungi. In; *Biochemical Aspects of Plant-Parasite Relationships*, ed. J. Friend & D.R. Threlfall, pp. 43–78. Academic Press: London.

Jarvis, M.C., Threlfall, D.R. & Friend, J. (1981). Potato cell wall polysaccharides: degradation with enzymes from *Phytophthora infestans*. *Journal of Experimental Botany*, **32**, 1309–19.

Keen, N.T., Dahlbeck, D., Staskawicz, B. & Belser, W. (1984). Molecular cloning of pectate lyase genes from *Erwinia chrysanthemi* and their expression in *Escherichia coli*. *Journal of Bacteriology*, **159**, 825–31.

Keenan, P. (1984). The degradation of potato cell walls by *Phoma exigua* var. *foveata*. Ph.D. Thesis, University of Hull.

Kenning, L.A. & Hanchey, P. (1980). Ultrastructure of lesion formation in *Rhizoctonia*-infected bean hypocotyls. *Phytopathology*, **70**, 998–1004.

Keon, J.P.R. (1985). Cytological damage and cell wall modification in cultured apple cells following exposure to pectin lyase from *Monilinia fructigena*. *Physiological Plant Pathology*, **26**, 11–29.

Knee, M., Fielding, A.H., Archer, S.A. & Laborda, F. (1975). Enzymic analysis of cell wall structure in apple fruit cortical tissue. *Phytochemistry*, **14**, 2213–22.

Lee, S.-C. & West, C.A. (1981a). Polygalacturonase from *Rhizopus stolonifer* an elicitor of casbene synthetase activity in castor bean (*Ricinus communis*) seedlings. *Plant Physiology*, **67**, 633–9.

Lee, S.-C. & West, C.A. (1981b). Properties of *Rhizopus stolonifer* polygalacturonase, an elicitor of casbene synthetase activity in castor bean (*Ricinus communis*) seedlings. *Plant Physiology*, **67**, 640–5.

Lewis, D.H. (1974). Microorganisms and plants:the evolution of parasitism and mutualism. In: *Evolution in the Microbial World. Symposium of the Society of General Microbiology*, vol. 24, ed. M.J. Carlile & J.J. Skehel, pp. 367–92.Cambridge University Press: Cambridge.

Lindeberg, G. & Lindeberg, M. (1977). Pectinolytic ability of some mycorrhizal and saprophytic hymenomycetes. *Archives of Microbiology*, **115**, 9–12.

McClendon, J.H. (1979a). The active site of yeast endopolygalacturonase contains seven subsites. *Phytochemistry*, **18**, 765–9.

McClendon, J.H. (1979b). Subterminal polygalacturonase, a nonmacerating enzyme attacks pectate from the reducing end. *Plant Physiology*, **63**, 74–8.

McNeil, M., Darvill, A.G., Fry, S.C. & Albersheim, P. (1984). Structure and function of the primary cell walls of plants. *Annual Review of Biochemistry*, **53**, 625–63.

Mahler, H.R. & Cordes, E.H. (1969). *Biological Chemistry*, pp. 10–14. Harper & Row, New York; Evanston & London. (International edition, 6th printing.)

Marcus, L. & Schejter, A. (1983). Single–step chromatographic purification and characterization of the endopolygalacturonases and pectinesterases of the fungus *Botrytis cinerea* Pers. *Physiological Plant Pathology*, **23**, 1–14.

Markert, C.L. (1975). Biology of isozymes. In:*Isozymes. I. Molecular Structure,* ed. C.L. Markert, pp. 1–9. Academic Press: New York & London.

Matile, P.L. (1975). The lytic compartment of plant cells. In: *Cell Biology Monographs,* vol. 1, ed. M. Alfert, W. Beerman, W.W. Franke, G. Rudkin & P. Sitte, p. 183. Springer-Verlag: Vienna.

Miyairi, K., Okuno, T. & Sawai, K. (1985). Purification and properties of endopolygalacturonase I. EC 3.2.1.15 from *Stereum purpureum,* a factor inducing silver-leaf symptoms on apple trees. *Agricultural and Biological Chemistry,* 49, 1111–18.

Mussell, H.W. (1973). Endopolygalacturonase: evidence for involvement in Verticillium wilt of cotton. *Phytopathology,* 63, 62–70.

Mussell, H. & Strand, L.L. (1977). Pectic enzymes: involvement in pathogenesis and possible relevance to tolerance and specificity. In: *Cell Wall Biochemistry Related to Specificity in Host-Plant Pathogen Interactions,* ed. B. Solheim & J. Raa, pp. 31–61. Universitetsforlaget, Tromso, Oslo, Bergen.

Mussell, H.W. & Strouse, B. (1972). Characterization of two polygalacturonases produced by *Verticillium albo-atrum. Canadian Journal of Biochemistry,* 50, 625–32.

Nguyen-The, C., Bertheau, Y. & Coleno, A. (1984). Study of isoenzymes of the polygalacturonases and endoglucanases of *Rhizopus* spp., and *Mucor* spp. and differentiation of isolates in South-Eastern France. *Canadian Journal of Botany,* 62, 2670–6.

O'Connell, R.J., Bailey, J.A. & Richmond, D.V. (1985). Cytology and physiology of infection of *Phaseolus vulgaris* by *Colletotrichum lindemuthianum. Physiological Plant Pathology,* 27, 75–98.

Pegg, G.F. (1985). Life in a black hole – the micro-environment of the vascular pathogen. *Transactions of the British Mycological Society,* 85, 1–20.

Perombelon, M. & Hadley, G. (1964). Production of pectic enzymes by pathogenic and symbiotic *Rhizoctonia* strains. *New Phytologist,* 64, 144–51.

Pring, R.J. (1980). A fine-structural study of the infection of leaves of *Phaseolus vulgaris* by uredospores of *Uromyces phaseoli. Physiological Plant Pathology,* 17, 269–76.

Pring, R.J., Byrde, R.J.W. & Willetts, H.J. (1981). An ultrastructural study of the infection of pear fruit by *Monilinia fructigena. Physiological Plant Pathology,* 19, 1–6.

Quantick, P., Cervone, F. & Wood, R.K.S. (1983). Isoenzymes of polygalacturonate transeliminase produced by *Erwinia atroseptica* in potato tissue and in liquid culture. *Physiological Plant Pathology,* 22, 77–86.

Rexova-Benkova, L. & Markovic, O. (1976). Pectic enzymes. *Advances in Carbohydrate Chemistry and Biochemistry,* 33, 323–85.

Rexova-Benkova, L. & Slezarik, A. (1968). Molecular weight and amino acid composition of *Aspergillus niger* endopolygalacturonase. *Collection of Czechoslovak Chemical Communications,* 33 , 1965–7.

Sakai, T. & Okushima, M. (1982). Purification and crystallization of a protopectin-solubilizing enzyme from *Trichosporon penicillatum. Agricultural and Biological Chemistry,* 46, 667–76,

Sakai, T., Okushima, M. & Yoshitake, S. (1984). Purification, crystallization and some properties of polygalacturonase from *Kluyveromyces fragilis. Agricultural and Biological Chemistry,* 48, 1951–61.

Sakai, T. & Yoshitake, S. (1984). Purification and some properties of a protopectin-solubilizing enzyme from *Galactomyces reesii* strain. *Agricultural and Biological Chemistry,* 48, 1941–50.

Sexton, R., Durbin, M.L., Lewis, L.N. & Thomson, W.W. (1980). Use of cellulase antibodies to study leaf abscission. *Nature,* 283, 873–4.

Shaykh, M., Soliday, C. & Kolattukudy, P.E. (1977). Proof for the production of cutinase by *Fusarium solani* f.sp. *pisi* during penetration into its host, *Pisum sativum*. *Plant Physiology*, 60,170–2.

Stephens, C.J. & Wood, R.K.S. (1975). Killing of protoplasts by soft-rot bacteria. *Physiological Plant Pathology*, 5, 165–81.

Strand, L.L., Corden, M.E. & MacDonald, D.L. (1976). Characterisation of two endopolygalacturonase isozymes produced by *Fusarium oxysporum* f.sp. *lycopersici*. *Biochimica et Biophysica Acta*, 429, 870–83.

Swinburne, T.R. (1983). Quiescent infections in post-harvest diseases. In: *Post-Harvest Pathology of Fruits and Vegetables*, ed. C. Dennis, pp. 1– 21.Academic Press: London & New York.

Tanabe, H., Kobayashi, Y., Matuo, Y., Nishi, N. & Wada, F. (1984). Isolation and fundamental properties of endo-pectate lyase pI–isozymes from *Erwinia carotovora*. *Agricultural and Biological Chemistry*, 48, 2113–20.

Tribe, H.T. (1955). Studies in the physiology of parasitism. XIX. On the killing of plant cells by enzymes from *Botrytis cinerea* and *Bacterium aroideae*. *Annals of Botany (N.S.)*, 19, 351–68.

Trocha, P. & Daly, J.M. (1974). Cell walls of germinating uredospores. *Plant Physiology*, 53, 527–32.

Turner, J.G. & Hoffman, R.M. (1985). Effect of the polygalacturonase inhibitor from pea on the hydrolysis of pea cell walls by the endopolygalacturonase from *Ascochyta pisi*. *Plant Pathology*, 34, 54–60.

Wade, M., Thomson, D.M. & Miflin, B.J. (1980). Saccharopine: an intermediate of L-lysine biosynthesis and degradation in *Pyricularia oryzae*. *Journal of General Microbiology*, 120, 11–20.

Wang, M.C. & Keen, N.T. (1970). Purification and characterization of endopolygalacturonase from *Verticillium albo-atrum*. *Archives of Biochemistry and Biophysics*, 141, 749–57.

Wijesundera, R.L.C., Bailey, J.A. & Byrde, R.J.W. (1984). Production of pectin lyase by *Colletotrichum lindemuthianum* in culture and in infected bean (*Phaseolus vulgaris*) tissue. *Journal of General Microbiology*, 130, 285–90.

Willetts, H.J., Byrde, R.J.W., Fielding, A.H. & Wong, A.-L. (1977). The taxonomy of the brown rot fungi (*Monilinia* spp.) related to their extracellular cell wall–degrading enzymes. *Journal of General Microbiology*, 103, 77–83.

Yamazaki, N., Fry, S.C., Darvill, A.G. & Albersheim, P. (1983). Host-pathogen interactions. XXIV. Fragments isolated from suspension-cultured sycamore cells inhibit the ability of the cells to incorporate ^{14}C leucine into proteins. *Plant Physiology*, 72, 864–9.

Yoder, O.C., Weltring, K., Turgeon, B.G. & Van Etten, H.D. (1986). Technology for molecular cloning of fungal virulence genes. In: *Biology and Molecular Biology of Plant Pathogen Interactions*. ed. J.A. Bailey. Plenum Press: New York, pp. 371–84.

Zink, R.T. & Chatterjee, A.K. (1985). Cloning and expression in *Escherichia coli* of pectinase genes of *Erwinia carotovora* subsp. *carotovora*. *Applied and Environmental Microbiology*, 49, 714–17.

8

The role of fungal toxins in plant disease

HERMAN W. KNOCHE[1] and
JONATHAN P. DUVICK[2]

[1]*Department of Agricultural Biochemistry, University of Nebraska, Lincoln, Nebraska 68583-0718, USA, and* [2]*Department of Biotechnology Research, Pioneer Hi-Bred International Inc., Johnston, Iowa 50131-0085, USA*

Introduction

Unquestionably toxins are a significant factor in a number important plant diseases. The epiphytotic caused by race T of *Helminthosporium maydis (Cochliobolus heterostrophus)* in the USA during 1970-71 illustrates the potential impact of toxin-associated virulence in a formerly minor pathogen of maize (Miller & Koeppe, 1971). Collectively, the plant pathogenic *Alternaria* species, many of which produce host-selective as well as non-selective toxins, are significant problems in fruit and vegetable production (Nishimura & Kohmoto, 1983*a,b*).

This chapter will present an overview of how fungal toxins are thought to participate in some plant diseases. The overall theme is one of diversity. Fungi have evolved many mechanisms to allow them to function as pathogens, but in terms of structural variety and modes of action, toxins may be the most freewheeling of disease agents. This diversity however exists within a framework of genetics and metabolism (host and pathogen) that responds to common research strategies, thus understanding one toxin aids in understanding others. The application of state-of-the-art techniques to toxin research will help to elucidate their role in host–pathogen biology (Macko, 1983, Gilchrist & Yoder, 1984).

Definitions of toxins

The lack of consensus on a definition for toxins in plant pathology reflects our incomplete knowledge of how most pathogen-produced compounds act in disease. A useful definition by Scheffer (1983) is that toxins are products of microorganisms that cause obvious damage to plant tissues and are known with confidence to be involved in plant diseases. The last criterion, known involvement in disease, is often difficult to establish.

Most compounds called toxins are low molecular weight compounds that may cause necrosis, chlorosis, wilting, or a combination of these symptoms in susceptible hosts. However, other compounds may fit these definitions to varying degrees. For example, hormones often play a part in disease, and fusicoccin is considered to be a toxin but also has growth-regulator activity (Ballio, 1981). Pathogen-produced enzymes may participate in host cell death, by methods that are not always well understood (Bateman & Basham, 1976). The roles of high molecular weight, wilt-inducing compounds have in some cases become suspect (Van Alfen & McMillan, 1982), but there is no doubt that some play a role, and perhaps should be included under the broad heading of toxins. Some mycotoxins may participate in plant disease development, but research in this area has been sparse (Rudolph, 1976). Nevertheless, the more narrowly defined toxins will be emphasised here because their involvement in disease appears to be well established.

Toxins may be classified as host-selective (host-specific) or non-selective (non-specific) (Rudolph, 1976; Scheffer, 1983). A line of demarcation does not exist, but in general, host-selective toxins are toxic to those plant species or cultivars that serve as hosts for the toxin-producing pathogen and lack toxicity towards nonhosts. A non-selective toxin may exhibit differential toxicity towards various plant species but toxicity is not highly correlated with the toxin-producer's host range.

Concepts that define a role for toxins in plant disease

Before a role in a particular plant disease can be defined for a fungal toxin produced as a cultural metabolite, the following questions must be answered:

1. Is the toxic compound a disease agent, an artifact of culture, or an incidental metabolite appearing only in the late stages of the disease?
2. At what stage(s) of the disease is the toxin produced?
3. What specific metabolic function(s) in the host is affected by the toxin and what is the molecular mechanism(s) by which the toxin elicits its metabolic effect?
4. How does the toxin's effect on host metabolism aid in disease development?
5. Does the toxin contribute to the host- or pathovar-selectivity of the pathogen?
6. How is toxin synthesis accomplished and regulated during disease development?
7. What is the impact of toxin production on the pathogen's existence outside the plant?

We are, needless to say, nowhere near answering all of these questions for any toxin. Any understanding of the toxin's structure is a key point since it paves the way for the study of structure-function relationships, probes for target sites, and routes of biosynthesis and degradation. Excellent progress has been made in the area of host-selective toxins and these will be discussed in some detail.

Toxins or incidental metabolites

Plant pathogens produce a variety of secondary compounds in culture which show phytotoxic activity, but only a small proportion of these have a demonstrated role in plant disease (Turner, 1971).

For toxins that display host-selectivity, correlations between toxin sensitivity and susceptibility to infection can be obtained for a range of potential hosts (Rudolph, 1976). Reproduction of typical disease symptoms by the toxin observed at the macro- and ultrastructural level is an excellent criterion for assessing the role of toxin involvement in a disease. Since many fungal metabolites are toxic to plant cells in sufficiently high concentrations, it is important to determine physiologically relevant concentrations (Yoder, 1981, 1983).

Once a putative toxin of known chemical structure is identified, the production of toxin in host tissue should be determined. Rudolph (1976) lists pathogens for which toxins have been detected *in vivo*. The list is short, and may reflect difficulties in detecting low concentration of toxins. Advances in high performance liquid chromatography should help alleviate this problem.

Recently, antisera raised against toxins have been used in combination with sensitive immunoassays to allow detection of minute amounts of toxin in tissue (Gendloff *et al.,* 1984; Tietjen, Hammer & Matern, 1985). Immunocytological analysis of infected and surrounding tissue can give quantitative and spatial evidence for the presence of disease-related metabolites (Hahn, Bonhoff & Grisebach, 1985). Nuclear magnetic resonance spectroscopy could be applied to detect minute amounts of toxin in intact tissue (Gadian, 1982). Molecular genetic analysis can provide definitive evidence for the involvement of a toxin in a disease, although this technology is only beginning to have an impact on the study of fungal plant pathogens (Gilchrist & Yoder, 1984).

Cercosporin, produced by many plant pathogenic species in the genus *Cercospora*, is an example of a toxic metabolite whose role in disease has been inferred, but not convincingly demonstrated. Cercosporin (Fig. 8.1), a perylene quinone possessing photodynamic activity, probably kills cells indirectly through its production of toxic oxygen radicals in the light (Daub, 1982*a*,*b*, 1984; Daub & Hangarter, 1983). Correlations between light intensity and severity of symptom expression have been observed for several diseases involving *Cercospora* spp. (Calpouzos & Stallknecht, 1976), and at least one ultrastructural study has shown numerous similarities between toxin- and fungus-induced lesions in host tissue (Steinkamp *et al.*, 1981). Production of cercosporin in culture is extremely dependent on the culture medium used (Lynch & Geoghegan, 1977), thus casting doubts about its production *in vivo*. Toxic compounds co-chromatographing with cercosporin have been extracted from lesions induced by the fungus in some cases (Fajola, 1978; Venkataramani, 1985; Kuyama & Tamura, 1957), and without success in at least one report (Mumma, Lukezic & Kelly, 1973) but there have been no measurements of toxin concentrations during lesion development. Since cercosporin is rapidly degraded by light in aqueous environments (Balis & Payne, 1981), accurate estimates of the small amounts likely to be present early in infection may be difficult. Thus, definitive work on the role of this toxin in disease remains to be done.

Stage of disease

Fungal toxins could be involved during any or all stages of infection: initial penetration, establishment, spread through host tissue, and senescence of infected tissue. There have been many attempts to distinguish between toxins that are required for infection versus those that act after the disease is established and thus add to the severity of symptoms (Yoder, 1980). Such a distinction depends, in part, on the

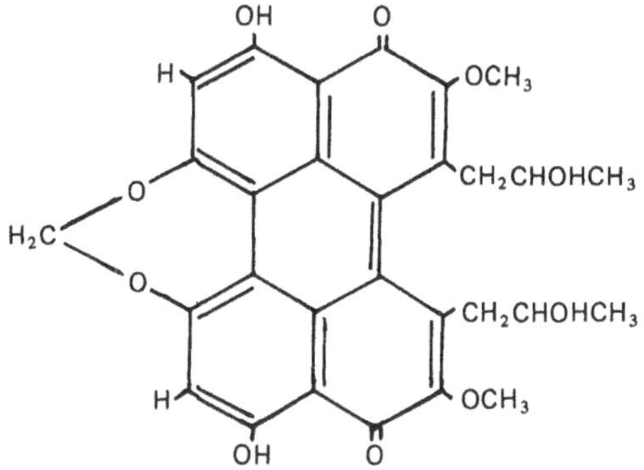

Fig. 8.1. Cercosporin, a nonselective toxin produced by numerous species in the genus *Cercospora* (Yamazaki & Ogawa, 1972).

definition of a successfion infection, which may occur without symptoms (Wheeler, 1981).

Studies on the progress of infection at the light or electron microscope level have provided evidence for a toxin's role in some cases. For the race T of *H. maydis*, ultrastructural studies on the penetration of resistant and susceptible maize lines led Wheeler (1977) to conclude that its toxin (HmT-Toxin) was not an important disease determinant during the early stages of infection. Comstock & Scheffer (1973) found that pretreatment of susceptible tissue with this toxin permitted infections by virulent pathogens. However, pretreatment does not occur in diseases; hence such studies do not prove the early involvement of a toxin.

Germinating spores of several *Alternaria* pathotypes produce their own characteristic toxins (Ueno, Nakashima & Fukarni, 1983; Yamamoto *et al.*, 1984), and chemical or heat treatment of germinating spores suppresses toxin production along with pathogenicity (Tsuge & Nishimura, 1984; Tsuge *et al.*, 1985; Otani *et al.*, 1985a). These studies suggest that the *Alternaria* toxins may be involved in the initial stages of disease development, but unnatural treatments must be interpreted with caution. Sensitive, quantitative studies on the presence of toxins during disease development are needed to provide more definitive answers.

Mechanism of action

Molecular mechanisms of toxicity and specificity, which may or may not involve identical processes, cannot be approached without a knowledge of chemical structures. At the beginning of this decade the structures for only two host-selective toxins had been established (Kono, Knoche & Daly, 1981a) and today that list has been extended to eleven. Each of the toxins identified has been shown to have more than one toxic component, but HmT-toxin and PM-toxin (see Fig. 8.2) represent extreme cases with ten or more closely related toxic components each. The structures of the major component of each host-selective toxin identified to date are shown in Fig. 8.2. The diversity in structure does not suggest common mechanisms of action as a group. The toxins from *H. maydis* and *P. maydis* however appear to be exceptions.

Current information on structures, structure-activity relationships and specific biochemical processes affected for certain host-selective toxins, provides some insights into possible mechanisms for toxin action. The earlier literature on structure and modes of action has been reviewed and will not be repeated here (Daly & Deverall, 1983; Durbin, 1981; Daly & Knoche, 1982; Daly *et al.*, 1983; Ueno *et al.*, 1983).

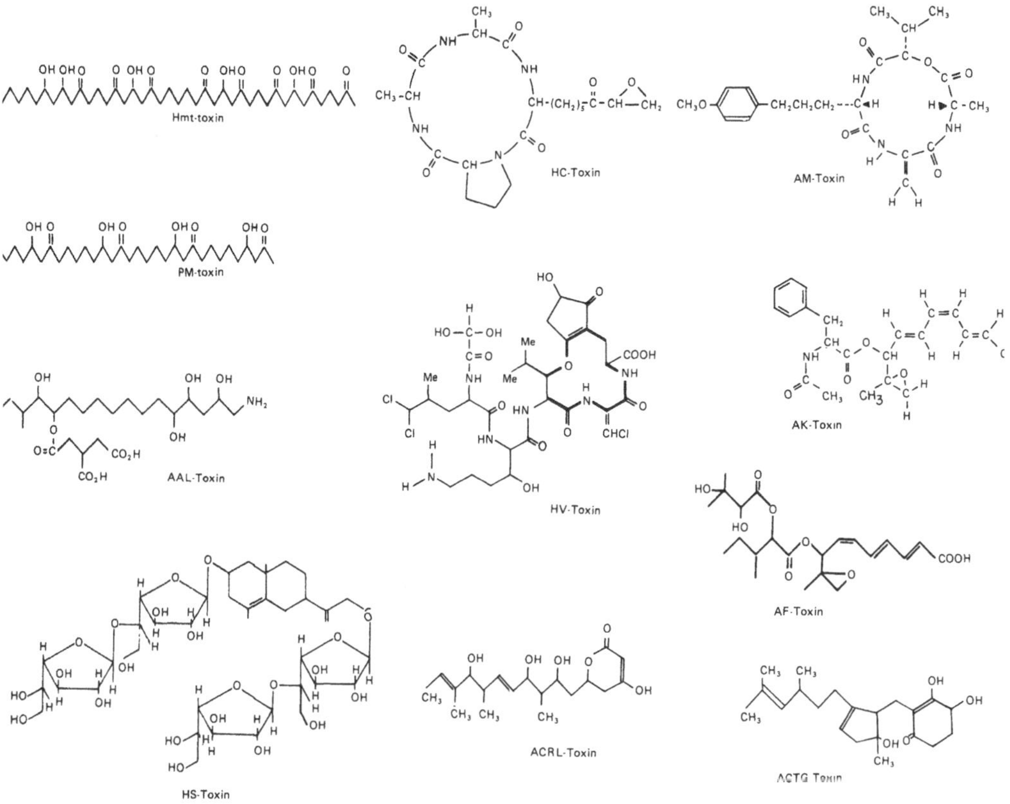

Fig. 8.2. Structures of host-selective toxins. All native toxins consist of more than one component, but only one representative component is shown for each toxin. HmT-Toxin, Band 1 from *Helminthosporium maydis,* race T, maize pathogen (Kono & Daly, 1979; Kono *et al.,* 1980a,b). PM-Toxin, component A from *Phyllosticta maydis,* maize pathogen (Kono *et al.,* 1983; Danko *et al.,* 1984). AA1-Toxin, compound T_A1 from *Alternaria alternata* f.sp. *lycopersici,* tomato pathogen (Bottini & Gilchrist, 1981; Bottini, Bowen & Gilchrist, 1981). HS-Toxin, component A from *Helminthosporium sacchari,* sugar cane pathogen (Macko *et al.,* 1981, 1983). HC-Toxin, major component from *Helminthosporium carbonum,* race 1, maize pathogen (Liesch *et al.,* 1982; Gross *et al.,* 1982; Walton, Earle & Gibson, 1982; Pope *et al.,* 1983). AM-Toxin, component I from *Alternaria mali,* apple pathogen (Okuno *et al.,* 1974a,b, 1975; Ueno *et al.,* 1975a,b,c). AK-Toxin, component I from *Alternaria kikuchiana,* pear pathogen (Nakashima, Ueno & Fukami, 1982; Nakashima *et al.,* 1982, 1985). ACRL-Toxin, Band 1 from *Alternaria citri,* rough lemon pathogen (Kono *et al.,* 1985b,c; Gardner *et al.,* 1985a,b). ACTG-Toxin, component A, from *Alternaria citri,* tangerine pathogen (Kono, Gardner & Takeuchi, 1986). AF-Toxin, component I, from *Alternaria fragariae,* strawberry pathogen (Kohomoto, pers. comm.). HV-Toxin, component C from *Helminthosporium victoriae,* oat pathogen (Macko *et al.,* 1985; Wolpert *et al.,* 1985).

Host selective toxins

HmT-toxin and PM-toxin

Helminthosporium maydis (Cochliobolus heterostrophus) race T causes southern corn leaf blight. Race T of this pathogen produces a host-selective toxin called HmT-toxin which consists of a family of about ten linear polyketols (Kono & Daly, 1979; Kono *et al.*, 1980*a*, 1981*b*). All components have an odd number of carbon atoms, vary in length from C_{35} to C_{45}, and also in their proportion of hydroxy and ketone groups. PM-toxin, produced by another maize pathogen, *Phyllosticta maydis*, also consists of a family of linear polyketols with odd numbers of carbon atoms (Kono *et al.*, 1983; Danko *et al.*, 1984). Compared to HmT-toxin, PM-toxin components have slightly shorter chain lengths and fewer oxygen functions. HmT-toxin and PM-toxin exhibit identical specificities toward T-maize (maize line carrying the Texas cytoplasmic male sterility trait) and their potencies are comparable, with PM-toxin being slightly more toxic on a molar basis (Danko *et al.*, 1984). Both toxins are active at $10^{-9}M$ against T-maize, but not against N-maize (lines with normal cytoplasm) or lines with other male sterile cytoplasmic types (C,S). Although HmT-toxin has been investigated more extensively than PM-toxin, no differences in biological effects (toxicity and selectivity) have ever been noted for these toxins. The conclusion that both act at the same site(s) and by the same mechanism(s) appears justified.

Organic synthesis of analogues and mimics of HmT- and PM-toxins has permitted extensive studies on structure-activity relationships for these toxins (Suzuki *et al.*, 1982*a,b*,1983,1984). Analogues with 15 to 26 carbon atoms were specifically toxic to T-maize, but at concentrations 100-fold higher than that for HmT-toxin. One component of the HmT-toxin family called 6 and 3, has four sets of (dihydroxyoxo) or (hydroxyoxo) groups and the analogues have similar sets of oxygen functions, but only two instead of four; thus they resemble half-molecules. Kono & Daly (1979) noted that HmT-toxin had a chain length appropriate to span a lipid bilayer, and Suzuki *et al.* (1982*a,b*) suggested that intermolecular associations between shorter toxin analogues might form comparable structures. A C_{41}-analogue with four sets of oxygen functions was as toxic as HmT-toxin in several bioassays (inhibition of dark CO_2-fixation, of photosynthesis and NADH oxidation by isolated mitochondria).

More recently Suzuki *et al.* (1985) have synthesised PM-toxin B (the toxin component of the PM-toxin family, labelled B) which is fully as active as native PM-toxin B. Using PM-toxin A as a model, Suzuki *et al.* (1987) prepared a series of mimics that varied by the length of the methylene bridge between the oxygen functions. From mitochondrial assays, the conclusions were that if the methylene bridges were as long as those in PM-toxin A or slightly longer, the mimics had activities comparable to PM-toxin. Shorter bridges and the introduction of double bonds which would increase chain rigidity reduced activity. However, these mimics still showed high potency. The least active mimic gave about 50% inhibition of malate oxidation at 5×10^{-7}M, compared to 50% inhibition for PM-toxin A at 2×10^{-9}M. Suzuki *et al.* (1987) have proposed that the toxin molecules may form circular structures (cage structures) with hydrogen bonding between oxygen functional groups.

Native toxins have several chiral centres (Kono *et al.*, 1985*a*). However, the synthetic compounds mentioned above were stereoisomeric mixtures, and their activities appeared to depend on the location and spacing of groups, not their configurations. For example, synthetic PM-toxin B was as potent and specific in its action as was native PM-toxin B (Suzuki *et al.*, 1985).

The selective insecticide methomyl, a cholinesterase inhibitor (Fig. 8.3), shows the same biological specificity and affects the same processes as HmT- and PM-toxins but structurally bears little resemblance to these toxins (Humaydan & Scott, 1977; Koeppe, Cox & Malone, 1978; Klein & Koeppe, 1985; Berville *et al.*, 1984). In mitochondrial assays, however, comparable effects are observed at concentrations of 4×10^{-3} M for methomyl and 1.4×10^{-8} M for the HmT-toxin (Klein & Koeppe, 1985). Aranda *et al.* (1983) have shown that the configuration of the $C = N$ bond in methomyl has an effect on biological activity. Both isomers are equally potent (5 mM) but, surprisingly, the E isomer is nonspecific. Kraus, Silveira & Danko (1984) also synthesised HmT-toxin and methomyl analogues that showed toxicity comparable to methomyl.

The high potency of many compounds, synthetic and native, and the lack of molecular stereospecificity all argue against the existence of a

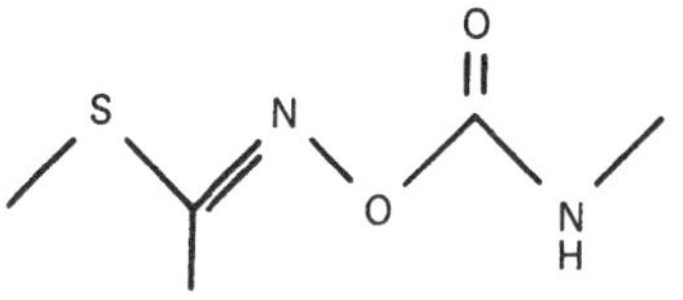

methomyl

Fig. 8.3. Structure of methomyl, the active ingredient in the insecticide lannate, which shows selective toxicity towards maize carrying Texas male sterile cytoplasm.

classical molecular receptor for these toxins. Since the report by Miller & Koeppe (1971) that NADH, succinate, and malate oxidation by isolated mitochondria from T-maize were selectively affected by the HmT-toxin, the mitochondrion has been a prime target for mechanistic studies. Daly (1981) reviewed earlier studies on the disruption of mitochondrial function *in vivo* and *in vitro* and cytological evidence for the involvement of mitochondria.

Several general conclusions may be drawn.

1. The HmT toxin and methomyl cause a stimulation of NADH oxidation in mitochondria from T-maize but not in mitochondria from N-maize, other plant species, or animals. The stimulation appears to be due to an uncoupling of oxidative phosphorylation, but HmT-toxin does not behave as a classic uncoupler such as 2,4-dinitrophenol.
2. HmT toxin and methomyl cause an inhibition of malate oxidation by T-mitochondria, contrary to its apparent action as an uncoupler.
3. The oxidation of organic acids is generally inhibited by the toxin and methomyl.

These effects on different mitochondrial processes are difficult to reconcile with action at only one site. Bouthyette, Spitsberg & Gregory (1985) have presented evidence that mitochondrial ATPase is affected, but uncoupling of oxidative phosphorylation does not explain all the effects of the toxin. Frantzen (1985) investigated the binding of HmT-toxin analogues to mitochondria. Analogues were prepared by reducing the ketone groups of HmT-toxin and PM-toxin C with tritium-labelled sodium borohydride. High specific radioactivities (8 Ci mol^{-1}) were obtained and labelled molecules showed comparable biological activities at concentrations within an order of magnitude of those for the native toxins. Binding studies conducted with concentrations comparable with those used in bioassays showed that mitochondria had a high capacity to bind the labelled toxin. However, no significant differences in ligand- displaceable binding could be detected between intact T-maize, N-maize, and soybean mitochondria. Compared to intact mitochondria, disrupted mitochondria had equal binding capacities, but enzymatic specificity between T- and N-maize mitochondria was lost. These results argue against a receptor hypothesis and suggest that toxin permeates the membranes of T- and N-maize mitochondria alike.

There are conflicting reports for methomyl- or toxin-induced mitochondrial swelling (Berville *et al.*, 1984; Klein & Koeppe, 1985), but it appears that the inner membrane remains intact (Koeppe *et al.*,1978; Berville, 1984; Klein & Koeppe, 1985). However, leakage of NAD^+ by treated mitochondria has been noted (Matthews, Gregory & Gracen, 1979; Berville *et al.*, 1984), and addition of NAD^+ to the reaction medium partially restored oxidation rates when substrates were isocitrate, α-keto-glutarate and malate. However, NADH oxidation effects are immediate, whereas NAD^+ leakage appears to be a slower response. Berville *et al.* (1984) also showed that addition of coenzyme A could relieve the inhibition of α-ketoglutarate oxidation by toxin-treated mitochondria. These results suggest that the inner membrane becomes permeable to some enzyme cofactors, but again this may not be a single, primary effect.

Pursuing Mitchell's (1979) chemiosmotic theory, Berville *et al.* (1984) found that the proton motive force collapsed when T-mitochondria were treated with toxin or methomyl. No collapse was observed for N-maize; measurements of transmembrane pH were not highly sensitive, but indicated a small, inconsistent pH gradient. Thus, the transmembrane potential was the major component of the proton motive force in T-mitochondria.

The suggestion that HmT-toxin may behave as an ionophore (Suzuki *et al.*, 1982*a*) has received support by Kimber & Sze (1984) and Holden & Sze (1984). They showed that HmT-toxin inhibited Ca^{2+} transport by T-mitochondria but had no or slight effect on Ca^{2+} transport by N-mitochondria and microsomal vesicles from T-maize. A Ca^{2+} ionophore, A23187, gave results identical to that of HmT-toxin on T-mitochondria, but it also caused the same effect on N-maize. More recently Holden, Colombini & Sze (1985) have reported that HmT-toxin at low concentrations (10^{-9} M) caused the formation of a cation-selective channel in phospholipid bilayers.

Thus, if HmT- or PM-toxins act as Ca^{2+} ionophores in T-mitochondria and synthetic membranes, and if toxin is taken up by N- and T-mitochondria, then mitochondria from nonsensitive organisms must have something in their membranes that blocks the toxin's action as an ionophore. This could be a polypeptide that interacts with the toxin to prevent or disrupt a cage formation.

Besides being a differentially permeable membrane, the inner mitochondrial membrane is a structure for the organisation of many enzymes. Indeed the physical environment around such enzymes affects their catalytic activities. Thus, membrane permeability and/or alteration of the physical nature of the membrane could be involved in the action of HmT- and PM-toxins. However specificity must be based on differences between

T- and N-mitochondria.

Forde, Oliver & Leaver (1978) and Forde & Leaver (1980) have reported that organelle translation products of T-mitochondria include a unique 13-kD polypeptide and lack a 21-kD polypeptide that is translated by N-mitochondria. Genes that restore male fertility in T-maize at least partially suppress the synthesis of the 13-kD polypeptide. The mitochondrial genome contains the genetic information for male sterility (Levings, 1983; Leaver *et al.*, 1985) and this trait has never been separated from sensitivity to HmT-toxin. Recently, Dewey, Levings & Timothy (1986) have characterised and sequenced a DNA fragment from T-mitochondria which appeared to contain genetic information for male sterility. The fragment contained an open reading frame for a 13-kD polypeptide from which a transcript was detected only in T-mitochondria. Whether or not this fragment codes for the 13-kD polypeptide or HmT-toxin sensitivity has not been shown.

Daly (1981) reviewed a number of studies that implicated the action of HmT-toxin at a chloroplastic site, and it is difficult to envisage how such light-mediated effects could be initiated by a mitochondrial site(s) of action. Restriction endonuclease profiles of chloroplast DNA from T- and N-maize have been investigated but differences such as those noted for mitochondria were not found (see Leaver *et al.*, 1985). Jigeng & Yinong (1983) noted differences between the chloroplast DNA from N-maize and lines carrying the cytoplasmic male sterility trait, but one of the lines tested was insensitive to the HmT-toxin.

HC-toxin

The HC-toxin produced by race 1 of *Helminthosporium (Cochliobolus) carbonum* (Fig. 8.2) affects certain lines of maize and is usually bioassayed by measuring ion leakage or root growth inhibition. However, no specific organelle has been shown to be involved in its action.

Recently *H. carbonum* has been shown to produce minor toxins. Kim *et al.* (1985) characterised one which differs from HC-toxin by the replacement of the D-alanine residue with glycine. A report by Scheffer's group (Tanis *et al.*, 1987) indicates that in addition to the glycine analogue, another cultural analogue is produced in which the proline residue is replaced by hydroxyproline. Structure-activity relationships for HC-toxin indicate that unlike the HmT- and PM-toxins, the structural requirements for HC-toxin activity are quite precise, and the unusual amino acid residue

Aoe (2-amino-8-oxo-9,10-epoxydecanoic acid) appears to be a critical feature.

Ciuffetti *et al.* (1983) observed that the epoxy function hydrolyses slowly upon storage and yields an 8-oxo-9,10-dihydroxy function in the Aoe residue. This conversion product showed no detectable activity at concentrations 100 times greater than HC-toxin and had no competitive effect with the native toxin. Walter & Earle (1983) treated HC-toxin with acid and also found it to be deactivated. The chirality of the epoxy function has been established and appears to be important, since Kawai & Rich (1983) synthesised HC-toxin containing both epoxide epimers and which in their assays were only about 60% as active as native HC-toxin. The ketone group of Aoe appears to be necessary for activity also. Kim (1986) reduced the ketone of Aoe with $NaBH_4$ and obtained the corresponding 8-hydroxy-9,10-epoxy function in the Aoe residue. The diastereoisomers produced were separated and found to be inactive in root growth inhibition assays either alone or in combination with one another.

Thus, the oxoepoxy function of Aoe appears to be a determinant for biological activity. However, other features of the molecule are important as well. The glycine-containing analogue represents the substitution of one hydrogen atom for a methyl group on the D-alanine residue, and the analogue's activity was found to be reduced by a factor of about 35 (Kim *et al.*, 1985). The conformation of the 12-membered peptide ring may be another important structural feature. In organic solvents, HC-toxin has the bis-γ-turn conformation (Kawai, Rich & Walton, 1983; Rich, Kawai & Jaronski, 1983; Mascagni *et al.*, 1983; Pope *et al.*, 1984).

Several fungi produce cyclic tetrapeptides that are surprisingly similar to the HC-toxin. All contain (a) the unusual amino acid Aoe, (b) either proline or pipecolic acid, and (c) at least one D-amino acid. These tetra-peptides also show toxicity in plant, animal or microbial systems. Cyl-2 was isolated from the plant pathogen *Cylindrocladium scoparium* (Hirota *et al.*, 1973). Kim (1986) also identified Cyl-2 in cultures of *Cylindro-cladium macrosporium*, a fungus not known to be a pathogen. An animal cytostatic agent called chlamydocin has been isolated from the soilborne fungus *Diheterospora chlamydosporia* the epoxy-ketone function of which is necessary for its biological activity (Closse & Hugeunin, 1974; Staehelin & Trappmacher, 1974). Kim (1986) found in the same organism another cyclic tetrapeptide which differed from chlamydocin by the substitution of an alanine residue for an α-aminoisobutyric acid residue. The cyclic tetrapeptide called WF-3161 has antitumour activity and is produced by the non-pathogenic fungus *Petriella guttulata* (Umehara *et al.*, 1983).

Reciprocal biological activities of some of these cyclic tetrapeptides have been studied by Walton *et al.* (1985). HC-toxin had cytostatic activity only slightly lower that that of chlamydocin. In root growth inhibition studies, chlamydocin and Cyl-2 required about twice the concentration of the HC-toxin for 50% inhibition in a susceptible line of maize. However, chlamydocin and Cyl-2 are not nearly as specific since both caused 50% inhibition in a resistant line at concentrations only two to three times higher than that needed for the susceptible line, while 100 times as much HC-toxin is required for similar effects in resistant maize (Pringle & Scheffer, 1967). Similar results with an ion leakage bioassay were observed by Dunkle & Kim (personal communication) in that chlamydocin and Cyl-2 were active on both resistant and susceptible maize seedlings at concentrations that were effective for HC-toxin. The ratios of responses between the susceptible and resistant seedlings were about 2 or 3 to 1.

Comstock & Scheffer (1973) first demonstrated toxin-assisted pathogenesis by establishing infections in susceptible host plants that were inoculated with a nonpathogen plus toxin. Similar experiments have been performed by Dunkle (personal communication) using races 1 and 2 of *H. carbonum*. Race 1 produces toxin whereas race 2 does not and race 2 is much less pathogenic than race 1. Inoculation of a susceptible maize line with race 2 plus HC-toxin yielded symptoms typical for race 1, but inoculation of race 2 plus either Cyl-2 or chlamydocin did not intensify infections beyond that typical for race 2.

Although little evidence is available, the mechanism of action for HC-toxin may involve disruption of genetic information. Unlike the immediate effects of HmT- and PM-toxins on isolated mitochondria, the bioassays for HC-toxin require hours for measurable responses (Ciuffetti *et al.*, 1983), suggesting a longer chain of events for expression of cellular dysfunction. Epoxides have a propensity to alkylate proteins and nucleic acids (Jerina & Daly, 1974) and a positive correlation between mutagenicity and ability of epoxides to alkylate the nucleophile 4-(*p*-nitrobenyl)-pyridine was shown by Van Duuren *et al.* (1967). Further, it was noted that electron-withdrawing groups, such as an adjacent ketone, increased mutagenicity.

The ability of HC-toxin to alkylate bionucleophiles *in vitro* as measured by fluorescence spectroscopy was shown by Kim (1986). It was found that guanosine, GMP, poly (dG-dC), and maize DNA were alkylated by HC-toxin. There was no specificity in the *in vitro* alkylation between susceptible and resistant maize DNA, and reaction conditions were not those found *in vivo*. However, these results show that, as expected, HC-toxin can alkylate maize DNA. Specificity of action could be due to hydrolysis of the epoxide group of HC-toxin by the resistant host or to differential permeabilities of membranes, but at present these are pure speculations.

AAL-toxin

The host-selective pathogen, *Alternaria alternata* f.sp. *lycopersici*, causes stem canker disease of tomato (Gilchrist & Grogan, 1976), and produces four closely related toxin components, all of which are biologically active (Bottini & Gilchrist, 1981; Bottini, Bowen & Gilchrist, 1981).

Gilchrist (1983) has summarised evidence for a molecular mechanism of action for this toxin which affects an allosteric enzyme, aspartate carbamyl transferase (ACTase) involved in pyrimidine biosynthesis. ACTase catalyses the condensation of aspartate and carbamyl phosphate, and uridine 5'-monophosphate (UMP) is a natural feedback inhibitor for the enzyme. The toxin also inhibits the enzyme, either uncompetitively or noncompetitively depending upon the substrate. Enzyme isolated from a susceptible genotype exhibited an apparent K_i of 2.4 mM compared to 4.2 mM for enzyme isolated from a resistant genotype. In the presence of UMP, AAL-toxin competitively inhibits the binding of carbamyl phosphate. More importantly, the toxin and UMP act synergistically, so that toxin plus UMP completely inhibit the enzyme at UMP concentrations that normally are not inhibitory for the susceptible genotype. A forty-fold difference in the toxin concentration needed to produce the synergistic effect was observed between enzymes from susceptible and resistant genotypes.

Gilchrist (1983), however, pointed out that although the concentration of toxin needed to demonstrate the *in vitro* effect was considerably higher than the concentration of toxin applied to a plant, this does not indicate what the concentration might be within a particular cell. Moreover metabolites of the pyrimidine pathway, e.g., L-aspartate and orotate, are potent protective compounds (Gilchrist, 1983). Thus, the evidence strongly suggests that susceptible plants produce an ACTase that differs from the enzyme of resistant plants and toxicity and selectivity of AAL-toxin are expressed by the toxin's effect on this enzyme.

AM-toxin

A. mali, the pathotype of *A. alternata* causing foliar necrosis in apple cultivars, produces three related toxins, AM-toxins I, II, and III.

AM-toxin I represents a landmark because it was the first host-selective toxin to have its structure elucidated and confirmed. As shown in Fig. 8.2, AM-toxin I has a 12-membered ring system formed by one ester and three peptide linkages. The AM-toxins and a number of analogues have been synthesised (Lee *et al.*, 1976; Shimohigashi *et al.*, 1977, 1978; Kammera *et al.*, 1981; Mihara *et al.*, 1983, 1984; Kozono *et al.*, 1984; Aoyagi *et al.*, 1985), and this has permitted studies of structure-activity relationships. The aromatic residue appears to be an important structural feature. Replacement of the methoxy group of AM-toxin I with a hydrogen atom (AM-toxin II) or a hydroxyl group (AM-toxin III) reduces potency by at least a factor of ten. Changes in the length of the methylene bridge between the phenyl group and α-carbon atom to CH_{24}, $(-CH_2-)_2$, and $(-CH_2-)$ lead to reductions in potencies of about 1000, 100, and 10,000, respectively. These results led to the proposal by Mihara *et al.* (1984) that the supposed cellular receptor has a highly specific hydrophobic pocket for the aromatic side chain of this amino acid residue.

Alteration of the residue at the L-alanine position appears to have less-dramatic effects on activity. Compared to AM-toxin I, the replacement of the methyl group by an ethyl group had little effect, but replacement by an isobutyl group reduced potency by 10 to 20 times, and replacement by a benzyl group reduced activity about 100- to 200-fold.

Factors that affect the configuration of the 12-membered ring appear to be the most important determinants of activity (Ueno, Nakashima & Fukami, 1983). Replacement of the 2-amino-2-propenoic acid residue with alanine (effectively a reduction of the double bond) creates another asymmetric centre in the molecule and reduces potency by 1000 or more times. However, the analogue with a D-alanine replacement was 10 to 20 times more active than the analogue with an L-alanine replacement. Proton NMR studies indicated that the D-alanine analogue had a ring conformation that was similar to that for the native toxin. Thus, AM-toxin has fairly strict structural requirements for activity, suggesting that the mode of action involves a highly specific interaction between the toxin and an active site. It is interesting that HC-toxin also has a 12-membered ring (all peptide linkages) and its ring conformation also appears to be quite important for activity. However, the conformations of HC-toxin and AM-toxin are not the same.

AK-toxin

A pathotype of *Alternaria alternata*, formerly called *A. kikuchiana* Tanaka, a pathogen on Nisisseiki pears causes necrotic lesions on leaves and fruit and produces the AK-toxin. As shown in Fig. 8.2 AK-toxin I appears to be derived from phenylalanine and an unusual 11-carbon fatty acid. Like the 10-carbon amino acid of HC-toxin (Aoe), it has an oxygen function at position 8 and a 9,10-epoxy group. The double bonds of the acid have the *trans, cis, trans* configuration. The addition of a methyl group on the phenylalanine residue (AK-toxin II) appears to reduce activity by about 10-fold compared to the AK-toxin I (Nakashima *et al.*, 1985).

AF-toxin

A strawberry pathotype, formerly called *A. fragariae*, of *A. alternata* produces the AF-toxin, which consists of three related compounds (Maekawa *et al.*,1984). Very recent structural studies (Namiki *et al.*, 1986*b*; Nakatsuka *et al.*, 1986*b*), show that AF-toxin contains the same C_{11}, epoxy-containing acid as AK-toxin. However the toxins differ in the double bond configurations of their acid moieties.

Biological activities of AM-, AK- and AF-toxins

In all cases, sensitivity of host plants to the toxin is correlated with pathogenicity of the particular fungal pathotype. Similarly toxin production by germinating spores also is correlated with pathogenicity. These three toxins show quite similar biological effects, leaf necrosis and K^+ leakage. All have been implicated in disruption of cellular membrane function (Ueno *et al.*, 1983; Nishimura & Kohmoto, 1983*a,b*; Otani *et al.*, 1985*b*; Maekawa *et al.*, 1984; Yamamoto *et al.*, 1984; Namiki *et al.*, 1986*a*), but precise molecular mechanisms of action are not yet evident.

ACRL-toxin

One pathotype of *A. citri* is pathogenic to rough lemon and rangpur lime and produces host-selective toxins. By a series of elegant structural studies, Kono & Gardner and co-workers have provided structures, including stereoisomeric assignments, for the ACRL-toxin complex which consists of a major ACRL-toxin I and five minor components (Gardner *et al.*, 1985*a,b*; Kono *et al.*, 1985*b,c*). The structure of the major toxin is shown in Fig. 8.2. The minor toxins are similar but contain an α-pyrone ring (instead of a dihydropyrone ring) and vary by the length of the side chain and the numbers of hydroxyl and methyl groups on the side chain. The same structure for the major ACRL-toxin apparently was reported by Nakatsuku *et al.* (1986*a*). Minor toxins are about five times less potent than the major toxin (Gardner *et al.*, 1985*b*), which may suggest an involvement of the pyrone ring in cellular interactions.

Ultrastructural studies on toxin-treated leaves indicate that ACRL-toxin affects mitochondria in susceptible tissue (Kohmoto *et al.*, 1984). ACRL-toxin-induced inhibition of malate oxidation and stimulation NADH of oxidation by isolated mitochondria appears to be a specific effect, which is surprisingly similar to the action of HmT-toxin (Kohmoto *et al.*, 1985).

ACTG-toxin

The pathotype of *A. citri* pathogenic to dancy tangerine also produces a host-selective toxin; three structurally related compounds have been identified by Kono, Gardner & Takeuchi (1986). ACTG-toxin has a structure somewhat similar to ACRL-toxin (Fig. 8.2) and one of the compounds, ACTG-toxin-C, is considerably less active than the other two.

HS-toxin

In culture *Helminthosporium sacchari*, causal agent of eyespot disease of sugar cane, produces a collection of toxic as well as nontoxic sesquiterpene galactofuranosides (Macko *et al.*, 1981, 1983; Livingstone & Scheffer, 1981). The major toxic component (Fig. 8.2), which has four galactofuranoside units per molecule, occurs in three isomeric forms differing only in the position of a double bond in the aglycone. These isomers differ in their activity as measured by two separate assays of leaf tissue from susceptible sugar cane clones (Duvick *et al.*, 1984; Livingstone & Scheffer, 1984*a*).

Compounds possessing the same sesquiterpenoid structure but having fewer than four galactofuranose units per molecule are also found in *H. sacchari* culture filtrates. These are not toxic to most of the sugar cane clones tested, but can protect against the toxic isomers when added to leaf tissue in molar excess (Livingstone & Scheffer, 1981, 1984*a,b*; Macko, 1983; Duvick *et al.*, 1984). Methyl and benzyl D-galactofuranosides were found to be protective as well, although much higher concentrations were required. Corresponding galactopyranosides were not protective (Duvick, unpublished). Thus, both the saccharide and aglycone moieties contribute to the specificity of HS-toxin. Interestingly, several clones of sugar cane showed sensitivity to one isomer of the normally nontoxic trigalactofuranoside (Livingstone & Scheffer, 1984*b*), indicating a degree of heterogeneity in whatever host component(s) interact with the toxin.

It is intriguing that nontoxic lower homologues appear to be generated by galactofuranosidases that are produced by the pathogen (Livingstone & Scheffer, 1983). Whether this enzyme activity, detected in culture fluid, is also expressed in eyespot lesions is not known. In addition, a 'latent' hexaglycoside that has a glucosyl residue linked to the 2-position of each of the two terminal galactofuranose residues of HS-toxin can also be found in culture filtrates and converted to active toxin by glycosidases (Macko, 1983). These observations illustrate the concept that pathogens may possess catabolic factors that may limit or increase the degree of expression of their virulence. Could this be a form of 'fine-tuning' of the pathogen-host interaction?

Although a plasma membrane-binding site for HS-toxin has been proposed (Strobel, 1973), to date there is no convincing evidence for such a receptor (Daly, 1981). There is evidence, however, for very rapid toxin-induced depolarisation of sugar cane leaf cell plasma membranes by HS-toxin (Schroter & Novacky, 1985). The structural data in hand, together with sensitive assays for activity (Macko, 1983), should allow toxin derivatives to be produced which could be used to probe for potential binding sites. For example, reduction of specific sites on the aglycone or saccharide portion of the molecule yields derivatives with toxicity close to that of the native toxin (Duvick, unpublished). Such derivatives could be prepared with a tritium label to give high specific radioactivity probes, or linked to a resin to make an affinity column that might be useful for purifying binding proteins. Using biosynthetic labelling, Nakajima & Scheffer (personal communication) fed tritium-labelled aglycone to cultures of *H. sacchari* and extracted a highly labelled glycosylated HS-toxin. Unfortunately, sugar cane is usually mixoploid, and genetic crosses are difficult to make. Thus it has not been possible to obtain genetically similar lines differing only in toxin sensitivity, which would be desirable for binding-site studies.

HV-toxin

The recent publication of proposed structures for the major species of HV-toxin (also called victorin), produced by *Helminthosporium (Cochliobolus) victoriae* (Wolpert et al., 1985; Macko et al., 1985), paves the way for mode of action studies. Oat cultivar 'Victoria', which carries a dominant crown rust-resistance gene (Rpc), is 10 000-fold more sensitive to HV-Toxin than are cultivars that lack the gene (Wheeler & Luke, 1963). The major toxic species are cyclic peptides containing some unusual features, including three chlorine substitutions (Fig. 8.2). As is the case with HmT-toxin and HS-toxin, chemical alteration of functional groups not essential for toxicity allows the possibility of sensitive probes for binding sites (Macko & Wolpert, personal communication).

Recently it was reported that purified HV-toxin preparations are sensitive elicitors of the oat phytoalexin avenalumin (Mayama *et al.*, 1986). Since this phytoalexin has been implicated in resistance to crown rust in Rpc lines (Mayama *et al.*, 1982), the observation provides a potential link between the rust resistance and toxin sensitivity that are both conferred by what is apparently the same locus. Resistance to HV-toxin has been obtained by selection *in vitro* using oat callus heterozygous at the Rpc locus (Rines & Luke, 1985). Regenerated plants were also resistant to the toxin and the fungus and susceptible to *Puccinia coronata,* thus increasing the likelihood that these two traits are conferred by a single locus.

Non-selective toxins

The structure-activity relationships of many nonhost-selective toxins have been reviewed by Ballio (1981). Although the molecular sites of several non-selective toxins produced by plant pathogenic bacteria are fairly well characterized (Mitchell, 1984), definitive sites of action for nonselective fungal toxins have yet to be found. Several examples will be discussed briefly here.

Tentoxin, a chlorosis-inducing cyclic tetrapeptide produced by *Alternaria* spp., has been shown to inactivate chloroplast coupling factor (CF1) *in vitro* (Steele *et al.*, 1976). A variety of structural analogues have been prepared and used to determine structural features important for toxin action (reviewed by Ballio, 1981). Vaughn & Duke (1981) presented evidence for a second site of action in tissue prior to chloroplast maturation, which involves loss of polyphenol oxidase (PPO) activity from toxin-sensitive tissue. A toxin-induced lack of processing of PPO precursor protein may explain this effect (Vaughn & Duke, 1984), but how this relates to the visible and ultrastructural effects of the toxin is still unclear. Tentoxin has been isolated from *A. mali* which also produces the host-selective AM-toxin. The structural similarities between tentoxin and AM-toxin are noteworthy and have been discussed by Stossel (1981).

Ophiobolin A is a phytotoxic sesquiterpene produced by a number of fungi, notably *Helminthosporium oryzae* from which it was first isolated (Nakamura & Ishibashi, 1958). Rapid, ophiobolin-induced alteration of plasma membrane potential has been observed (Cocucci *et al.*, 1983), but the mode of action is unknown. In a recent paper, Leung *et al.* (1985) demonstrated an interaction between ophiobolin A and maize calmodulin *in vitro*, and proposed calmodulin as the target site for the toxin. The relative activity of ophiobolin and several derivatives was similar and of the same order of magnitude in both phytotoxicity assays and *in vitro* assays for inactivation of calmodulin. The relationship between inactivation of calmodulin and the metabolic effects of ophiobolin has yet to be established.

Cercosporin, discussed earlier, represents a non-specific toxin which probably does not have a single, specific cellular target site. The photo-induced oxygen radicals or their by-products can react with and damage many cell components, including nucleic acids, proteins and lipids. The plasmalemma is likely to be a major target, since the toxin is highly lipophilic, and electrolyte leakage is observed within minutes after addition of the toxin to leaf tissue in the light (Daub, 1982a). There is also direct evidence for free radical-induced membrane damage in cercosporin-treated leaf tissue (Daub & Briggs, 1983).

While cercosporin produces both singlet oxygen and superoxide under illumination *in vitro* (Daub & Hangarter, 1983), it is difficult to assess which of these species (or both) is responsible for cercosporin's toxic effect on cells. Cavallini *et al.* (1979), on the basis of the inability of superoxide-quenching compounds to mitigate cercosporin's ability to peroxidate lipids *in vitro*, concluded that superoxide was not involved. However, suitable electron donors needed for superoxide production (Daub & Hangarter, 1983) may not have been present. According to at least one report (Macri & Vianello, 1981), cercosporin, in micromolar concentrations, can also disrupt plasma membrane functions of plant cells in the dark. The relationship of this activity to the toxin's photo-sensitising ability is not known. Cercosporin may also serve as an iron chelator (Cavallini *et al.*, 1979).

With this apparent diversity of activity by cercosporin, its precise mode of action may be difficult to ascertain. Although structural modifications of the toxin molecule, including isomerisation, have been reported (Yamazaki & Ogawa, 1972), the product's activities in causing cell damage have not been tested.

The involvement of toxins in disease development

Mechanisms of toxin action are important in explaining how gross symptoms arise, but may shed little light on how a pathogen benefits from those metabolic changes. The complexity of integration between biochemical processes may not permit an accurate prediction of the overall effects when one process is deranged, unless the whole system is understood. Diseases involving host-selective toxins have been considered excellent models for studying host-parasite interactions. However, the involvement of a toxin may be just another factor that must be integrated into an already complex system. Penetration devices, hydrolytic enzymes, specific recognition factors, and local chemical messages undoubtedly function in some toxin-related diseases. To understand thoroughly the disease state requires a knowledge of the interrelations of these processes.

The idea that toxin-producing pathogens merely kill cells in advance to permit them to live as saprophytes on dead tissue has not received experimental support (Scheffer, 1983). The induction of leakage of electrolytes and nutrients from toxin-damaged cells would benefit a pathogen capable only of intercellular growth in host tissue, but there is no reason to conclude that this is a toxin's only function, or even the primary one, in all cases.

Some toxins may act as suppressors of host-defence mechanisms. Hayani *et al.* (1982) claim to have isolated an inducer which activates host defence mechanisms. The inducer, isolated from germinating spores of *A. kikuchiana*, is thought to be a glycoprotein, and protects susceptible pear leaves against the pathogen, when leaves are treated with it 4-6 h prior to inoculation. Simultaneous treatment with inducer and toxin yielded no protection. Hayani *et al.* (1982) argued against the involvement of a phytoalexin in this induced-resistance response. These results could be interpreted as representing the induction of a self-repair mechanism (Wheeler, 1981) in susceptible hosts, a mechanism that is constitutive in resistant hosts.

While cell death is an obvious response of susceptible host tissue to many toxins, it is also the end result of hypersensitive resistance to certain pathogens. For these phenomena the role of cell death is not well understood. In some diseases, the outcome of infection may depend on whether the host or pathogen controls how and when cell death occurs (Mansfield, 1984).

Role in specificity

By definition, a role in the specificity or selectivity is implied for host-selective toxins and the list of such toxins extends beyond those given in Fig. 8.2 (Macko, 1983; Kono, Knoche & Daly, 1981*a*). In fact, the list continues to grow (Sugawara *et al.*, 1985; Vidhyasekaran, Borromeo & Mew, 1986).

Regardless of classification, specificity relationships are not clear in all cases. Scheffer & Livingstone (1980) have presented data comparing susceptibility to eyespot disease with sensitivity of leaf tissue to HS-toxin for a range of sugar cane varieties. The results indicate that toxin sensitivity and susceptibility to *H. sacchari* are not always well correlated. The authors concluded that the toxin may determine pathogenicity to some sugar cane clones but not to others. For many toxins, including HS-toxin, there are multiple forms with sometimes different or even antagonistic activities. Reproducing the array of molecules that contribute to toxicity and specificity may be a critical condition in such experiments.

Three *Alternaria* pathogens and their toxins, AM-, AK-, and AF-toxins exhibit some interesting cross reactions. AM-toxin shows selectivity among apple cultivars and some pear cultivars (Maeno, Kohmoto & Otani, 1984). Although many of the pear cultivars sensitive to AK-toxin are also sensitive to AM-toxin, the correlation is not perfect, and generally, higher concentrations of AM-toxin are needed for a sensitive reaction on pear cultivars. The reciprocal, apple cultivars sensitive to AK-toxin, has not been reported. Likewise, Nijisseiki, a pear cultivar sensitive to AK-toxin, is also sensitive to AF-toxins at the same concentration as that which affects the susceptible cultivar of strawberry, Marioka-16 (Maekawa *et al.*, 1984). The epoxy-containing acid that is present in both AK- and AF-toxins may explain some of the cross-activity, but the sensitivity and susceptibility of Nijisseiki to three pathogens and their toxins are puzzling.

The selection of bioassays may affect the apparent specificity of toxin action. Using four different bioassays methods for ACRL-toxin, Gardner, Kono & Chandler (1986) found a correlation between host-selectivity and toxin sensitivity except for one nonhost (Cleopatra mandarin) which showed an intermediate response to toxin in two of the four bioassays. The ACTG-toxin shows poor correlation between host-selectivity and host-sensitivity when the same four bioassays are used. These results and those cited for HS-toxin suggest that general toxicity as well as specific toxicity may be expressed by a particular toxin.

It has often been pointed out that the paradigm of toxin and receptor as determinants of disease represents the exact opposite of a gene-for-gene specificity based on complementary genes conferring incompatibility (Ellingboe, 1976). Complications often emerge, however, when gene-for-gene systems are analysed in detail, as in the example of the Sr6 allele in wheat stem rust resistance (Daly, 1984). Toxins likewise often fail to conform to the simple models we assign them in disease. These facts may merely indicate that our current models are inadequate to explain the action of determinant genes in the host and pathogen. The merging of classical and recombinant genetic analysis of both pathogen and host (Yoder, 1983) should permit the more accurate identification of relevant genes. Initial efforts are under way in several systems to clone genes for toxin production in *H. maydis* (Yoder, 1983) and *H. carbonum* (Briggs, Bass & Jones, 1985), and the tools are available for cloning corresponding genes for resistance or susceptibility in several host species, notably maize (Federoff, Furtek & Nelson, 1984).

Toxin production by the pathogen

While a few toxins have well-defined biosynthetic pathways, the majority, particularly the host-selective compounds, are not well known. This question is a key one in understanding how toxins may have evolved, as well as for elucidating how they are regulated in disease.

For pathogens amenable to genetic analysis, differences between toxin + and toxin − strains appear to be controlled by a single locus (Leach *et al.*, 1982; Scheffer, Nelson & Ullstrup, 1967). Several genes are likely to be required for production of most toxins, however. Evidence for multiple genes for HmT-toxin production is suggestive but incomplete (Yoder, 1980). In general, little is known about the genetics of toxin production for most fungi.

Using *Alternaria* as an example, Nishimura & Kohmoto (1983*a*) have advanced the idea that selection pressure in large-scale monocultures may be sufficient for the conversion of saprophytic to parasitic forms by becoming toxin producers. Does this mean that the saprophytic forms carry genes for AM-, AK-, and AF-toxins, but only express them under certain environmental conditions? If not, the probability of a saprophyte becoming a toxin producer, and hence a parasite, seems remote unless the genes for secondary products are particularly mutable, or perhaps have unusual control mechanisms. That toxins are the result of certain metabolic processes being 'out of control' is suggested by the amount of toxin that may be produced in culture. It seems unlikely that a pathogen needs to convert 2% of its metabolisable energy into a toxin that is active at 10^{-9} M (Daly, 1981). On the other hand, toxin producers grown continuously in culture have been known suddenly to cease to produce toxins.

Yoder (1980) hypothesised that some toxins could represent intermediates in a normal metabolic pathway. Mutations that blocked the pathway, e.g., produced a defective enzyme, would lead to an accumulation of the intermediate (toxin). He also pointed out that such a mutation would be lethal unless the end product of the normal pathway was not essential or could be produced by another pathway. Another alternative might be for the fungus to obtain an intermediate or end-product from its environment.

Using *H. maydis* as an example, Yoder reasoned that if the pathway is present in other corn pathogens, the potential for them to produce the same toxin exists. At that time, before the structure of PM-toxin was established, he suggested that *P. maydis* might be such an example. Now we know the structures of PM- and HmT-toxins are not identical, but their similarities do suggest they could arise from a common, normal pathway. The differences in the number of oxygen functions could be due to differences in the position of the enzymatic blocks. Also a catabolic process may alter the true intermediate before excretion. The critical question regarding this hypothesis is: What is that normal pathway? Since the toxins have an odd number of carbon atoms and lack methyl branches, it could be one similar to that used to synthesise waxes (Goodwin & Mercer, 1983).

Although unproven, such metabolic errors could be responsible for toxin production by some organisms, but that would be an imprudent assumption for all toxin producers. Consider the biosynthesis of HC-toxin. Unlike PM- and HmT-toxins, HC-toxin does not look like a metabolic intermediate, and although only a few fungi have been surveyed to date, the number of species that produce cyclic tetrapeptides with similar structural features suggests that they may have an important role in fungal metabolism. Perhaps their toxicities to plant tissues are incidental to their primary function. These speculations only point out the need to investigate biosynthetic pathways for toxins, which are final expressions of fungal genetics.

Toxin production outside host

What is the effect of toxin production on a pathogen's existence outside the host? For nonspecific toxins, a role in general antibiosis during the saprophytic stage could be imagined, but toxin-producers must be able to resist the effects of their own toxin. This has been addressed in detail for phaseolotoxin-producing strains of the bacterial plant pathogen *Pseudomonas syringae*, which have an altered target enzyme that is insensitive to the toxin (Staskawicz, Panopoulos & Hoogenroad, 1980). The mechanism by which *Cercospora* species resist or escape the adverse effects of cercosporin is not known, but may involve higher levels of protective carotenoids (Daub & Payne, 1985). An unknown factor correlated with fungal cell wall synthesis appears also to be involved in resistance (Gwinn & Daub, 1985).

The concept of stabilising selection (Person, Groth & Mylyk, 1976) predicts that genes involved in pathogenicity will not persist in a pathogen population unless a susceptible host is present. Observations on the low incidence of toxin producers among natural isolates of *Alternaria alternata* even in the presence of susceptible hosts (Nishimura & Kohmoto, 1983a) lead one to suspect that toxin genes may not persist outside the host. Klittich & Brosnan (1986) detected differences in fitness of near isogenic strains of *H. maydis* differing in HmT-toxin production, and concluded that in the absence of a susceptible host (i.e., T-cms maize), the toxin-producing strain was at a competitive disadvantage in the field. Similar studies need to be done using other pathogens to determine to what extent this observation is generally true. Mutants derived by insertional mutagenesis or recombinant excision of the appropriate DNA coding sequences would be most desirable for such studies.

Conclusion

In many cases a disease-causing 'toxin' is, in fact, a complex mixture of molecules of differing structure and sometimes of differing activity and specificity. Our knowledge of fungal toxins is in many respects at the forefront of research in disease physiology and biochemistry, yet we lack precise descriptions of how they participate in their respective diseases. Nevertheless, with the tools for structural analysis, biochemistry and molecular genetics, the next decade should yield critical details that will explain, in molecular terms, the disease role for some of the more significant toxins.

References

Aoyagi, H., Mihara, H., Lee, S., Kato, T., Ueno, T. & Izumiya, N. (1985). Cyclic Peptides XVIII. Syntheses of AM-toxin I analogs containing bulky L-amino acid residues instead of an L-alanine. *International Journal of Peptide and Protein Research*, **25**, 144–8.

Aranda, G., Gauvrit, C., Cesario, M., Guilhem, J., Pascard, C. & Tran Huu Dau, M.E. (1983). Biological activity of the two geometrical isomers of methomyl on maize mitochondria. *Phytochemistry*, **22**, 2431–5.

Balis, C. & Payne, M.B. (1981). Triglycerides and cercosporin from *Cercospora beticola*: fungal growth and cercosporin production. *Phytopathology*, **61**, 1477–84.

Ballio, A. (1981). Structure-activity relationships. In: *Toxins in Plant Disease*, ed. R. Durbin, pp. 395–448. London: Academic Press.

Bateman, D.F. & Basham, H.G. (1976). Degradation of plant cell walls and membranes by microbial enzymes. In *Physiological Plant Pathology*, ed. R. Heitefuss & P.H. Williams, pp. 316–55. Berlin: Springer-Verlag.

Berville, A., Ghazi, A., Charbonnier, M. & Bonavent, F.-F. (1984). Effects of methomyl and *Helminthosporium maydis* toxin on matrix volume, proton motive force, and NAD accumulation in maize (*Zea mays L.*) mitochondria. *Plant Physiology*, 76, 508–17.

Bottini, A.T. & Gilchrist, D.G. (1981). Phytotoxins I.A 1-aminodimethylheptadecapentol from *Alternaria alternata* f.sp. *lycopersici*. *Tetrahedron Letters*, 22, 2719–22.

Bottini, A.T., Bowen, J.R. & Gilchrist, D.G. (1981). Phytotoxins II. Characterization of phytotoxic fractions of *Alternaria alternata* f.sp. *lycopersici*. *Tetrahedron Letters*, 22, 2723–6.

Bouthyette, P.Y., Spitsberg, V. & Gregory, P. (1985). Mitochondrial interaction with *Helminthosporium maydis* race T toxin: blocking by dicyclohexyl carbodiimide. *Journal of Experimental Botany*, 36, 511–28.

Briggs, S.P., Bass, R.W. & Jones, S.L. (1985). Molecular genetics of pathogenicity in *Cochliobolus carbonum*. *Journal of Cellular Biochemistry*, supplement 9C, 196 (abstract).

Calpouzos, L. & Stallknecht, G. (1976)). Symptoms of *Cercospora* leaf spot of sugar beet influenced by light intensity. *Phytopathology*, 57, 799–800.

Cavallini, L., Bindoli, A., Macri, F. & Vianello, A. (1979). Lipid peroxidation induced by cercosporin as a possible determinant of its toxicity. *Chemical-Biological Interactions*, 28, 139–46.

Ciuffetti, L.M. (1983). Purification, structure, and detection of the *Helminthosporium carbonum* toxin and its inactive conversion product. *Ph.D. Dissertation*, Purdue University, West Lafeyette, Indiana.

Ciuffetti, L.M., Pope, M.R., Dunkle, L.D., Daly, J.M. & Knoche, H.W. (1983). Isolation and structure of an inactive product derived from the host-specific toxin produced by *Helminthosporium carbonum*. *Biochemistry*, 22, 3507–10.

Closse, A.& Huguenin, R. (1974). Isolierung und strukturaufklarung von Chlamydocin. *Helvica Chemie Acta*, 57, 533–45.

Cocucci, S.M., Morgutti, S., Cocucci, M. & Gianani, L. (1983). Effects of ophiobolin A on potassium permeability, transmembrane electrical potential and proton extrusion in maize roots. *Plant Science Letters*, 32, 9–16.

Comstock, J.C. & Scheffer, R.P (1973). Role of host-selective toxin in colonization of corn leaves by *Helminthosporium carbonum*. *Phytopathology*, 63, 24–9.

Daly, J.M. (1981). Mechanisms of action. In: *Toxins in Plant Disease*, ed. R.D. Durbin, pp. 331–94. London: Academic Press.

Daly, J.M. (1984). The role of recognition in plant disease. *Annual Review of Phytopathology*, 22, 273–307.

Daly, J.M. & Deverall, B.J. (1983). *Toxins and Plant Pathogenesis*. Sydney: Academic Press.

Daly, J.M. & Knoche, H.W. (1982). The chemistry and biology of pathotoxins exhibiting host-selectivity. In: *Advances in Plant Pathology*, vol.I., ed. D.S. Ingram & P.H. Williams, pp. 83–138. London: Academic Press.

Daly, J.M., Kono, Y., Suzuki, Y. & Knoche, H.W. (1983). Biological activities and structures of host-selective pathotoxins. In: *Pesticide Chemistry Human Welfare and the Environment*, vol.2, ed. J. Miyamoto & P.C. Kearney, pp. 11–20. New York: Pergamon Press.

Danko, S.J., Kono, Y., Daly, J.M., Suzuki, Y., Takeuchi, S. & McCrery, D.A. (1984). Structure and biological activity of a host-specific toxin produced by the fungal corn pathogen *Phyllosticta maydis*. *Biochemistry*, 23, 759–66.

Daub, M.E. (1982*a*). Cercosporin, a photosensitizing toxin from *Cercospora* species. *Phytopathology*, 72, 370–4.

Daub, M.E. (1982*b*). Peroxidation of tobacco membrane lipids by the photosensitizing toxin, cercosporin. *Plant Physiology*, 69, 1361–4.

Daub, M.E. (1984). A cell culture approach for the development of disease resistance: studies on the phytotoxin cercosporin. *Horticultural Science*, 19, 18–23.

Daub, M.E. & Briggs, S.P. (1983). Changes in tobacco cell membrane composition and structure caused by cercosporin. *Plant Physiology*, 71, 763–6.

Daub, M.E. & Hangarter, R.P. (1983). Light-induced production of singlet oxygen and superoxide by the fungal toxin, cercosporin. *Plant Physiology*, 73, 855–7.

Daub, M.E. & Payne, G.A. (1985). The role of carotenoids in resistance of fungi to cercosporin. Abstracts of the 1985 Meeting of the American Phytopathological Society, Abstract No. 166.

Dewey, R.E., Levings, C.S. & Timothy, D.H. (1986). Novel recombinations in the maize mitochondrial genome produce a unique transcriptional unit in Texas male-sterile cytoplasm. *Cell*, 44, 439–49.

Durbin, R.D. (1981). *Toxins in Plant Disease*. London: Academic Press.

Durbin, R.D. & Uchytil, T.F. (1977). A survey of plant insensitivity to tentoxin. *Phytopathology*, 67, 602–3.

Duvick, J.P., Daly, J.M., Kratky, Z., Macko, V., Acklin, W. & Arigoni, D. (1984). Biological activity of the isomeric forms of *Helminthosporium sacchari* toxin and of homologs produced in culture. *Plant Physiology*, 74, 117–22.

Ellingboe, A.H. (1976). Genetics of host-parasite interactions. In: *Physiological Plant Pathology*, ed. R. Heitefuss & P.H. Williams, pp. 761–78. Berlin: Springer-Verlag.

Fajola, A.O. (1978). Cercosporin, a phytotoxin from *Cercospora* spp. *Physiological Plant Pathology*, 13, 157–64.

Federoff, N.V., Furtek, D.B. & Nelson, O.E. (1984). Cloning of the *bronze* locus in maize by a simple and generalizable procedure using the transposable controlling element Activator (Ac). *Proceedings of the National Academy of Science, U.S.A.*, 81, 3825–9.

Forde, B.G. & Leaver, C.J. (1980). Nuclear and cytoplasmic genes controlling synthesis of variant mitochondrial polypeptides in male-sterile maize. *Proceedings of the National Academy of Science, U.S.A.*, 77, 418–22.

Forde, B.G., Oliver, R.J.C. & Leaver, C.J. (1978). Variation in mitochondrial translation products associated with male-sterile cytoplasms in maize. *Proceedings of the National Academy of Science, U.S.A.*, 75, 3841–5.

Frantzen, K.A. (1985). The binding of the host-specific toxins from *Helminthosporium maydis*, race T and *Phyllosticta maydis* to mitochondria isolated from *Zea mays*. *Ph.D. Dissertation*, University of Nebraska, Lincoln.

Gadian, D.B. (1982). *Nuclear Magnetic Resonance and its Applications to Living Systems*. Oxford: Oxford University Press.

Gardner, J.M., Kono, Y. & Chandler, J.L (1986). Bioassay and host-selectivity of *Alternaria citri* toxins affecting rough lemon and mandarins. *Physiological and Molecular Plant Pathology*, 29, 293–304.

Gardner, J.M., Kono, Y., Tatum, J.H., Suzuki, Y. & Takeuchi, S. (1985a). Structure of the major component of ACRL-toxins, host-specific pathotoxic compounds produced by *Alternaria citri*. *Agricultural and Biological Chemistry*, 49, 1235–8.

Gardner, J.M., Kono, Y., Tatum, J.H., Suzuki, Y. & Takeuchi, S. (1985b). Plant pathotoxins from *Alternaria citri*: the major toxin specific for rough lemon plants. *Phytochemistry*, 24, 2861–7.

Gendloff, E., Pestka, J., Swanson, S. & Hart, L.P. (1984). Detection of T-2 toxin in *Fusarium sporotrichoides*-infected corn by enzyme-linked immunosorbent assay. *Applied and Environmental Microbiology*, 47, 1161–3.

Gilchrist, D.G. (1983). Molecular modes of action. In: *Toxins and Plant Pathogenesis*, ed. J. Daly & B. Deverall, pp. 81–136. Sydney: Academic Press.

Gilchrist, D. & Grogan, R. (1976). Production and nature of a host-specific toxin from *Alternaria alternata* f.sp. *lycopersici*. *Phytopathology*, **66**, 165–71.

Gilchrist, D.G. & Yoder, O.C. (1984). Genetics of host-parasite systems: a prospectus for molecular biology. In: *Plant-Microbe Interactions, Molecular and Genetic Perspectives*, volume 1, ed. T. Kosuge & E. Nester, pp. 69–90. New York: McMillan.

Goodwin, T.W. & Mercer, E.I. (1983). *Introduction to Plant Biochemistry*, 2nd edn., pp. 312–15. Oxford: Pergamon Press.

Gross, M.L., McCrery, D., Crow, F., Tomer, K.B., Pope, M.R., Ciuffetti, L.M., Knoche, H.W., Daly, J.M. & Dunkle, L.D. (1982). The structure of the toxin from *Helminthosporium carbonum*. *Tetrahedron Letters*, **23**, 5381–4.

Gwinn, K.D. & Daub, M.E. (1985). Relationship of fungal cell wall regeneration to cercosporin resistance. Abstract of the 1985 Meeting of the American Phytopathyological Society, Abstract No. 165.

Hahn, M.G., Bonhoff, A. & Grisebach, H. (1985). Quantitative localization of the phytoalexin glyceollin I in relation to fungal hyphae in soybean roots infected with *Phytophthora megasperma* f.sp. *glycinea*. *Plant Physiology*, **77**, 591–601.

Hayani, C., Otaini, H., Nishimura, S. & Kohmoto, K. (1982). Induced resistance in pear leaves by spore germination fluids of nonpathogens to *Alternaria alternata* Japanese pear pathotype and suppression of the induction by AK-toxin. *Journal of the Faculty of Agriculture of Tottori University*, **20**, 9–18.

Hirota, A., Suzuki, A., Aizawa, K. & Tamura, S. (1973). Structure of Cyl-2, a novel cyclotetrapeptide from *Cylindrocladium scoparium*. *Agricultural and Biological Chemistry*, **37**, 955–6.

Holden, M.J. & Sze, H. (1984). *Helminthosporium maydis* T toxin increased membrane permeability to Ca^{2+} in susceptible corn mitochondria. *Plant Physiology*, **75**, 235–7.

Holden, M.J., Colombini, M. & Sze, H. (1985). Channel formation in phospholipid bilayer membranes by toxin of *Helminthosporium maydis*, race T. *Journal of Membrane Biology*, **87**, 151–8.

Humaydan, H.S. & Scott, E.W. (1977). Methomyl insecticide selective phytotoxicity on sweet corn hybrids and inbreds having the Texas male sterile cytoplasm. *Horticultural Science*, **12**, 312–13.

Jerina, D.M. & Daly, J.W. (1974). Arene oxides: a new aspect of drug metabolism. *Science*, **185**, 573–82.

Jigeng, L. & Yi-nong, L. (1983). Chloroplast DNA and cytoplasmic male sterility. *Theoretical and Applied Genetics*, **64**, 231–8.

Kammera, T., Aoyagi, H., Waki, M., Kato, T., Izumiya, N., Noda, K. & Ueno, T. (1981). Synthesis of AM-toxin III and its analogs using the Hoffmann degradation. *Tetrahedron Letters*, **22**, 3625–8.

Kawai, M. & Rich, D.H. (1987). Total synthesis of the cyclic tetrapeptide, HC-toxin. *Tetrahedron Letters*, **24**, 5309–12.

Kawai, M., Rich, D.H. & Walton, J.D. (1983). The structure and conformation of HC-toxin. *Biochemical and Biophysical Research Communications*, **111**, 398–403.

Kim, S.-D. (1986). Chemistry and biological activities of HC-toxin and related cyclic tetrapeptides. *Ph.D. Dissertation*, University of Nebraska, Lincoln.

Kim, S.-D., Knoche, H.W, Dunkle, L.D., McCrery, D.A & Tomer, K.B. (1985). Structure of an amino acid analog of the host specific toxin from *Helminthosporium carbonum*. *Tetrahedron Letters*, **26**, 969–72.

Kimber, A. & Sze, H. (1984). *Helminthosporium maydis* T toxin decreased calcium transport into mitochondria of susceptible corn. *Plant Physiology*, **74**, 804–9.

Klein, R.R. & Koeppe, D.E. (1985). Mode of methomyl and *Bipolaris maydis* (race T) toxin in uncoupling Texas male-sterile cyctoplasm corn mitochondria. *Plant Physiology*, **77**, 912–16.

Klittich, C.J.R. & Bronson, C.R. (1986). Reduced fitness associated with Tox 1 locus of *Cochliobolus heterostrophus. Phytopathology* **76**, 1294–8.

Koeppe, D.E., Cox, J.K. & Malone, C.P. (1978). Mitochondrial heredity: a determinant in the toxic response of maize to the insecticide methomyl. *Science*, **201**, 1227–9.

Kohmoto, K., Kohguchi, T., Kondoh, Y., Otani, H., Nishimura, S., Nakatsuka, S. & Goto, T. (1985). The mitochondrion: the prime site for a host-selective toxin (ACR-toxin I) produced by *Alternaria alternata* pathogenic to rough lemon. *Proceedings of the Japan Academy*, **61B**, 269–72.

Kohmoto, K., Kondoh, Y., Kohguchi, T., Otani, H., Nishimura, S. & Scheffer, R.P. (1984). Ultrastructural changes in host leaf cells caused by host-selective toxin of *Alternaria alternata* from rough lemon. *Canadian Journal of Botany*, **62**, 2485–92.

Kono, Y. & Daly, J.M. (1979). Characterization of the host-specific pathotoxin produced by *Helminthosporium maydis*, race T, affecting corn with Texas male sterile cytoplasm. *Bioorganic Chemistry*, **8**, 391–7.

Kono, Y., Danko, S.J., Suzuki, Y., Takeuchi, S. & Daly, J.M. (1983). Structure of the host-specific pathotoxins produced by *Phyllosticta maydis. Tetrahedron Letters*, **24**, 3803–6.

Kono, Y., Gardner, J.M., Kobayashi, K., Suzuki, Y., Takeuchi, S. & Sakurai, T. (1985*b*). Plant pathotoxins from *Alternaria citri*: stereochemistry of the major and minor toxins. *Phytochemistry*, **25**, 69–72.

Kono, Y., Gardner, J.M., Suzuki, Y. & Takeuchi, S. (1985*c*). Plant pathotoxins from *Alternaria citri*: the minor ACRL toxins. *Phytochemistry*, **24**, 2869–74.

Kono, Y., Gardner, J.M. & Takeuchi, S. (1986). Structure of the host-selective toxins produced by a pathotype of *Alternaria citri* causing brown spot disease of mandarins. *Agricultural and Biological Chemistry*, **50** (in press).

Kono, Y., Knoche, H.W. & Daly, J.M. (1981*a*). Structure: fungal host specific. In: *Toxins in Plant Disease*, ed. R.D. Durbin, pp. 221–57. London: Academic Press.

Kono, Y., Suzuki, Y., Takeuchi, S., Knoche, H.W. & Daly, J.M. (1985*a*). Studies on the host-specific pathotoxins produced by *Helminthosporium maydis*, race T and *Phyllosticta maydis*: absolute configuration of PM-toxins and HmT-toxins. *Agricultural and Biological Chemistry*, **49**, 559–62.

Kono, Y., Takeuchi, S., Kawarada, A., Daly, J.M. & Knoche, H.W. (1980*a*). Structure of the host-specific pathotoxins produced by *Helminthosporium maydis*, race T. *Tetrahedron Letters*, **21**, 1537–40.

Kono, Y., Takeuchi, S., Kawarada, A., Daly, J.M. & Knoche, H.W. (1980*b*). Studies on the host-specific pathotoxins produced by *Helminthosporium maydis*, race T. *Agricultural and Biological Chemistry*, **44**, 2613–22.

Kono, Y., Takeuchi, S., Kawarada, A., Daly, J.M. & Knoche, H.W. (1981*b*). Studies on the host-specific pathotoxins produced in minor amounts by *Helminthosporium maydis*, race T. *Bioorganic Chemistry*, **10**, 206–18.

Kozono, T., Mihara, H., Aoyagi, H., Kato, T. & Izumiya, N. (1984). Cyclic peptides XVII. Synthesis of AM-toxin II by cyclization of linear tetradepsipeptide containing a dehydroalanine residue. *International Journal of Peptide and Protein Research*, **24**, 402–6.

Kraus, G.A., Silveira, M. & Danko, S.J. (1984). Synthesis and testing of analogues of *Helminthosporium maydis* race T toxin. *Journal of Agricultural and Food Chemistry*, **32**, 1265–8.

Kuyama, S. & Tamura, T. (1957). Cercosporin. A pigment of *Cercospora kikuchii* Matsumoto and Tomoyasu. I. Cultivation of the fungus, isolation and purification of pigment. *Journal of the American Chemical Society*, **79**, 5725–6.

Leach, J., Tegtmeier, K., Daly, J.M. & Yoder, O.C. (1982). Dominance at the Tox 1 locus controlling T-toxin production by *Cochliobolus heterostrophus*. *Physiological Plant Pathology*, **21**, 327–33.

Leaver, C.J., Dawson, A.J., Isaac, P., Jones, V.P. & Hack, E. (1985). Structure and expression of plant mitochondrial genes. In: *Molecular Form and Function of the Plant Genome*, ed. L. van Vloten-Doting, G.S.P. Groat & T.C. Hall, pp. 399–411. New York: Plenum.

Lee, S., Aoyagi, H., Shimohigashi, Y., Izumiya, N., Ueno, T. & Fukami, H. (1976). Synthesis of cyclotetradepsipeptides, AM-toxin I and its analogs. *Tetrahedron Letters*, **11**, 843–6.

Leung, P.C., Taylor, W.A., Wang, J.H. & Tipton, C.L. (1985). Role of calmodulin inhibition in the mode of action of ophiobolin A. *Plant Physiology*, **77**, 303–8.

Levings, C.W. (1983). Cytoplasmic male sterility. In *Genetic Engineering of Plants*, ed. T. Kosuge, C.P. Meredith & A. Hollander, pp. 81–92. New York: Plenum.

Liesch, J.M., Sweeley, C.C., Staffeld, G.D., Anderson, M.S., Weber, D.J. & Scheffer, R.P. (1982). Structure of HC-toxin, a cyclic tetrapeptide from *Helminthosporium carbonum*. *Tetrahedron*, **38**, 45–8.

Livingstone, R.S & Scheffer, R.P. (1981). Isolation and characterization of host-selective toxin from *Helminthosporium sacchari*. *Journal of Biological Chemistry*, **256**, 1705–10.

Livingstone, R.S & Scheffer, R.P. (1983). Conversion of *Helminthosporium sacchari* toxin to toxoids by β-galactosidase from *Helminthosporium*. *Plant Physiology*, **72**, 530–4.

Livingstone, R.S & Scheffer, R.P. (1984a). Selective toxins and analogs produced by *Helminthosporium sacchari*. *Plant Physiology*, **76**, 96–102.

Livingstone, R.S. & Scheffer, R.P. (1984b). Toxic and protective effects of analogues of *Helminthosporium sacchari* toxin on sugarcane tissues. *Physiological Plant Pathology*, **24**, 133–42.

Lynch, F.J. & Geoghegan, M.J. (1977). Production of cercosporin by *Cercospora* species. *Transactions of the British Mycological Society*, **69**, 496–7.

Macko, V. (1983). Structural aspects of toxins. In: *Toxins in Plant Pathogenesis*, ed. J.M. Daly & B.J. Deverall, pp. 41–80. Sydney: Academic Press.

Macko, V., Acklin, W., Hildenbrand, C., Weibel, F. & Arigoni, D. (1983). Structure of three isomeric host-specific toxins from *Helminthosporium sacchari*. *Experientia*, **39**, 343–7.

Macko, V., Goodfriend, K., Wachs, T., Renwick, J.A.A., Acklin, W. & Arigoni, D. (1981). Characterization of the host-specific toxins produced by *Helminthosporium sacchari*, the causal organism of eyespot disease of sugarcane. *Experientia*, **37**, 923–4.

Macko, V., Wolpert, T.J., Acklin, W., Jaun, B., Swibl, J., Meili, J. & Arigoni, D. (1985). Characterization of victorin C, the major host-selective toxin from *Cochliobolus victoriae*: structure of degradation products. *Experientia*, **41**, 1366–70.

Macri, F. & Vianello, A. (1981). Inhibition of K^+ uptake and H^+ extrusion caused by non-irradiated cercosporin. *Plant Science Letters*, **11**, 29–37.

Maekawa, N., Yamamoto, M., Nishimura, S., Kohmoto, K., Kuwada, M. & Watanabe, Y. (1984). Studies on host-specific AF-toxins produced by *Alternaria alternata* strawberry pathotype causing Alternaria black spot of strawberry. (1) Production of host-specific toxins and their biological activities. *Annals of the Phytopathological Society of Japan*, **50**, 600–9.

Maeno, S., Kohmoto, K. & Otani, H. (1984). Different sensitivities among apple and pear cultivars to AM-toxin produced by *Alternaria alternata* apple pathotype. *Journal of the Faculty of Agriculture of Tottori University*, 19, 8–19.

Mansfield, J.W. (1984). Plant cell death during infection by fungi. *Society for Experimental Biology*, Series 25, pp. 322–45. Cambridge, U.K.: Cambridge University Press.

Mascagni, P., Pope, M., Gibbons, W.A., Ciuffetti, L.M. & Knoche, H.W. (1983). The backbone and side chain conformations of the cyclic tetrapeptide, HC-toxin. *Biochemical and Biophysical Research Communications*, 113, 10–17.

Matthews, D.E., Gregory, P. & Gracen, V.E. (1979). *Helminthosporium maydis* race T toxin induces leakage of NAD^+ from T cytoplasm corn mitochondria. *Plant Physiology*, 63, 1149–53.

Mayama, S., Matsuura, Y., Iida, H. & Tani, T. (1982). The role of avenalumin in the resistance of oat to crown rust, *Puccinia coronata* f.sp. *avenae*. *Physiological Plant Pathology*, 20, 189–99.

Mayama, S., Tani, T., Ueno, T., Midland, S.L., Sims, J.J. & Keen, N.T. (1986). The purification of victorin and its phytoalexin elicitor activity in oat leaves. *Physiological Plant Pathology*, 29, 1–19.

Mihara, H., Aoyagi, H., Kato, T., Ueno, T. & Izumiya, N. (1983). Synthesis of AM-toxin I analogs containing a lower or higher homolog of L-2-amino-5-(p-methoxyphenyl) pentanoic acid. *Chemistry Letters*, 811–14.

Mihara, H., Aoyagi, H., Lee, S., Waki, M., Kato, T. & Izumiya, N. (1984). Cyclic peptides XVII. Syntheses of AM-toxin I analogs containing lower or higher homologs of the component L-2-amino-5-(p-methoxyphenyl) pentanoic acid residue. *International Journal of Peptide and Protein Research*, 23, 447–53.

Miller, R.M. & Koeppe, D.E. (1971). Southern corn leaf blight: susceptible and resistant mitochondria. *Science*, 173, 67–9.

Mitchell, P. (1979). Compartmentation and communication in living systems. Ligand conduction: a general catalytic principle in chemical osmotic and chemiosmotic reaction systems. *European Journal of Biochemistry*, 95, 1–20.

Mitchell, R.E. (1984). The relevance of non-host-specific toxins in the expression of virulence by pathogens. *Annual Review of Phytopathology* 22, 215–45.

Mumma, R.O., Lukezic, F.L. & Kelly. M.G. (1973). Cercosporin from *Cercospora hayii*. *Phytochemistry*, 12, 917–22.

Nakamura, M. & Ishibashi, K. (1958). On the new antibiotics 'ophiobolin', produced by *Ophiobola miyabeanus*. *Journal of the Agricultural Chemistry Society of Japan*, 32, 739–44.

Nakashima, T., Ueno, T. & Fukami, H. (1982). Structure elucidation of AK-toxins, host-specific phytotoxic metabolites produced by *Alternaria kikuchiana* Tanaka. *Tetrahedron Letters*, 23, 4469–72.

Nakashima, T., Ueno, T., Fukami, H., Taga, T., Masuda, H., Osaki, K., Otani, H., Kohmoto, K. & Nishimura, S. (1985). Isolation and structures of AK-toxin I and II, host-specific phytotoxin metabolites produced by *Alternaria alternata* Japanese pear pathotype. *Agricultural and Biological Chemistry*, 49, 807–15.

Nakatsuka, S., Goto, T., Kohmoto, K. & Nishimura, S. (1986a). Host-specific phytotoxins. In: *Natural Products and Biological Activities, a NATO Foundation Symposium*, ed. H. Imura, T. Goto, T. Murachi & T. Terumi, pp. 11–18. Tokoyo: University of Tokoyo Press.

Nakatsuka, S., Ueda, K., Goto, T., Yamamoto, M., Nishimura, S. & Kohmoto, K. (1986b). Structure of AF-toxin II, one of the host-specific toxins produced by *Alternaria alternata* strawberry pathotype. *Tetrahedron Letters*, in press.

Namiki, F., Okamoto, H., Katou, K., Yamamoto, M., Nishimura, S., Nakatsuka, S., Goto, T., Kohmoto, K., Otani, H. & Novacky, A. (1986a). Studies on host-specific AF-toxins produced by *Alternaria alternata* strawberry pathotype causing alternaria black spot of strawberry (5) effect of toxins on membrane potential of susceptible plants by means of electrophysiological analysis. *Annals of the Phytopathological Society of Japan*, in press.

Namiki, F., Yamamoto, M., Nishimura, S., Nakatsuka, S., Goto, T., Kohmoto, K. & Otani, H. (1986b). Studies on host-specific AF-toxins produced by *Alternaria alternata* strawberry pathotype causing alternaria black spot of strawberry (4) protective effect of pre-treatment with AF-toxin II on AF-toxin I-induced toxic action and fungal infection against strawberry tissues. *Annals of the Phytopathological Society of Japan*, in press.

Nishimura, S. & Kohmoto, K. (1983a). Roles of toxins in pathogenesis. In: *Toxins and Plant Pathogenesis*, ed. J.M. Daly & B.J. Deverall, pp. 137–57. Sydney: Academic Press.

Nishimura, S. & Kohmoto, K. (1983b). Host-specific toxins and chemical structures from *Alternaria* species. *Annual Review of Phytopathology*, 21, 87–116.

Okuno, T., Ishita, Y., Sawaii, K. & Matsumoto, T. (1974b). Characterization of alternariolide, a host-specific toxin produced by *Alternaria mali* Roberts. *Chemistry Letters*, pp. 635–38.

Okuno, T., Ishita, Y., Sugawara, A., Mori, Y., Sawai, K. & Matsumoto, T. (1975). Structure of the biologically active cyclopeptides produced by *Alternaria mali* Roberts. *Tetrahedron Letters*, 5, 335–6.

Okuno, T., Ishita, Y., Nakayama, S., Sawai, K., Fujita, T. & Sawamura, K. (1974a). Isolation of a host-specific toxin produced by *Alternaria mali* Roberts. *Annals of the Phytopathological Society of Japan*, 40, 375–6.

Otani, H., Haramoto, M., Kohmoto, K. & Nishimura, S. (1985a). Two different phases in host cell damage induced by AK-toxin of *Alternaria alternata* Japanese pear pathotype. *Journal of the Faculty of Agriculture of Tottori University*, 20, 8–17.

Otani, H., Kohmoto, K., Nishimu, S., Nakashima, T., Ueno, T. & Fukami, H. (1985b). Biological activities of AK-toxins I and II, host-specific toxins from *Alternaria alternata* Japanese pear pathotype. *Annals of the Phytopathological Society of Japan*, 51, 285–93.

Peron, C., Groth, J.V. & Mylyk, O.M. (1976). Genetic change in host-parasite populations. *Annual Review of Phytopathology*, 14, 177–88.

Pope, M.R., Ciuffetti, L.M., Knoche, H.W., McCrery, D., Daly, J.M. & Dunkle, L.D. (1983). Structure of the host-specific toxin produced by *Helminthosporium carbonum*. *Biochemistry*, 22, 3502–6.

Pope, M., Mascagni, P., Gibbons, W.A., Ciuffetti, L.M. & Knoche, H.W. (1984). Fourier transform IR and NMR studies of hydrogen bonding in *Helminthosporium carbonum* toxin. *Journal of the American Chemical Society*, 106, 3863–5.

Pringle, R.B. & Scheffer, R.P. (1967). Isolation of the host-specific toxin and a related substance with nonspecific toxicity from *Helminthosporium carbonum*. *Phytopathology*, 57, 1169–72.

Rich, D.H., Kawai, M. & Jaronski, R.D. (1983). Conformational studies of cyclic tetrapeptides. *International Journal of Peptide Protein Research*, 21, 35–42.

Rines, H.W. & Luke, H.H. (1985). Selection and regeneration of toxin-insensitive plants from tissue cultures of oats (*Avena sativa*) susceptible to *Helminthosporium victoriae*. *Theoretical and Applied Genetics*, 71, 16–21.

Rudolph, K. (1976). Non-specific toxins. In: *Encyclopaedia of Plant Physiology, vol. 4, Physiological Plant Pathology*, ed. R. Heitefuss & P.H. Williams, pp. 270–315. Berlin: Springer-Verlag.

Scheffer, R.P. (1983). Toxins as chemical determinants of plant disease. In: *Toxins and Plant Pathogenesis*, ed. J.M. Daly & B.J. Deverall, pp. 1–40. Sydney: Academic Press.

Scheffer, R.P. & Livingstone, R.S. (1980). Sensitivity of sugarcane clones to toxin from

Helminthosporium sacchari as determined by electrolyte leakage. *Phytopathology*, **70**, 400–4.

Scheffer, R.P., Nelson, R.R. & Ullstrup, A.J. (1967). Inheritance of toxin production and pathogenicity in *Cochliobolus carbonum* and *Cochliobolus victoriae*. *Phytopathology*, **57**, 1288–91.

Schroter, H. & Novacky, A. (1985). Effect of *Helminthosporium sacchari*-toxin on cell membrane potential of susceptible sugarcane. *Physiological Plant Pathology*, **26**, 165–74.

Shimohigashi, Y., Lee, S., Kato, T. & Izumiya, N. (1978). Cyclic peptides. IV. Synthesis of diastereomeric dihydro-AM-toxin I and its analogs. *Bulletin of the Chemical Society of Japan*, **51**, 584–8.

Shimohigashi, Y., Lee, S., Kato, T., Izumiya, N., Ueno, T. & Fukami, H. (1977). Synthesis and necrotic activity of dihydro-AM-toxin I. *Agricultural and Biological Chemistry*, **41**, 1533–4.

Staehelin, H. & Trippmacher, A. (1974). Cytostatic activity of chlamydocin, a rapid inactivated cyclic tetrapeptide. *European Journal of Cancer*, **10**, 801–8.

Staskawicz, B.J., Panopoulos, N.J. & Hoogenroad, N.J. (1980). Phaseolotoxin-insensitive ornithine carbamoyltransferase of *Pseudomonas syringae* pv. *phaseolicola*: basis for immunity to phaseolotoxin. *Journal of Bacteriology*, **142**, 720–3.

Steele, J.A., Uchytil, T.F., Durbin, R.D., Bhatnagar, P. & Rich, D.H. (1976). Chloroplast coupling factor 1: a species-specific receptor for tentoxin. *Proceedings of the National Academy of Science, U.S.A..*, **73**, 2245–8.

Steinkamp, M.P., Martin, S.S., Hoefert, L.L. & Ruppel, E.G. (1981). Ultrastructure of lesions produced in leaves of *Beta vulgaris* by cercosporin, a toxin from *Cercospora beticola*. *Phytopathology*, **15**, 13–26.

Stossel, A. (1981). Structure and biogenetic relations: fungal nonhost-specific. In: *Toxins in Plant Disease*, ed. R.D. Durbin, pp. 109–219. London: Academic Press.

Strobel, G.A. (1973). Biochemical basis of the resistance of sugarcane to eyespot disease. *Proceedings of the National Academy of Science, U.S.A.*, **70**, 1693–6.

Sugawara, F., Strobel, G., Fisher, L., Van Duyne, G. & Clardy, J. (1985). Bipolaroxin, a selective phytotoxin produced by *Bipolaris cynodontis*. *Proceedings of the National Academy of Science, U.S.A.*, **82**, 8291–4.

Suzuki, Y., Coleman, L.W., Daly, J.M., Kono, Y., Knoche, H.W. & Takeuchi, S. (1987). Synthesis and biological activities of mimics of PM-toxins, the host-specific corn pathotoxins produced by *Phyllosticta maydis*. *Phytochemistry*, **26**, 687–96.

Suzuki, Y., Danko, S.J., Daly, J.M., Kono, Y., Knoche, H.W. & Takeuchi, S. (1983). Comparison of activities of the host-specific toxin of *Helminthosporium maydis*, race T, and a synthetic C_{42} analog. *Plant Physiology*, **73**, 440–4.

Suzuki, Y., Danko, S.J., Kono, Y., Daly, J.M. & Takeuchi, S. (1985). Synthesis of a stereoisomeric mixture of $(\pm)$-PM-Toxin B, a host-specific corn pathotoxin produced by *Phyllosticta maydis*. *Agricultural and Biological Chemistry*, **49**, 149–58.

Suzuki, Y., Danko, S.J., Kono, Y., Takeuchi, S., Daly, J.M. & Knoche, H.W. (1984). Synthesis of $(\pm)$-8,16,26,34-[tetrahydroxy-6,10,14,18,24,28,32,36]- -octaoxohentetracontane as a potent synthetic analog of race T toxin produced by *Helminthosporium maydis*. *Agricultural and Biological Chemistry*, **48**, 2321–9.

Suzuki, Y., Knoche, H.W. & Daly, J.M. (1982a). Analogs of host-specific phytotoxin produced by *Helminthosporium maydis*, race T, I. Synthesis. *Bioorganic Chemistry*, **11**, 300–12.

Suzuki, Y., Tegmeier, K.J., Daly, J.M. & Knoche, H.W. (1982b). Analogs of host-specific phytotoxin produced by *Helminthosporium maydis*, race T: II. Biological activities. *Bioorganic Chemistry*, **11**, 313–21.

Tanis, S.P., Horenstein, B., Scheffer, R.P. & Rasmussen, J.B. (1987). A new host-specific toxin from *Helminthosporium carbonum*. *Heterocycles,* **24**, 3423-31.

Tietjen, K.G., Hammer, D. & Matern, U. (1985). Determination of toxin distribution in *Alternaria* leaf spot diseased tissue by radioimmunoassay. *Physiological Plant Pathology*, **26**, 241-57.

Tsuge, T. & Nishimura, S. (1984). Metabolic regulation of host-specific toxin production in *Alternaria alternata* pathogen. (1) Suppression of toxin production from germinating spores under high temperature stress. *Annals of the Phytopathological Society of Japan*, **50**, 189-96.

Tsuge T., Nishimura, A., Omura, S., Kohmoto, K. & Otani, H. (1985). Metabolic regulation of host-specific toxin production in *Alternaria alternata* pathogens. (2) Suppression of toxin production from germinating spores by chemical treatments. *Annals of the Phytopathological Society of Japan*, **51**, 277-84.

Turner, W.B. (1971). *Fungal Metabolites*. New York: Academic Press.

Ueno, T.M., Nakashima, T. & Fukami, H. (1983). Chemical basis of host recognition by *Alternaria* species. In: *Pesticide Chemistry Human Welfare and the Environment*, vol. 2, ed. J. Miyomoto & P.C. Kearney, pp. 75-80. New York: Pergamon Press.

Ueno,T., Hayashi, Y., Nakashima, T., Fukami, H., Nishimita, S., Kohmoto, K. & Sekiguchi, A. (1975*a*). Isolation of AM-toxin I. A new phytotoxic metabolite from *Alternaria mali*. *Phytopathology*, **65**, 82-3.

Ueno, T., Nakashima, T., Hayashi, Y. & Fukami, H. (1975*b*). Structures of AM-toxin I and II. Host-specific phytotoxic metabolites produced by *Alternaria mali*. *Agricultural and Biological Chemistry*, **39**, 1115-22.

Ueno, T., Nakashima, T., Hayashi, Y. & Fukami, H. (1975*c*). Isolation and structure of AM-toxin III. Host-specific phytotoxic metabolites produced by *Alternaria mali*. *Agricultural and Biological Chemistry*, **39**, 2081-2.

Umehara, K., Nakahara, K., Kiyoto, S., Iwami, M., Okamoto, M., Tanaka, H., Kohsaka, M., Aoki, H. & Imanaka, H. (1983). Studies on WF-3161, a new antitumour antibiotic. *The Journal of Antibiotics*, **36**, 478-83.

Van Alfen, N.K. & McMillan, B.D. (1982). Macromolecular plant-wilting toxins: artifacts of the bioassay? *Phytopathology*, **72**, 132-5.

Van Duuren, V.L., Longseth, L., Goldsmith, B.M. & Orris, L. (1967). Carcinogenicity of epoxides, lactones and peroxy compound VI structure and carcinogenic activity. *Journal of the National Cancer Institute*, **39**, 217-28.

Vaughn, K.C. & Duke, S.O. (1981). Tentoxin-induced loss of plastidic polyphenol oxidase. *Physiologia Plantarum*, **53**, 421-8.

Vaughn, K.C. & Duke, S.O. (1984). Tentoxin stops the processing of polyphenol oxidase into active protein. *Physiologia Plantarum*, **60**, 257-61.

Venkataramani, K. (1985). Isolation of cercosporin from *Cercospora personata*. *Phytopathologische Zeitschrift*, **58**, 379-81.

Vidhyasekaran, P., Borromeo, E.S. & Mew, T.W. (1986). Host-specific toxin production by *Helminthosporium oryzae*. *Phytopathology*, **76**, 261-6.

Walton, J.D. & Earle, E.D. (1983). The epoxide in HC-toxin is required for activity against susceptible maize. *Physiological Plant Pathology*, **22**, 371-6.

Walton, J.D., Earle, E.D. & Gibson, B.W. (1982). Purification and structure of the host-specific toxin from *Helminthosporium carbonum* race 1. *Biochemical and Biophysical Research Communications*, **107**, 785-94.

Walton, J.D., Earle, E.D., Staehelin, H., Grieder, A., Hirota, A. & Suzuki, A. (1985). Reciprocal biological activities of the cyclic tetrapeptides chlamydocin and HC-toxin. *Experientia*, **41**, 348-50.

Wheeler, H. (1977). Ultrastructure of penetration by *Helminthosporium maydis*. *Physiological Plant Pathology*, **11**, 171-8.

Wheeler, H. (1981). Role in pathogenesis. In: *Toxins in Plant Disease*, ed. R.D. Durbin, pp. 477–94. London: Academic Press.

Wheeler, H. & Luke, H.H. (1963). Microbial toxins in plant disease. *Annual Review of Microbiology*, 17, 232–43.

Wolpert, T.J., Macko, V., Acklin, W., Jaun, B., Seibl, J., Meili, J. & Argoni, D. (1985). Structure of victorin C, the major host-selective toxin from *Cochliobolus victoriae*. *Experientia*, 41, 1524–9.

Yamamota, M., Nishimura, S., Kohmoto, K. & Otani, H. (1984). Studies on host-specific AF-toxins produced by *Alternaria alternata* strawberry pathotype causing *Alternaria* black spot of strawberry. (2) Role of toxins in pathogenesis. *Annals of the Phytopathological Society of Japan*, 50, 610–19.

Yamazaki, S. & Ogawa, T. (1972). The chemistry and stereochemistry of cercosporin. *Agricultural and Biological Chemistry*, 36, 1707–17.

Yoder, O.C. (1980). Toxin in pathogenesis. *Annual Review of Phytopathology*, 18, 103–29.

Yoder, O.C. (1981). Assay. In: *Toxins in Plant Disease*, ed. R. Durbin, pp. 45–78. New York: Academic Press.

Yoder, O.C. (1983). Use of pathogen-produced toxins in genetic engineering of plants. In: *Genetic Engineering of Plants*, ed. T. Kosuge, C. Meredith & A. Hollaender, pp. 335–53. New York: Plenum Press.

9

Antifungal substances from herbaceous plants

PAUL J. KUHN[1] and
JOHN A. HARGREAVES[2]

[1]*Shell Research Ltd., Sittingbourne Research Centre, Sittingbourne, Kent ME9 8AG, UK*, [2]*Long Ashton Research Station, Department of Agricultural Sciences, University of Bristol, Bristol BS18 9AF, UK*

Introduction

Antifungal agents may be implicated in both active and constitutive defence, though caution should be exercised in ascribing such a role, where evidence is based solely on crude extracts from plants. This is especially so in instances of putative constitutive defence, where comparisons between healthy and infected plant tissues may be less useful than in cases of active defence. The observation that a crude extract, when incorporated into an agar or liquid medium leads to reduced fungal growth relative to that on, or in an unamended substrate, does not imply the agency of an antifungal substance per se. Rather it might reflect nonspecific alterations in the physiological environment (e.g. pH, osmotic pressure), with consequent adverse effects on growth. The reduction in growth could also be due to a substance(s) that is formed or released as a result of extraction procedures, and which may be inaccessible to an infecting fungus *in vivo*.

It may be helpful to consider briefly some of the basic terminology used. The qualifier 'active' in active defence refers to the fact that this form of defence is only expressed during fungal infection. Generally, antifungal agents are formed from remote precursors via a number of biosynthetic steps. These may depend on enhanced activities and increased rates of synthesis of participating enzymes. Constitutive mechanisms on the other hand, reflect features typical of the intact healthy plant. Here, the antifungal principle either already exists in plant tissues prior to infection, or is formed by simple modification (e.g. hydrolysis or oxidation) of an immediate precursor following tissue damage or attempted ingress by the fungus. The term 'antifungal substance' as used in this chapter is defined as '*any compound present in a plant, before or after infection, which exhibits an inhibitory effect on fungal growth'.* In general, these substances are products of secondary metabolism.

Constitutive defence

Low molecular weight compounds

In view of the diversity and large number of secondary products found in healthy plants, it is not unexpected that some exhibit antifungal activity *in vitro*. Indeed it would seem likely that the list of known constitutive inhibitors would be longer had the appropriate bioassays been carried out during the course of phytochemical investigations. Although many antifungal substances undoubtedly exist in uninfected plant tissues, few have been characterised in full and fewer still can be claimed with justification to have a role in disease resistance.

Table 9.1 presents a number of examples of constitutive antifungal materials from herbaceous plants and, although non-exhaustive illustrates the wide range of structural types found. Chemical classes represented include phenolics (catechol and protocatechuic acid), isoflavones (luteone and wighteone), terpenoids (sclareol and episclareol) and alkaloid and triterpenoid saponins (tomatine and avenacin, respectively). Further examples can be found in reviews by Harborne & Ingham (1978), Ingham (1973), Mansfield (1983) and Schönbeck & Schlösser (1976).

Heterogeneity in chemical structure is paralleled by the range of plant tissues from which constitutive inhibitors have been isolated. The distribution of individual inhibitors within source plants has not been examined systematically in all cases, but where the information is available, it shows that compounds may be more or less widely distributed (Table 9.1). The inhibitor content may also depend on the age of plant involved. It is self-evident that the localisation, distribution, level of activity, and spectrum of antifungal effect of a compound, are critical factors in determining its value in disease resistance. Thus a compound distributed throughout a plant, and exhibiting high intrinsic activity and low specificity, could be of prime importance in protection against fungal infection.

Antifungal activity, specificity, and mode of action

Comparing antifungal activities using values drawn from a wide range of literature sources is difficult particularly where chemically-unrelated inhibitors are involved. Table 9.1 gives some indication of the variation in antifungal activities exhibited by different constitutive inhibitors. All of the compounds appear primarily to exert a direct effect on fungal growth and development. Of the compounds included in Table 9.1, hordatine A and avenacin are probably the most active. The figures for the former compound, however, refer only to effects on spore germination, since hordatine A has no effect in agar plate tests (Stoessl & Unwin, 1970).

Most of the compounds show a low degree of specificity *in vitro*, i.e. their activity is generally not restricted to any particular fungus or group of fungi. An exception, however, is the comparative insensitivity of members of the Pythiaceae to saponins, reflecting the absence of sterols in their membranes. Low specificity may be associated with general or unsophisticated modes of action. This is true of saponins, which by virtue of their surfactant properties and ability to bind sterols are general membranolytic agents. The antifungal activity of the tulipalins may be due to the inactivation of fungal SH-enzymes (Overeem, 1976). The terminal oxidase inhibitor HCN is another example of a constitutive compound that would not be expected to display selectivity.

Despite the low degree of specificity, pathogens of inhibitor-producing plants have been shown to be less sensitive to those inhibitors than other fungi. Examples are *Gaeummanomyces graminis* var. *avenae* to avenacin (Turner, 1961); *Stemphylium loti* to HCN (Millar & Higgins, 1970); and *B. tulipae* to tulipalin B (Schönbeck & Schroder, 1972). In some cases, relative insensitivity may be attributed to the pathogen's ability to metabolise and detoxify the inhibitor.

Role in disease resistance

Constitutive inhibitors are often present in high concentrations where they are localised, or where they are more widely distributed but predominate in particular tissues, e.g. avenacins at 8 µg per root tip (Crombie, Crombie & Whiting, 1985) and tuliposides at up to 32.3% of pistil dry weight (Kämmerer, 1967). This, coupled with their demonstrable antifungal activity, might lead to the conclusion that their *raison d'être* is as resistance mechanisms against disease. Unfortunately, such evidence is convincing in only the minority of cases; the role of catechol and protocatechuic acid against *C. circinans*, tuliposides against *B. tulipae*, and the avenacins against *G. graminis* being the best examples (Walker & Stahmann, 1955; Mansfield, 1980, 1983). It is noteworthy in each case that resistant and susceptible interactions could be compared using closely related plants (coloured v. uncoloured onion bulbs), or parasites *B. cinerea* v. *B. tulipae*, *G. graminis* v. *G. graminis* var. *avenae*. Evidence of an *in vitro* role for most other constitutive inhibitors is much more equivocal. In at least one instance where follow-up studies have been

Table 9.1 *Examples of constitutive inhibitors from herbaceous plants*

Plant	Compound	Source tissue and distribution	Antifungal activity	Reference
Allium cepa (onion)	Catechol and protocatechuic acid	Outer scales of coloured bulbs	Active against *Colletotrichum circinans*, *Diplodia natalensis* and *Botrytis* spp.	Walker & Stahmann (1955)
Lupinus albus (white lupin)	Luteone and wighteone	Leaf surface and within leaves	Radial mycelial growth of *Helminthosporium carbonum* inhibited by luteone with ED_{50} of 99–113 μM	Harborne *et al.* (1976)
Nicotiana glutinosa (tobacco)	Sclareol and episclareol	Distributed throughout aerial parts with high concentrations in leaf surface droplets. Only traces in roots.	Radial mycelial growth of 16 fungi inhibited at 17–331 μM. *Septoria nodorum* and *Phytophthora cinnamomi* insensitive. No effect on spore germination.	Bailey, Vincent & Burden (1974)
Avena sativa (oat)	Avenacin	Roots and particularly root tips.	Growth of 15 fungi completely inhibited at 3–46 μM.	Maizel, Burkhardt & Mitchell (1964)

Hordeum vulgare (barley)	Hordatines A and **B**	Young shoots.	Spore germination of six fungi completely inhibited by hordatine A at 7.5–30 μM.	Stoessl & Unwin (1970)
Lycopersicon esculentum (tomato)	Tomatine	Predominantly in leaves with lower concentrations in stems and roots.	Mycelial growth of 14 fungi completely inhibited at 130–2000 μM.	Arneson & Durbin (1968)
Tulipa gesneriana (tulip)	Tuliposides A and **B**; active forms tulipalins A and **B**	Widely distributed with highest concentrations in flower pistils	Pistils inhibitory to a wide range of fungi. Tulipalin **B** at c.105 μM inhibited mycelial growth of *Botrytis cinerea* and *B. tulipae* by 81 and 4% respectively.	Schönbeck & Schroeder (1972)
Lotus corniculatus (birds foot trefoil)	Linamarin and lotaustralin; active form HCN	Leaves	KCN at 75 μM inhibited spore germination in seven test fungi. *Stemphylium loti* was less sensitive, germination occurring at 185 μM KCN.	Millar & Higgins (1970)

done, e.g. with the white lupin isoflavones luteone and wighteone, the previously suggested involvement in disease resistance was not supported (Hargreaves, Brown & Holloway, 1982). There is clearly much scope for further work on low molecular weight constitutive inhibitors, in both the isolation and characterisation of novel compounds, and in establishing their function *in vivo*. Unfortunately, the study of these inhibitors is currently not an active area of research.

High molecular weight compounds

Several high molecular weight substances might also be termed constitutive inhibitors. Firstly there are polymeric materials that compose the cuticle (Kolattukudy & Köller, 1983) and cell wall (Ride, 1983) both of which represent structural barriers to infection. Lectins, which by virtue of their binding properties may be involved in determining disease reactions (Sequeira, 1978) could also be included in this category. Finally proteins associated with the cell walls of dicotyledonous plants, and shown to inhibit endo-polygalacturonase activity (Albersheim & Anderson-Prouty, 1975), may have an indirect effect on fungal growth and development and, therefore, disease.

Active defence

Active or inducible defence mechanisms result from a confrontation between a host cell or group of host cells and the infection structures or metabolites (elicitors) produced by a fungus attempting infection. The resistance mechanisms lead to the development of a highly localised hostile environment that prevents further colonisation by the fungus. It seems unlikely that a single factor is responsible for the arrest of fungal growth, and it is more probable that a number of related or unrelated metabolites are involved. Central to the concept of active defence is the ability of host cells to perceive the presence of a foreign organism, and following this recognition process, to trigger the co-ordinated organisation of resistance mechanisms.

Low molecular weight compounds (phytoalexins)

Following the pioneering work by Müller & Börger (1940), Müller (1956) subsequently defined phytoalexins as, 'antibiotics which are the result of two different metabolic systems, the host and the parasite, and which inhibit the growth of microorganisms pathogenic to plants.' Phytoalexins have recently been redefined as 'low molecular weight antimicrobial compounds that are both synthesised by and accumulate in plants after exposure to microorganisms' (Paxton, 1980). Although the latter definition makes clear the *de novo* nature of biosynthesis, and the antimicrobial activity of phytoalexins *in vitro*, it does not require that compounds are present at physiologically active concentrations in plant cells or tissues resisting attack, or that their site of accumulation is in the immediate vicinity of the attempted infection. Strictly, therefore, a compound could satisfy the criteria laid down in this definition, without necessarily performing a role in *in vivo* resistance. Notwithstanding this, however, the fact that in many host-parasite interactions phytoalexin accumulation coincides with morphological events such as hypersensitivity and local lesion formation associated with the cessation of fungal growth (Mansfield, 1982, 1983), suggests that in at least some instances phytoalexins are major determinants of resistance.

Space limitation in this chapter precludes a comprehensive treatment of phytoalexins; the discussion has therefore focused on three central questions: (1) can phytoalexin accumulation be considered a general disease resistance mechanism in plants? (2) does the fungus come into contact with the phytoalexin(s) of the host during a resistant interaction? (3) is the amount of phytoalexin to which a fungus is exposed sufficient to prevent the establishment of a parasitic relationship?

For other reviews on phytoalexins see those by Bailey & Mansfield (1982), Kuć (1984), Mansfield (1983, 1986) and Stoessl (1983).

Phytoalexin accumulation as a general resistance mechanism in plants
Although phytoalexins do not appear to be produced by nonvascular plants, and occur only rarely in gymnosperms (see Chapter 10), it is tempting to speculate that phytoalexin accumulation is ubiquitous among angiosperms. Although the search has not been exhaustive, representatives from seventeen families of higher plants, mostly herbaceous species, have now been shown to exhibit a phytoalexin response (Coxon, 1982; Stoessl, 1983). Furthermore the Gramineae, previously believed to lack induced antifungal compounds, have recently been shown to include phytoalexin-producing species (Mayama *et al.*, 1981; Kono *et al.*, 1985).

Phytoalexins are secondary metabolites, mainly synthesised via three metabolic pathways: the acetate-malonate, the acetate-mevalonate, and the shikimate pathways (Stoessl, 1980, 1982), according to the compound and the plant species involved. In many cases phytoalexins accumulate in plants exposed to various types of stress in the absence of infection (Bailey, 1982*a*). This feature has led to the suggestion that phytoalexin synthesis may be part of a co-ordinated response to injury (Ward, 1986), representing an integral part of a wound healing or cellular repair mechanisms following cell death or damage.

Although members of a particular plant family typically produce phytoalexins of a single structural type, phytoalexins as a whole exhibit a wide diversity of structure in different families (see Bailey & Mansfield, 1982). In addition to this diversity in basic structure, there is also considerable variation in the types and positions of substituent groups.

Exceptions to the above, however, are species of the genera *Lens* and *Vicia*. Their principal induced antimicrobial compounds are furanoacetylenic derivatives (Hargreaves, Mansfield & Rossall, 1977; Robeson & Harborne, 1980), rather than isoflavonoids characteristic of most members of the Leguminosae (Ingham, 1982). In the case of *V. faba*, the isoflavonoid phytoalexin medicarpin is produced together with the furanoacetylenic compounds albeit at low levels (Hargreaves, Mansfield & Coxon, 1976). This raises the interesting possibility that isoflavonoid phytoalexin accumulation in this legume may have been superceded in favour of furanoacetylenic compounds. A particular compound may occur as a phytoalexin in one plant but as a constitutive inhibitor in another. For example, medicarpin and maackiain, two common isoflavonoid phytoalexins, are also found as preformed metabolites in the heartwood of some tropical legumes (Wong, 1975).

Accumulation and localisation of phytoalexins

Detailed studies on the distribution of phytoalexins in infected plant tissues have shown that phytoalexin production and accumulation are highly localised events (Sato & Tomiyama, 1969; Sato, Kitazawa & Tomiyama, 1971; Mansfield & Deverall, 1974; Yoshikawa, Yamauchi & Masago, 1978). Their synthesis and accumulation are closely associated, both temporally and spatially, with the appearance of hypersensitive cell death and necrosis (Bailey, Rowell & Arnold, 1980; Mansfield, 1980; Bell *et al.*, 1984; Cramer *et al.*, 1985) and also with the *in vivo* distribution of infection hyphae (Yoshikawa *et al.*, 1978; Hahn, Bonhoff & Grisebach, 1985). These correlations strongly suggest that phytoalexin synthesis is a co-ordinated response associated with cell damage, and secondly, that the accumulation of phytoalexins has a role in impeding fungal growth. The results of Hahn *et al.* (1985) however, show that phytoalexins may not be solely responsible for the initial inhibition of *Phytophthora megasperma* f.sp. *glycinea* in the resistant reaction in soybean.

It is difficult to envisage how rapidly dying cells could possess the necessary degree of organisation or integrity to mobilise the complex biosynthetic pathways required for phytoalexin synthesis. Moreover, in a number of cases it appears that cells are killed by the infecting fungus before the onset of phytoalexin accumulation (Tomiyama *et al.*, 1979; Bailey, 1982*b*) (Fig 9.1). A more likely site for phytoalexin synthesis, therefore, is in live cells immediately adjacent to those cells that have died or are damaged (Bailey, 1982*b*). Microspectrofluorimetric studies of

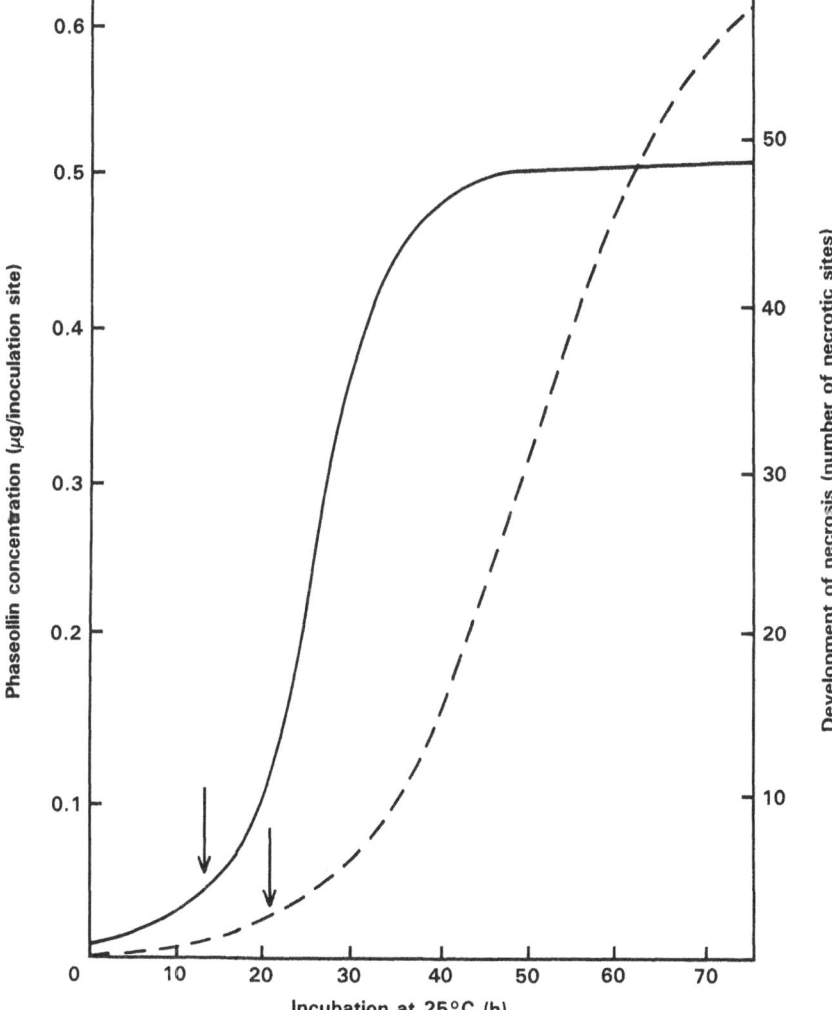

Fig. 9.1 The development of host cell necrosis (—) and the accumulation of phaseollin (--) during the temperature-dependent hypersensitive reaction between French bean (*Phaseolus vulgaris*) and *Colletotrichum lindemuthianum*. Arrows indicate when necrotic cells were visible (—) and when phaseollin was detected (--). (For details see Bailey, Rowell & Arnold, 1980.)

the epidermis of *V. faba* have demonstrated the presence of wyerone acid and other wyerone derivatives in live cells at the edge of necrotic lesions (Mansfield, Hargreaves & Boyle, 1974; Mansfield, 1980) (Fig. 9.2). The synthesis of isoflavonoid and sesquiterpenoid phytoalexins in French bean (*Phaseolus vulgaris*) (Hargreaves & Bailey, 1978) and potato (*Solanum tuberosum*) (Nakajima, Tomiyama & Kinukawa, 1975; Sakai, Tomiyama & Doke, 1979) respectively, also appears to conform to this pattern. In considering the location of phytoalexin synthesis within infected tissues, it should be remembered that resistance is a dynamic event. Thus, cells that respond initially to infection by producing phytoalexins, may ultimately die, either through the action of the fungus or as a consequence of the innate phytotoxicity of the phytoalexins themselves (Lyon & Mayo, 1978; Hargreaves, 1980).

Phytoalexin accumulation in cells surrounding those initially damaged by the fungus could itself provide an effective barrier to further colonisation of host tissues. During a resistant response, however, infection hyphae are normally restricted to hypersensitive cells (Fig. 9.3) or necrotic tissues. In addition, during the early stages of infection, it is uncommon for hyphae to penetrate the plasmalemma of host protoplasts. Hyphae, for example haustoria of rusts and mildews either occupy the paramural space between the plasmalemma and cell wall, or grow in the cell walls or intercellular spaces (Hancock & Huisman, 1981). Unless the cells producing phytoalexins therefore die and 'leak' the newly synthesised compounds, some form of export from the cells must be essential.

The ability to detect many phytoalexins by the drop diffusate technique (Bailey, 1982a) and in at least one case in the absence of cell death (Paxton, Goodchild & Cruikshank, 1974), illustrates that phytoalexins can move out from the cells in which they are produced. Transport from live cells most probably accounts for the fact that the highest concentrations of phytoalexin are often found in neighbouring necrotic cells (Hargreaves & Bailey, 1978; Ersek, Holliday & Keen, 1982; Mayama & Tani, 1982).

Although many phytoalexins are able to diffuse into inoculum droplets, the broad bean phytoalexin precursor, wyerone, cannot be detected in inoculum droplets from leaves or pods (Mansfield *et al.*, 1974; Hargreaves *et al.*, 1977). Wyerone may be adsorbed onto the plant cell wall (Hargreaves *et al.*, 1977), and be subsequently converted to the more active and water-soluble wyerone acid by an extracellular esterase produced by the infecting fungus. The adsorption of phytoalexins on cell walls (Hargreaves & Bailey, 1978) could be important where the infecting fungus

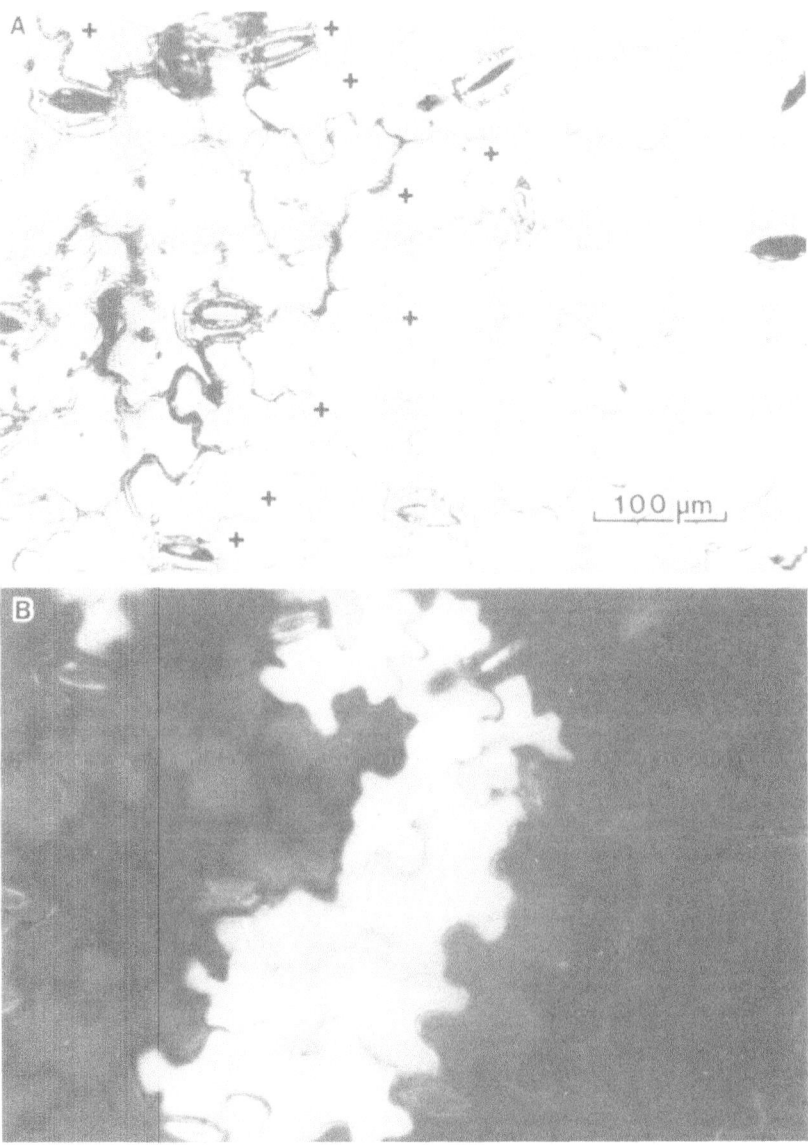

Fig. 9.2. Fluorescence indicating the production of the phytoalexin, wyerone acid, by live cells in the epidermis of a broad bean (*Vicia faba*) leaf infected with *Botrytis fabae*. An epidermal strip was viewed under tungsten illumination (A) and with transmitted UV light (B). In (A) the fluorescent cells are marked with crosses. The infection hypha produced by *B.faba* was at the centre of the lesion, to the left of the micrograph. (From Mansfield, 1980. Courtesy of J.W. Mansfield, Biological Sciences Department, Wye College, UK. Reproduced by permission of Academic Press Inc. (London) Limited).

penetrates or grows within cell walls. This would be particularly true for phytoalexins like kievitone, that can inactivate extracellular cell wall degrading enzymes (Bull & Smith, 1981). Induced structural changes, e.g. cell wall appositions and encapsulations, at sites of fungal penetration, could also be important regions for the accumulation and adsorption of phytoalexins.

The fact that many phytoalexins presumably in water soluble form, can be detected in drop diffusates outside plant cells, might suggest that these compounds could move systemically within the host plant. This does not appear to be so. The inability to detect phytoalexins at sites remote from the infection court could be due to their irreversible binding to plant cells at the locus of infection, or to their metabolism by these or closely adjacent cells. Both mechanisms would lower the amount of phytoalexin to which surrounding tissues might be exposed, and phytotoxicity would thereby be reduced.

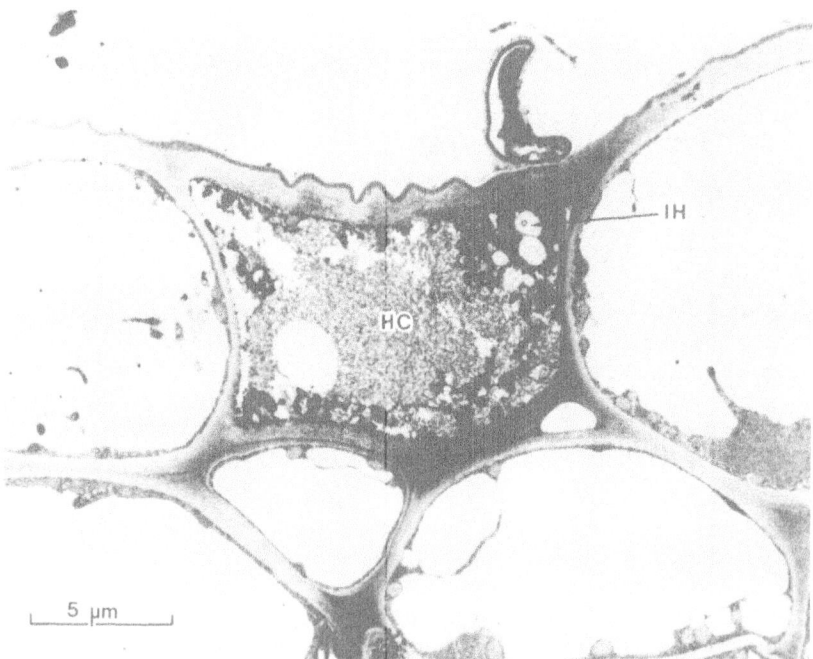

Fig. 9.3. A transmission electron micrograph of a hypersensitive epidermal cell (HC) from a hypocotyl of French bean (*Phaseolus vulgaris*) infected by an incompatible race of *Colletotrichum lindemuthianum*. The invading hypha (IH) is restricted within the hypersensitive cell, and does not penetrate the surrounding live cells. (Courtesy of R.J. O'Connell, Long Ashton Research Station, University of Bristol, UK.)

It is clear that for a phytoalexin to have a significant role in resistance, it must eventually interact with hyphae of the infecting fungus. Circumstantial evidence of this is provided when localised accumulation of phytoalexins precedes or coincides with the cessation of fungal growth. The isolation of phytoalexin metabolites, resulting from intracellular fungal enzyme activities is additional definitive evidence that phytoalexin and fungus come into contact (Stoessl, Unwin & Ward, 1973; Van Etten & Smith, 1975; van den Heuvel & Grootveld, 1980; Mansfield, 1980).

Toxicity of phytoalexins in vivo

Although many types of assay system have been used to determine the antifungal activities of phytoalexins it is unlikely that the conditions at the infection site can be completely reproduced by any bioassay *in vitro*. In plant tissues resisting attack, for example, the fungus is exposed to increasing concentrations of a number of fungitoxic compounds, whereas most bioassays *in vitro* typically employ only a single phytoalexin at a specific concentration. *In vivo*, it is possible that exposure to initially low non-inhibitory levels of toxicant allows the fungus to adapt and grow in the presence of phytoalexin concentrations shown by bioassay *in vitro* to be completely inhibitory (Skipp & Carter, 1978). Non-lethal levels of phytoalexin might also allow the induction of detoxifying enzymes by the fungus (Van Etten, Matthews & Smith, 1982). A related problem is that in bioassays *in vitro* phytoalexin concentrations may decrease with time due either to metabolism or adsorption of the compound by the fungus (Skipp & Bailey, 1976). The composition and pH of the bioassay medium may also influence phytoalexin toxicity (Deverall & Rogers, 1972). It is thus clear that bioassays *in vitro* can give only an approximation of the concentrations of a phytoalexin required to limit fungal growth in plant tissues.

The type of toxicity exhibited by a phytoalexin, i.e. fungistatic or fungicidal is an important consideration when discussing activity *in vivo*. A number of phytoalexins have been shown to kill individual fungal cells (Skipp & Bailey, 1976; Harris & Dennis, 1977; Rossall, Mansfield & Hutson, 1980), while others cause only a temporary cessation of growth (Higgins, 1978). Even where fungi are exposed to apparently fungicidal levels of a phytoalexin, some cells may survive and subsequently regrow (Skipp & Bailey, 1976; Rossal et al., 1980) (Fig. 9.4). This may reflect adaptation in the fungus, perhaps by the induction of an energy-dependent efflux system (cf. de Waard & Van Nistelrooy, 1980), or by a reduction in the permeability of the fungal cell to the toxicant.

In some cases where a phytoalexin has been found to be fungicidal *in vitro* it also appears to kill the fungus *in vivo*. For instance, Rossall *et al.* (1980) showed that wyerone acid and related compounds can kill germinated sporelings of *Botrytis cinerea in vitro* and that most of the infection hyphae of this pathogen were dead in leaf epidermal cells accumulating high levels of these compounds. By contrast, although phaseollin is fungicidal towards germlings of *Colletotrichum lindemuthianum* (Skipp & Bailey, 1976) this fungus remains viable though restricted in hypersensitive cells (Bailey & Rowell, 1980) containing an estimated phaseollin level greater than 3000 μg cm^{-3} cell volume (Bailey & Deverall, 1971); a quantity far in excess of the amount required to kill the fungus *in vitro*. In this case it would appear that the pathogen

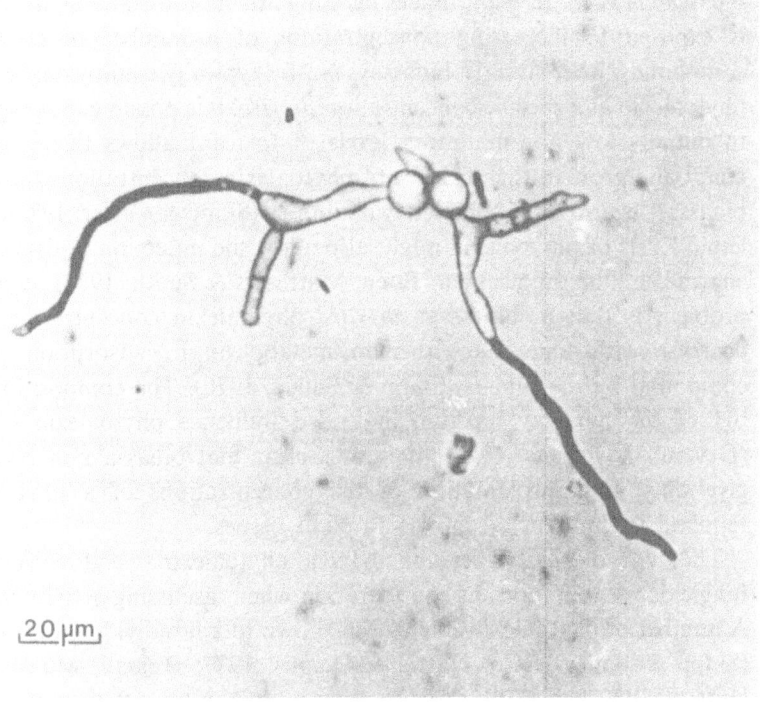

20 μm

Fig. 9.4. Sporelings of *Botrytis cinerea* treated with the phytoalexin wyerone acid at 15 μg/ml and stained with trypan blue; dead cells have accumulated the stain. The apical cells have been killed and regrowth has occurred from the sub-apical cells and also one conidium. (From Rossall, Mansfield & Hutson, 1980). (Courtesy of J.W. Mansfield, Biological Sciences Department, Wye College, UK. Reproduced by permission of Academic Press Inc. (London) Limited).

has become tolerant of the phytoalexin through adaptation, or that the phytoalexin is not wholly accessible to the fungus. The finding that *Aphanomyces euteiches*, a fungus sensitive to the phytoalexin pisatin *in vitro*, is able to cause spreading lesions in pea stems despite the presence of high concentrations of pisatin (c.800 μg cm^{-3} lesion tissue) (Pueppke & Van Etten, 1974) is further evidence that the accumulation of phytoalexins in infected plants to levels that prevent growth *in vitro* does not necessarily indicate that a similar concentration will lead to the arrest of fungal growth *in vivo*.

Although there is no general correlation between pathogenicity and the ability of a pathogen metabolically to detoxify a phytoalexin, there is some recent evidence to suggest that such a relationship may exist. It has been shown that isolates of *Nectria haematococca* that are able to demethylate and tolerate pisatin (Fig. 9.5) are also virulent pathogens of pea. Conversely, those isolates that are unable to detoxify pisatin are sensitive to the phytoalexin and are avirulent (Van Etten *et al.*, 1980). Genetic investigations showed that these traits (pisatin demethylation, pisatin tolerance and virulence) were inherited together following crosses between non-virulent and virulent isolates (Tegtmeier & Van Etten, 1982).

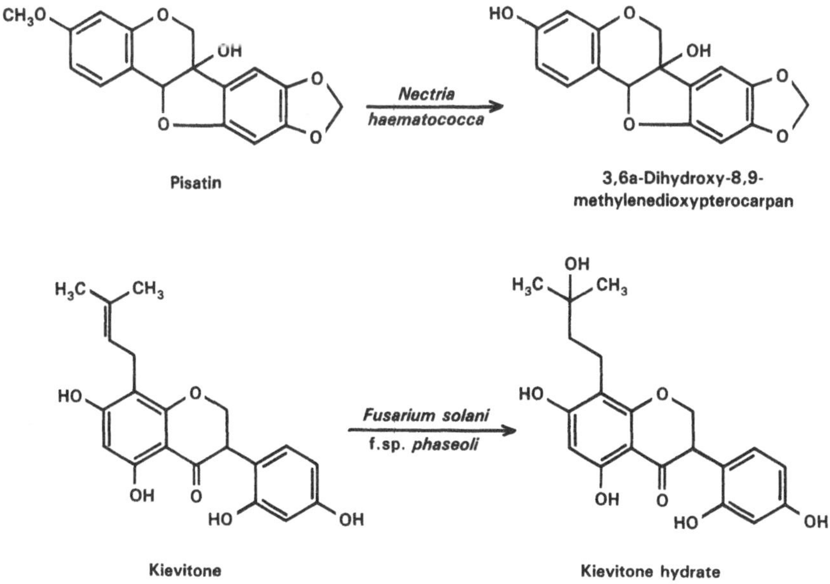

Pisatin

Nectria haematococca →

3,6a-Dihydroxy-8,9-methylenedioxypterocarpan

Kievitone

Fusarium solani f.sp. *phaseoli* →

Kievitone hydrate

Fig. 9.5. Metabolic detoxification of the phytoalexin pisatin (Matthews & VanEtten, 1981) and kievitone (Kuhn, Smith & Ewing, 1977) by *N. haematococca* and *F.solani* f.sp. *phaseoli*, respectively.

Smith, Harrer & Cleveland (1982) examined the pathogenicity of a range of isolates of *Fusarium solani* towards French bean, and the ability of the same isolates to produce the enzyme kievitone hydratase (Kuhn & Smith, 1979) which detoxifies the bean phytoalexin kievitone by metabolism to kievitone hydrate (Kuhn, Smith & Ewing, 1977) (Fig. 9.5). Although kievitone hydratase could be formed by all of the isolates tested, only those that secreted the enzyme extracellularly were highly virulent on *P. vulgaris*. This correlation suggests a possible link between pathogenicity and production of extracellular kievitone hydratase.

The discussion on phytoalexins presented here highlights the difficulty in extrapolating from results of bioassays *in vitro* to the interaction of a host with a potential pathogen. In order to demonstrate a role for phytoalexins in disease resistance it is necessary to conduct detailed and often complex experiments. Although to date only a few host-parasite interactions have been studied in sufficient depth, it is clear that phytoalexins have a key role in the resistance of certain plants to fungal pathogens.

High molecular weight compounds

A number of high molecular weight compounds found in plant tissues following infection may have an inhibitory effect on fungal growth and may, therefore, function in active defence. Such substances include lignin, suberin, oxidised polyphenolic compounds, glycoproteins and enzymes. Unlike phytoalexins, satisfactory methodology for the extraction and purification of some of these materials is not yet available, making difficult an assessment of their effects on fungal growth and development *in vitro*. Furthermore, because the precise chemical nature of a number of these polymers is unknown, it is difficult to make quantitative measurements or to determine their localisation.

Among the induced polymeric compounds, the accumulation of 'lignin-like' substances appears to be a common response of herbaceous plants during the expression of resistance (Vance, Kirk & Sherwood, 1980; Ride, 1983). The deposition of lignin is a highly localised event, often associated with modifications to the host cell wall at the initial infection site (Sherwood & Vance, 1976; Ride & Pearce, 1979). Lignin is probably laid down as an apposition layer on the cell wall polymers, perhaps including those composing the walls of infection hyphae. It may also be associated with polymers (e.g. callose) that are deposited as cell wall appositions or papillae beneath penetration sites (Ride & Pearce, 1979). Finally, in some instances there may be lignification of whole cells and their contents, particularly where cells react to infection in a hypersensitive manner (Maule & Ride, 1982; Beardmore, Ride & Granger, 1983).

In vivo, lignin could influence the progress of a fungus in a number of ways. It might for example be deposited in the cell wall at the tip of an infection hypha, reducing the elasticity and plasticity of the wall, and thereby directly preventing hyphal extension (Ride, 1978). In addition the low molecular weight precursors (e.g. coniferyl alcohol) of lignin, which are themselves antifungal, may accumulate at sites of infection and function as phytoalexins (Hammerschmidt & Kuć, 1982). As well as these direct effects, lignin might impede fungal growth by forming a physical barrier to the movement of nutrients and water.

Not all structural changes associated with resistance involve lignin deposition however. For instance, lignin does not accumulate in barley leaf epidermal cells where the infection hyphae of *Leptosphaeria nodorum* are restricted to the outer layers of the cell wall (Keon & Hargreaves, 1984) (Fig. 9.6). In this case it appears that the major components of modified cell walls are unidentified phenolic materials in association with glycoproteins.

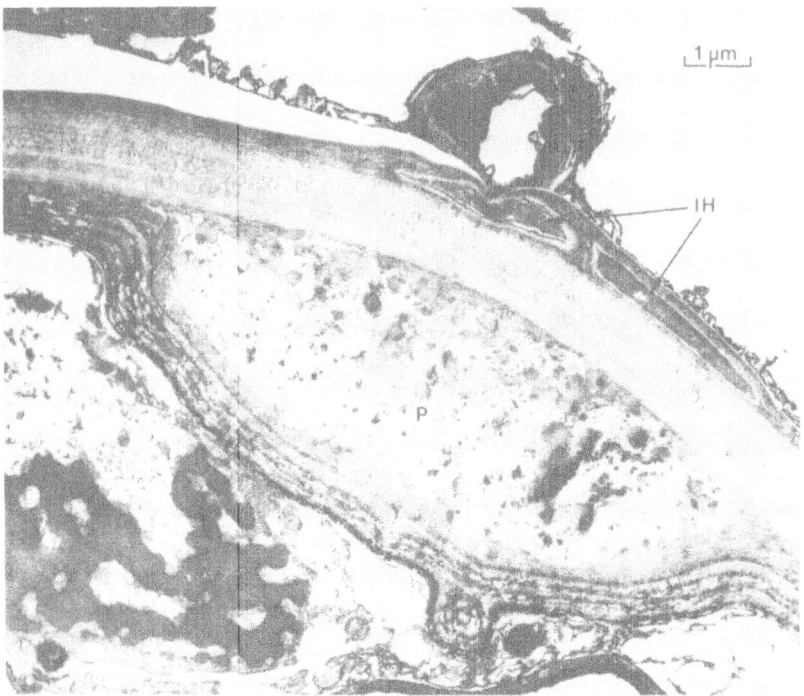

Fig. 9.6. A transmission electron micrograph of an unsuccessful attempted penetration of a barley (*Hordeum vulgare*) leaf epidermal cell by *Leptosphaeria nodorum*. Although a cell wall apposition (papilla, P) has been deposited beneath the site of attempted penetration, the infection hyphae (IH) are restricted to the outer layers of the cell wall.

Another class of macromolecules, hydroxyproline-rich glycoproteins associated with cell walls, accumulate in tissues following fungal infection (Esquerré-Tugayé & Lamport, 1979; Hammerschmidt, Lamport & Muldoon, 1984), and in elicitor-treated plant cell suspension cultures (Bolwell, Robbins & Dixon, 1985). Although no direct effect of these compounds on fungal growth has so far been demonstrated, they might function as agglutinins (Touzé & Esquerré-Tugayé, 1982). Alternatively, their role in resistance may reflect their ability to act as matrices upon which lignin or other substances are deposited (Roberts & Northcote, 1972; Whitmore, 1978).

As well as polymers that are deposited in association with cell walls, disease development may also be influenced by some of the enzymic activities that increase following attempted fungal invasion. The highly complex nature of the interrelationships probably involved in active defence mechanisms is illustrated by one example of this type, the accumulation in *Fusarium solani*-infected pea tissues of glycosidic enzymes that can degrade components of hyphal cell walls (Nichols, Beckman & Hadwiger, 1980). Chitosan, one of the fragments released by the plant glycosidic enzyme, chitinase, is not only a potent inducer of the pea phytoalexin, pisatin, but it is also itself toxic to macroconidia of *F. solani* (Hadwiger & Beckman, 1980; Hadwiger, Fristensky & Riggleman, 1984). This example emphasises the complexity of interrelated biochemical events that may be typical of incompatible interactions between plants and fungi.

Phytoalexins as exogenous plant protectants

A limited number of studies have explored the possibilities of using naturally occurring inhibitors as fungicides. Sclareol at 100 μg cm^{-3} gave better than 90% control of rust diseases on French bean, broad bean and wheat (Bailey *et al.*, 1975), while the pepper phytoalexin, capsidiol at 118 μg cm^{-3} reduced by nearly 90% the number of lesions on tomato caused by *Phytophthora infestans* (Ward, Unwin & Stoessl, 1975). Although these figures look impressive, by comparison with conventional systemic fungicides, disease control required high concentrations of the active compound. Furthermore, in a study by Rathmell & Smith (1980), a range of seven isoflavonoid phytoalexins at concentrations up to 100 μg cm^{-3} failed to give useful control of six diseases. Likewise, analogues of the cowpea phytoalexin, vignafuran, exhibited only limited activity when applied as foliar protectant sprays at 100 μg cm^{-3} against chocolate spot and rust of broad bean, and powdery mildew on wheat (Carter,

Chamberlain & Wain, 1978). As well as having low *in vivo* activity, phytoalexins or constitutive inhibitors would lack systematicity and present possible problems with phytotoxicity.

In a limited number of cases conventional fungicides may function through the control of endogenous anti-fungal substances, particularly with some indirectly acting fungicides which are either inactive *in vitro*, e.g. fosetyl-A1 (Bompeix *et al.*, 1980), or where activity *in vitro* is insufficient to account for the observed levels of disease control e.g. the experimental fungicide 2,2-dichloro-3,3-dimethylcyclopropane carboxylic acid (Langcake, Cartwright & Ride, 1981). For both these fungicides, disease control is thought to involve the participation of host defence mechanisms including the formation of phytoalexins. In some instances phytoalexins have also been shown to be produced following treatment with directly acting fungicides (Ward, 1984), though their role under these circumstances is more difficult to establish unequivocally.

Concluding remarks

In 1976 Overeem claimed: 'It is not very difficult to discuss antimicrobial substances in plants because there are so many. It is much more difficult to discuss the role that these substances may play in resistance of plants to fungal or bacterial attack because so much has been suggested and so little has been convincingly proved'. Although the situation has changed somewhat in the intervening years, it might be argued that the sentiment contained in this view is still largely valid today.

The use of appropriate bioassays, together with detailed *in vivo* studies of fungal development related to localisation and accumulation of antifungal compounds should elucidate the role of these substances in disease resistance.

In order to establish more firmly a role *in vivo*, a number of approaches can be adopted, some of which are summarised here:

(i) Comparison of disease reactions in plant cultivars or closely related plant species that contain or produce differing levels of inhibitor. This approach has been used to assess the role of phytoalexins in race-specific resistance (Bailey & Deverall, 1971; Yoshikawa *et al.*, 1978) and in the study of constitutive inhibitors from woody plants (Noveroske, Kuć & Williams, 1964; Zaki *et al.*, 1980). A similar method can be employed where inhibitor content changes with the stage of plant development (Schönbeck, 1976).

(ii) Application of antibiotics, e.g. blasticidin S (Yoshikawa, 1978), or other chemicals, e.g. glyphosate (Keen, Holliday & Yoshikawa,

1982), artificially to manipulate host–parasite interactions by suppressing a normally incompatible response. Results must be interpreted with caution, however, since the antimetabolites used to date would probably have multiple biochemical effects. The availability of specific inhibitors of the biosynthesis of antifungal compounds would permit more definitive experiments to be undertaken. Host-parasite interactions may also be modified by environmental factors, e.g. temperature (Bailey, 1974; Ward & Lazarovits, 1981), providing another useful means of studying the role of antifungal compounds in disease resistance.

(iii) Selection of fungal mutants with altered virulence, and sensitivity to the relevant antifungal compounds (Hutson & Mansfield, 1980; Defago & Kern, 1983; Defago, Kern & Sedlar, 1983). If enhanced virulence is positively correlated with decreased sensitivity, then this would support a role for the inhibitor in disease resistance. Clearly, however, where mutants differ from the wild-type in several respects, conclusions must be equivocal. A related method involves comparison of naturally occurring fungal isolates in terms of virulence, sensitivity to phytoalexin, and ability metabolically to detoxify that phytoalexin (see under 'Toxicity of phytoalexins *in vivo*'). Where crosses can be made, relationships between these parameters may be studied by genetic analysis.

(iv) Transfer from a pathogen to a non-pathogen of the gene(s) enabling inhibitor detoxification, with assessment of consequent effects on disease reaction. Should the transformation confer pathogenicity towards the inhibitor-producing plant, this would imply that the inhibitor has a vital function in natural disease.

By applying techniques such as these, either singly or in combination, it should be possible to provide convincing evidence for or against a role in disease resistance. In this way, the deficiency highlighted in the quotation given earlier hopefully will be redressed.

References

Albersheim, P. & Anderson-Prouty, A.J. (1975). Carbohydrates, proteins, cell surfaces, and the biochemistry of pathogenesis. *Annual Review of Plant Physiology,* **26,** 31–52.

Arneson, P.A. & Durbin, R.D. (1968). The sensitivity of fungi to α-tomatine. *Phytopathology,* **58,** 536–7.

Bailey, J.A. (1974). The relationship between symptom expression and phytoalexin concentration in hypoctyls of *Phaseolus vulgaris* infected with *Colletotrichum lindemuthianum. Physiological Plant Pathology,* **4,** 477–88.

Bailey, J.A. (1982a). Mechanisms of phytoalexin accumulation. In: *Phytoalexins*, ed. J.A. Bailey & J.W. Mansfield, pp. 289–318. Glasgow: Blackie.

Bailey, J.A. (1982b). Physiological and biochemical events associated with expression of resistance to disease. In: *Active Defence Mechanisms in Plants*, ed. R.K.S. Wood, pp. 39–65. New York: Plenum Press.

Bailey, J.A., Carter, G.A., Burden, R.S. & Wain, R.L. (1975). Control of rust diseases by diterpenes from *Nicotiana glutinosa*. *Nature*, **255**, 328–9.

Bailey, J.A. & Deverall, B.J. (1971). Formation and activity of phaseollin in the interaction between bean hypocotyls (*Phaseolus vulgaris*) and physiological races of *Colletotrichum lindemuthianum*. *Physiological Plant Pathology*, 1, 435–49.

Bailey, J.A. & Mansfield, J.W. (1982). *Phytoalexins*. Glasgow: Blackie.

Bailey, J.A. & Rowell, P.M. (1980). Viability of *Colletotrichum lindemuthianum* in hypersensitive cells of *Phaseolus vulgaris*. *Physiological Plant Pathology*, 17, 341–5.

Bailey, J.A., Rowell, P.M. & Arnold, G.M. (1980). The temporal relationship between host cell death, phytoalexin accumulation and fungal inhibition during hypersensitive reactions of *Phaseolus vulgaris* to *Colletotrichum lindemuthianum*. *Physiological Plant Pathology*, 17, 329–39.

Bailey, J.A., Vincent, G.G. & Burden, R.S. (1974). Diterpenes from *Nicotiana glutinosa* and their effect on fungal growth. *Journal of General Microbiology*, **85**, 57–64.

Beardmore, J., Ride, J.P. & Granger, J.W. (1983). Cellular lignification as a factor in the hypersensitive resistance of wheat to stem rust. *Physiological Plant Pathology*, **22**, 209–20.

Bell, J.N., Dixon, R. A., Bailey, J.A., Rowell, R.A. & Lamb, C.J. (1984). Differential induction of chalcone synthase mRNA activity at the onset of phytoalexin accumulation in compatible and incompatible plant-pathogen interactions. *Proceedings of the National Academy of Science, USA*, **81**, 3384-8.

Bolwell, G.P., Robbins, M.P. & Dixon, R.A. (1985). Metabolic changes in elicitor-treated bean cells. Enzymic responses associated with rapid changes in cell wall components. *European Journal of Biochemistry*, **148**, 571–8.

Bompeix, G., Ravisé, A., Raynal, G., Fettouche, F. & Durand, M.C. (1980). Modalités de l'obtention des nécroses bloquantes sur feuilles détachées de tomate par l'action du tris-O-éthyl phosphonate d'aluminium (phosethyl d'aluminium), hypothèses sur son mode d'action *in vivo*. *Annales de Phytopathologie*, **12**, 337–51.

Bull, C. A. & Smith, D. A. (1981). Pectic enzyme inhibition by the phytoalexin kievitone. *Phytopathology*, **71**, 206 (Abstr.).

Carter, G.A., Chamberlain, K. & Wain, R.L. (1978). Investigations on fungicides. XX. The fungitoxicity of analogues of the phytoalexin 2-(2'-methoxy-4'-hydroxyphenyl)-6-methoxybenzofuran (vignafuran). *Annals of Applied Biology*, **88**, 57–64.

Coxon, D.T. (1982). Phytoalexins from other families. In: *Phytoalexins*, ed. J.A. Bailey & J.W. Mansfield, pp. 106–32. Glasgow: Blackie.

Cramer, C.L., Bell, J.N., Ryder, T.B., Bailey, J.A., Schuch, W., Bolwell, G.P., Robbins, M.P., Dixon, R.A. & Lamb, C.J. (1985). Co–ordinated synthesis of phytoalexin biosynthetic enzymes in biologically-stressed cells of bean (*Phaseolus vulgaris* L.). *EMBO Journal*, **4**, 285–9.

Crombie, L., Crombie, W.M.L. & Whiting, D.A. (1985). The avenacins. Natural fungicides from oat roots active against 'take-all' attack. In: *Fungicides for Crop Protection 100 years of Progress*, ed. I.M. Smith, pp. 267–71. Croydon: BCPC Publications.

Défago, G. & Kern, H. (1983). Induction of *Fusarium solani* mutants insensitive to tomatine, their pathogenicity and aggressiveness to tomato fruits and pea plants. *Physiological Plant Pathology*, **22**, 29–37.

Défago, G., Kern, H. & Sedlar, L. (1983). Genetic analysis of tomatine insensitivity, sterols content and pathogenicity for green tomato fruits in mutants of *Fusarium solani*. *Physiological Plant Pathology*, 22, 39–43.

Deverall, B.J. & Rogers, P.M. (1972). The effect of pH and composition of test solutions on the inhibitory activity of wyerone acid towards germination of fungal spores. *Annals of Applied Biology*, 59, 375–87.

de Waard, M.A. & Van Nistelrooy, J.G.M. (1980). An energy-dependent efflux mechanism for fenarimol in wild-type strain and fenarimol-resistant mutants of *Aspergillus nidulans*. *Pesticide Biochemistry and Physiology*, 13, 255–66.

Ersek, T., Holliday, M. & Keen, N. T. (1982). Association of hypersensitive cell death and autofluorescence with a gene for resistance to *Peronospora manshurica* in soybean. *Phytopathology*, 72, 628–31.

Esquerré-Tugayé, M-T. & Lamport, D.T.A. (1979). Cell surfaces in plant-microorganism interactions. I. A structural investigation of cell wall hydroxyproline-rich glycoproteins which accumulate in fungus-infected plants. *Plant Physiology*, 64, 314–19.

Ford, J.E., McCance, D.J. & Drysdale, R.B. (1977). The detoxification of α-tomatine by *Fusarium oxysporum* f.sp. *lycopersici*. *Phytochemistry*, 16, 545–46.

Hadwiger, L.A. & Beckman, J.M. (1980). Chitosan as a component of pea–*Fusarium solani* interactions. *Plant Physiology*, 66, 205–11.

Hadwiger, L.A., Fristensky, B. & Riggleman, R.C. (1984). Chitosan, a natural regulator in plant-fungal pathogen interactions, increases crop yields. In: *Chitin, Chitosan and Related Enzymes*, ed. J.P. Zikakis, pp. 291–302. London: Academic Press.

Hahn, H.G., Bonhoff, A. & Grisebach, H. (1985). Quantitative localization of the phytoalexin glyceollin I in relation to fungal hyphae in soybean roots infected with *Phytophthora megasperma* f.sp. *glycinea*. *Plant Physiology*, 77, 591–601.

Hammerschmidt, R. & Kuć, J. (1982). Lignification as a mechanism for induced systemic resistance in cucumber. *Physiological Plant Pathology*, 20, 61–71.

Hammerschmidt, R., Lamport, D.T.A. & Muldoon, E.P. (1984). Cell wall hydroxyproline enhancement and lignin deposition as an early event in the resistance of cucumber to *Cladosporium cucumerinum*. *Physiological Plant Pathology*, 24, 43–7.

Hancock, J.G. & Huisman, O.C. (1981). Nutrient movement in host-pathogen systems. *Annual Review of Phytopathology*, 19, 309–31.

Harborne, J.B. & Ingham, J.L. (1978). Biochemical aspects of the coevolution of higher plants with their fungal parasites. In: *Biochemical Aspects of Plant and Animal Coevolution* ed. J.B. Harborne, pp. 343–405. London: Academic Press.

Harborne, J.B., Ingham, J.L., King, L. & Payne, M. (1976). The isopentenyl isoflavone luteone as a pre-infectional antifungal agent in the genus *Lupinus*. *Phytochemistry*, 15, 1485–7.

Hargreaves, J.A. (1980). A possible mechanism for the phytotoxicity of the phytoalexin phaseollin. *Physiological Plant Pathology*, 16, 351–7.

Hargreaves, J.A., Brown, G.A. & Holloway, P.J. (1982). The structural and chemical characteristics of the leaf surface of *Lupinus albus* L. in relation to the distribution of antifungal compounds. In: *The Plant Cuticle*, ed. D.F. Cutler, K.L. Alvin & C.E. Price, pp. 331–40. London: Academic Press.

Hargreaves, J. A., Mansfield, J. W. & Coxon, D. T. (1976). Identification of medicarpin as a phytoalexin in the broad bean plant (*Vicia faba* L.). *Nature*, 262, 318–19.

Hargreaves, J.A., Mansfield, J.W. & Rossall, S. (1977). Changes in phytoalexin concentrations in tissues of the broad bean plant (*Vicia faba* L.) following inoculation with species of *Botrytis*. *Physiological Plant Pathology*, 11, 227–42.

Harris, J.E. & Dennis, C. (1977). The effect of post-infectional potato tuber metabolites and surfactants on zoospores of Oomycetes. *Physiological Plant Pathology*, 11, 163–9.

Higgins, V.J. (1978). The effects of some pterocarpanoid phytoalexins on germ tube elongation of *Stemphylium botryosum*. *Phytopathology*, 68, 339–45.

Hutson, R.A. & Mansfield, J.W. (1980). A genetic approach to the analysis of mechanisms of pathogenicity in *Botrytis/Vicia faba* interactions. *Physiological Plant Pathology*, 17, 309–17.

Ingham, J.L. (1973). Disease resistance in higher plants. The concept of pre-infectional and post-infectional resistance. *Phytopathologische Zeitschrift*, 78, 314–35.

Ingham, J. L. (1982). Phytoalexins from the Leguminosae. In: *Phytoalexins*, ed. J.A. Bailey & J.W. Mansfield, pp. 21–80. Glasgow: Blackie.

Kämmerer, F.-J. (1967). Über die antibiotischen Substanzen aus der Tulpe (*Tulipa hybrida*). Dissertation, Bonn.

Keen, N.T., Holliday, M.J. & Yoshikawa, M. (1982). Effects of glyphosate on glyceollin production and the expression of resistance to *Phytophthora megasperma* f.sp. *glycinea* in soybean. *Phytopathology*, 72, 1467–70.

Keon, J.P.R. & Hargreaves, J.A. (1984). The response of barley leaf epidermal cells to infection by *Septoria nodorum*. *New Phytologist*, 98, 387–98.

Kolattukudy, P.E. & Köller, W. (1983). Fungal penetration of the first line defensive barriers of plants. In: *Biochemical Plant Pathology*, ed. N.A. Callow, pp. 79–100. Chichester: Wiley.

Kono, Y., Takeuchi, S., Kodama, O., Sekido, H. & Akatsuka, T. (1985). Novel phytoalexins (oryzalexins A, B and C) isolated from rice blast leaves infected with *Pyricularia oryzae*. Part II: Structural studies of oryzalexins. *Agricultural and Biological Chemistry*, 49, 1695–701.

Kuć, J. (1984). Phytoalexins and disease resistance mechanisms from a perspective of evolution and adaption. In: *Origins and Development of Adaption*, Ciba Foundation Symposium, 102, pp. 100–98. London: Pitman.

Kuhn, P.J. & Smith, D.A. (1979). Isolation from *Fusarium solani* f.sp. *phaseoli* of an enzymic system responsible for kievitone and phaseollidin detoxification. *Physiological Plant Pathology*, 14, 179–90.

Kuhn, P.J., Smith, D.A. & Ewing, F.F. (1977). 5,7,2',4'-tetrahydroxy-8-(3'-hydroxy-3' methyl-butyl) isoflavanone, a metabolite of kievitone produced by *Fusarium solani* f.sp. *phaseoli*. *Phytochemistry*, 16, 296–7.

Langcake, P., Cartwright, D.W. & Ride, J.P. (1981). The dichlorocyclopropanes and other fungicides with indirect mode of action. In: *Systemic Fungicides and Antifungal Compounds*, ed. H. Lyr & C. Polter, pp. 199–210. Berlin: Akademie der Landwirtschaftwissen-schaften der DDR.

Langcake, P., Kuhn, P. J. & Wade, M. (1983). The mode of action of systemic fungicides. In: *Progress in Pesticide Biochemistry and Toxicology*, Volume 3, ed. D.H. Hutson & T. R. Roberts, pp. 1–109. Chichester: Wiley.

Lyon, G.D. & Mayo, M.A. (1978). The phytoalexin rishitin affects the viability of isolated plant protoplasts. *Phytopathologische Zeitschrift*, 92, 298–304.

Maizel, J.V., Burkhardt, H.J. & Mitchell, H.K. (1964). Avenacin, an antimicrobial substance isolated from *Avena sativa*. I. Isolation and antimicrobial activity. *Biochemistry*, 3, 424–31.

Mansfield, J.W. (1980). Mechanisms of resistance to *Botrytis*. In: *The Biology of Botrytis*, ed. J.R. Coley-Smith, K. Verhoeff & W.R. Jarvis, pp. 181–218. London: Academic Press.

Mansfield, J.W. (1982). The role of phytoalexins in disease resistance. In: *Phytoalexins*, ed. J.A. Bailey & J.W. Mansfield, pp. 253–88. Glasgow: Blackie.

Mansfield, J.W. (1983). Antimicrobial compounds. In: *Biochemical Plant Pathology*, ed. J.A. Callow, pp. 237–65. Chichester: Wiley.

Mansfield, J.W. (1986). Induced antimicrobial systems in plants. In: *Induced Antimicrobial Systems*, ed. R.M. Cooper. Bath: Bath University Press (In press).

Mansfield, J.W. & Deverall, B.J. (1974). Changes in wyerone acid concentrations in leaves of *Vicia faba* after infection by *Botrytis cinerea* or *B. fabae*. *Annals of Applied Biology*, **77**, 227, 235.

Mansfield, J.W., Hargreaves, J.A. & Boyle, F.C. (1974). Phytoalexin production by live cells in broad bean leaves infected with *Botrytis cinerea*. *Nature*, **252**, 316–17.

Matthews, D.E. & Van Etten, H.D. (1981). Demethylation of pisatin by a cell-free preparation from *Nectria haematococca*. *Abstracts of the Annual Meeting of the American Society for Microbiology 1981*, 163 (Abstr).

Maule, A.J. & Ride, J.P. (1982). Ultrastructure and autoradiography of lignifying cells in wheat leaves wound-inoculated with *Botrytis cinerea*. *Physiological Plant Pathology*, **20**, 235–41.

Mayama, S. & Tani, T. (1982). Microspectrophotometric analysis of the location of avenalumin accumulation in oat leaves in response to fungal infection. *Physiological Plant Pathology*, **21**, 141–50.

Mayama, S., Tani, T., Matsuura, Y., Ueno, T. & Fukami, H. (1981). The production of phytoalexins by oat in response to crown rust, *Puccinia coronata* f.sp. *avenae*. *Physiological Plant Pathology*, **19**, 217–26.

Millar, R.L. & Higgins, V.J. (1970). Association of cyanide with infection of birdsfoot trefoil by *Stemphylium loti*. *Phytopathology*, **60**, 104–10.

Müller, K.O. (1956). Einige einfache Versuche zum Nachweis von Phytoalexinen. *Phytopathologische Zeitschrift*, **27**, 237–54.

Müller, K.O. & Börger, H. (1940). Experimentelle Untersuchungen uber die *Phytophthora*–Resistenz der Kartoffel. Zugleich ein Beitrage zum Problem der 'Erworbenen Resistenz' in Pflanzenreich. *Arbeiten aus der biologischen Reichsamstalt für Land und Forstwirtschaft Berlin-Dahlem*, **23**, 189–231.

Nakajima, T., Tomiyama, K. & Kinukawa, M. (1975). Distribution of rishitin and lubimin in potato tuber tissue infected by an incompatible race of *Phytophthora infestans* and the site where rishitin is synthesised. *Annals of the Phytopathological Society of Japan*, **41**, 49–55.

Nichols, E.J., Beckman, J.M. & Hadwiger, L.A. (1980). Glycosidic enzyme activity in pea tissue and pea-*Fusarium solani* interactions. *Plant Physiology*, **66**, 199–204.

Noveroske, R.L., Kuć, J. & Williams, E.B. (1964). Oxidation of phloridzin and phloretin related to resistance of *Malus* to *Venturia inaequalis*. *Phytopathology*, **54**, 92–7.

Overeem, J.C. (1976). Pre-existing antimicrobial substances in plants and their role in disease resistance. In: *Biochemical Aspects of Plant-Parasite Relationships*, ed. J. Friend & D.R. Threlfall, pp. 195–206. London: Academic Press.

Paxton, J. (1980). A new working definition of the term 'phytoalexin'. *Plant Disease*, **64**, 734.

Paxton, J., Goodchild, D.J. & Cruikshank, I.A.M. (1974). Phaseollin production by live bean endocarp. *Physiological Plant Pathology*, **4**, 167–72.

Pueppke, S.G. & Van Etten, H.D. (1974). Pisatin accumulation and lesion development in peas infected with *Aphanomyces euteiches*, *Fusarium solani* f.sp. *pisi*, or *Rhizoctonia solani*. *Phytopathology*, **64**, 1433–40.

Rathmell, W.G. & Smith, D.A. (1980). Lack of activity of selected isoflavonoid phytoalexins as protectant fungicides. *Pesticide Science*, **11**, 568–72.

Ride, J.P. (1978). The role of cell wall alterations in resistance to fungi. *Annals of Applied Biology*, **89**, 302–6.

Ride, J.P. (1983). Cell walls and other structural barriers in defence. In: *Biochemical Plant Pathology*, ed. J.A. Callow, pp. 215–36. Chichester: Wiley.

Ride, J.P. & Pearce, R.B. (1979). Lignification and papilla formation at sites of attempted penetration of wheat leaves by non–pathogenic fungi. *Physiological Plant Pathology*, 15, 79–92.

Roberts, K. & Northcote, D.H. (1972). Hydroxyproline: observations on its chemical and autoradiographical localisation in plant cell wall protein. *Planta*, 107, 43–51.

Robeson, D.J. & Harborne, J.B. (1980). A chemical dichotomy in phytoalexin induction within the tribe Viciae of the Leguminosae. *Phytochemistry*, 19, 2359–65.

Rossall, S., Mansfield, J.W. & Hutson, R.A. (1980). Death of *Botrytis cinerea* and *B. fabae* following exposure to wyerone derivatives *in vitro* and during infection development in broad bean leaves. *Physiological Plant Pathology*, 16, 135–46.

Sakai, R., Tomiyama, K. & Doke, N. (1979). Synthesis of a sesquiterpenoid phytoalexin rishitin in non-infected tissue from various parts of potato plants immediately after slicing. *Annals of the Phytopathological Society of Japan*, 45, 705–11.

Sato, N., Kitazawa, K. & Tomiyama, K. (1971). The role of rishitin in localising the invading hyphae of *Phytophthora infestans* in infection sites at the cut surfaces of potato tubers. *Physiological Plant Pathology*, 1, 289–95.

Sato, N. & Tomiyama, K. (1969). Localised accumulation of rishitin in potato-tuber tissue infected by an incompatible race of *Phytophthora infestans. Annals of the Phytopathological Society of Japan*, 35, 202–17.

Schönbeck, F. (1976). Role of preformed factors in specificity. In: *Specificity in Plant Diseases*, ed. R.K.S. Wood & A. Graniti, pp. 237–50. New York: Plenum Press.

Schönbeck, F. & Schlösser, E. (1976). Preformed substances as potential protectants. In: *Encyclopedia of Plant Physiology, New Series, vol. 4, Physiological Plant Pathology*, ed. R. Heitefuss & P. Williams, pp. 653–78. Heidelberg: Springer-Verlag.

Schönbeck, F. & Schroeder, C. (1972). Role of antimicrobial substances (tuliposides) in tulips attacked by *Botrytis* spp. *Physiological Plant Pathology*, 2, 91–9.

Sequeira, L. (1978). Lectins and their role in host-pathogen specificity. *Annual Review of Phytopathology*, 16, 453–81.

Sherwood, R.T. & Vance, C.P. (1976). Histochemistry of papillae formed in reed canary-grass leaves in response to non-infecting pathogenic fungi. *Phytopathology*, 66, 503–10.

Skipp, R.A. & Bailey, J.A. (1976). The effect of phaseollin on the growth of *Colletotrichum lindemuthianum* in bioassays designed to measure fungitoxicity. *Physiological Plant Pathology*, 9, 253–63.

Skipp, R.A. & Carter, G.A. (1978). Adaption of fungi to isoflavonoid phytoalexins. *Annals of Applied Biology*, 89, 366–69.

Smith, D.A., Harrer, J.M. & Cleveland, T.E. (1982). Relation between production of extracellular kievitone hydratase by isolates of *Fusarium* and their pathogenicity on *Phaseolus vulgaris. Phytopathology*, 72, 1319–23.

Stoessl, A. (1980). Phytoalexins – a biogenetic perspective. *Phytopathologische Zeitschrift*, 99, 251-72.

Stoessl, A. (1982). Biosynthesis of phytoalexins. In: *Phytoalexins*, ed. J.A. Bailey & J.W. Mansfield, pp. 133–80. Glasgow: Blackie.

Stoessl, A. (1983). Secondary plant metabolites in pre–infectional and post-infectional resistance. In: *The Dynamics of Host Defence*, ed. J. A. Bailey & B. J. Deverall, pp. 71–122. Sydney: Academic Press.

Stoessl, A. & Unwin, C.H. (1970). The antifungal factors in barley. V. Antifungal activity of the hordatines. *Canadian Journal of Botany*, 48, 465–70.

Stoessl, A., Unwin, C.H. & Ward, E.W.B. (1973). Post-infectional fungus inhibitors from plants: fungal oxidation of capsidiol in pepper fruit. *Phytopathology*, 63, 1225–31.

Tegtmeier, K.J. & Van Etten, H.D. (1982). The role of pisatin tolerance and degradation in the virulence of *Nectria haematococca*. A genetic analysis. *Phytopathology*, 72, 608–12.

Tomiyama, K., Doke, N., Nozue, M. & Ishiguri, Y. (1979). The hypersensitive response of resistant plants. In: *Recognition and Specificity in Plant Host-Parasite Interactions*, ed. J.M. Daly & I. Uritani, pp. 69–84. Tokyo: Japan Scientific Society Press.

Touzé, A. & Esquerré-Tugayé, M.-T. (1982). Defence mechanisms of plants against varietal non-specific pathogens. In: *Active Defence Mechanisms in Plant*, ed. R. K. S. Wood, pp. 103–18. New York: Plenum Press.

Turner, E.M.C. (1961). An enzymic basis for pathogenic specificity in *Ophiobolus graminis*. *Journal of Experimental Botany*, 12, 169–75.

Vance, C.P., Kirk, T.K. & Sherwood, R.T. (1980). Lignification as a mechanism of disease resistance. *Annual Review of Phytopathology*, 18, 259–88.

Van Etten, H.D. & Smith, D.A. (1975). Accumulation of antifungal isoflavonoids and 1a-hydroxyphaseollone, a phaseollin metabolite, in bean tissue infected with *Fusarium solani* f.sp. *phaseoli*. *Physiological Plant Pathology*, 5, 225–37.

Van Etten, H.D., Matthews, D.E. & Smith, D.A. (1982). Metabolism of phytoalexins. In: *Phytoalexins*, ed. J.A. Bailey & J.W. Mansfield, pp. 181–217. Glasgow: Blackie.

Van Etten, H.D., Matthews, P.S., Tegtmeier, K.J., Dietert, M.F. & Stein, J.I. (1980). The association of pisatin tolerance and demethylation with virulence on pea in *Nectria haematococca*. *Physiological Plant Pathology*, 16, 257–68.

van den Heuvel, J. & Grootveld, D. (1980). Formation of phytoalexins within and outside lesions of *Botrytis cinerea* in French bean leaves. *Netherlands Journal of Plant Pathology*, 86, 27–35.

Walker, J.C. & Stahmann, M.A. (1955). Chemical nature of disease resistance in plants. *Annual Review of Plant Physiology*, 6, 351–66.

Ward, E.W.B. (1984). Suppression of metalaxyl activity by glyphosate: evidence that host defence mechanisms contribute to metalaxyl inhibition of *Phytophthora megasperma* f.sp. *glycinea* in soybeans. *Physiological Plant Pathology*, 25, 381–6.

Ward, E.W.B. (1986). Biochemical mechanisms involved in resistance of plants to fungi. In: *Biology and Molecular Biology of Plant-Pathogen Interactions*, ed. J.A. Bailey, pp. 107–32. New York & London: Plenum Press.

Ward, E.W.B. & Lazarovits, G. (1981). Influence of temperature on the disease reaction of soybean hypocotyls inoculated with *Phytophthora megasperma* var. *sojae*. *Phytopathology*, 71, 911 (Abstr.).

Ward, E.W.B., Unwin, C.H. & Stoessl, A. (1975). Experimental control of late blight of tomatoes with capsidiol, the phytoalexin from peppers. *Phytopathology*, 65, 168–9.

Whitmore, F.W. (1978). Lignin-protein complex catalysed by peroxidase. *Plant Science Letters*, 13, 241–45.

Wong, E. (1975). Isoflavonoids. In: *The Flavonoids*, ed. J.B. Harborne, T.J. Mabry & H. Mabry, pp. 743–800. London: Chapman and Hall.

Yoshikawa, M. (1978). *De novo* messenger RNA and protein synthesis are required for phytoalexin mediated disease resistance in soybean hypocotyls. *Plant Physiology*, 61, 314–17.

Yoshikawa, M., Yamauchi, K. & Masago, H. (1978). Glyceollin: its role in restricting fungal growth in resistant soybean hypocotyls infected with *Phytophthora megasperma* var. *sojae*. *Physiological Plant Pathology*, 12, 73–82.

Zaki, A.I., Zentmyer, G.A., Pettus, J., Sims, J.J., Keen, N.T. & Sing, V.O. (1980). Borbonol from *Persea* spp. – chemical properties and antifungal activity against *Phytophthora cinnamomi*. *Physiological Plant Pathology*, 16, 205–12.

10
Antimicrobial defences in secondary tissues of woody plants

R. B. PEARCE

Oxford Forestry Institute, Department of Plant Sciences, University of Oxford, South Parks Road, Oxford OX1 3RB, UK

Our understanding of the physiology of disease in trees lags considerably behind that of herbaceous plants, especially agriculturally important crop species. Prominent amongst the factors contributing to this is the generally low economic value of forest trees and the comparatively unsophisticated approaches to silviculture, which have not encouraged or necessitated the use of disease resistance in forest management. Moreover, the size of trees and the time-scale of the development of both trees and their diseases present severe practical difficulties for the would-be investigator. Although relatively few detailed investigations of disease resistance in the primary tissues of woody plants have been carried out, sufficient examples have been reported to indicate that the major defence mechanisms known for herbaceous species occur also in woody angiosperms. Some, at least, of these are also known in gymnosperms.

The most obvious difference between herbaceous and woody plants is the extensive development of secondary tissues in the latter. The secondary xylem constitutes the wood, while the phloem, cortex and periderms are collectively referred to as the bark. The secondary tissues of trees are more massive and much longer-lived than the comparatively ephemeral organs of herbaceous species. In this chapter emphasis is given to the defence mechanisms of these secondary tissues.

Although major differences exist between gymnospersms and angiosperms (c.f. Esau, 1977), there are sufficient parallels between these groups to permit many aspects of their host-parasite interactions to be considered together. However, the paucity of information on the physiology of disease resistance in woody plants makes most such generalisations of questionable validity. Where it is evident that important differences may exist between the two groups they will be discussed separately.

Bark

An important but largely neglected means of constitutive defence of woody plants is their secondary surface, comprising a periderm or a rhytidome composed of a series of sequent periderms. In addition to the suberised walls of the phellem cells which may present a structural barrier to infection, these tissues also contain substantial amounts of low molecular weight compounds which may have antimicrobial properties. Few fungi appear able to penetrate these secondary surfaces directly. Most pathogens normally infecting the underlying tissues, bypass this barrier, gaining access to their host *via* wounds (e.g. *Endothia parasitica* on *Castanea dentata* (Anderson & Rankin, 1914)), or other infection courts where the periderm is incomplete (e.g. *Lachnellula willkommii* on *Larix decidua* (Buczacki, 1973)). Only a few fungi appear able to infect through intact periderms. Direct penetration by the rhizomorphs of *Armillaria* spp. has long been known (Thomas, 1934) and several other root–infecting fungi (e.g. *Heterobasidion annosum* (Peek, Liese & Parameswaran, 1972) and *Rosellinia necatrix* (Tourvieille de Labrouhe, 1982)) also appear able to breach the periderm barrier.

An important response to wounding and microbial infection in cortical and phloem tissues is the restoration of a periderm around the damaged site, isolating it from the body of the plant by a suberised barrier. This process, resulting in the formation of a necrophylactic periderm was described in conifers by Mullick (1977). An early event in this process was the formation of a zone, termed by Mullick the non-suberised impervious tissue (NIT) overlying the site of eventual necrophylactic periderm differentiation. A similar process has been observed in the bark of woody angiosperms at the margins of wounds and fungal lesions (Biggs, Davis & Merrill, 1983; Biggs, 1984; Biggs, 1985*a*), although the tissue corresponding to the NIT is suberised in these plants (Biggs, 1985*b*). Recent work with improved histochemical methods has also demonstrated the presence of suberin in the 'NIT' of the conifer *Picea sitchensis* (S. Woodward & R.B. Pearce, unpublished). This process of periderm restoration leads to the formation of a durable suberised barrier to the growth of potential pathogens. Successful pathogens may be able to circumvent this barrier and invade new host tissues, although direct penetration of a fully formed necrophylactic periderm does not normally occur (Biggs *et al.*, 1983; Hebard, Griffin & Elkins, 1984).

Other defence mechanisms may be especially important in the early stages of infection before periderm restoration has occurred. In *Picea sitchensis* several weeks may elapse before the completion of necrophylactic periderms (S. Woodward & R.B. Pearce, unpublished). In this species the antifungal stilbenes *iso*-rhapontigenin and astringenin accumulate around bark wounds inoculated with potentially pathogenic fungi. This occurs concomitantly with the disappearance of the corresponding stilbene glucosides rhaponticin and astringin, which are markedly less inhibitory (Fig. 10.1a) (S. Woodward & R.B. Pearce, unpublished). These compounds thus appear to provide a rapid response to infection that may inhibit fungal growth in the bark tissues pending the development of durable structural barriers. As these antifungal stilbenes appear to be released from closely related constitutive precursors, they cannot be regarded as phytoalexins; indeed doubt still remains whether gymnosperms have true phytoalexins.

Phytoalexins have been detected in the bark of a number of woody

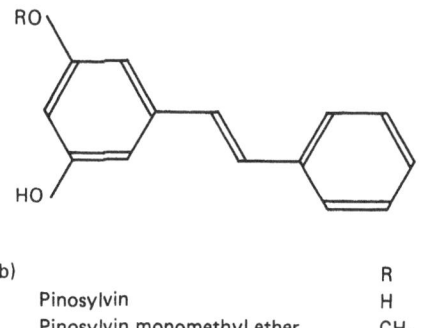

(a)

	R^1	R^2
Rhaponticin	CH_3	Glucosyl-
Astringin	H	Glucosyl-
Iso-rhapontigenin	CH_3	H
Astringenin	H	H

(b)

	R
Pinosylvin	H
Pinosylvin monomethyl ether	CH_3

Fig. 10.1. Antifungal stilbenes from bark and sapwood tissues of coniferous trees.

angiosperms, including *Populus tremuloides* (Flores & Hubbes, 1979; 1980), and *Morus alba* (Shirata, 1978; Takasugi *et al.*, 1978). Bark tissues are also frequently rich in constitutive phenolic compounds, which may have a defensive function (Tattar & Rich, 1973; Parker, 1977). Other possible defence mechanisms operating in bark tissues include cell necrosis, equatable with the hypersensitive response (see also de Wit, Chapter 1), which has been reported in interactions between pines and rust fungi (Miller *et al.*, 1976) and lysis of fungal hyphae. The enzymes chitinase and β-1,3-glucanase are present in the phloem of *Quercus* and *Acer* species, where a role for them in defence has been suggested (Wargo, 1975). Such defences may be of particular importance during the early stages of host-parasite interaction, slowing down pathogen growth and allowing time for the development of durable necrophylactic periderm barriers (c.f. Biggs *et al.*, 1983; Hebard *et al.*, 1984).

Sapwood

Sapwood comprises that part of the secondary xylem containing living parenchyma cells which is functional in conduction and storage (Esau, 1977). In the living tree sapwood is more resistant to degradation by microorganisms than heartwood, whereas after death, and in wood products, the converse is true. Studies suggesting the occurrence of active defence mechanisms in living sapwood date back many years (Swarbrick, 1926; Hepting & Blaisdell, 1936), but until recently, have been largely ignored. In consequence the resistance of living trees to decay has often been regarded as an essentially passive process (e.g. Peace, 1962). More recently the restriction of decay in living trees has been explained by schemes involving active resistance.

The most straightforward, termed the 'Reaction zone' model (Shain, 1967, 1971, 1979) proposes reaction zones formed in advance of the fungal infection front, which act in a dynamic mechanism of host resistance. Healthy sapwood initially forms a drier 'transition zone' containing metabolically active parenchyma cells. These cells produce the antimicrobial compounds, and eventually die, forming the reaction zone itself (Shain, 1979). The accumulated inhibitors may impede, but not necessarily completely arrest, the advance of the pathogen, as many microorganisms have the ability to degrade the aromatic compounds that accumulate in reaction zones (Tattar & Rich, 1973; Johanssen & Stenlid, 1985).

A second scheme was proposed on the basis of the distribution of decay observed in the vicinity of major wounds on living trees, and has been termed the 'Compartmentalisation of decay in trees' (CODIT) model

(Shigo & Marx, 1977). It proposes that the spread of decay fungi and their pioneers is limited by barriers of unspecified mechanism within the xylem tissues, termed walls 1-4. Wall 1 limits the axial spread of decay, wall 2 limits the radial spread of decay inwards, and wall 3 its lateral spread. These three walls all act in wood extant at the time of wounding and are essentially equivalent to reaction zones. Wall 4 is distinctive, comprising tissues laid down by the cambium subsequent to wounding, and prevents the outward radial spread of decay into the wood formed after wounding. This is the most important of these barriers, protecting the cambium and the youngest xylem tissues from invasion. As originally formulated the CODIT hypothesis did not seek to explain the mechanisms by which these proposed walls operated. However, subsequent studies have indicated likely mechanisms for these barriers (see below).

As an alternative to these models in which fungal colonisation is limited by active host responses, Rayner and co-workers (Boddy & Rayner, 1983; Cooke & Rayner, 1984) have suggested that the distribution of decay in living wood can be explained primarily in terms of wood moisture content. According to this hypothesis the resistance of living wood to colonisation by decay and other wood-inhabiting fungi is attributable to the high water content of functional sapwood, which can be inhibitory to fungal growth. The distribution of decay observed in association with wounds would therefore correspond to the distribution of sapwood rendered non-functional by the damage. Thus the limits of fungal colonisation, corresponding to the compartment walls 1-4, would be defined by anatomical characteristics of the wood and the vessel-occluding responses that maintain xylem function in the tree. The essentially passive nature of this process more closely parallels the classical 'heartrot' concept of tree decay than the active host-parasite interactions envisaged in the reaction zone and CODIT models. Although this provides an explantation for some of the apparent weaknesses of active defence models (e.g. the durability of the compartmentalisation wall 3), it fails to provide a full explanation for other observations, such as: the existence of well defined defence responses in decay margin tissues, and the absence of fungal colonisation in reaction zones, which are frequently dryer than functional sapwood (Shain, 1971). Further work is needed to clarify the role of environmental limitations in the host-parasite interaction, but some evidence suggesting the importance of hydrostatic pressure has been presented (Leben, 1985).

Antimicrobial defence mechanisms
Structural and chemical defence mechanisms have been described from living sapwood tissues. Both probably operate synergistically,

although individual research programmes have tended to concentrate on a single aspect only.

In many woody species constitutive components of the sapwood, in particular phenolic compounds, may act as pre-infectional microbial inhibitors. These may be active against wood-decaying pathogens; for example a range of sapwood extractives of *Pinus densiflora* are inhibitory to the growth of *Heterobasidion annosum* (Dumas, Hubbes & Strunz, 1983). Such compounds may be utilised or modified and detoxified by certain wood-inhabiting fungi (Tattar & Rich, 1973; Johanssen & Stenlid, 1985). The successions of microorganisms colonising wounded xylem tissues may result from such changes in the inhibitory compounds present in the wood (Shortle & Cowling, 1978). The antimicrobial activities of wood phenolics have generally been assessed by bioassay *in vitro* (e.g. Shortle, Tattar & Rich, 1971; Shain & Hillis, 1971); this may not, however, truly reflect their activities in the host-parasite interaction, where their ability to inactivate degradative enzymes may further arrest the progress of infection (c.f. Wood, 1967; Swain, 1977). Little work however, has been carried out on antimicrobial substances in relation to the enzymes of wood decay.

Inducible defence mechanisms in sapwood include both chemical and structural responses to infection. Several instances of phytoalexins or phytoalexin-like compounds accumulating in sapwood in response to fungal infection have been documented (Kemp & Burden, 1986).

Apart from an isolated report of phytoalexin-like compounds in the leaves of *Gingko biloba* (Christensen & Sproston, 1972), the principal candidates for consideration as gymnosperm phytoalexins are the stilbenes pinosylvin and pinosylvin monomethyl ether in pines (Fig. 10.1b). These accumulate in the sapwood of *Pinus* species in response to various fungi including *Heterobasidion annosum* (Shain, 1967; Prior, 1976) and *Amylostereum chailletii* (Hillis & Inoue, 1968), and evidence for their *de novo* synthesis from remote precursors in abiotically stressed sapwood tissues has been presented (Rudolf & Jorgensen, 1963). Although frequently accepted as phytoalexins (Hart, 1981; Kemp & Burden, 1986), further work is needed to clarify their status in microbially challenged tissues, particularly since the stilbene glucosides astringin and rhaponticin are known to occur constitutively in the sapwood of *Picea sitchensis* (C.D. Toscano, S. Woodward & R.B. Pearce, unpublished). Other candidates for consideration as gymnosperm phytoalexins include resin components which typically accumulate at lesions after wounding or infection, although their precise role in defence remains uncertain (Prior, 1976).

The occurrence of phytoalexins in the sapwood of woody angiosperms is more positively established. The sesquiterpene phytoalexins mansonones E and F (Fig. 10.2) have been identified in *Ulmus glabra* (Wych elm) challenged with *Ceratocystis ulmi*, along with six other phytoalexins including (-)-7- hydroxycalamenene (Burden & Kemp, 1984), which has

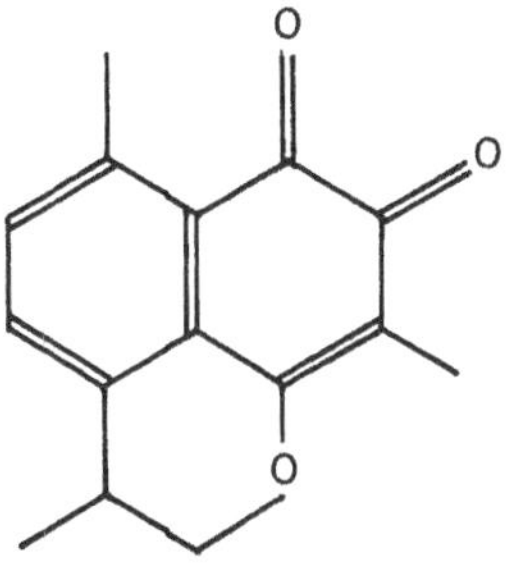

mansonone E 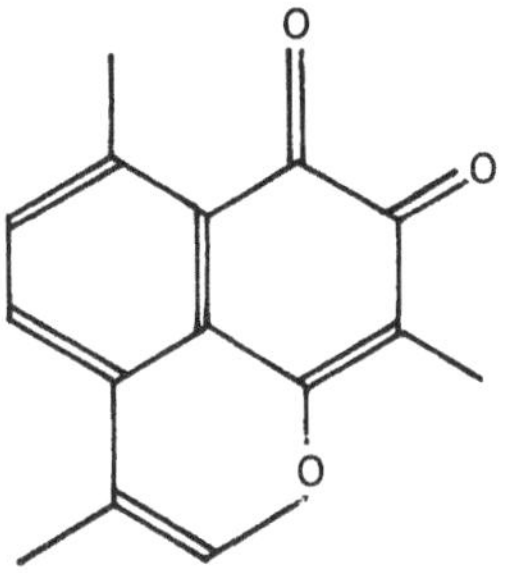   mansonone F

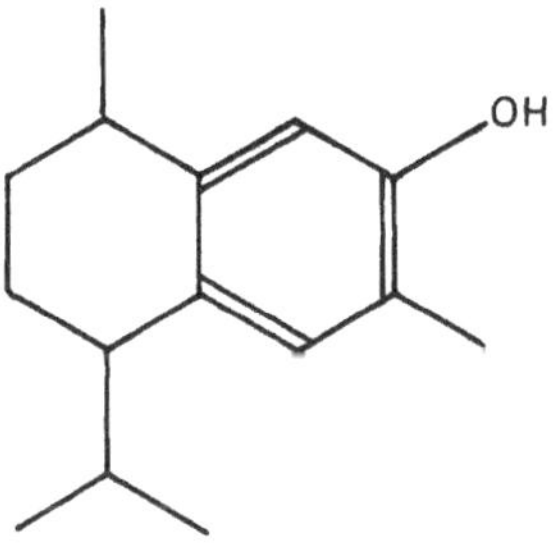

(−)-7-hydroxycalamenene

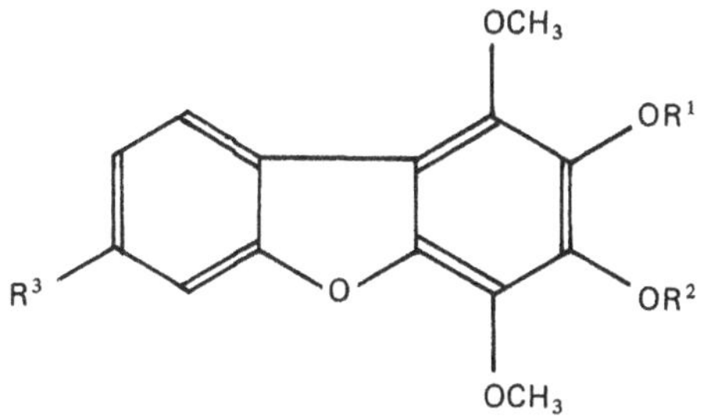

	R^1	R^2	R^3
α-pyrufuran	H	CH_3	H
β-pyrufuran	CH_3	H	H
γ-pyrufuran	H	CH_3	OH

Fig. 10.2. Sapwood phytoalexins from woody angiosperms.

also been found in reaction zones in the sapwood of *Tilia europaea* (European lime) decayed by *Ganoderma applanatum* (Burden & Kemp, 1983) (Fig. 10.2). Further examples of sapwood phytoalexins include the α-, β- and γ- pyrufurans induced in the sapwood of *Pyrus communis* (pear) after inoculation with *Chondrostereum purpureum* (Kemp, Burden & Loeffler, 1983; Kemp & Burden, 1984) and the uncharacterised compounds detected in reaction zones and compartmentalisation wall 4 barrier tissues of *Acer saccharinum* (Pearce & Woodward, 1986). In most cases these induced antifungal compounds accumulate in the reaction zones at the margin between healthy and invaded tissues, where they may provide, or contribute to, the inhibitory environment arresting the spread of the pathogen. Typically, the tissues comprising a functional reaction zone are not themselves colonised by the pathogen (see below).

Cell wall alterations in xylem tissues may lead to the formation of structural barriers to fungal spread. Probably the most important of these is the suberisation response induced in oak sapwood by fungal colonisation (Pearce & Rutherford, 1981; Pearce & Holloway, 1984), and many other species (R.B. Pearce, unpublished, c.f. Tables 10.1, 10.2). The response occurs in both angiosperms and gymnosperms, but is less well developed in the latter, perhaps as a consequence of the fewer living parenchyma cells in gymnosperm xylem. Suberised xylem parenchyma cells are highly resistant to degradation by wood-decaying fungi and may present an effective barrier to fungal penetration (Pearce & Rutherford, 1981). Additionally, tyloses and vessel linings may be suberised (Pearce & Holloway, 1984; Parameswaran, Knigge & Liese, 1985).

Brown, probably polyphenolic compounds, are frequently deposited on the walls of sapwood cells (Pearce & Woodward, 1986). These deposits, which mask the normal staining of suberin in the infected sapwood of many species may also contribute to host defence. Gummy deposits commonly found in the vessels of some angiosperms in association with sapwood infections can also have a polyphenolic composition (Wong & Preece, 1978*a,b*), and may have a vessel-occluding function similar to tyloses.

In addition to a possible role as chemical inhibitors, resins may also contribute to defence in conifers by forming mechanical barriers in resin-soaked tissues (Prior, 1976). Further work is, however, necessary to elucidate their importance in disease resistance.

Although our current understanding of the host-parasite interaction in woody plant tissues is still generally inadequate to allow precise roles

Table 10.1 *Association of xylem suberisation with compartmentalisation wall 4 barriers*

Wall 4 barrier suberised[a]	Wall 4 barrier not suberised
Acer saccharinum	*Fraxinus excelsior*
Carpinus betulus	*Sophora japonica*
Celtis occidentalis	*Taxus baccata*
Fagus sylvatica	*Thuja plicata*
Populus × *deltoides*	
Quercus robur	
Quercus rubra	
Zelkova carpinifolia	

[a]Suberin detected by staining with Sudan IV after extraction with chlorine dioxide and acetone (Pearce & Woodward, 1986).

to be ascribed to individual resistance mechanisms it is probable that these active defences are important in the establishment of the clearly demarcated boundaries of fungal infection in living sapwood. Passive factors, including wood anatomy and the high water content of healthy sapwood, may also be involved (c.f. Boddy & Rayner, 1983; Cooke & Rayner, 1984).

Infection boundaries

The best understood infection boundary in sapwood is wall 4 of the CODIT model. This comprises a wood zone of altered anatomy produced by the cambium in response to major wounds. Typically, the first xylem tissue formed after wounding is a traumatic axial parenchyma, ranging in thickness from a few, to in excess of 50 cell layers (Sharon, 1973; McGinnes *et al.*, 1977; Moore, 1978; Pearce & Rutherford, 1981; Pearce & Woodward, 1986). In coniferous species, similar parenchyma tissues often accompanied by abundant resin ducts, have been reported (Tippett & Shigo, 1980, 1981; Shigo & Tippett, 1981). These traumatic tissues have variously been termed 'barrier zones' without much consideration of their mode of action (Moore, 1978). In *Quercus robur*, however, the parenchyma tissues overlying decay are suberised and highly resistant to degradation by the sapwood-colonising decay fungus *Stereum gausapatum* (Pearce & Rutherford, 1981; Pearce & Holloway, 1984). Similar suberisation responses have been observed in the wall 4 parenchyma of some, but not all, other hardwood species examined (Table 10.1). In the absence of suberisation, antifungal compounds may be important in the mechanism of the barrier and have been found in conjunction with suberisation in the wall 4 barrier of *Acer saccharinum* (Pearce & Woodward, 1986). A role for the passive restriction of fungal growth (c.f. Boddy & Rayner, 1983; Cooke & Rayner, 1984) cannot be excluded, however, in the absence of investigation.

Wall 4 barriers are variable in extent. In some cases distinctive traumatic tissues may form around the greater part of the cambial circumference near wounds (Tippett & Shigo, 1980); in *Q. robur* however there was little development beyond the lateral limits of the wounds (R.B. Pearce, unpublished). Axial development was more extensive, paralleling staining in the underlying wood. At first, the barrier zone comprised living parenchyma cells with normal walls. Later, only those overlying sparsely colonised sapwood became suberised (Fig. 10.3), suggesting that the suberisation response (but not the formation of the traumatic parenchyma tissue) was fungally elicited.

Where development of a wall 4 barrier is limited in extent, its continued effectiveness in the CODIT concept is dependent on the durability of walls 1 and 3 which, together with wall 2, are formed in wood already fully differentiated before infection, thus differing fundamentally from wall 4. It has been recognised that these, in essence, comprise reaction zones (Shortle, 1979; Shigo, 1984).

In contrast to wall 4, reaction zones do not normally form a structurally homogeneous barrier, and consist of diverse cell types, including non-living vessels, fibres, etc.

Reaction zones frequently appear as brightly coloured margins to decay, ranging from less than 1 mm to greater than 10 mm in thickness, usually with no fungal hyphae present (Shain, 1967; Pearce & Woodward, 1986). The mechanism of fungal containment has been attributed to the accumulation of inhibitory compounds (Shain, 1979). Indeed most phytoalexin-like compounds described in woody plants have been characterised from reaction zones (Kemp & Burden, 1986). Cell well alterations may also contribute to the resistance of these zones. In *Quercus robur* (Pearce & Rutherford, 1981) (Fig. 10.4) and a range of other tree species (Table 10.2) suberisation occurs either on the walls of xylem parenchyma cells, or as a vessel lining or on tyloses. Such deposits, although not forming a continuous barrier, may greatly hinder fungal growth in the xylem by blocking the easiest routes of fungal spread: axially along vessels and radially in rays (c.f. Greaves & Levy, 1965). In *Picea abies,* suberisation in a circumferential reaction zone formed in response to infection by *Heterobasidion annosum* was localised in the region of the ray parenchyma end wall pits (Fig. 10.5).

The antimicrobial defence responses of living xylem tissues, perhaps together with environmental conditions, appear able to restrict the growth of wood-inhabiting fungi at the site of reaction zones. The difficulty of conducting *in vivo* studies in trees limits our knowledge of the dynamics of reaction zones.

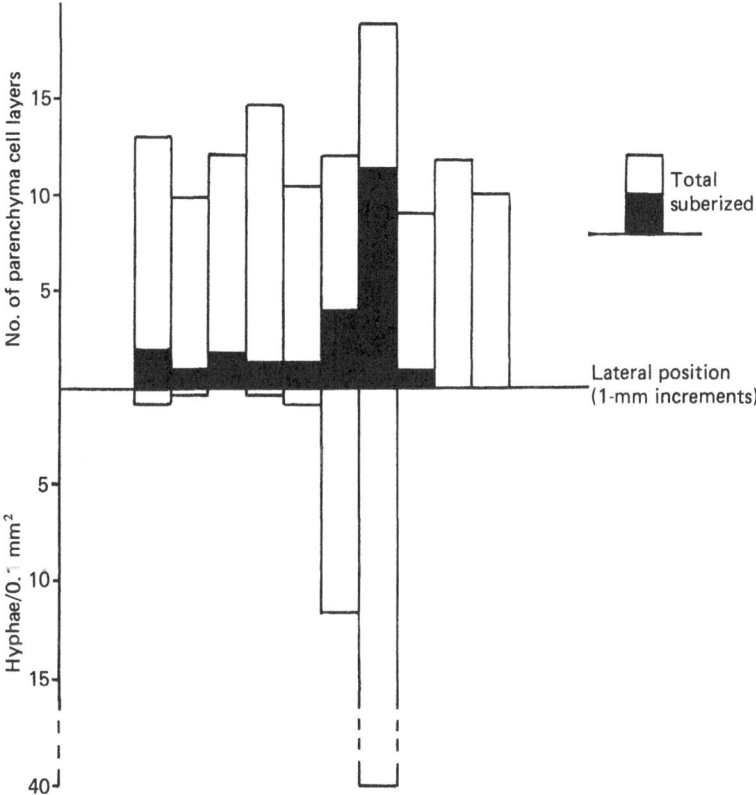

Fig. 10.3. Suberisation of parenchyma cells in a wall 4 barrier zone in *Quercus robur* in relation to fungal colonisation of the underlying sapwood, 32 weeks after wounding and inoculation with *Stereum gausapatum*. Trees were wounded in June by the removal of a square of bark tissue having an area of c. 100 cm^2. Wounds were inoculated by insertion of a fungally colonised dowel, at the lower wound margin. After felling and dissection of the tree, sections were cut from xylem tissues 1.5 cm below the site of inoculation. Serial sections were stained for suberisation with Sudan IV (Pearce & Rutherford, 1981) or for fungal hyphae with rhodamine B and methyl green (Pearce, 1984). The total thickness of the traumatic axial parenchyma comprising the compartment wall 4; the thickness of this tissue that had suberised and the number of hyphae cut per 0.1 mm^2 of the underlying sapwood tissues, were determined at lateral increments of 1 mm along the wall 4 barrier. The results shown are means obtained from three serial replicate pairs of sections.

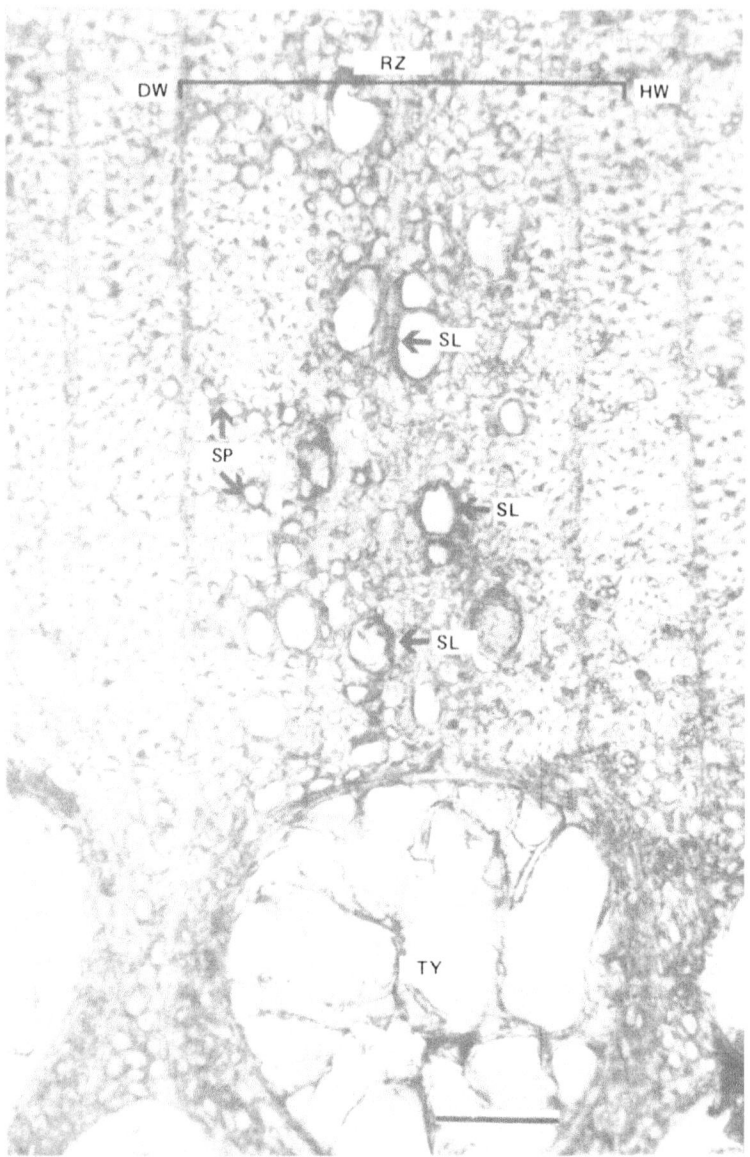

Fig. 10.4. Radially aligned reaction zone in *Quercus robur*. Transverse section stained with Sudan IV after extraction with acetone under reflux, to show suberisation (Pearce & Rutherford, 1981). In the reaction zone region (RZ) between healthy wood (HW) and discoloured wood (DW), suberised tyloses (TY) in early wood vessels, vessel lining material (SL), and parenchyma cells (SP) stain strongly. Scale bar represents 100 μm.

Table 10.2 *Occurrence of xylem suberisation in reaction zones*

Suberisation present[a]	Suberisation absent
Aesculus hippocastanum	*Acer pseudoplatanus*
Celtis occidentalis	*A. saccharinum*
Crataegus monogyna	*Betula pendula*
Fagus sylvatica	*Fraxinus excelsior*
Ilex aquifolium	*Ligustrum vulgare*
Malus pumila	*Picea sitchensis*
Picea abies	*Prunus avium*
Populus deltoides	*Sophora japonica*
Quercus ilex	
Quercus robur	
Quercus rubra	
Robinia pseudoacacia	
Tilia × *europaea*	
Zelkova carpinifolia	

[a]Suberin detected by staining with Sudan IV after extraction with chlorine dioxide and acetone (Pearce & Woodward, 1986).

Although some wood-inhabiting fungi may degrade reaction zone products (Shain, 1967; Johanssen & Stenlid, 1985), some reaction zones, particularly those acting as compartmentalisation walls (especially wall 3), appear to provide static and durable barriers to decay. This supports the proposition that decay is arrested not by active defence mechanisms, but by the normal physiological properties of sapwood (c.f. Cooke & Rayner, 1984).

Reaction zones however were initially envisaged as areas of dynamic response, being degraded by the fungus from within, while healthy sapwood externally was being converted to a reaction zone *via* a 'transition zone' (Shain, 1967; Shain, 1979). Fungi generally leave few structural or chemical relics of degraded reaction zones in decayed wood. Such evidence however is seen in *Acer pseudoplatanus* infected by the canker-causing pathogen *Dichomera saubinetii* (Cooke & Rayner, 1984) or by *Ustulina deusta* (Pearce, unpublished). In the latter a green-coloured circumferential reaction zone, 10mm wide, surrounded the decayed wood, within which were visible the remains of previous reaction zones (Fig. 10.6). These were in various stages of degradation. In places, reaction zone tissue, apparently healthy and still free of fungal invasion, was surrounded by decay. Higher levels of phenolic materials were present in the defeated reaction zones than in surrounding decayed wood (Table

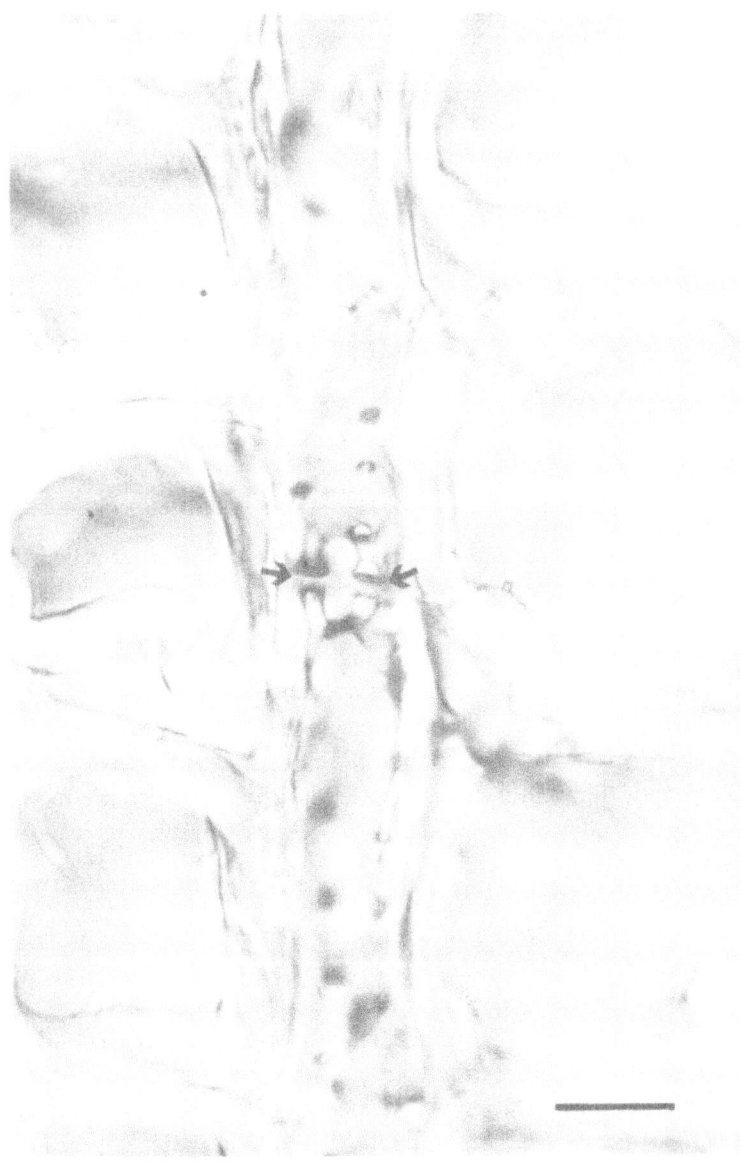

Fig. 10.5. Suberisation in a circumferential reaction zone in *Picea abies*, formed in response to attack by *Heterobasidion annosum*. Transverse section stained with Sudan IV after extraction with chlorine dioxide and acetone (Pearce & Woodward, 1986). Stained material (suberin) is concentrated in the region of the ray parenchyma end-wall pits (arrowed). Scale bar represents 10 μm.

10.3). Such evidence suggests that, in this species at least, the transition from sapwood to reaction zone to decayed wood is not a continuous process. Instead, the circumferential reaction zone appears to form a static barrier, able to persist for an extended period of time, but which is ultimately breached. Following this a rapid fungal advance may occur, which is eventually halted, accompanied by the formation of a new reaction zone.

Further evidence for this behaviour of reaction zones has been obtained from *Acer saccharinum*, decayed by *Ganoderma adspersum*. Pearce & Woodward, 1986). No relics of defeated reaction zones remained in the decayed wood but two distinct types of decay margin were evident. A typical green but discontinuous reaction zone, devoid of fungal colonisation and containing high levels of phytoalexin-like substances surrounded the decayed wood. Locally the decay margin appeared orange on exposure rather than green and levels of phytoalexin-like compounds were lower. Here hyphae could be seen invading apparently healthy ray parenchyma cells. These regions were considered to be where active fungal

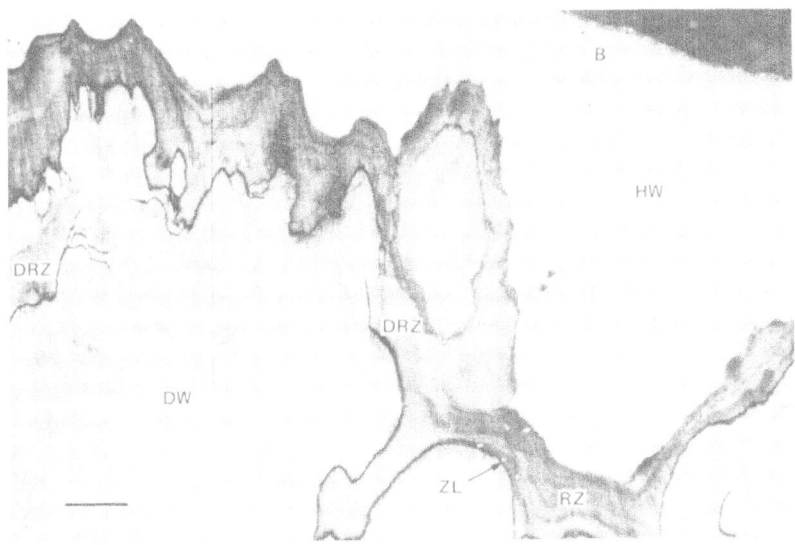

Fig. 10.6. Transverse section through wood of *Acer pseudoplatanus*, decayed by *Ustulina deusta*. Healthy bark (B) and sapwood (HW) overlie a broad green reaction zone (RZ). The margin between this and the decayed wood (DW) is marked by a fungal 'zone line' (ZL): mycelium does not normally extend beyond this into the reaction zone tissues. Within the decayed wood the remains of defeated reaction zones (DRZ) can be seen in various stages of degradation. Scale bar represents 10 mm.

Table 10.3 *Phenolic content of reaction zones, defeated reaction zones, healthy wood and decayed wood of* Acer pseudoplatanus, *decayed by* Ustulina deusta

Site	Phenol content (mg g^{-1} f. wt)[a]
Healthy sapwood	3.59
Functional reaction zone	17.40
Decayed wood between functional and defeated reaction zones	3.46
Defeated reaction zone	8.81
Decayed wood behind defeated reaction zone	3.51

[a]Expressed as gallic acid equivalents. Wood tissues extracted twice with methanol for 24 h in the cold, and assayed with the Folin–Ciocalteau reagent using a method based on that of Wong & Preece (1978*b*).

advance was occurring following the defeat of the reaction zone and before the establishment of a new, static, line of defence.

Studies on reaction zones to date have considered events in a transverse plane. In a tree, however, the major channels for the spread of pathogens are axial, and future research must concentrate in this dimension if reaction zone defences are to be elucidated.

Heartwood

Despite claims to the contrary Shigo & Shortle (1979), it is doubtful whether active defence mechanisms can occur in heartwood in the absence of living cells. Although residual enzymic activity in heartwood is possible, its capacity for active responses, as opposed to abiotic chemical changes consequent upon alterations in oxygen or moisture availability associated with decay, must be very limited.

Constitutive defences, rendering heartwood an unfavourable environment for the growth of microorganisms, however, are well known. A wide range of antifungal compounds are known to accumulate in heartwood, contributing to its resistance to decay (see: Kuć & Shain, 1977; Hart, 1981). Phytoalexin-like compounds found in infected sapwood tissues often appear to be identical to constitutive substances in healthy heartwood, e.g. (-)-7-hydroxycalamenene in *Ulmus* species (Burden & Kemp, 1984), and pinosylvin and pinosylvin monomethyl ether in *Pinus* species (Erdtman & Misiorny, 1952; Shain, 1967) (c.f. Figs. 10.1, 10.2). In the heartwood of *Quercus robur*, tyloses and vessel linings may be suberised (Pearce & Holloway, 1984; Parameswaran *et al.*, 1985). This closely parallels the response of sapwood to infection and could function similarly by obstructing fungal spread.

The similarities between the normal changes associated with heartwood formation, and those occurring in reaction zone responses to fungal infection in sapwood are striking (c.f. Johanssen & Stenlid, 1985; Kemp & Burden, 1986). One interpretation is that during heartwood formation the plants' inducible defence mechanisms are activated thereby rendering the future moribund xylem resistant to microbial colonisation. In *Pinus radiata* the ratio of pinosylvin to pinosylvin monomethyl ether differs significantly between heartwood and reaction zone tissues (Hillis & Inoue, 1968), suggesting intrinsic differences between reaction zone and heartwood formation. The relationship of these processes warrants further investigation.

The role of suberin

Defence mechanisms in both bark and wood, often need to remain effective for many years. In such circumstances structural defences are likely to prove more durable than chemical inhibitors which may be labile or diffusible. Induced suberisation of xylem tissues appears important in many woody species. Suberin is particularly suited to a defensive role in xylem tissues: it is primarily an aliphatic polymer (Kolattukudy, 1984) differing markedly from other xylem cell wall components. Wood-inhabiting fungi may thus be poorly equipped to deal with it, although *Armillaria mellea* has been shown to degrade it (Zimmermann & Seemuller, 1984). A major function of suberin in plants is as a permeability barrier, particularly in association with wounds (Kolattukudy, 1984). In sapwood, suberisation may thus minimise the extent of disruption to xylem transport resulting from damage caused by microbial colonisation.

References

Anderson, P.J. & Rankin, W.H. (1914). *Endothia* canker of chestnut. *Bulletin of the Cornell University Agricultural Experiment Station*, No. 347, 533–618.

Biggs, A.R. (1984). Boundary zone formation in peac' bark in response to wounds and *Cytospora leucostroma* infection. *Canadian Journal of Botany*, **62**, 2814–21.

Biggs, A.R. (1985*a*). Suberized boundary zones and the chronology of wound response in tree bark. *Phytopathology*, **75**, 1191–5.

Biggs, A.R. (1985*b*). Detection of impervious tissue in tree bark with selective histochemistry and fluorescence microscopy. *Stain Technology*, **60**, 299–304.

Biggs, A.R., Davis, D.D. & Merrill, W. (1983). Histopathology of cankers on *Populus* caused by *Cytospora chrysosperma*. *Canadian Journal of Botany*, **61**, 563–74.

Boddy, L. & Rayner, A.D.M. (1983). Origins of decay in living deciduous trees: the role of moisture content and a re-appraisal of the expanded concept of tree decay. *New Phytologist*, **94**, 623–41.

Buczacki, S.T. (1973). Observations on the infection biology of larch canker. *European Journal of Forest Pathology*, **3**, 228–32.

Burden, R.S. & Kemp, M.S. (1983). (-)-7-hydroxycalamenene, a phytoalexin from *Tilia europaea*. *Phytochemistry*, **22**, 1039–40.

Burden, R.S. & Kemp, M.S. (1984). Sesquiterpene phytoalexins from *Ulmus glabra*. *Phytochemistry*, **23**, 383–5.

Christensen, T.G. & Sproston, T. (1972). Phytoalexin production in *Ginkgo biloba* in relation to inhibition of fungal penetration. *Phytopathology*, **62**, 493–4.

Cooke, R.C. & Rayner, A.D.M. (1984). *Ecology of Saprotrophic Fungi*. London: Longmans.

Dumas, M.T., Hubbes, M. & Strunz, G.M. (1983). Identification of some compounds associated with resistance of *Pinus densiflora* to *Fomes annosus*. *European Journal of Forest Pathology*, **13**, 151–60.

Erdtman, H. & Misiorny, A. (1952). Constituents of pine heartwood XXXI. The content of pinosylvin phenols in Swedish pines. *Svensk Papperstidning*, **55**, 605–8.

Esau, K. (1977). *Anatomy of Seed Plants*, 2nd edn. New York: John Wiley.

Flores, G. & Hubbes, M. (1979). Phytoalexin production by aspen (*Populus tremuloides* Michx.) in response to infection by *Hypoxylon mammatum* (Wahl.) Mill and *Alternaria* spp. *European Journal of Forest Pathology*, **9**, 280–8.

Flores, G. & Hubbes, M. (1980). The nature and role of phytoalexins produced by aspen (*Populus tremuloides* Michx.). *European Journal of Forest Pathology*, **10**, 95–103.

Greaves, H. & Levy, J.F. (1965). Comparative degradation of the sapwood of Scots pine, beech and birch by *Lenzites trabea*, *Polystictus versicolor*, *Chaetomium globosum* and *Bacillus polymixa*. *Journal of the Institute of Wood Science*, **15**, 55–63.

Hart, J.H. (1981). Role of phytostilbenes in decay and disease resistance. *Annual Review of Phytopathology*, **19**, 437–58.

Hebard, F.V., Griffin, G.J. & Elkins, J.R. (1984). Developmental histopathology of cankers incited by hypovirulent and virulent isolates of *Endothia parasitica* on susceptible and resistant chestnut trees. *Phytopathology*, **74**, 140–9.

Hepting, G.H. & Blaisdell, D.J. (1936). A protective zone in red gum fire scars. *Phytopathology*, **26**, 62–7.

Hillis, W.E. & Inoue, T. (1968). The formation of polyphenols in trees IV. The polyphenols formed in *Pinus radiata* after *Sirex* attack. *Phytochemistry*, **7**, 13–22.

Johansson, M. & Stenlid, J. (1985). Infection of roots of Norway spruce (*Picea abies*) by *Heterobasidion annosum*. I. Initial reactions in sapwood by wounding and infection. *European Journal of Forest Pathology*, **15**, 32–45.

Kemp, M.S. & Burden, R.S. (1984). Isolation and structure determination of γ-pyrufuran, a third induced antifungal dibenzofuran from the wood of *Pyrus communis* L. infected with *Chondrostereum purpureum* (Pers. ex Fr.) Pouzar. *Journal of the Chemical Society. Perkin Transactions, Part 1*, 1441–3.

Kemp, M.S. & Burden, R.S. (1986). Phytoalexins and stress metabolites in the sapwood of trees. *Phytochemistry*, **25**, 1261–9.

Kemp, M.S., Burden, R.S. & Loeffler, R.S.T. (1983). Isolation, structure determination and total synthesis of the dibenzofurans α and β-pyrufuran, new phytoalexins from the wood of *Pyrus communis* L. *Journal of the Chemical Society. Perkin Transactions, Part 1*, 2267–72.

Kolattukudy, P.E. (1984). Biochemistry and function of cutin and suberin. *Canadian Journal of Botany*, **62**, 2918–33.

Kuć, J. & Shain, L. (1977). Antifungal compounds associated with disease resistance in plants. In: *Antifungal Compounds*, vol. 2, ed. M.R. Siegel & H.D. Sisler, pp. 497–535. New York: Marcel Dekker.

Leben, C. (1985). Wound occlusion and discolouration columns in red maple. *New Phytologist*, **99**, 485–90.

McGinnes, E.A Jr., Phelps, J.E., Szopa, P.S. & Shigo, A.L. (1977). *Wood anatomy after tree injury − a pictorial study. Research Bulletin of the Missouri Agricultural Experiment Station*, No. 1025.

Miller, T., Cowling, E.B., Powers, H.R. Jr. & Blalock, T.E. (1976). Types of resistance and compatibility in slash pine seedlings infected by *Cronartium fusiforme*. *Phytopathology*, **66**, 1229–35.

Moore, K.E. (1978). Barrier-zone formation in wounded stems of sweetgum. *Canadian Journal of Forest Research*, **8**, 389–97.

Mullick, D.B. (1977). The non-specific nature of defence in bark and wood during wounding, insect and pathogen attack. In: *The Structure Biosynthesis and Degradation of Wood*, ed. F.A. Loewus & V.C. Runeckles, pp. 395–442. New York: Plenum Press.

Parameswaran, N., Knigge, H. & Liese, W. (1985). Electron microscope demonstration of a suberized layer in the tylosis wall of beech and oak. *IAWA Bulletin. New Series*, **6**, 269–71.

Parker, J. (1977). Phenolics in black oak bark and leaves. *Journal of Chemical Ecology*, **3**, 489–96.

Peace, T.R. (1962). *Pathology of Trees and Shrubs*. Oxford: Oxford University Press

Pearce, R.B. (1984). Staining fungal hyphae in wood. *Transactions of the British Mycological Society*, **82**, 564–7.

Pearce, R.B. & Holloway, P.J. (1984). Suberin in the sapwood of oak (*Quercus robur* L.): its composition from a compartmentalisation barrier and its occurrence in tyloses in undecayed wood. *Physiological Plant Pathology*, **24**, 71–81.

Pearce, R.B. & Rutherford, J. (1981). A wound-associated suberized barrier to the spread of decay in the sapwood of oak (*Quercus robur* L.). *Physiological Plant Pathology*, **19**, 359–69.

Pearce, R.B. & Woodward, S. (1986). Compartmentalisation and reaction zone barriers at the margin of decayed sapwood in *Acer saccharinum* L. *Physiological and Molecular Plant Pathology*, **29**, 197–216.

Peek, R-D., Liese, W. & Parameswaran, N. (1972). Infektion ünd Abbau der Wurzelrinde von Fitche durch *Fomes annosus*. *European Journal of Forest Pathology*, **2**, 104–15.

Prior, C. (1976). Resistance by Corsican pine to attack by *Heterobasidion annosum*. *Annals of Botany*, **40**, 261–79.

Rudolf, E. von & Jorgensen, E. (1963). The biosynthesis of pinosylvin in the sapwood of *Pinus resinosa* Ait. *Phytochemistry*, **2**, 297–304.

Shain, L. (1967). Resistance of sapwood in stems of loblolly pine to infection by *Fomes annosus*. *Phytopathology*, **57**, 1034–45.

Shain, L. (1971). The response of sapwood of Norway spruce to infection by *Fomes annosus*. *Phytopathology*, **61**, 301–7.

Shain, L. (1979). Dynamic responses of differentiated sapwood to injury and infection. *Phytopathology*, **69**, 1143–7.

Shain, L. & Hillis, W.E. (1971). Phenolic extractives in Norway spruce and their effects on *Fomes annosus. Phytopathology*, **61**, 841–5.

Sharon, E.M. (1973). Some histological features of *Acer saccharum* wood formed after wounding. *Canadian Journal of Forest Research*, **3**, 83–9.

Shigo, A.L. (1984). Compartmentalisation: a conceptual framework for understanding how trees grow and defend themselves. *Annual Review of Phytopathology*, **22**, 189–214.

Shigo, A.L. & Marx, H.G. (1977). *Compartmentalisation of decay in trees. U.S.D.A. Forest Service Agriculture Information Bulletin*, No. 405.

Shigo, A.L. & Shortle, W.C. (1979). Compartmentalisation of discoloured wood in heartwood of red oak. *Phytopathology*, **69**, 710–11.

Shigo, A.L. & Tippett, J.T. (1981). *Compartmentalisation of decayed wood associated with Armillaria mellea in several tree species.* U.S.D.A. Forest Service Research Paper, NE-488.

Shirata, A. (1978). Production of phytoalexin in cortex tissue of mulberry shoot. *Annals of the Phytopathological Society of Japan*, **44**, 485–92.

Shortle, W.C. (1979). Mechanisms of compartmentalisation of decay in living trees. *Phytopathology*, **69**, 1147–51.

Shortle, W.C. & Cowling, E.B. (1978). Interaction of live sapwood and fungi commonly found in discoloured and decayed wood. *Phytopathology*, **68**, 617–23.

Shortle, W.C., Tattar, T.A. & Rich, A.E. (1971), Effects of some phenolic compounds on the growth of *Phialophora melinii* and *Fomes connatus. Phytophathology*, **61**, 552–5.

Swain, T. (1977). Secondary compounds as protective agents. *Annual Review of Plant Physiology*, **28**, 479–501.

Swarbrick, T. (1926). The healing of wounds in woody stems. *Journal of Pomology and Horticultural Science*, **5**, 98–114.

Takasugi, M., Nagao, T., Masamune, T., Shirata, A. & Takahashi, K. (1978). Structure of moracin A and B, new phytoalexins from diseased mulberry. *Tetrahedron Letters*, (1978), 797–8.

Tattar, T.A. & Rich, A.E. (1973). Extractable phenols in clear, discoloured and decayed woody tissues and bark of sugar maple and red maple. *Phytopathology*, **63**, 167–9.

Thomas, H.E. (1934). Studies on *Armillaria mellea* (Vahl.) Quel.: Infection, parasitism and host resistance. *Journal of Agricultural Research*, **48**, 187–218.

Tippett, J.T. & Shigo, A.L. (1980). Barrier zone anatomy in red pine roots invaded by *Heterobasidion annosum. Canadian Journal of Forest Research*, **10**, 224–32.

Tippett, J.T. & Shigo, A.L. (1981). Barriers to decay in conifer roots. *European Journal of Forest Pathology*, **11**, 51–9.

Tourvieille de Labrouhe, D. (1982). Penetration de *Rosellinia necatrix* (Hart) Berl. dans les racines du pommier en conditions de contamination artificielle. *Agronomie*, **2**, 553–60.

Wargo, P.M. (1975). Lysis of the cell wall of *Armillaria mellea* by enzymes from forest trees. *Physiological Plant Pathology*, **5**, 99–105.

Wong, W.C. & Preece, T.F. (1978a). *Erwinia salicis* in cricket bat willows: histology and histochemistry of infected wood. *Physiological Plant Pathology*, **12**, 321–32.

Wong, W.C. & Preece, T.F. (1978b). *Erwinia salicis* in cricket bat willows: phenolic constituents in healthy and diseased wood. *Physiological Plant Pathology*, **12**, 349–57.

Wood, R.K.S. (1967). *Physiological Plant Pathology.* Blackwell Scientific Publications, Oxford, Edinburgh.

Zimmerman, W. & Seemuller, E. (1984). Degradation of raspberry suberin by *Fusarium solani* f.sp. *pisi* and *Armillaria mellea. Phytopathologische Zeitschrift*, **110**, 192–9.

11
Comparisons between vesicular-arbuscular mycorrhizal fungi with respect to the development of infection and consequent effects on plant growth

S. AL-NAHIDH and F. E. SANDERS

Department of Plant Sciences, University of Leeds, Leeds LS2 9JT, UK.

Although there is no absolute specificity in the relations between vesicular-arbuscular mycorrhizal (VAM) fungi and their hosts (see Chapter 2), quantitative differences can be found between the responses of host plants to infection by different species (Harley & Smith, 1983). This chapter reports the results of experiments designed to investigate the effectiveness or comparability of a number of VAM associations, and illustrates the problems involved in comparing such associations.

VAM fungi are often compared on the basis of their 'effectiveness' in stimulating the growth of a host (Smith, 1985) but this may vary with host species, fertility of the soil and other environmental conditions. Inocula which may contain propagules with different germination behaviour and be at different densities can influence early levels of infection and the development of growth responses to different fungi (Carling *et al.*, 1979). The relation between a VAM fungus and its host may also be considered in terms of the 'compatibility' of the symbionts. Where compatibility is high and conditions suitable, it is to be expected that colonisation of the root system will be rapid and extensive, and this may lead to substantial improvement in rates of phosphorus uptake and growth of the host. However, the possibility of a depression of host growth by the formation of a compatible association, which may occur in soils high in phosphorus (Koide, 1985), is not excluded.

Glomus *spp. compared in the field*
Inoculation trials were carried out in successive years on the same field site (calcareous loam of the Wothersome series (Crompton &

Table 11.1 *Results of field trials in 1981 and 1982 comparing the growth responses of onion inoculated with three species of VAM fungus (from Al-Nahidh, 1985)*

Inoculant	1981	1982
Shoot dry weight at harvest[a], g plant–¹		
Control	0.18a	0.21a
Glomus caledonium	0.57b	0.23a
G. clarum	0.51b	0.37a
G. occultum	0.46b	0.32a

For each year, means followed by the same letter do not differ significantly at $P = 0.05$.

% mycorrhizal infection at harvest (spores g^{-1} in inoculum)			
Control	3	(0)	12 (0)
Glomus caledonium	33	(10)	19 (1)
G. clarum	18	(1)	54 (80)
G. occultum	42	(125)	36 n.a.

[a]Harvests were at 68 and 70 d after sowing in 1981 and 1982 respectively.
[b]Differences between % infections were clear and significant.
n.a. Not applicable.

Matthews, 1970); pH 7.2; 20 mg kg⁻¹ NaHCO₃-extractable P). In 1981 the soil had been previously fumigated with methyl bromide (120 g m⁻²); in 1982 the soil had been treated with dazomet (40 g m⁻²). Both fumigation treatments reduced considerably, but did not eliminate, indigenous VAM fungi.

In both trials, 1 m² plots were inoculated with *Glomus caledonium* (Nicol and Gerd, Gerd and Trappe), *G. clarum* (Nicol and Schenk) or *G. occultum* (Walker), or left uninoculated. There were five replicates of each treatment as part of a randomised block design. Inocula were produced by growing mycorrhizal maize in sand culture and consisted of spores and mycorrhizal root fragments mixed in sand. Ten kilograms of the appropriate inoculum or sterile sand was incorporated into the soil of each plot (1981) or applied in drills 10 cm apart (1982). Onion (*Allium cepa* cv. Hygro F1) was sown in rows 10 cm apart and thinned to 100 plants m⁻² in both years.

In 1981 there was a significant response to inoculation with each of the three fungi (Table 11.1). The magnitude of the response was not related to the level of infection, though this varied from 18% (*G. clarum*) to 42% (*G. occultum*). In contrast, in spite of substantial differences obtained in 1982 between the fungi in the level of infection, there was no significant response to inoculation. There was some relationship between per cent infection and numbers of spores applied in the inoculum but it was impossible to draw any general conclusions regarding the relative effectiveness of the different species.

Glomus *spp. compared under controlled conditions*

Because screening in the field proved unsatisfactory, an experiment was designed to compare the responses of four species of host plant inoculated with each of three species of VAM fungus. To ensure that inherent differences between the species were not obscured, levels of inoculum were sufficiently high to avoid any limitation of the rate of mycorrhizal infection (Al-Nahidh, 1985). The mycorrhizal fungi selected could be easily distinguished from each other on the basis of spore characteristics.

Experimental technique

Soil was collected from the field site, air dried, sieved to 3 mm and heat treated at 105°C for 48 h to eliminate indigenous VAM fungi. Spores of the mycorrhizal fungi were extracted from the crude sand-based inocula (as used in the field trial), concentrated by wet-sieving and mixed with heat-sterilised sand to form inocula for immediate use.

Mycorrhizal development and plant response were studied in all host/ fungus combinations. Pots were therefore given one of 16 possible treatments (4 hosts x (3 fungi + 1 uninoculated)). Numbers of replicates were sufficient to allow single pots of each treatment to be harvested on ten occasions.

Inoculum of each fungus or sterile sand was mixed with heat-treated soil in the proportion 1:2 by weight. Pots contained 1.5 kg of the appropriate mixture. Final spore densities were 3.5, 20 and 42 spores g^{-1} medium in pots given inoculum of *G. caledonium*, *G. clarum* or *G. occultum* respectively.

Seeds of onion (*Allium cepa* L. cv. Hygro F1), carrot (*Daucus carota* L. cv. Chantenay), lucerne (*Medicago sativa* L. cv. Europe) or red clover (*Trifolium patense* L. cv. Hungaropoly) were sown and pots were covered with plastic beads to retard evaporation. Pots were placed randomly in a growth chamber (21°C day, 18°C night, 200 μmol m^{-2} s^{-1} PAR, 16 h photoperiod) and watered regularly with deionised water. After emergence, seedlings were thinned to seven per pot for onion, six for carrot and five for lucerne and clover. Nitrogen (as NH_4NO_3) and potassium (as KCl) were applied to the pots at rates of 500 and 250 mg kg^{-1} of N and K respectively.

Single pots of the sixteen treatments were harvested on ten occasions. Roots were cleared of soil and their length was measured by the grid line intersect method (Marsh, 1971). Either the entire root systems (3 or 4 harvests) or random subsamples (subsequent harvests) were cleared and stained for assessment of infection, again by a line intersect method (Giovanetti & Mosse, 1980). Fresh weights of roots were also obtained. Dry weights of roots were estimated from subsamples. Phosphorus concentration in the dry matter was measured using the vanado-molybdate method after wet digestion in nitric/perchloric acid (Cavell, 1955).

Phosphorus inflows were calculated as $Lt^{-1}.dP/dt$ (Sanders *et al.*, 1977),

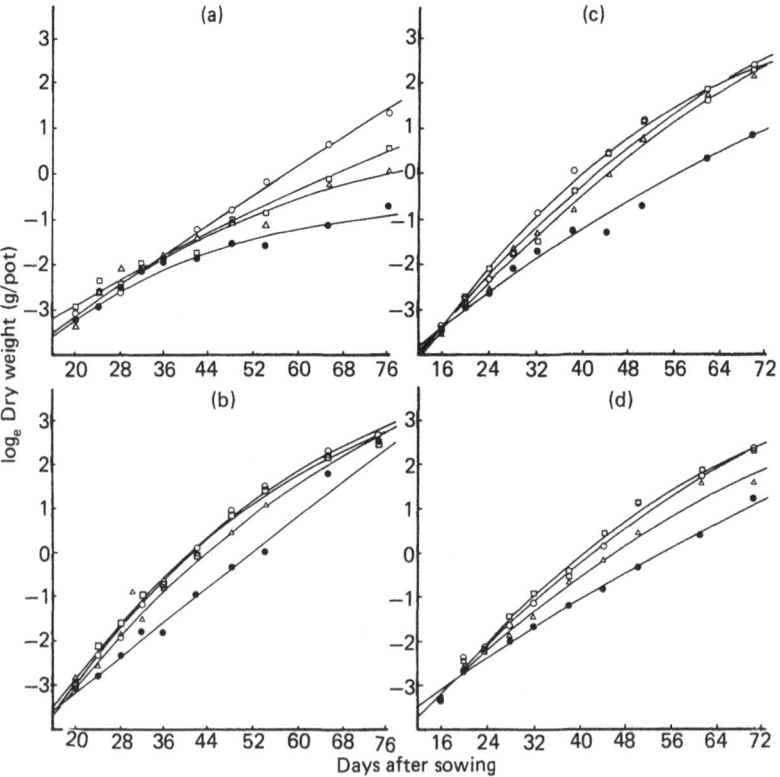

Fig. 11.1. Changes in plant dry weight with time. a, onion; b, carrot; c, lucerne; d, clover. $-\bullet-$ uninoculated controls; inoculated with $-\bigcirc-$ *Glomus caledonium*; $-\square-$ *G. clarum*; $-\triangle-$ *G. occultum*.

where Lt was total root length and dP/dt was the rate of change in the phosphorus content of the plants. Values of Lt and dP/dt were obtained from the equations of exponential or Gompertz functions fitted to the data using the statistical programming package MLP (Ross, 1980).

Growth of hosts and fungi

Plant growth was stimulated by inoculation in all host/fungus combinations (Fig. 11.1,a–d). Responses were largest with *G. caledonium* and smallest with *G. occultum.* Root:shoot ratios (dry weight basis) tended to increase with time in lucerne and clover and decrease in onion and carrot but there were no consistent differences between inoculation treatments.

Root lengths (Fig. 11.2,a–d) were greater in inoculated plants than in controls, with marked differences between fungi being in onion. Unino-

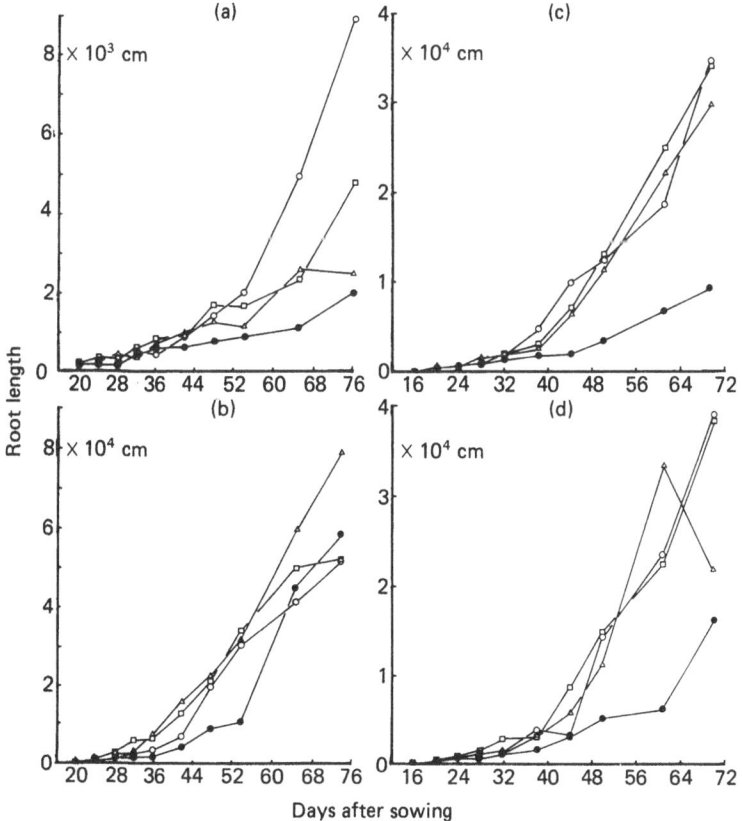

Fig. 11.2. Changes in root length with time. a, onion; b, carrot; c, lucerne; d, clover. Symbols as in Fig. 11.1.

culated carrot plants were found to be slightly (infection $\langle 10\%\rangle$) mycorrhizal at the last four harvests and data from these were therefore unreliable.

Fractional infection (infected root length/total root length) usually increased sigmoidally with time (Fig. 11.3; similar results obtained for lucerne and clover). *G. caledonium* and *G. clarum* behaved similarly in lucerne, clover and carrot. Fractional infection was usually lowest with *G. occultum*. The behaviour of *G. clarum* in onion was similar to that of *G. occultum*. *G. occultum* produced much shorter lengths of infected root than the other fungi except in the case of onion when *G. clarum* behaved similarly.

Uptake of phosphorus

At early harvests inflows to mycorrhizal root systems were frequently lower than to non-mycorrhizal roots. It is not clear whether

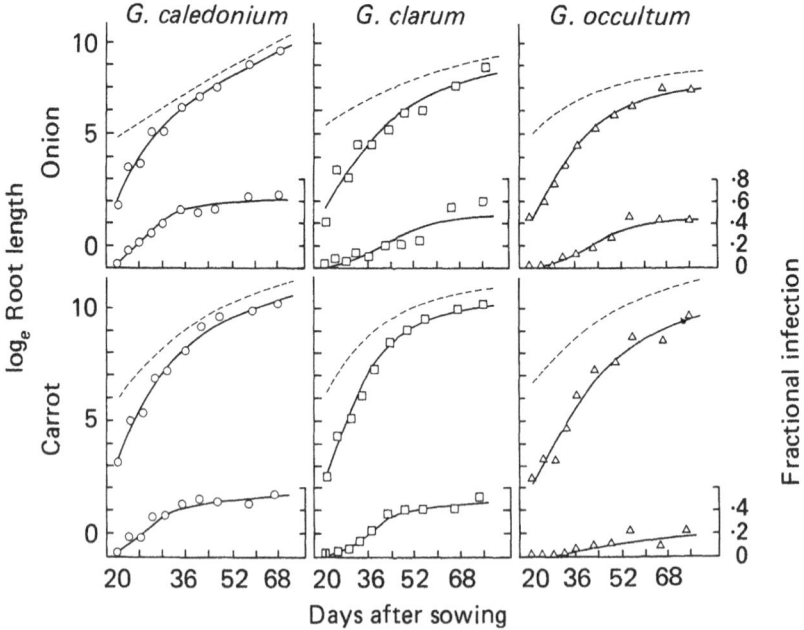

Fig. 11.3. Changes in total root length, mycorrhizal root length and fractional infection with time. ---- fractions fitted to data on total root length; – computer predictions. Symbols as in Fig. 11.1.

this was a real effect, arising perhaps from competition between the symbionts for phosphorus during the early stages of colonisation of the root system, or an artefact resulting from errors in curve-fitting. At later harvests mycorrhizal plants took up greater quantities of phosphorus than the uninoculated controls, in the order, control $<G.$ *occultum* $<G.$ *clarum* $= G.$ *caledonium.* Concentrations of phosphorus in the dry matter of mycorrhizal plants were also generally higher than in the uninoculated controls (Fig. 11.4, a–d). Phosphorus concentrations changed in a similar way with time in all host/fungus combinations, usually decreasing over the first few harvests, increasing to a maximum and then decreasing over the last few harvests. Maximum concentrations were reached at different

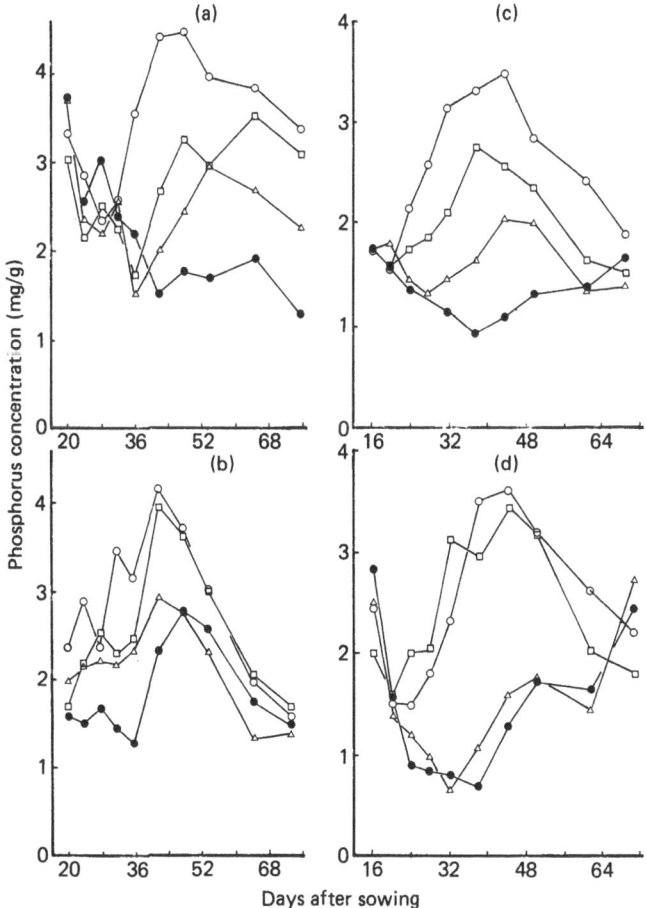

Fig. 11.4. Changes with time in concentrations of phosphorus in dry matter. a, onion; b, carrot; c, lucerne; d, clover. Symbols as for Fig. 11.1.

times depending on the host/fungus combination. In lucerne and onion, concentrations were highest with *G. caledonium* and lowest with *G. occultum* compared to the controls. In clover, the effects of *G. caledonium* and *G. clarum* were the same. Plants with *G. occultum* did not differ from the controls. In carrot, *G. caledonium* and *G. clarum* were also similar in their effects.

Values of phosphorus inflow to roots were calculated from each harvest time, using the equations of curves fitted to the data on phosphorus content and root length (see Sanders *et al.*, 1977). Inflow values could be assumed to be least accurate towards the beginning and end of the experimental period but it was not possible, due to the method of calculation, to assign precise confidence limits. Phosphorus inflows per unit root length (Fig. 11.5) were in general increased by mycorrhizal infection and changed with time in a manner resembling the changes in phosphorus concentration described above. Graphs for lucerne and clover were similar to each other. *G. caledonium* produced the largest increases in inflow and *G. occultum* the smallest. In both hosts, inflows to non-mycorrhizal plants rose towards the end of the experiment. This effect has occurred frequently and may be due to the solubilisation of phosphorus resulting from an increased production of organic exudates by the roots in phosphorus-stressed plants (Ratanayake *et al.*, 1978). In carrot, only *G. caledonium* caused a substantial increase in inflow of phosphorus. In onion, compared to *G. caledonium*, *G. clarum* and *G. occultum* increased inflow only slightly.

Hyphal inflow

An attempt was made to calculate hyphal inflows using the procedure of Sanders *et al.* (1977). Total inflows from hyphae to mycorrhizal roots, i.e. differences between control and mycorrhizal inflows, declined from their maxima after plants had taken up more than 3 mg of phosphorus from the pots (Fig. 11.6). Such declines in inflow have been reported previously (Sanders *et al.*, 1977; Smith, 1982), and have been attributed to the formation of overlapping zones of phosphorus depletion around the rapidly extending roots and external mycelium. Data from these pots were therefore ignored for the purpose of calculation of hyphal inflow.

In view of the undoubtedly low precision of the estimates of hyphal inflow in lucerne or clover, there was probably little difference between the abilities of the external fungal mycelia to aid the uptake of phosphorus (Table 11.2). Estimates for carrot could not be made in the case of two

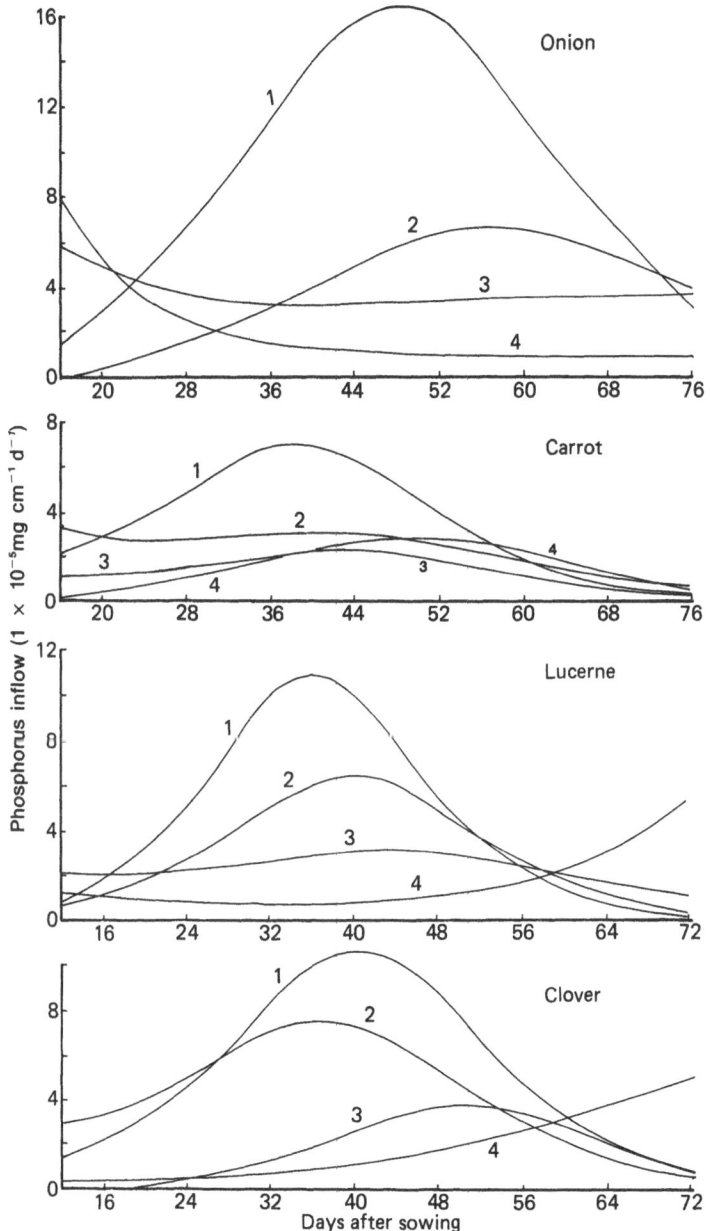

Fig. 11.5. Changes in phosphorus inflow per unit length of root with time. Inoculated with: 1, *Glomus caledonium*; 2, *G. clarum*; 3, *G. occultum*; 4, uninoculated control.

of the fungi, so comparisons were not possible. Only in onion were differences between hyphal inflows sufficiently large to be meaningful, and *G. caledonium* and *G. clarum* seemed to be considerably more efficient in combination with this host than was *G. occultum*.

With *G. caledonium* and *G. clarum*, hyphal inflows seemed much larger in onion than in the other plant species. This may have arisen because onion roots have a greater diameter and can probably accommodate a greater biomass of fungus per unit length, which may in turn be associated with greater lengths of external mycelium.

The differences between species in their effects on the growth of the four hosts probably arose, therefore, both from different abilities to increase the rate of phosphorus uptake per unit length of mycorrhizal root and from differences in rates of colonisation.

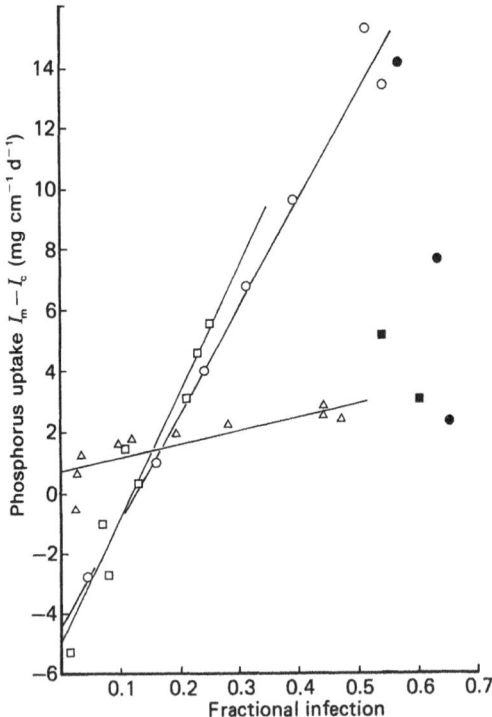

Fig. 11.6. Differences between mycorrhizal and control inflows of phosphorus per unit length of root ($I_m - I_c$) plotted against fractional infection. $-\bigcirc-$ *Glomus caledonium*; $-\square-$ *G. clarum*; $-\triangle-$ *G. occultum*, means of all hosts. Solid symbols are for plots where phosphorus uptake by the plants had exceeded 3 mg.

Table 11.2. *Estimated hyphal inflows of phosphorus (1×10^{-5} mg cm^{-1} d^{-1}) in the various host/fungus combinations*

Inoculant	Lucerne	Clover	Onion	Carrot
G. caledonium	18	14	42	6
G. clarum	12	7	36	?
G. occultum	9	12	5	?

Modelling the spread of VAM infection

When comparing VAM fungi it may be possible to measure parameters which relate to the compatibility or effectiveness of the association but are independent of difficult–to-control experimental variables, such as density of viable propagules in the inoculum, or the time of their germination.

Modelling the spread of VAM infection in growing root systems may allow identification of suitable parameters. A model originally developed by Buwalda *et al.* (1982) and later modified (Sheikh, 1984; Sanders, 1986) is described here. In this model the precise mechanism of spread is ignored and it is assumed only that the rate of increase of infected root length (dL/dt) is proportional to the length already infected (Li) and to the probability that regions of the root system into which the fungus may extend are not yet infected ($1-Li/\text{fs}.Lt$), such that

$$Li^{-1}.dLi/dt = S(1 - Li/fs.Lt)$$

where S and fs are constants. S has been termed the 'inherent spread rate' and is the proportional growth rate of Li when the fraction of root length infected is very small. fs may be defined as the fraction of root length which is susceptible to infection and takes account of the possibility that a proportion of the root system cannot, for various reasons, become mycorrhizal (Sanders, 1986). If root length increases exponentially, the above equation can be solved directly (Sanders, 1986) but more usually a numerical solution must be obtained using a computer.

The degree of compatibility between host and VAM fungus may be reflected in S, which should be independent of inoculum density. The value of S can easily be estimated by plotting the proportional rate of increase in length of infected root against fractional infection (Sanders, 1986). If the model correctly represents the process of colonisation such plots must give straight lines. The intercepts of these lines with the axes provide estimates of the values of S and fs.

Plots for each host/fungus combination used in the experiments described above are shown in Fig. 11.7 and corresponding estimates of S are in Table 11.3. There is unfortunately no easy method for assessing the statistical significance of differences between values of S. Nevertheless, there were obvious differences between the plots for the different host/fungus combinations. In some plots, points seem not to lie on a straight line which may cause the values of S to be underestimated (Sanders, 1986).

Therefore, as a final check, computer solutions to the Buwalda equation were obtained (Sheikh, 1984; Al-Nahidh, 1985) for each host/fungus

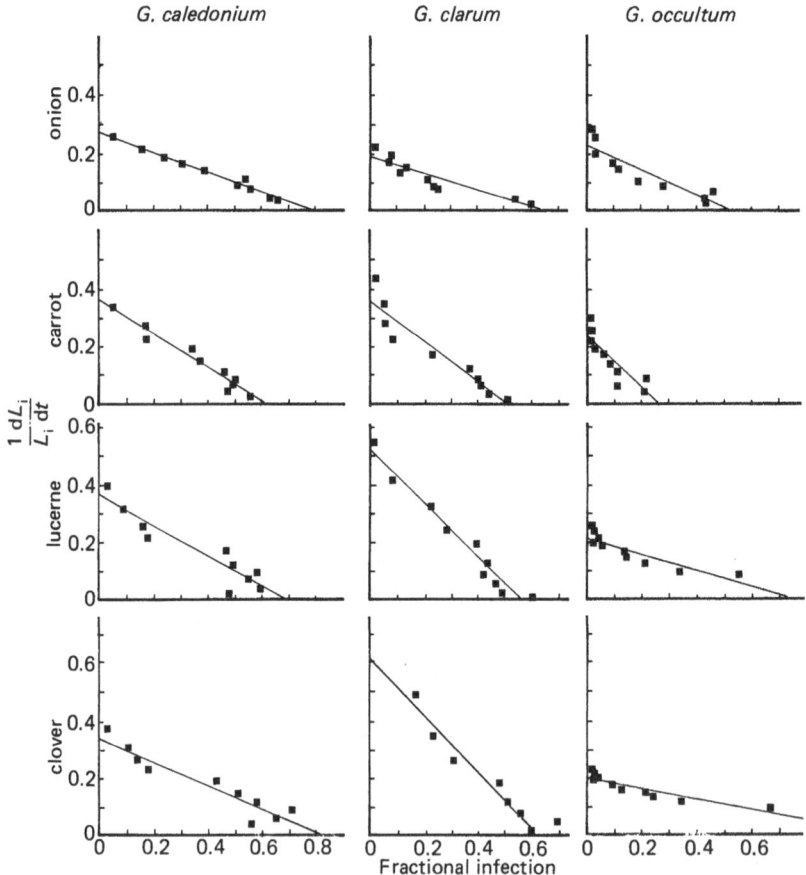

Fig. 11.7. Plots of proportional rates of increase in length of mycorrhizal root (cm cm^{-1}d^{-1}) against fractional infection. Intercept of vertical axis defines s, the inherent spread rate; intercept of horizontal axis defines fs, the fraction of root length susceptible to infection.

Table 11.3. *Estimated values of the inherent spread rate* (S) *of three VAM species in four host plants.*

Fungus	Host			
	Lucerne	Clover	Carrot	Onion
Glomus clarum	0.53	0.61	0.37	0.20
G. caledonium	0.37	0.34	0.37	0.28
G. occultum	0.22	0.20	0.25	0.24

combination using values of S and fs estimated from the plots shown in Fig. 11.7, appropriate initial values for infected root length and the equations describing the increases in total root length with time. These solutions are *predictions* of what will occur given certain parameter values, and this use of the model thus differs from that of Buwalda *et al.* (1982a).

In Fig. 11.3 predictions are shown alongside the experimental data. Agreement is satisfactory for both carrot and onion (shown) and lucerne and clover (not shown), indicating that for practical purposes the model describes the colonisation process well. This agreement also lends weight to the validity of estimated values of S (Table 11.3).

The poor effectiveness of *G. occultum* could therefore be explained on the basis of low inherent spread rate and small hyphal inflow. *G. caledonium* was the most effective of the three fungi, probably mainly because of its high inherent spread rates.

Varying the level of phosphorus in the soil has little effect on S (Buwalda *et al.*, 1982b). On the other hand, since the fungus colonises the roots by external as well as internal spread, S may be affected by unfavourable soil conditions such as water stress. It therefore remains to be established whether S is a characteristic of a given host/fungus combination that is generally independent of soil conditions. Nevertheless, models such as that of Buwalda *et al.* (1982a) have enabled us to clarify our ideas on what determines effectiveness in VAM associations and will probably remain an essential tool for future research in this area.

References

Al–Nahidh, S. (1985). Host-endophyte compatibility in mycorrhizal crop plants. *Ph.D. Thesis, University of Leeds.*

Buwalda, J.G., Ross, G.J.S., Stribley, D.P. & Tinker, P.B. (1982a). The development of endomycorrhizal root systems. III. The mathematical representation of the spread of vesicular-arbuscular mycorrhizal infection in root systems. *New Phytologist*, **91**, 669–82.

Buwalda, J.G., Ross, G.J.S., Stribley, D.P. & Tinker, P.B. (1982b). The development of endomycorrhizal root systems. IV. The mathematical analysis of effects of phosphorus on the spread of vesicular-arbuscular mycorrhizal infection in root systems. *New Phytologist*, **92**, 391–9.

Carling, D.E., Brown, M.F. & Brown, R.A. (1979). Colonisation rates and growth responses of soybean plants infected by vesicular-arbuscular mycorrhizal fungi. *Canadian Journal of Botany*, **57**, 1769–71.

Cavell, A.J. (1955). The colorimetric determination of phosphorus in plant materials. *Journal of the Science of Food and Agriculture*, **6**, 479–80.

Crompton, A. & Matthews, B. (1970). *Soils of the Leeds District*. Memoirs of the Soil Survey of Great Britain.

Giovanetti, M. & Mosse, B. (1980). An evaluation of techniques for measuring vesicular-arbuscular mycorrhizal infection in roots. *New Phytologist*, **84**, 489–500.

Harley, J.L. & Smith, S.E. (1983). *Mycorrhizal Symbiosis*. Academic Press: London.

Koide, R. (1985). The nature of growth depressions in sunflower caused by vesicular-arbuscular mycorrhizal infection. *New Phytologist*, **99**, 449–62.

Marsh, B. A'B. (1971). Measurement of length in a random arrangement of lines. *Journal of Applied Ecology*, **8**, 265–7.

Ratanayake, M., Leonard, R.T. & Menge, J. (1978). Root exudation in relation to supply of phosphorus and its possible relevance to mycorrhizal formation. *New Phytologist*, **81**, 543–52.

Ross, G.J.S. (1980). *Maximum Likelihood Programme*. Statistics Department, Rothamsted Experimental Station.

Sanders, F.E. (1986). Quantitative approaches to the analysis of the development of mycorrhizal root systems. In: *Physiological and Genetical Aspects of Mycorrhizae*, ed. V. Gianinazzi-Pearson & S. Gianinazzi. INRA: Paris.

Sanders, F.E., Tinker, P.B., Black, R.L. & Palmerley, S.M. (1977). The development of endomycorrhizal root systems. I. Spread of infection and growth promoting effects with four species of vesicular-arbuscular endophyte. *New Phytologist*, **78**, 257-68.

Sheikh, N.A. (1984). The dynamics of spread of vesicular-arbuscular mycorrhizal infection and temperature responses. *Ph.D. Thesis, University of Leeds*.

Smith, S.E. (1982). Inflow of phosphate into mycorrhizal and non-mycorrhizal plants of *Trifolium subterraneum* at different levels of soil phosphate. *New Phytologist*, **90**, 293–303.

Smith, S.E. (1985). *Proceedings of the 6th North American Conference on Mycorrhizae*. ed. R. Molina. Corvallis, Oregon: Oregon State University Press.

12
Fruiting and successions of ectomycorrhizal fungi

P. A. MASON, F. T. LAST, J. WILSON[1],
and
J. W. DEACON, L. V. FLEMING and F. M. FOX[2]

[1]Institute of Terrestrial Ecology, Bush Estate, Penicuik, Midlothian EH26 0QB, UK and [2]Microbiology Department, School of Agriculture, West Mains Road, Edinburgh EH9 3JG, UK

Changes in the structure of fungal communities are often difficult to verify because hyphae, and other vegetative structures, can seldom be identified to species or even genus level without the preparation of pure cultures. The sequential appearance of reproductive structures in natural environments needs to be interpreted with care (Cooke & Rayner, 1984) because different fungi produce their reproductive structures at different ages (Harper & Webster, 1964). The presence of reproductive structures is always associated with the presence of mycelia, but their absence does not necessarily signify the absence of mycelia.

The (sheathing) ectomycorrhizas of birch (*Betula* spp.), pine (*Pinus* spp.) and spruce (*Picea* spp.), fortunately can be categorised by gross morphology and microscopical features such that the associated mycorrhizal fungi can be attributed at least to the genus (Last *et al.*, 1985). By coupling records of fruitbodies with direct observations of mycorrhizas it is possible, therefore, to study successional events with some degree of confidence for at least some tree species.

Concepts of fungal succession

Critical reviews of successions of saprotrophic fungi have recently been made by Frankland (1981) and Cooke & Rayner (1984). Earlier, Park (1968) drew analogies with the development of assemblages of higher plants and distinguished between (i) seral successions and (ii) substratum successions. The former equate to the series of changes leading to climax communities as has been described for fungi colonising sand dunes (Brown, 1958). Substratum successions, on the other hand, relate to events on a smaller scale within a large environment; for example, on individual twigs or fallen leaves.

Seral successions doubtless occur among mycorrhizal fungi. For example, the mycorrhizal fungi associated with a seral succession of higher plants, progressing from herbaceous communities to climax woodland, will inevitably change from vesicular–arbuscular mycorrhizal fungi colonising herbs (Rose, 1980) to ectomycorrhizal fungi predominating on climax or sub-climax trees such as birch, pine, oak and beech. Successions can also occur in association with individual trees; these are substratum successions. In these instances, fungi differentially colonise roots at different times and at different distances from the tree base. Changes in populations of mycorrhizal fungi may be rapid, particularly in the early years of a tree's life. On a woodland scale, therefore, it is probable that both substratum and seral successions of mycorrhizal fungi occur, as trees exert strong influences both on their aerial and edaphic environments.

Successions of ectomycorrhizal fungi

Temporal aspects

Among mycorrhizal fungi, *Tricholoma fulvum* is characteristically associated with birch, *Suillus luteus* with pines, and *Suillus grevillei* with larch, but such associations are seldom exclusive, most tree species being associated with a spectrum of fungi, some host-specific and some of broad host-range as described in Chapter 2. A few tree species such as red alder (*Alnus rubra*), which develops in pure stands, tend to be more fungus-specific (Malajczuk, Molina & Trappe, 1982; Molina & Trappe, 1982). Apart from such exceptions, mycorrhizal successions, at least in previously treeless sites, appear to proceed through three broad phases: (1) a relatively narrow range of pioneer fungi of broad host-range develop in young stands, for example species of *Hebeloma*, *Laccaria* and *Inocybe*; (2) an increase in species-diversity occurs as the stand develops (see Table 12.1); (3) one or a few fungi become dominant in, or at least characterise, old stands, these fungi sometimes being host-specific and best adapted to the host in the particular edaphic conditions pertaining to a site. This view has evolved as a result of observations of mycorrhizal fungi associated with a variety of coniferous and broadleaved trees of different ages in many parts of the world (Chu-Chou, 1979; Chu-Chou & Grace, 1981; Mason *et al.*, 1982). A detailed study of the fungal fruitbodies arising annually in a young birch plantation (Table 12.1) revealed that certain species of *Inocybe*, *Hebeloma*, *Laccaria* and *Thelephora* appeared in the first four years after planting, followed largely in later years by species of *Lactarius*, *Leccinum*, *Cortinarius* and *Russula* (Mason *et al.*, 1982).

Table 12.1. *Succession of fruitbodies of proven or suspected ectomycorrhizal fungi appearing in a stand of birches planted at Bush estate, near Edinburgh (lat. 55°52′N) (after* Last et al.*, 1983)*

Year after planting	Fungi with fruit bodies occurring for the first time
1	Nil
2	*Hebeloma crustuliniforme*
	Laccaria proxima
3	*Laccaria tortilis*
	Thelephora terrestris
4	*Hebeloma fragilipes*
	H. sacchariolens
	H. mesophaeum
	Inocybe lanuginella
	Lactarius pubescens
6	*Cortinarius* sp.
	Hebeloma leucosarx
	Hymenogaster tener
	Inocybe petiginosa
	Leccinum roseofractum
	L. scabrum
	L. versipelle
	Peziza badia
	Ramaria sp.
7	Other *Cortinarius* spp.
	Other *Hebeloma* spp.
	Lactarius glyciosmus
	Leccinum subleucophaeum
10	*Hebeloma vaccinum*
	Russula betularum
	R. grisea
	R. versicolor
14	*Laccaria laccata*
	Lactarius spinosulus
	Russula atropurpurea

Observations of fruitbodies beneath a number of other tree species growing in young stands revealed similarities to the populations of mycorrhizal fungi on young birch. J.H. Warcup (pers. comm.), studying four-year-old *Pinus contorta* in Scotland recorded *Hebeloma fragilipes, Thelephora terrestris, Amphimena byssoides, Rhizopogon luteolus* and *Suillus luteus* . This range of species is comparable to species of *Hebeloma, Rhizopogon* and *Suillus* found in young stands of *Pinus radiata, Pseudotsuga menziesii,* and *Eucalyptus* spp. growing in New Zealand (Chu–Chou, 1979; Chu-Chou & Grace, 1981, 1982, 1983) and *Picea sitchensis* in the UK (Thomas, Rogers & Jackson, 1983).

Watling (1981) similarly found species of *Laccaria* and *Hebeloma* pioneer fungi which colonised newly planted woodland and open land invaded by sapling trees; whereas in 19 established woodland sites in Perthshire he recorded fruitbodies of 45 species of *Cortinarius*, 38 of *Lactarius*, 57 of *Russula* and 7 of *Amanita* (Watling, 1984). Richardson (1970) found that the mycorrhizal flora of a mature (55-years-old) plantation of Scots pine (*Pinus sylvestris*) was dominated by *Russula (Russula emetica)* but also had significant numbers of fruitbodies of *Amanita (A. inaurata, A. rubescens* and *A. vaginata*) and *Lactarius (L. rufus* and *L. turpis)*. In a stand of *Betula pendula* (72 years old), Miles (1985) found that 32% and 28% of the fruitbodies of mycorrhizal fungi were of *Amanita muscaria* and *Tricholoma columbetta*, respectively (Table 12.2).

Circumstantial evidence for the third phase mentioned above is provided by the relatively narrow range of fruitbodies often reported for ageing stands. Harvey, Larsen & Jurgensen (1976) found in a 250 year-old forest of Douglas fir and larch that roots of larch were associated with a single dominant mycorrhizal fungus, *Suillus cavipes*; whereas the fir roots were 'dominated' by *Russula brevipes*. Malajczuk *et al.* (1982) suggested that not only does a succession of mycorrhizal fungi occur in natural stands of eucalypts and *Pinus radiata*, but also that this succession, over time, tends towards dominance by host-specific fungi. Such uniformity does not occur in all mature woodlands, even if these contain only one tree species. The relict birchwood at Struan, Perthshire, is ancient and yet has a diverse community of ectomycorrhizal fungi (Fleming, Deacon & Last, 1986).

Spatial aspects

In addition to the age-related sequence, clear evidence of a spatial sequence of fruitbodies was noted in a young stand of birch (Ford, Mason & Pelham, 1980; Mason *et al.*, 1982). The mean distance of fruitbodies of different fungi from the bases of trees tended to increase in successive years, with later-appearing species usually fruiting, in the first instance, close to the trees, as did earlier appearing species before them (Ford *et al.*, 1980). However, while most species were associated with the fine roots in rings or arcs at particular distances from the trunk, some (e.g. *Laccaria tortilis*) developed a radial, spoke-like distribution as if following the main structural roots. The existence of these spatial distributions was corroborated by Deacon, Donaldson & Last (1983) who examined cores of soil at different distances from the bases of trees. For each tree, the predominant mycorrhizas typical of the early-fruiting species occurred furthest from the base.

Table 12.2. *Influence of age of natural stands of birch* (Betula pendula) *on the occurrence of fruitbodies of fungi known to form ectomycorrhizas with* B. pendula

	B. pendula age (yr)		
	20	29	72
	Occurrence of fruitbodies (per cent)		
Boletus impolitus	3	Nil	Nil
Cortinarius spp.	78	18	Nil
Hebeloma spp.	Nil	52	Nil
Lactarius pubescens	16	5	5
Russula atropurpurea	3	Nil	Nil
Amanita muscaria	Nil	8	32
Boletus edulis	Nil	3	Nil
Leccinum scabrum	Nil	3	5
Inocybe sp.	Nil	Nil	5
Laccaria amethystina	Nil	Nil	10
Lactarius tabidus	Nil	Nil	10
Russula aeruginea	Nil	3	5
Tricholoma columbetta	Nil	Nil	28
T. nudum	Nil	8	Nil

Data from Kerrow, near Cannich, Invernesshire (Miles, 1985).

The transition from nursery to outplanting sites

Several studies have involved seedlings transplanted from nursery or glasshouse environments to field sites, a transfer involving a major change of environment. Ectendomycorrhizal fungi, such as the 'E-strain' group, are typical of nursery soils but often disappear from the root system in the first few years after transplanting to field sites (Thomas *et al.*, 1983).

Teleomorphs of the 'E-strain' fungi are unknown but evidence suggests that these fungi may be ascomycetous, possibly species of *Geospora, Trichophaea, Sphaerosporella* or *Humaria* (Danielson, 1982). The Ascomycotina, as a group, has tended to be overlooked in studies of mycorrhizal successions because the fruitbodies are, for the most part, inconspicuous. Danielson (1984) and Danielson, Zak & Parkinson (1984) have drawn attention to the presence of ascomycetes associated with saplings in field soils. *Sphaerosporella brunnea*, for example, is a carbonicolous operculate discomycete that may play an important role in mycorrhizal succession in forest stands frequently subjected to fire (Danielson, 1984). It remains to be seen whether ascospores of this fungus have a degree of heat-resistance and are heat-activated, properties well recognised in *Rhizina undulata*, the important pathogen that causes 'group dying' of conifers on recently burned sites (Jalaluddin, 1967).

Effects of different genotypes

Different genotypes of one host may significantly modify at least the early stages of mycorrhizal successions, as shown for *Betula pubescens* (Last *et al.*, 1984). After planting rooted cuttings of two clones of *B. pubescens* (9.3D and 9.3G) in two soils, fruitbodies of *Hebeloma* spp. and *Inocybe petiginosa* developed around both clones in one soil; in the other, however, fruitbodies of *Laccaria tortilis* developed around 9.3G while *Hebeloma* spp. were found in association with 9.3D. It is not known if these early differences persist or affect the later stages of successions.

It seems that variation within the host is paralleled by similar variation within species of mycorrhizal fungi, as shown in *Amanita muscaria* by Mason (1975). Fries (1987) has recognised two morphologically distinct field populations of *Paxillus involutus* in Scandinavia; one associated with conifers and the other found with broadleaved trees such as birch.

Effects of different soils

Soil and site characteristics markedly influence the types of mycorrhizal fungi that develop on tree roots, though as yet there is insufficient information to enable generalisations to be made except where extreme differences occur. *Cenococcum geophilum* typically forms mycorrhizas in dry locations (Mikola, 1949; Pigott, 1982), whereas *Paxillus, Scleroderma* and *Pisolithus* are common on coal spoil and other mining wastes (Marx, 1980; Gardner & Malajczuk, 1985; Ingleby, Last & Mason, 1985). Gardner & Malajczuk (1985) noted that a fungal succession occurred in association with eucalypts on a bauxite mine, with *Pisolithus, Scleroderma* and *Laccaria* representing the early colonisers while *Cortinarius, Paxillus, Ramaria* and *Russula* spp. represented the later colonisers.

As trees develop on formerly treeless sites numerous alterations are caused to the soil. The environment becomes less favourable for litter breakdown and nutrient remobilisation by saprotrophic fungi (Swift, Heal & Anderson, 1979; Vogt *et al.*, 1983), and substantial amounts of organic matter accumulate. In mature forests, ectomycorrhizas tend to be concentrated in the upper soil horizons, particularly the litter and raw humus layers. In a Douglas fir/larch forest, Harvey *et al.* (1976) found 95% of active mycorrhizas associated with the organic soil fraction and only 5% with the underlying mineral soil.

Tyler (1985) noted the occurrence of fruitbodies in woodlands on a range of soil types and found that some mycorrhizal fungi were more prevalent in organic than in mineral soils. Whereas *Laccaria laccata* fruited mainly at sites with small amounts of organic matter (<13%) *Amanita citrina, Cantharellus tubaeformis* and five *Cortinarius* spp. fruited mainly, if not exclusively, at sites with 30–93% organic matter. Such an observation is supported by Reddell & Malajczuk (1984) who grew jarrah (*Eucalyptus marginata*) seedlings in samples of mineral soil and forest litter taken from jarrah forests. Whereas black mycorrhizas (attributed to *Cenococcum geophilum*) were virtually confined to the mineral soils, white and brown mycorrhizas, attributed to basidiomycetes, were almost exclusively associated with seedlings growing in the litter.

Experiments have shown that mycorrhiza formation by *Paxillus involutus* is markedly affected by soil type. When spores of this fungus from fruitbodies occurring on coal spoil were added to samples of coal spoil and a brown earth in pots, birch seedlings developed many mycorrhizas of *Paxillus* in the coal spoil but only few mycorrhizas of this type (and on few seedlings) in the brown earth (Fox, 1986*a*). Last *et al.* (1985) found that *P. involutus*, after establishment on birch seedling roots in axenic culture, could persist and dominate the root system for at least a year after the seedlings were planted into pots of four contrasting (2 peat and 2 mineral) soils. However, by the end of the second year, *P. involutus* accounted for 44% of mycorrhizas in sphagnum peat of pH 3.4, and only 19% in a pH 5.4 mineral soil. In the same experiment, birch seedlings inoculated with *Hebeloma sacchariolens* had almost 100% of their roots tips colonised by this fungus after two years in both mineral soils (pH 5.1 and 5.4) and in sedge peat (pH 4.7) but the fungus completely disappeared from root systems in the first year in Sphagnum peat (pH 3.4). These observations from controlled experiments are corroborated by records made in the field. Thus, Dighton, Poskitt & Howard (1986) observed *Paxillus* and *Laccaria* in association with saplings of *Pinus contorta* and *Picea sitchensis* at a peat site, while Mason *et al.* (1982) found fruitbodies of *Hebeloma*, but not *Paxillus*, in conjunction with birch growing in a mineral soil.

In a detailed study made over four years, Fleming *et al.* (1984) grew *Betula pendula* for eight weeks from seed in two propagating soils; soils were taken from beneath the canopy ('mycorrhizal soil') of a birch tree and *c.* 11m from its periphery ('non-mycorrhizal'). Seedlings were then grown for a further two years in 'non-mycorrhizal' soil. Distinct assemblages of mycorrhizas developed on the seedlings during this time, with fruitbodies of *Hebeloma, Thelephora terrestris* and *Peziza badia* developing in association with the seedlings propagated initially in pots of 'non-mycorrhizal' soil, whereas *Inocybe lanuginella* and *Laccaria* fruited in those propagated in 'mycorrhizal' soil. After two years of glasshouse cultivation, the seedlings were outplanted to a treeless site about 100m from where the propagating soils were collected. The periphery of the root zone of all plants became dominated by mycorrhizas of *Inocybe*. However, the two batches of plants still retained different types of mycorrhiza in the older parts of the root zone after a total of four years growth. The significance of this lies in the fact that the majority of mycorrhizal root tips occurred in these older regions of the root zone. Thus, the type of soil used for seedling propagation during the first few weeks of plant growth can have a significant influence on the mycorrhizal status of plants for at least four years.

Irrespective of host and soil effects, a broad distinction can be made between 'early' and 'late' fungal colonisers. The only important qualification that needs to be made is that these categories refer to mycorrhizal successions on previously tree-less sites, because from evidence presented in the next section it will be clear that early-stage fungi may be bypassed if seedlings develop in the root zone of old trees.

Experimental analysis of mycorrhizal succession

Succession in most, if not all, communities involves sequential colonisation and species-replacement.

Sequential colonisation

All fungi, whether early- or late-stage, that have so far been investigated, can establish mycorrhizas on very young seedlings from mycelial inocula in axenic culture (e.g. Mason, 1980); none has proved to be incompatible with seedlings. However, when equivalent inocula were added to unsterile (natural) soils, only early-stage fungi were able to establish mycorrhizas on seedling roots (Deacon *et al.*, 1983; Danielson, Visser & Parkinson, 1984; Last *et al.*, 1985). When Fox (1986a) added spores of late-stage fungi to unsterile soils, she found that birch seedlings either developed mycorrhizas of early-stage fungi, such as *Thelephora* or *Hebeloma*, from naturally occurring sources of inoculum, or the seedlings remained non-mycorrhizal. This suggests that competition from other mycorrhizal fungi is not the major factor preventing the establishment of late-stage fungi on seedlings in soil.

In glasshouse and growth chamber experiments, Fox (1983, 1986a) added large numbers of spores of different mycorrhizal fungi to pots of mineral soil from a treeless site and found that only the early-stage fungi in successions established mycorrhizas on birch seedlings in this soil. All tested species of *Hebeloma (H. crustuliniforme, H. leucosarx* and *H. sacchariolens), Inocybe (I. geophylla, I. lacera, I. lanuginella)* and *Laccaria (L. proxima* and *L. tortilis)* were successful in establishing mycorrhizas from spores, whereas all species of *Cortinarius (C. bulbosus* and *C. debilitus), Lactarius (L. blennius, L. pubescens, L. rufus, L. spinosulus, L. turpis* and *L. vietus), Leccinum (Lecc. rosefractum* and *Lecc. scabrum)* and *Russula (R. cyanoxantha* and *R. grisea)* were unsuccessful. Fox's results suggest that under her conditions *genera* rather than individual species can be categorised as being able or unable to establish mycorrhizas from spores on seedling roots. Other workers have also reported major differences in the ability of fungi to establish mycorrhizas from spores in soil (Theodorou, 1971; Lamb & Richards, 1974a,b; Donald, 1975; Marx, 1976; Mullette, 1976; Marx, 1980). Marx (1980) reported that *Pisolithus* and *Thelephora* formed mycorrhizas from spores on pine seedlings in soil, whereas three species of *Amanita*, three of *Lactarius*, and *Suillus luteus, Russula emetica, Laccaria laccata* and *Paxillus involutus* did not.

A partial explanation of these results may lie in differences in spore germination between mycorrhizal fungi, though there have been few comparative studies of germination, and none that relates directly to mycorrhizal succession. On agar media, germination frequencies of 0.1% or less are reported for *Amanita muscaria, Leccinum scabrum, Lactarius helvus* and *Paxillus involutus*, even in the presence of activated charcoal or 'stimulatory' *Rhodotorula glutinis* (Fries, 1978). However, the relevance of this type of study is questionable since plant roots can stimulate the germination of basidiospores of mycorrhizal fungi (Melin, 1959; Birraux & Fries, 1981) and, also, because vegetative hyphae can stimulate germination, an effect that is highly taxon-specific for *Leccinum* spp. (Fries, 1981) but not for *Laccaria laccata* (Fries, 1983). It is not known whether homokaryotic mycelia derived from basidiospores can initiate mycorrhizal infection, or if dikaryotisation is necessary before infection can occur. Resolution of this question may be important because spores are the primary agents responsible for initiating mycorrhizal development – whether of early- or late-stage fungi – in sites that do not have a more or less continuous history of woodland cover. Nevertheless, it should be remembered that the behaviour of spore inocula in soils is paralleled by that of dikaryotic mycelial inocula produced in laboratory conditions and derived initially from fruitbody material.

In relation to mycorrhizal establishment, Fleming *et al.* (1986) found that 12 of 99 birch seedlings grown in pots of soil from an ancient birchwood (Struan wood) developed mycorrhizas of *Amanita* and *Leccinum*, and many seedlings grown in soil samples from beneath fruitbodies of *Lactarius tabidus, Tricholoma fulvum* or *Cortinarius* spp. developed mycorrhizas of these respective fungi. These results contrast markedly with previous work on soil samples from a young birch stand (Deacon *et al.*, 1983) where only early-stage fungi developed mycorrhizas on birch seedlings, even if the soil samples were taken from beneath fruitbodies of late-stage fungi. They also contrast with results from a sweet chestnut coppice with oak standards, soil samples from which gave rise to only early-stage mycorrhizal types on birch seedlings in a glasshouse, whereas the predominant mycorrhizal type on tree roots in the field was typical of the Boletaceae (late-stage) (Fleming, 1983). The coppice woodland was at least 50 years old, as indicated by the size of the oak standards, but was on a shallow clay-with-flints soil overlying chalk and did not have a substantial accumulation of organic matter. The available but sparse evidence suggests that litter layers and humus-rich woodland soils may be more 'conducive' to the establishment of late-stage mycorrhizal types on seedlings than are predominantly mineral soils, a deduction possibly explaining the zonation of mycorrhizal types outwards from the tree base towards the periphery of a root system in spaced trees on previously tree-less sites.

Competition between primary mycorrhizal colonisers (early-stage), or at least niche exclusion by them, has been demonstrated in glasshouse studies with soils supplemented with spores (Fox, 1983, 1986*a*). When non-mycorrhizal birch, pine or spruce seedlings were grown in soil from a treeless site or in vermiculite-peat mixtures, mycorrhizas developed from spores of early-stage fungi. In contrast, when the soils were supplemented with spores of an early-stage inoculant fungus, sucn as *Hebeloma, Laccaria* or *Inocybe,* the same number of mycorrhizas developed as in the non-supplemented soil but were usually solely attributable to the inoculant fungus, which thus excluded the 'contaminants'. The same was observed when axenically grown birch seedlings, colonised by mycelial cultures of *Hebeloma sacchariolens*, were transferred to unsterile soils (Mason *et al.*, 1983). This ineffectiveness of introduced inoculum may have implications for the commercial inoculation of trees with mycorrhizal fungi.

Species replacement

The possibility that early-stage mycorrhizal fungi facilitate the subsequent establishment of late-stage fungi was investigated by Fleming (1985). Birch seedlings were inoculated with *Hebeloma sacchariolens* or

Laccaria proxima, or left uninoculated, before planting in a brown earth supplemented with mycelial inocula of *Amanita muscaria, Leccinum scabrum* or *Lactarius pubescens*. For the 441 d duration of the experiment, no mycorrhizas of *Amanita* or *Leccinum* developed. *Lactarius* mycorrhizas developed on roots, but independently of inoculation with *L. pubescens*, suggesting that they arose from naturally-occurring inoculum. At the end of the experiment *Lactarius* mycorrhizas represented up to 30% of the total number of mycorrhizas occurring on the originally non-mycorrhizal seedlings and those inoculated with *Hebeloma*, whereas none occurred on seedlings initially mycorrhizal with *Laccaria*. There was, therefore, no evidence of facilitated replacement of early-stage fungi by late-stage fungi, but there was evidence of inhibition of the development of the late-stage fungus, *Lactarius pubescens*, by *Laccaria* previously established on the root systems. In some respects these results parallel those of Fleming *et al.* (1984) who found that *L. pubescens* from natural inoculum sources colonised birch saplings if these had previously become colonised by *Hebeloma* sp. but not if they were colonised by *Inocybe* sp.

Two other experimental approaches have yielded information relating to replacement phenomena. In the first (Fleming, 1985), mycelial cultures of three early-stage fungi (*Paxillus involutus, Hebeloma sacchariolens* and *Thelephora terrestris*) were either inoculated onto seedlings in aseptic conditions (primary inoculants) or added to soil as 'secondary inoculants'. The seedlings were then planted in soil so that the different combinations of primary and secondary inoculants were compared together with uninoculated controls. In such an experiment in which two 'soils' were used – a brown earth and a peat – *Paxillus* did not persist as a primary inoculant in any soil or soil treatment, even if no secondary inoculant was present. Instead, in the absence of a secondary inoculant the root systems were colonised by *Thelephora* in brown earth and *Inocybe lacera* in peat, from presumably low levels of naturally-occurring inoculum. *Paxillus* colonised all seedlings in all treatments if it was used as a secondary inoculant, and it then replaced the primary inoculants. In contrast to *Paxillus, Thelephora* persisted as a primary inoculant in both soils in the absence of secondary inoculants, but was replaced by *Paxillus* secondary inoculum in both soils and by *Hebeloma* secondary inoculum in brown earth. Yet another pattern was displayed by *Hebeloma*, which persisted as a primary inoculant in brown earth and not in peat, but was replaced by *Paxillus* (secondary inoculant) in both soils. Thus, in addition to demonstrating the over-riding importance of soil type, the experiment showed unexpectedly that *Paxillus involutus* was a successful colonist when added to soil as a secondary inoculant but failed to persist when inoculated to seedlings as a primary inoculant. *Hebeloma* and *Thelephora* experienced no such difficulty as primary inoculants in soils favourable to their development.

In the second approach, non-mycorrhizal birch seedlings were planted among the roots of old trees (Fleming, 1983, 1984; Fleming *et al.*, 1986). The early stages of mycorrhizal succession were bypassed, because even young seedlings developed late-stage mycorrhizas or a mixture of early- and late-stage types. However, if the seedlings were planted into soil which had previously been cored or trenched to separate the roots in it from contact with the parent tree, then late-stage mycorrhizal fungi developed poorly, if at all, on the seedling roots. It seems, therefore, that the ability of late-stage fungi to colonise the roots of seedlings growing in unsterile soils is enhanced if they have uninterrupted links with their food bases represented by their 'parent' tree, a situation comparable to the behaviour of rhizomorphs of the pathogen *Armillaria mellea* and strands of *Serpula lacrymans*, the dry rot fungus (Garrett, 1956).

In terms of replacement, fungi that form mycelial strands or copious networks of unaggregated hyphae from the mycorrhizal sheath might be expected progressively to dominate root systems once they are established in a succession. Many late-stage mycorrhizal fungi do, indeed, produce mycelial strands, which may serve roles in addition to infection, such as the transport of water and mineral nutrients (Read, Francis & Finlay, 1985). In contrast, many early-stage fungi do not produce conspicuous mycelial strands and, instead, rely on spore inocula for colonisation of new flushes of roots. Persistence of some of these early-stage fungi may be facilitated by the production of sclerotia, as in *Hebeloma sacchariolens* (Fox, 1986*b*) and *Paxillus involutus* (Fox, 1986*c*).

Concluding remarks

For long periods the existence and value of mycorrhizas was taken for granted by foresters. From observations of fungal fruitbodies, and more recently supported by inoculation experiments, it has been possible to identify some of the factors which together control a well-ordered, age-related sequence (succession) of mycorrhizal fungi colonising the roots of trees. With the recognition of the phenomena of succession and host specificity (Chapter 2), and the need to take account of the influence of soil type, it is clear that the stage is set for a rational approach to the manipulation of mycorrhizal successions in forests. Bearing in mind the increasing areas felled and replanted, it is now imperative to broaden our focus to include the ecology of mycorrhizal fungi in second-rotation forests, which may not be the same as in the first rotation.

References

Birraux, D. & Fries, N. (1981). Germination of *Thelephora terrestris* basidiospores. *Canadian Journal of Botany*, **59**, 2062–4.

Brown, J.C. (1958). Soil fungi of some British sand dunes in relation to soil type and succession. *Journal of Ecology*, **46**, 641–64.

Chu-Chou, M. (1979). Mycorrhizal fungi of *Pinus radiata* in New Zealand. *Soil Biology and Biochemistry*, **11**, 557–62.

Chu-Chou, M. & Grace, L.J. (1981). Mycorrhizal fungi of *Pseudotsuga menziesii* in the north island of New Zealand. *Soil Biology and Biochemistry*, **13**, 247–9.

Chu-Chou, M. & Grace, L.J. (1982). Mycorrhizal fungi of *Eucalyptus* in the north island of New Zealand. *Soil Biology and Biochemistry*, **14**, 133–7.

Chu-Chou, M. & Grace, L.J. (1983). Characterisation and identification of mycorrhizas of Douglas-fir in New Zealand. *European Journal of Forest Pathology*, **13**, 251–60.

Cooke, R.C. & Rayner, A.D.M. (1984). *Ecology of Saprotrophic Fungi*. London: Longman.

Danielson, R.M. (1982). Taxonomic affinities and criteria for identification of the common ectendomycorrhizal symbiont of pines. *Canadian Journal of Botany*, **60**, 7–18.

Danielson, R.M. (1984). Ectomycorrhiza formation of the operculate discomycete *Sphaerosporella brunnea* (Pezizales). *Mycologia*, **76**, 454–61.

Danielson, R.M., Zak, J.C. & Parkinson, D. (1984). Mycorrhizal inoculum in a peat deposit formed under a white spruce stand in Alberta. *Canadian Journal of Botany*, **63**, 2557–60.

Danielson, R.M., Visser, S. & Parkinson, D. (1984). The effectiveness of mycelial slurries of mycorrhizal fungi for the inoculation of container-grown Jack pine seedlings. *Canadian Journal of Forest Research*, **14**, 140–2.

Deacon, J.W., Donaldson, S.J. & Last, F.T. (1983). Sequences and interactions of mycorrhizal fungi on birch. *Plant and Soil*, **71**, 257–62.

Dighton, J., Poskitt, J.M. & Howard, D.M. (1986). Changes in occurrence of basidiomycete fruitbodies during forest stand development: with specific reference to mycorrhizal species. *Transactions of the British Mycological Society*, **87**, 163–71.

Donald, D.G.M. (1975). Mycorrhizal inoculation of pines. *Journal of the South African Forestry Association*, **92**, 27–9.

Fleming, L.V. (1983). Succession of mycorrhizal fungi on birch: infection of seedlings planted around mature trees. *Plant and Soil*, **71**, 263–7.

Fleming L.V. (1984). Effects of soil trenching and coring on the formation of ectomycorrhizas on birch seedlings grown around mature trees. *New Phytologist*, **98**, 143–53.

Fleming, L.V. (1985). Experimental study of sequences of ectomycorrhizal fungi on birch (*Betula* sp.) seedling root systems. *Soil Biology and Biochemistry*, **17**, 591–600.

Fleming, L.V., Deacon, J.W., Last, F.T. & Donaldson, S.J. (1984). Influence of propagating soil on the mycorrhizal succession of birch seedlings transplanted to a field site. *Transactions of the British Mycological Society*, **82**, 707–11.

Fleming, L.V., Deacon, J.W. & Last, F.T. (1986). Ectomycorrhizal succession in a Scottish birch wood. In: *Physiological and Genetical Aspects of Mycorrhizae* (ed. V. Gianinazzi-Pearson & S. Gianinazzi), pp. 259–64. Paris: INRA-Presse.

Ford, E.D., Mason, P.A. & Pelham, J.P. (1980). Spatial patterns of sporophore distribution around a young birch tree in three successive years. *Transactions of the British Mycological Society*, **75**, 287–96.

Fox, F.M. (1983). Role of basidiospores as inocula of mycorrhizal fungi of birch. *Plant and Soil*, **71**, 269–73.

Fox, F.M. (1986a). Groupings of ectomycorrhizal fungi of birch and pine, based on establishment of mycorrhizas on seedlings from spores in unsterile soils. *Transactions of the British Mycological Society*, **87**, 371–80.

Fox, F.M. (1986b). Ultrastructure and infectivity of sclerotium-like bodies of the ectomycorrhizal fungus *Hebeloma sacchariolens*, on birch (*Betula* spp.). *Transactions of the British Mycological Society*, **87**, 359–69.

Fox, F.M. (1963c). Ultrastructure and infectivity of sclerotia of the ectomycorrhizal fungus *Paxillus involutus* on birch (*Betula* spp.). *Transactions of the British Mycological Society*, **87**, 627–31.

Frankland, J.C. (1981). Mechanisms in fungal successions. In: *The Fungal Community: its Organisation and Role in the Ecosystem*, ed. D.T. Wicklow & G.C. Carroll, pp. 403–26. New York: Marcel Dekker.

Fries, N. (1978). Basidiospore germination in some mycorrhiza-forming hymenomycetes. *Transactions of the British Mycological Society*, **70**, 319-24.

Fries, N. (1981). Recognition reactions between basidiospores and hyphae in *Leccinum*. *Transactions of the British Mycological Society*, **77**, 9–14.

Fries, N. (1983). Spore germination, homing reaction, and inter-sterility groups in *Laccaria laccata* (Agaricales). *Mycologia*, **75**, 221–7.

Fries, N. (1987). Ecological and evolutionary aspects of spore germination in the higher basidiomycetes. *Transactions of the British Mycological Society*, **88**, 1–7.

Gardner, J.N. & Malajczuk, N. (1985). Succession of ectomycorrhizal fungi associated with eucalypts on rehabilitated bauxite mines in South Western Australia. In: *Proceedings of the 6th North American Conference on Mycorrhizae*, ed. R. Molina, p. 265. Corvallis. Oregon: Oregon State University Press.

Garrett, S.D. (1956). *Biology of Root-infecting Fungi*. Cambridge: Cambridge University Press.

Harper, J.E. & Webster, J. (1964). An experimental analysis of the coprophilous fungus succession. *Transactions of the British Mycological Society*, **47**, 511–30.

Harvey, A.E., Larsen, M.J. & Jurgensen, M.F. (1976). Distribution of ectomycorrhizae in a mature Douglas fir-larch forest soil in western Montana. *Forest Science*, **22**, 393–8.

Ingleby, K., Last, F.T. & Mason, P.A. (1985). Vertical distribution and temperature relations of sheathing mycorrhizas of *Betula* spp. growing on coal spoil. *Forest Ecology and Management*, **12**, 279–85.

Jalaluddin, M. (1967). Studies on *Rhizina undulata*. I. Mycelial growth and ascospore germination. *Transactions of the British Mycological Society*, **50**, 449–59.

Lamb, R.J. & Richards, B.N. (1974a). Inoculation of pines with mycorrhizal fungi in natural soils. I. Effects of density and time of application of inoculum and phosphorus amendment on mycorrhizal infection. *Soil Biology and Biochemistry*, **6**, 167–71.

Lamb, R.J. & Richards, B.N. (1974b). Inoculation of pines with mycorrhizal fungi in natural soils. II. Effects of density and time of application of inoculum and phosphorus amendment on seedling yield. *Soil Biology and Biochemistry*, **6**, 163–7.

Last, F.T., Mason, P.A., Wilson, J. & Deacon, J.W. (1983). Fine roots and sheathing mycorrhizas: their formation, function and dynamics. *Plant and Soil*, **71**, 9–21.

Last, F.T., Mason, P.A., Pelham, J. & Ingleby, K. (1984). Fruitbody production by sheathing mycorrhizal fungi: effect of 'host' genotypes and propagating soils. *Forest Ecology and Management*, **9**, 221–7.

Last, F.T., Mason, P.A., Wilson, J., Ingleby, K., Fleming, L.V. & Deacon, J.W. (1985). 'Epidemiology' of sheathing (ecto-) mycorrhizas in unsterile soils: a case study of *Betula pendula*. *Proceedings of the Royal Society of Edinburgh*, **85B**, 299–315.

Malajczuk, N., Molina, R. & Trappe, J.M. (1982). Ectomycorrhiza formation in *Eucalyptus*. 1. Pure culture synthesis, host specificity and mycorrhizal compatibility in *Pinus radiata*. *New Phytologist*, **91**, 467–82.

Marx, D.H. (1976). Synthesis of ectomycorrhizae on loblolly pine seedlings with basidiospores of *Pisolithus tinctorius*. *Forest Science*, **91**, 467–82.

Marx, D.H. (1980). Ectomycorrhizal fungus inoculation: a tool for improving forestation practices. In: *Tropical Mycorrhiza Research*, ed. P. Mikola, pp. 13–71. Oxford: Clarendon Press.

Mason, P.A. (1975). The genetics of mycorrhizal associations between *Amanita muscaria* and *Betula verrucosa*. In: *The Development and Function of Roots*, ed. J.G. Torrey & D.T. Clarkson, pp. 567–74. London: Academic Press.

Mason, P.A. (1980). Aseptic synthesis of sheathing (ecto-) mycorrhizas. In: *Tissue Culture Methods for Plant Pathologists*, ed. D.S. Ingram & J.P. Helgeson, pp. 173–8. Oxford: Blackwell Scientific.

Mason, P.A., Last, F.T., Pelham, J. & Ingleby, K. (1982). Ecology of some fungi associated with an ageing stand of birches (*Betula pendula* and *B. pubescens*). *Forest Ecology and Management*, **4**, 13–39.

Mason, P.A., Wilson, J., Last, F.T. & Walker, C. (1983). The concept of succession in relation to the spread of sheathing mycorrhizal fungi on inoculated tree seedlings growing in unsterile soils. *Plant and Soil*, **71**, 247–56.

Melin, E. (1959). Physiological aspects of mycorrhizae of forest trees. In: *Tree Growth*, ed. T.T. Kozlowski, pp. 247–63. New York: Ronald Press.

Mikola, P. (1949). On the physiology and ecology of *Cenococcum graniforme*, especially as a mycorrhizal fungus of birch. *Communicationes Instituti Forestalis Fenniae*, **36**, 1–104.

Miles, J. (1985). Soil in the ecosystem. In: *Ecological Interactions in Soil: Plants, Microbes and Animals*, ed. A.H. Fitter, D. Atkinson, D.J. Read & M.B. Usher, pp. 407–27. British Ecological Society Special Publication 4. Oxford: Blackwell Scientific.

Molina, R. & Trappe, J.M. (1982). Applied aspects of ectomycorrhizae. In: *Advances in Agricultural Microbiology*, ed. N.S. Subba Rao, pp. 305–24. London: Butterworth Scientific.

Mullette, K.H. (1976). Studies on Eucalypt mycorrhizas. I. A method of mycorrhizal induction in *Eucalyptus gummifera* (Gaertn. & Hochr.) by *Pisolithus tinctorius* (Pers.) Coker & Couch. *Australian Journal of Botany*, **24**, 193–200.

Park, D. (1968). The ecology of terrestrial fungi. In: *The Fungi*, ed. G.C. Ainsworth & A.S. Sussmann, pp. 5–39. New York: Academic Press.

Pigott, C.D. (1982). Survival of mycorrhiza formed by *Cenococcum geophilum* Fr. in dry soils. *New Phytologist*, **92**, 513–17.

Read, D.J., Francis, R. & Finlay, R.D. (1985). Mycorrhizal mycelia and nutrient cycling in plant communities. In: *Ecological Interactions in Soil: Plants, Microbes and Animals*, ed. A.H. Fitter, D. Atkinson, D.J. Read & M.B. Usher, pp. 193–217. British Ecological Society Special Publication 4. Oxford: Blackwell Scientific.

Reddell, P. & Malajczuk, N. (1984). Formation of mycorrhizae by Jarrah (*Eucalyptus marginata* Donn ex Smith) in litter and soil. *Australian Journal of Botany*, **32**, 511–20.

Richardson, M.J. (1970). Studies on *Russula emetica* and other agarics in a Scots pine plantation. *Transactions of the British Mycological Society*, **55**, 217–29.

Rose, S.L. (1980). Mycorrhizal associations of some actinomycete nodulated nitrogen-fixing plants. *Canadian Journal of Botany*, **58**, 1449–54.

Swift, M.J., Heal, O.W. & Anderson, J.M. (1979). *Decomposition in Terrestrial Ecosystems.* Oxford: Blackwell Scientific.

Theodorou, C. (1971). Introduction of mycorrhizal fungi into soil by spore inoculation of seed. *Australian Forestry*, **34**, 183–91.

Thomas, G.W., Rogers, D. & Jackson, R.M. (1983). Changes in the mycorrhizal status of Sitka spruce following outplanting. *Plant and Soil*, **71**, 319–23.

Tyler, G. (1985). Macrofungal flora of Swedish beech forest related to soil organic matter and acidity characteristics. *Forest Ecology and Management*, **10**, 13–29.

Vogt, K.A., Grier, C.C., Meier, C.E. & Keyes, M.R. (1983). Organic matter and nutrient dynamics in forest floor of young and mature *Abies amabilis* stands in western Washington, as affected by fine-root input. *Ecological Monographs*, **53**, 139–57.

Watling, R. (1981). Relationships between macromycetes and the development of higher plant communities. In: *The Fungal Community: its Organization and Role in the Ecosystem*, ed. D.T. Wicklow & G.C. Carroll, pp. 427–58. New York: Marcel Dekker.

Watling, R. (1984). Larger fungi around Kindrogan, Perthshire. *Transactions of the Botanical Society of Edinburgh*, **44**, 237–59.

13

Formation and dispersal of propagules of endogonaceous fungi

CHRISTOPHER WALKER

Forestry Commission, Northern Research Station, Roslin, Midlothian EH25 9SY, UK

Introduction

Endogonaceous fungi are among the most common of organisms, occurring from north of the Arctic Circle through the temperate zones of both hemispheres, the tropics, and almost to the Antarctic. They are found in grassland, woodland, arid lands, bogs and swamps, and even in some aquatic environments. The group seems to have evolved early, possibly concurrently with land plants (Malloch, Pirozynski & Raven, 1980). Fossil remains which are virtually indistinguishable from some present forms have been found from as early as the Carboniferous Period (Wagner & Taylor, 1982). Even the much earlier Devonian deposits of Rhynie contain fossils which bear a strong resemblance to extant endogonaceous spores (Butler, 1939; Nicolson, 1975). Although there are currently only about 100 species formally described (Trappe & Schenck, 1982), it is likely that many more will be found as taxonomic and ecological investigations continue (Walker, 1985). The number of species known belies their ecological importance and in nature very few families of land plants do not depend on them as essential symbionts (Harley & Smith, 1983). The mycorrhizal relationships of these fungi have been reviewed extensively in recent years (e.g., Smith, 1980; Mosse, Stribley & LeTacon, 1981; Harley & Smith, 1983; Molina, 1985).

Much research on mycorrhizas has been concerned with the possibility that the fungi may be used to enhance plant growth. Unlike studies of plant pathology, in which it is usually necessary to know not only the species but also the strain of fungus involved, studies of endomycorrhizal Endogonaceae have often been made on the assumption that the identity and biology of the fungi are of little importance, an attitude possibly engendered by the apparent lack of host-fungus specificity. Although many of these fungi will form symbioses with most endomycorrhizal hosts, there is now no doubt that not only different species, but different strains within a species, produce very different plant-growth responses, depending on such factors as edaphic conditions and host genotype (Abbott & Robson, 1982a,b; Miller, Domoto & Walker, 1985). Experimental information on the formation and dispersal of propagules of endogonaceous fungi is sparse, perhaps mainly because most have not yet been cultured *in vitro*. Much of the knowledge that is available is therefore based on chance observation and on the results of survey work.

The Endogonaceae

Most endogonaceous fungi form vesicular-arbuscular or arbuscular mycorrhizas (VAM). Their current placing in a single family, the Endogonaceae in the Endogonales of the Zygomycotina (Benjamin, 1979), is more a matter of historical convenience than of known phylogeny. Fungi in the type genus *Endogone* are either ectomycorrhizal or saprotrophic whereas species of other genera in the family are all endomycorrhizal symbionts where their nutritional mode is known. Such fungi may be phylogenetically linked, but much more information is needed before credence can be given to theories about the relationships among the genera (Walker, 1985).

Together with the ectomycorrhizal or saprotrophic *Endogone*, there are six endomycorrhizal genera in the Endogonaceae: *Acaulospora, Entrophospora, Gigaspora, Scutellospora, Glomus* and *Sclerocystis*. The genera *Modicella* and *Complexipes*, formerly included in the family, were removed from it by Trappe & Schenck (1982) and in view of recent evidence of ascomycetous affinities (Gibson, 1984; Gibson, Kimbrough & Benny, 1985), *Glaziella* is now also excluded. The genus *Scutellospora* is the result of a recent taxonomic dichotomy of *Gigaspora* (Walker & Sanders, 1985, 1986). The genera in the Endogonaceae are principally separated on their mode of spore formation though there is some debate as to whether this taxonomic scheme is valid (Mosse *et al.*, 1981). It relies largely on spore morphology (Gerdemann & Trappe, 1974) and is little more than a phenetic system for cataloguing organisms which produce similar resting spores. Nevertheless, it is a workable system of classification. The most recent guides to the taxonomy of the group are by Hall (1984) and Trappe (1982). However, due to the erection of a new genus (Walker & Sanders, 1986) and recent descriptions of new species (Koske, Miller & Walker, 1983; Koske & Walker 1984, 1985; Morton & Walker, 1984; Walker, Reed & Sanders, 1984), these are already outdated.

Propagule formation
In both ecto- and endo-mycorrhizal Endogonaceae, the most obvious form of propagule is the spore. These are formed in a variety of ways, either singly, in clusters, or in small sporocarps. Except for one species of *Gigaspora,* each genus is known to form only one type of spore, but in the Endogonaceae the term 'spore' is conventionally used in a very broad sense. Discussion of characteristics such as wall structure have usually not distinguished between sporangia and spores (e.g. Gerdemann & Trappe, 1974), though for *Endogone* the distinction has recently been made (Berch & Fortin, 1982; Gibson, 1985). Although the convention has been followed here, the reader should be aware that the so-called spores' of such genera as *Acaulospora, Entrophospora,* and *Scutellospora* may prove in fact to be sporangia containing a single sporangiospore. Two genera form zygospores (see Note on p. 281), four form spores which are often (perhaps erroneously) termed 'azygospores', and the remaining two form chlamydospores. There are also a few organisms that do not precisely fit into these patterns. Questions have been raised about the relationships among these genera (Walker, 1985). When further information is available, it may be necessary to create more than one family from this assemblage of taxa.

Endogone
Spores of *Endogone* species are formed embedded in a matrix of glebal hyphae within sporocarps (Gerdemann & Trappe, 1974; Tandy, 1975). A zygospore develops from the conjugation of two gametangia by development of the gametangial walls into a zygosporangium (Berch & Fortin, 1983). The zygosporangium may bud from the apex of the larger of two gametangia, from the point of the gametangial union, or from two distinctly separated suspensor cells (Gerdemann & Trappe, 1974). The only known details of karyogamy in the genus are those described by Bucholtz (1912) for *Endogone flammicorona* (considered at the time to be *E. lactiflua* but later separated from that species (Trappe & Gerdemann, 1972). The zygote, formed as a terminal outgrowth from the female gametangium, contains two haploid nuclei, one female and one male. These do not fuse in the resting spore, but are assumed to fuse at the time of germination. Germination has been recorded and illustrated for only two species of the genus, *E. pisiformis* and *E. incrassata* (Berch & Fortin, 1982, 1983), but no cytological details were given. Tandy (1975) examined material of several species of *Endogone* and interpreted the mode of spore formation from observations of sporocarps of

presumed different ages collected from the field. Illustrations were presented of the mode of zygospore formation for *Endogone flammicorona* (Fig. 13.1A) and it was shown that *E. reticulata* forms its zygosporangium in a similar manner. Berch & Fortin (1982) illustrated mature spores of *E. pisiformis* and indicated that, in this species, gametangia develop separately, though closely and the zygosporangium forms between them. A more detailed study of the species by Gibson (1985) confirms this, but also reveals that both primary and secondary zygosporangia are formed (Fig. 13.1B). In one figure (Fig. 1.15, Gibson, 1985) Gibson illustrated a section which led him to suggest that nuclear fusion takes place within the zygospore, though he clearly pointed out the speculative nature of this observation.

Glomus and *Sclerocystis*

Both these genera produce chlamydospores, formed usually at the tip of a single hypha, but sometimes from two or more hyphae (Thaxter, 1922; Gerdemann & Trappe, 1974). In addition to being produced in sporocarps, spores of *Glomus* may be produced ectocarpically either singly or in loose clusters; those of *Sclerocystis* are apparently produced only in sporocarps and are arranged in an orderly manner around a sterile central hyphal plexus rather than randomly among the glebal hyphae (Gerdemann & Trappe, 1974). The distinction between these genera is not great. Recently described species of *Glomus* produce both ectocarpic and sporocarpic spores (Smith & Schenck, 1985), the latter forming around a hyphal plexus in a manner similar to those of *Sclerocystis.*

There is no published information about the ontogeny of spores in this group. Their wall structure and mode of occlusion is extremely variable. For example, spores of *Glomus deserticola* have a simple wall structure with only a single laminated wall (Trappe, Bloss & Menge, 1984), whereas those of *Glomus maculosum* have multiple walls which have the effect of producing an endospore (Miller & Walker, 1986). Madelin (1979) has pointed out that diverse forms of conidiogenesis in the Fungi Imperfecti may not indicate diverse phylogeny. Conversely in the Endogonaceae, superficial similarity of spore morphology does not necessarily indicate a common phylogeny.

Species of *Glomus* are the most commonly used organisms in endomycorrhizal research, yet little is known about the formation of their spores. Such information would be of great taxonomic interest and of value for the production of inoculum for use in horticulture, agriculture and forestry.

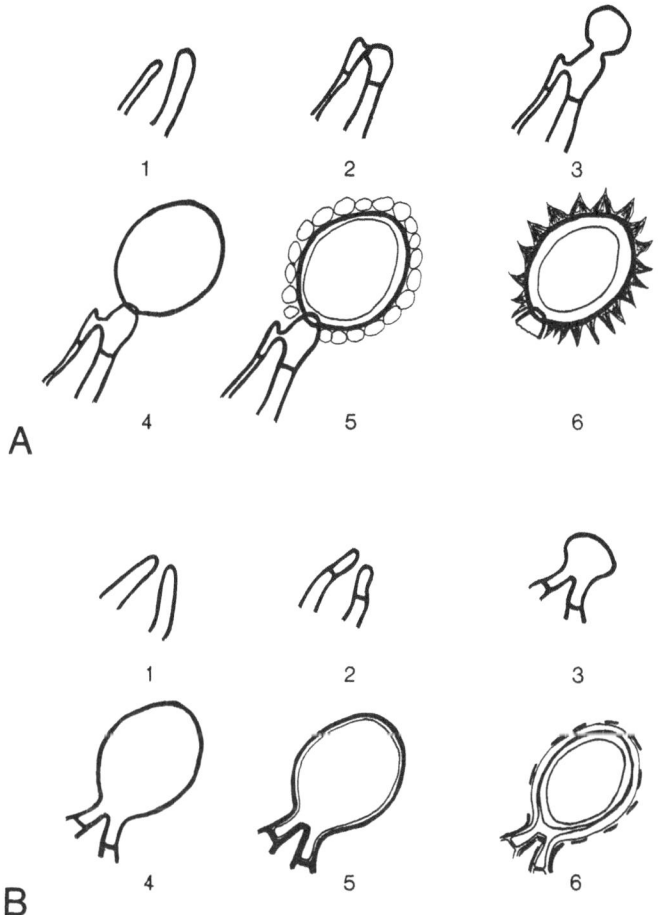

Fig. 13.1. Diagrammatic representation of formation of *Endogone* zygospores. Drawn from the plates of Berch & Fortin (1983), Bonfante-Fasola & Scannerini (1976), Gibson (1985), and Tandy (1975).
(A) *E. flammicorona*. (1) formation of progametangia. (2) gametangia are formed. (3) formation of a fusion pore between gametangia. (4) zygosporangium formed by larger gametangium. (5) zygospore formation. Hyphae around the zygosporangium are shown in this diagram. (6) mature zygospore. The surrounding hyphae collapse and their walls thicken to form 'flammenkronen', a feature typical of this species.
(B) *E. pisiformis*. (1) formation of progametangia. (2) gametangia are formed. (3) and (4) primary zygosporangium formation. (5) a secondary zygosporangium is formed. (6) the zygospore is formed and the primary sporangium disintegrates.

Acaulospora

This genus has a unique mode of spore formation (Fig. 13.2). Initially, the somatic hypha begins to enlarge at the tip and forms characteristic filiform hyphal appendages. The tip gradually enlarges to form a swollen saccule termed a sporiferous saccule (Walker, Reed & Sanders, 1984), mother spore (Mosse, 1970*a,b*), or swollen hyphal terminus (Schenck *et al.*, 1984). A swelling appears to the side of the saccule and a spore begins to form. The contents empty into the spore as it enlarges, after which the saccule collapses and the pore through which the contents passed becomes occluded. The spore walls thicken as maturity is reached. In some species, the saccule becomes completely detached, leaving a sessile spore with little evidence of its origin, except for a thickened collar around the occluded pore. In other species, the saccule or remnants of it usually remains attached to the spore and can be seen on microscopic examination.

The exact nature of the spore of *Acaulospora* is unknown. It is not clear whether the saccule wall is continuous with the outer spore wall, though it appears so in two species, *A. nicolsonii* (Walker *et al.*, 1984) and an unidentified member of the genus (Mosse, 1970*b*), thought to be *A. laevis* (Gerdemann & Trappe, 1974). With the possible exception of

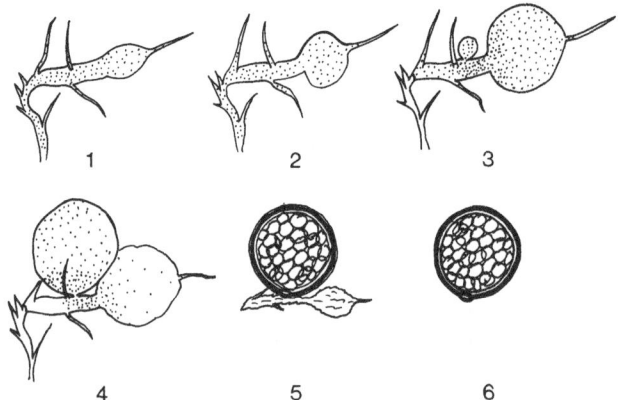

Fig. 13.2. Diagrammatic representation of the development of a typical spore of an *Acaulospora* sp. Drawn from the plates of Mosse (1970*a,b*) and Walker *et al.* (1984). (1) and (2) a hyphal tip begins to swell and forms a saccule. (3) and (4) a swelling appears on the neck of the saccule, and becomes enlarged to form a spore into which the contents of the saccule then empty. (5) the spore walls are laid down and thicken, the contents become vacuolate, and the saccule collapses. (6) the collapsed saccule often becomes completely detached to leave a sessile spore with only a small, raised collar to indicate the point of origin.

A. trappei, all species in this genus have inner, membranous walls (Walker, 1983) and the outer walls seem to act as a sporangium containing a single endospore. Details of karyogamy are unknown, but from cytological study the spores of one species are known to be multinucleate (Mosse, 1970*a*). Because of the presence of the filiform hyphal appendages and the enlarged sporiferous saccule, both of which could be considered to be forms of gametangia, it has been suggested that spores of *Acaulospora* might be zygospores (Mosse, 1970*a*). However, since conjugation of gametangia has not been demonstrated, they are generally referred to as 'azygospores' (Gerdemann & Trappe, 1974).

Entrophospora

Until recently, this was a monospecific genus of unknown mycorrhizal status. However, new species have now been named, and all described members of the genus form vesicular–arbuscular mycorrhizas (Schneck *et al.*, 1984; Sieverding & Toro, 1986). Spore production in this genus (Fig. 13.3) is somewhat similar to that in *Acaulospora,* though its detail is not as well elucidated. The spores (usually termed 'azygospores') (Ames & Schneider, 1979) are produced from a sporiferous saccule, but inside its neck, rather than from its side. The cytology and karyogamy of *Entrophospora* spp. and details of spore ultrastructure are unknown.

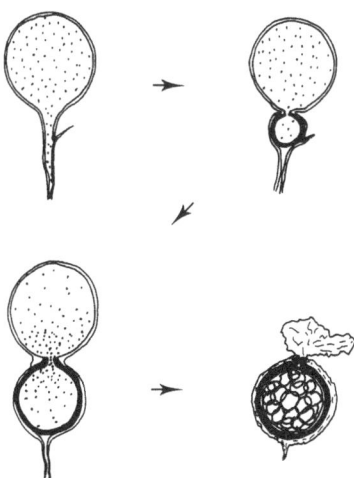

Fig. 13.3. The development of a spore of *Entrophospora infrequens.* Drawn from the plates of Ames & Schneider (1979). Apart from the formation of the spore in, rather than at the side of, the saccule neck, the details are similar to spore formation in *Acaulospora* spp.

Gigaspora and *Scutellospora*

These genera both produce very large single spores (in excess of 800 μm diameter in some species) on inflated hyphal cells in the soil (Fig. 13.4). These bulbous cells are termed 'suspensor-like cells' (Gerdemann & Trappe, 1974) and usually have one or more small peg-like processes projecting towards the spore, from which extend a thin filiform hypha that appears collapsed and empty at spore maturity. These projections seem not to be gametangial in nature (hence the term azygospore for the spores) and similar structures may be observed at irregular intervals along the sporogenic hypha. No detailed study of spore ontogeny exists and Fig. 13.4 is drawn from light-microscope observations of spores extracted from pot cultures, with some additional details deduced from the plates of Gibson (1985). No details of karyogamy are known, but the spores are multinucleate.

The spores of *Scutellospora* and *Gigaspora*, although superficially similar, seem to be fundamentally different. The former has an endospore formed by flexible walls in which develops a complex structure, the 'germination shield' (Koske, Miller & Walker, 1983; Walker & Sanders, 1986), from which the germ tubes arise before penetrating the outer spore wall. Spores of *Gigaspora* do not possess flexible walls and germination is simpler, germ tubes emerging directly through the spore wall.

Species in these genera form small (approximately 50 μm diameter) ornamented cells, called 'auxiliary cells', on their somatic hyphae in the soil (Fig. 13.5). Those of *Gigaspora* are echinulate, whilst those of *Scutellospora* are knobbly (Walker & Sanders, 1986). The function of these structures is unknown, though Gerdemann & Trappe (1974) suggest that they may be temporary food storage organs or modified subsporangial vesicles. They do not seem to be spores, although hyphae can be produced from their subtending hypha after detachment (Pons & Gianinazzi-Pearson, 1985).

The fine endophyte

One of the commonest forms of endomycorrhizal colonisation, particularly in upland and acidic soils, is that formed by the so-called fine endophyte. This type of symbiosis is characterised by having extremely fine hyphae, about 0.5 μm thick, and by producing similarly small arbuscules, vesicles and fan-shaped hyphal structures (Hall, 1977). It can be an efficient endophyte, enhancing plant growth in certain circumstances (Crush, 1973; Sparling & Tinker, 1978b; Powell, 1980). The fungus was originally named *Rhizophagus tenuis* by Greenall (1963), but was transferred to *Glomus* (Hall, 1977) after spores were found. Because of the difference in mycorrhizal anatomy, and the tenuous relationship with other members of *Glomus*, it merits separate treatment here. The fungus *(G. tenue)* produces minute, dark brown spores (presumably chlamydospores), 10-12 μm in diameter, in the soil. When immature, these spores are colourless and are indistinguishable from the 'vesicles' formed within

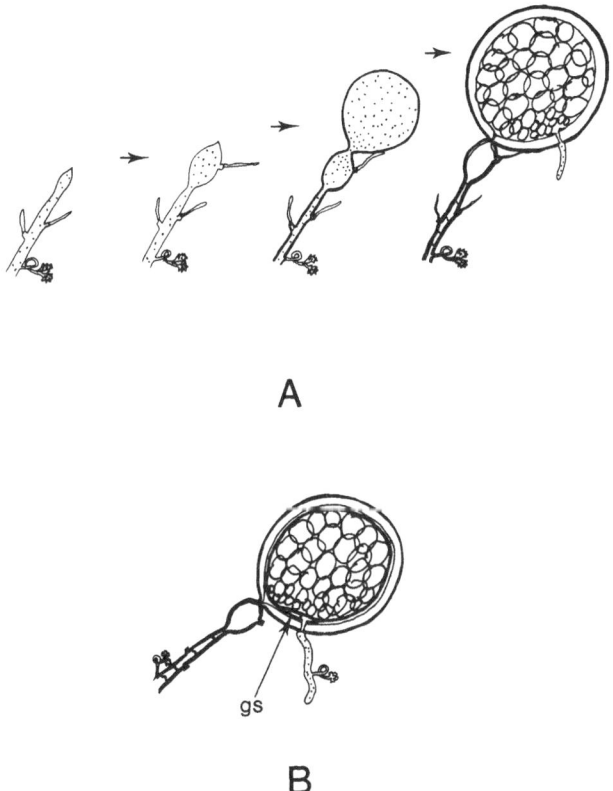

Fig. 13.4. The development of spores of (A) *Gigaspora* and (B) *Scutellospora* (inferred from light–microscope studies and the work of Gibson (1985)). A hyphal tip swells somewhat at the tip. A thin hypha continues to grow into the soil from a small, lateral peg-like projection. The walls of the hyphal tip develop to form a large swollen spore which becomes separated from the sporogenous hypha by cross-walls. Finally, in *Gigaspora* spp., the walls thicken, and germination takes place directly through the spore wall. In *Scutellospora*, the initial development appears to be the same as for *Gigaspora*, but inner, flexible walls form to create an 'endospore', and germination is by means of a germination shield (gs) formed upon it. The two different forms of auxiliary cells are illustrated on the drawings (see Fig. 13.5).

the mycorrhizas. The spores are described (see p. 350 in Hall, 1984) as having a subtending hypha 'swollen approximately into a sphere 1.5 μm thick and attached to fine endophyte infection', but unfortunately, these are not illustrated. Germination is said to be directly through the pore of the subtending hypha. It is quite possible that more than one species is involved and studies of different isolates from several parts of the world are required to clarify the position of the fine endophyte in relation to other VAM species.

Other types of propagule

Spores may be the most obvious form of propagule formed by endogonaceous fungi, but their survival and dissemination can also be achieved in other ways. Root fragments containing fungal structures and pieces of detached hyphae can also be effective. Indeed, according to some workers, there may well be non-sporing forms of endomycorrhizal Endogonaceae (Powell, 1977; Crush, 1978). Despite the general assumption that the mycorrhizal species are obligate symbionts, they may have some independent growth in soil. One isolate of *Glomus mosseae* demonstrated an independent existence for 50 d in soil, or sand with added soil. When sand alone was used, the fungus was unable to survive (Hepper & Warner, 1983).

Although little experimental evidence for non-sporulating forms has been offered, some endophytes of New Zealand forests do not readily sporulate in pot culture (Johnson, 1977). This was shown experimentally by carefully scrubbing roots of *Griselinia littoralis* free of all soil and attached mycelium and repotting in disinfected soil. After two years, three-quarters of the pots were still devoid of spores although the plants

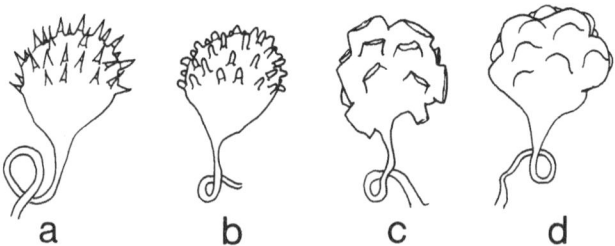

a   b   c   d

Fig. 13.5. Echinulate auxiliary cells (a and b) typical of *Gigaspora* spp., and knobby auxiliary cells (c and d) typical of *Scutellospora* spp. (Modified from Walker (1979) and Walker & Sanders (1986)).

were intensely endomycorrhizal. It should be pointed out, however, that the factors stimulating or inhibiting sporulation are virtually unknown and that the only treatment applied in this instance was drought stress. It has been suggested that there may be a soil water potential below which sporulation fails to take place (Walker, Mize & McNabb, 1982). If this is so, drought would have had the opposite effect to that desired. No evidence for non-sporing types was found in open pot culture surveys in Australia and the suggestion was made that lack of sporulation merely reflects a lack of stimulus for sporulation and not inability of the fungus to sporulate (Abbott & Robson, 1982b).

Inadequate extraction techniques and sampling limited by time and space may give the erroneous impression that spores are not formed in some soils. For example, while Read, Koucheki & Hodgson (1976) and Sparling & Tinker (1975, 1978a) claimed that spores are infrequent in upland soils in Yorkshire, sampling of similar locations in northern Britain using extraction techniques suitable for organic soils (Walker *et al.*, 1982), has always yielded abundant spores of many different species. Excepting the agitation, settling, decanting and filtration method, which is useful only in very sandy soils, (Koske & Walker, 1984), all the currently described techniques use sieves for extracting spores from soil. These are not capable of extracting spores of *Glomus tenue* (which are less than 15 μm in diameter) since the smallest sieve has 45-μm openings (Walker *et al.*, 1982). Indeed, most investigators use a smallest sieve of 106-μm mesh, with the consequence that many small-spored species are lost with the fine material.

Evidence has been reported for the role of hyphal fragments and particles of senescent roots in the dissemination of endogonaceous fungi (Biermann & Linderman, 1983). Mycorrhizas detached from the plant can produce regrowth of hyphae, but much variation exists among the genera. Species of *Gigaspora* and *Scutellospora* and the ectomycorrhizal *Endogone* do not form vesicles or, very rarely, spores in roots unlike most other endomycorrhizal Endogonaceae. Therefore, root fragments harbouring fungi in these genera are less efficient as propagules than those mycorrhizal with most species of *Glomus, Sclerocystis, Acaulospora,* and *Entrophospora* which produce such structures.

Although the auxiliary cells of *Gigaspora margarita* can germinate *in vitro*, there is no evidence to suggest that they operate as propagules *in vivo*. Indeed, attempts to use them as inoculum for the establishment of mycorrhizas in pot cultures have failed (Biermann & Linderman, 1983). The role of these structures is not understood.

Dispersal of propagules

Since the earliest observations that endogonaceous spores could be found, apparently unharmed, in the guts of voles and myriopods (Thaxter, 1922; Trappe & Maser, 1976), much evidence has been forthcoming to suggest that small mammals are perhaps the major means of dispersing sporocarpic members of the Endogonaceae. The large sporocarps of these fungi contain peridial and glebal hyphae which are more digestible than the resistant spores (Gerdemann & Trappe, 1974) and which for some species of small mammal may be of major dietary importance (Maser, Trappe & Nussbaum, 1978; Ure & Maser, 1982). One endogonaceous species, *Glomus botryoides*, has been described only from the stomachs of chipmunks and mice (Rothwell & Victor, 1984).

It seems unlikely that small mammals would deliberately ingest ectocarpic spores. Although such spores are very large compared with other fungal spores, they are produced in the soil and would probably be too small for even the smallest mammal to locate. However, other animals may play an important role in disseminating such propagules. Grasshoppers, crickets and millipedes ingest spores (Thaxter, 1922; Gerdemann & Trappe, 1974; Ponder, 1980) and it seems likely that other arthropods do the same. It is unknown whether such ingestion is incidental to feeding on roots or is deliberate, but spores defecated by such animals can still be viable (Ponder, 1980). Birds consume the animals and may fly considerable distances before depositing their excreta, presumably still containing viable spores. Soil incorporated into the nests of certain species of swallow included viable endogonaceous fungi (McIlveen & Cole, 1976). Such material could be transported in surface water as the nests disintegrate. There have been no investigations of the role of water in the transport of propagules in the Endogonaceae, but it can be assumed that, for example, spores brought to the soil surface by animal activity or erosion would be moved by surface water and could be washed into the soil later where they could germinate to create new centres of colonisation.

In view of the ubiquity of endogonaceous species, any animal which inadvertently carries soil will also carry with it viable propagules of endogonaceous fungi. Principal amongst carriers must be man who transports both soil and living plant material. Modern agricultural practice must be a major factor in spreading inoculum through arable lands. Burrowing activity by such animals as ants, worms, rabbits and gophers is of major importance in mixing propagules through the soil and in bringing them to the surface (McIlveen & Cole, 1976; Allen, MacMahon & Ianson, 1985). At the surface, they can be spread by the action of wind and rain (Warner, MacMahon & Allen, 1985).

In view of the importance to plants of the mycorrhizal Endogonaceae, it is surprising that there is so little known about the formation and dispersal of their propagules. Investigation of the introduction and spread of these organisms in agricultural and horticultural soils, as well as in such areas as spoil–heap reclamation, would be greatly aided by increased knowledge of their life-histories. Much of the work done in the past has been hindered either by a reluctance to consider the identity of the endophytes in VAM systems, or a tendency to misidentify the fungi involved. Failure to lodge herbarium specimens has meant that errors of the past cannot be rectified as taxonomic knowledge progresses. This matter has been discussed in detail in relation to the fungus *Glomus fasciculatum* (Walker, 1985, 1986).

Note added in proof. Gigaspora decipiens has recently been shown to have a perfect (zygospore) state (I. Tommerup pers. comm.). The zygospores are different from any other known spore type in the family, and are capable of germination.

References

Abbot, L.K. & Robson, A.D. (1982*a*). The role of vesicular arbuscular mycorrhizal fungi in agriculture and the selection of fungi for inoculation. *Australian Journal of Agricultural Research*, **33**, 389–408.

Abbott, L.K. & Robson, A.D. (1982*b*). Infectivity of vesicular-arbuscular mycorrhizal fungi in agricultural soils. *Australian Journal of Agricultural Research*, **33**, 1049–59.

Allen, M.F., MacMahon, J.A. & Ianson, D.C. (1985). Ecesis on Mount St. Helens: can mycorrhizal fungi spread from animal-dispersed inoculum? In: *Proceedings of the 6th North American Conference on Mycorrhizae*, ed. R. Molina, p. 291. Corvallis, Oregon: Forest Research Laboratory, Oregon State University Press.

Ames, R.N. & Schneider, R.W. (1979). *Entrophospora*, a new genus in the Endogonaceae. *Mycotaxon*, **8**, 347–52.

Benjamin, R.K. (1979). Zygomycetes and their spores. In: *The Whole Fungus*, vol. 2, ed. B. Kendrick, pp. 573–622. Ottawa, Canada: National Museums of Natural Sciences, National Museums of Canada, and the Kananaskis Foundation.

Berch, S.M. & Fortin, J.A. (1982). Germination of zygospores of *Endogone incrassata*. *Mycologia*, **74**, 861–4.

Berch, S.M. & Fortin, J.A. (1983). Germination of zygospores of *Endogone pisiformis*. *Mycologia*, **75**, 328–32.

Biermann, B. & Linderman, R.G. (1983). Use of vesicular-arbuscular mycorrhizal roots, intraradical vesicles and extraradical vesicles as inoculum. *New Phytologist*, **95**, 97–105.

Bonfante-Fasolo, P. & Scannerini, S. (1976). The ultrastructure of the zygospore in *Endogone flammicorona* Trappe & Gerdemann. *Mycopathologia*, **59**, 117–23.

Bucholtz, F. (1912). Beitrage zur Kenntris der gattung *Endogone* Link. *Beihefte sum Botanishchen Centralblatt Abteilung 2*, **39**, 147–224.

Butler, E.J. (1939). The occurrence and systematic position of the vesicular-arbuscular type of mycorrhizal fungi. *Transactions of the British Mycological Society*, **22**, 274–301.

Crush, J.R. (1973). The effect of *Rhizophagus tenuis* mycorrhizas on ryegrass, cocksfoot and sweet vernal. *New Phytologist*, **72**, 965–73.

Crush, J.R. (1978). Changes in effectiveness of soil endomycorrhizal fungal populations during pasture development. *New Zealand Journal of Agricultural Research*, **21**, 683–5.

Gerdemann, J.W. & Trappe, J.M. (1974). The Endogonaceae in the Pacific Northwest. *Mycologia Memoir No. 5*., 76 pp.

Gibson, J.L. (1984). *Glaziella aurantiaca* (Endogonaceae): Zygomycete or ascomycete? *Mycotaxon*, **20**, 325–8.

Gibson, J.L. (1985). Morphology, cytology and ultrastructure of selected species of Endogonaceae (Endogonales: Zygomycetes). *PhD dissertation, University of Florida*, Gainesville, Florida.

Gibson, J.L., Kimbrough, J.W. & Benny, G.L. (1985). Ultrastructural evidence for the ascomycete-like nature of *Glaziella aurantiaca*. In: *Proceedings of the 6th North American Conference on Mycorrhizae*, ed. R. Molina, p. 431. Corvallis, Oregon: Forest Research Laboratory, Oregon State University Press.

Greenall, J.M. (1963). The mycorrhizal endophytes of *Griselinia littoralis* (Cornaceae). *New Zealand Journal of Botany*, **1**, 389–400.

Hall, I.R. (1977). Species and mycorrhizal infections of New Zealand Endogonaceae. *Transactions of the British Mycological Society*, **68**, 341–56.

Hall, I.R. (1984). Taxonomy of VA mycorrhizal fungi. In: *VA Mycorrhiza*, ed. C. L. Powel & D. J. Bagyaraj, pp. 57–93. Florida: CRC Press.

Harley, J.L. & Smith, S.E. (1983). *Mycorrhizal symbiosis*. London: Academic Press.

Hepper, C.M. & Warner, A. (1983). Role of organic matter in growth of a vesicular-arbuscular mycorrhizal fungus in soil. *Transactions of the British Mycological Society*, **81**, 155–6.

Johnson, P.N. (1977). Mycorrhizal Endogonaceae in a New Zealand Forest. *New Phytologist*, **78**, 161–70.

Koske, R.E., Miller, D.D. & Walker, C. (1983). *Gigaspora reticulata*: a newly described endomycorrhizal fungus from New Zealand. *Mycotaxon*, **16**, 429–35.

Koske, R.E. & Walker, C. (1984). *Gigaspora erythropa*, a new species forming arbuscular mycorrhizae. *Mycologia*, **76**, 250–5.

Koske, R.E. & Walker, C. (1985). Species of *Gigaspora* (Endogonaceae) with roughened outer walls. *Mycologia*, **77**, 702–20.

McIlveen, W.D. & Cole, H., Jnr. (1976). Spore dispersal of endogonaceae by worms, ants, wasps and birds. *Canadian Journal of Botany*, **54**, 1486–9.

Madelin, M.F. (1979). An appraisal of the taxonomic significance of some different modes of producing blastic conidia. In: *The Whole Fungus*, ed. B. Kendrick, pp. 63–80. Ottawa, Canada: National Museums of Natural Sciences, National Museums of Canada and the Kananaskis Foundation.

Malloch, D.W., Pirozynski, K.A. & Raven, P.H. (1980). Ecological and evolutionary significance of mycorrhizal symbioses in vascular plants (a review). *Proceedings of the National Academy of Science, USA* **77**, 2113–18.

Maser, C., Trappe, J.M. & Nussbaum, R.A. (1978). Fungal-small mammal inter-relationships with emphasis on Oregon coniferous forests. *Ecology*, **59**, 799–809.

Miller, D.D., Domoto, P.A. & Walker, C. (1985). Colonization and efficacy of different endomycorrhizal fungi with apple seedlings at two phosphorus levels. *New Phytologist*, **100**, 393–402.

Miller, D.D. & Walker, C. (1986). *Glomus maculosum* sp. nov. (Endogonaceae): an endomycorrhizal fungus. *Mycotaxon*, **25**, 217–27.

Molina, R. (ed.) (1985). *Proceedings of the 6th North American Conference on Mycorrhizae*. Corvallis, Oregon: Forest Research Laboratory, Oregon State University Press.

Morton, J.B. & Walker, C. (1984). *Glomus diaphanum*: a new species in the Endogonaceae common in West Virginia. *Mycotaxon*, **21**, 431–40.

Mosse, B. (1970a). Honey-coloured, sessile *Endogone* spores. I. Life history. *Archiv fur Mikrobiologie*, **74**, 129–45.

Mosse, B. (1970b). Honey-coloured, sessile *Endogone* spores. III. Wall structure. *Archiv fur Mikrobiologie*, **74**, 146–59.

Mosse, B., Stribley, D.P. & Le Tacon, F. (1981). Ecology of mycorrhizae and mycorrhizal fungi. In: *Advances in Microbial Ecology*, vol. 5, ed. M. Alexander, pp. 137–210. London: Plenum Press.

Nicolson, T.H. (1975). Evolution of vesicular-arbuscular mycorrhizas. In: *Endomycorrhizas*, ed. F. E. Sanders, B. Mosse & P. B. Tinker, pp. 25–34. London: Academic Press.

Ponder, F. (1980). Rabbits and grasshoppers: vectors of endomycorrhizal fungi on new coal mine soil. *USDA Forest Service Research Note*, NC250.

Pons, F. & Gianinazzi-Pearson, V. (1985). Observations on extra-matrical vesicles of *Gigaspora margarita in vitro*. *Transactions of the British Mycological Society*, **84**, 168–70.

Powell, C. Ll. (1977). Mycorrhizas in hill-country soils. I. Spore-bearing mycorrhiza fungi in thirty-seven soils. *New Zealand Journal of Agricultural Research*, **20**, 53–7.

Powell, C. Ll. (1980). Mycorrhizal infectivity of eroded soils. *Soil Biology and Biochemistry*, **12**, 247–50.

Read, D.J., Koucheki, H.K. & Hodgson, J. (1976). Vesicular-arbuscular mycorrhiza in natural vegetation systems. I. The occurrence of infection. *New Phytologist*, **77**, 641–53.

Rothwell, F.M. & Victor, B.J. (1984). A new species of Endogonaceae: *Glomus botryoides*. *Mycotaxon*, **20**, 163–7.

Schenck, N.C., Spain, J.L., Sieverding, E. & Howeler, R.H. (1984). Several new and unreported vesicular-arbuscular mycorrhizal fungi (Endogonaceae) from Colombia. *Mycologia*, **76**, 685–99.

Sieverding, E. & Toro, S. (1986). The genus *Entrophospora* in Columbia. *Proceedings of the 1st European Symposium on Mycorrhiza*. INRA, Dijon. In Press.

Smith, G.S. & Schenck, N.C. (1985). Two new dimorphic species in the Endogonaceae: *Glomus ambisporum* and *Glomus heterosporum*. *Mycologia*, **77**, 566–74.

Smith, S.E. (1980). Mycorrhizas of autotrophic higher plants. *Biological Review*, **55**, 475–510.

Sparling, G.P & Tinker, P.B. (1975). Mycorrhizas in Pennine grassland. In: *Endomycorrhizas*, ed. F.E. Sanders, B. Mosse & P.B. Tinker, pp. 545–60. London: Academic Press.

Sparling, G.P. & Tinker, P.B. (1978a). Mycorrhizal infection in Pennine grassland. I. Levels of infection in the field. *Journal of Applied Ecology*, **15**, 943–50.

Sparling, G.P. & Tinker, P.B. (1978b). Mycorrhizal infection in Pennine grassland. II. Effects of mycorrhizal infection on the growth of some upland grasses on gamma-irradiated soils. *Journal of Applied Ecology*, **15**, 951–8.

Tandy, P.A. (1975). Sporocarpic species of Endogonaceae in Australia. *Australian Journal of Botany*, **23**, 849–66.

Thaxter, R. (1922). A revision of the Endogonaceae. *Proceedings of the American Academy of Arts and Sciences*, **57**, 290–351.

Trappe, J.M. (1982). Synoptic keys to the genera and species of zygomycetous mycorrhizal fungi. *Phytopathology*, **72**, 1102–8.

Trappe, J.M., Bloss, H.E. & Menge, J.A. (1984). *Glomus deserticola* sp. nov. *Mycotaxon*, **20**, 123–7.

Trappe, J.M. & Gerdemann, J.W. (1972). *Endogone flammicorona* sp. nov., a distinctive segregate from *Endogone lactiflua*. *Transactions of the British Mycological Society*, **59**, 403–7.

Trappe, J.M. & Maser, C. (1976). Germination of spores of *Glomus macrocarpus* (Endogonaceae) after passage through a rodent digestive tract. *Mycologia*, **68**, 433–6.

Trappe, J.M. & Schenck, N.C. (1982). Taxonomy of the fungi forming endomycorrhizae. A. Vesicular-arbuscular mycorrhizal fungi (Endogonales). In: *Methods and Principles of Mycorrizal Research*, ed. N. C. Schenck, pp. 1–9. St Paul, Minnesota: The American Phytopathological Society.

Ure, D.C. & Maser, C. (1982). Mycophagy of red-backed voles in Oregon and Washington. *Canadian Journal of Zoology*, **60**, 3307–15.

Wagner, C.A. & Taylor, T.N. (1982). Fungal chlamydospores from the Pennsylvanian region of North America. *Review of Palaeobotany and Palynology*, **37**, 317–28.

Walker, C. (1979). Hybrid poplar mycorrhizae and Endogonaceous spores in Iowa. *PhD dissertation, Iowa State University*, Ames, Iowa.

Walker, C. (1983). Taxonomic concepts in the Endogonaceae: spore wall characteristics in species descriptions. *Mycotaxon*, **18**, 443–55.

Walker, C. (1985). Taxonomy of the Endogonaceae. In: *Proceedings of the 6th North American Conference on Mycorrhizae*, ed. R. Molina, pp. 193–9. Corvallis, Oregon: Forest Research Laboratory.

Walker, C. (1986). Problems in the taxonomy of mycorrhizal fungi. In: *Proceedings of the 1st European Symposium on Mycorrhizae*, ed. V. Gianinazzi-Pearson, pp. 605–9. Paris: INRA.

Walker, C., Mize, C.W. & McNabb, H.S. Jnr. (1982). Populations of endogonaceous fungi at two locations in central Iowa. *Canadian Journal of Botany*, **60**, 2518–29.

Walker, C., Reed, L.E. & Sanders, F.E. (1984). *Acaulospora nicolsonii*, a new endogonaceous species from Great Britain. *Transactions of the British Mycological Society*, **83**, 360–4.

Walker, C. & Sanders, F.E. (1985). Germination shields in the genus *Gigaspora*. In: *Proceedings of the 6th North American Conference on Mycorrhizae*, ed. R. Molina, p. 433. Corvallis, Oregon: Forest Research Laboratory, Oregon State University Press.

Walker, C. & Sanders, F.E. (1986). Taxonomic concepts in the Endogonaceae: III. The separation of *Scutellospora* gen. nov. from *Gigaspora Gerd & Trappe*. *Mycotaxon*, **27**, 169–82.

Warner, N., MacMahon, J.A. & Allen, M.F. (1985). Dispersal of VA-mycorrhizal fungi in a disturbed semi-arid ecosystem. In: *Proceedings of the 6th North American Conference on Mycorrhiza*, ed. R. Molina, p. 292. Corvallis, Oregon: Forest Research Laboratory.

14
Programmed cortical senescence: a basis for understanding root infection

J. W. DEACON

Microbiology Department, School of Agriculture, West Mains Road, Edinburgh EH9 3JG, UK

Post-harvest pathologists have long been aware of the importance of tissue senescence, not least in enabling disease to develop from latent infections. Root pathologists, however, have been slow to appreciate its significance, perhaps because tissue senescence was thought to occur only late in the life of a root when it would have little relevance for primary pathogens. Holden (1975) forced a reappraisal of this view, bringing tissue senescence into centre stage in root pathology and root microbiology as a whole.

Programmed early senescence

Holden (1975) used nuclear staining with Feulgen reagent to assess rates of cell death in cereal roots. In 3-week-old seedlings grown in pathogen-free soil at 15°C the roots appeared white and healthy with no sign of cortical senescence. Specific staining showed, however, that 65% of cortical cells were dead in the older parts of wheat seminal root axes, and 41% were dead in barley. These findings were confirmed by Henry & Deacon (1981) and Lewis & Deacon (1982), using both nuclear and cytoplasmic stains. Early root cortex death (RCD) occurred even in sterile conditions, excluding the possible involvement of micro-organisms. It occurred while individual roots continued to grow and branch, and followed a definite pattern (Fig. 14.1) suggesting that it is a programmed phenomenon. At a variable distance behind the root tip, depending on growth conditions, nuclei disappear first from the epidermis and then from successive cortical cell layers until five of the six cell layers of the seminal root axes of wheat and barley are anucleate. Nuclei persist for a considerable time in the innermost cortical cell layer, next to the endodermis. They also persist temporarily in most cell layers of the root axes around the bases of lateral branches but, as laterals age, they too show RCD.

RCD occurs in plants grown under both glasshouse and field conditions (Deacon & Henry, 1981; Henry & Deacon, 1981); its advance is consistently slower in barley than in wheat, and is apparently even slower in oats and rye (Deacon & Mitchell, 1985). It occurs also in a range of British grasses (Kirk & Deacon, 1986) and in maize (Deacon, Drew & Darling, 1986). Dicotyledonous plants have received little attention in this respect; Macleod, Robson & Abbott (1986) observed RCD in rape, but not in subterranean clover after three weeks growth, whereas Tommerup (1984) did not detect it in 6-week-old rape plants.

Root cortical senescence was known long before Holden's work, but it had not been quantified and its occurrence on very young plants was not appreciated except in the work of Beckel (1956) who did not distinguish it clearly from aerenchyma formation. It occurs early enough to affect the establishment and subsequent growth of a whole range of root-associated microbes, including pathogens.

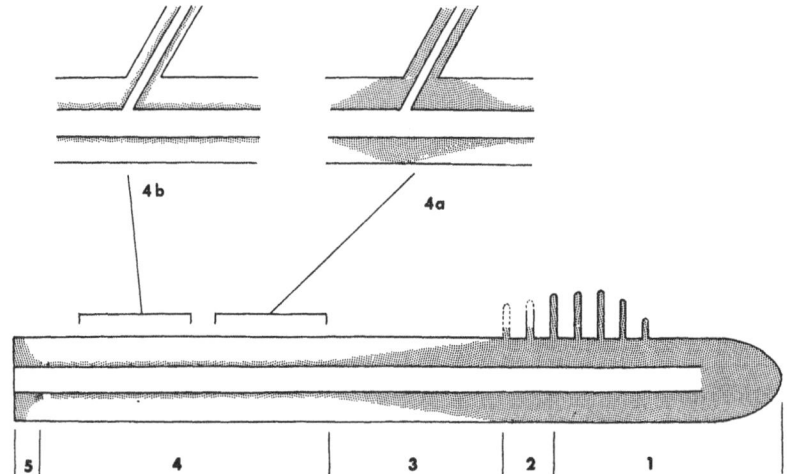

Fig. 14.1. Diagram of cortical cell death in wheat seminal root axes, not to scale. Stippling represents nucleate cortical cells. 1, Zone including root tip, regions of cell differentiation and elongation and living root hairs. 2, Zone in which nuclei are lost from root hairs and non–piliferous epidermal cells. 3, Zone of progressive death of cortical cell layers 2–5. 4, Zone in which cell layers 2–5 are anucleate but nuclei remain in the innermost (sixth) cortical cell layer, next to the endodermis. In young parts (4a), but not in older parts (4b), of this zone nuclei are present in several cell layers around the bases of laterals. 5, Short zone (up to 1 cm) of nucleate cortex immediately below the seed. (From Henry & Deacon, 1981.)

Implications for take-all, *Gaeumannomyces graminis*, and its biocontrol agents

Phialophora graminicola

Phialophora graminicola is a non-pathogenic parasite of cereal and grass roots. It is closely related to the take-all fungus, *Gaeumannomyces graminis*, and mycelia of both fungi develop similarly on roots, but *P. graminicola* does not penetrate the vascular system. Under some conditions it can give significant control of take-all (Deacon, 1981; Wong, 1981). In glasshouse experiments *P. graminicola* grew better if inoculated on older (6–12 d) than younger (3-d) regions of wheat roots (Deacon, 1973), and its pattern of growth changed as roots aged (Holden, 1976; Kirk & Deacon, 1987a). In young regions the dark runner hyphae of *P. graminicola* are confined to the root surface and the fungus penetrates very little into the cortex; on older regions the runner hyphae grow deep within the cortex, in the intercellular spaces (Fig. 14.2). Holden (1976) explained this behaviour in terms of RCD; he concluded that *P. graminicola* is a weak parasite (not merely a weak pathogen) that is restricted by host resistance and capable of invading only dead and dying cells.

Much evidence now supports Holden's conclusion. For example, it is now known that *P. graminicola* (like *Microdochium bolleyi*, which is discussed later) does not enhance the rate of cortical death in soil (Henry & Deacon, 1981). In naturally infected wheat roots in the field, there is

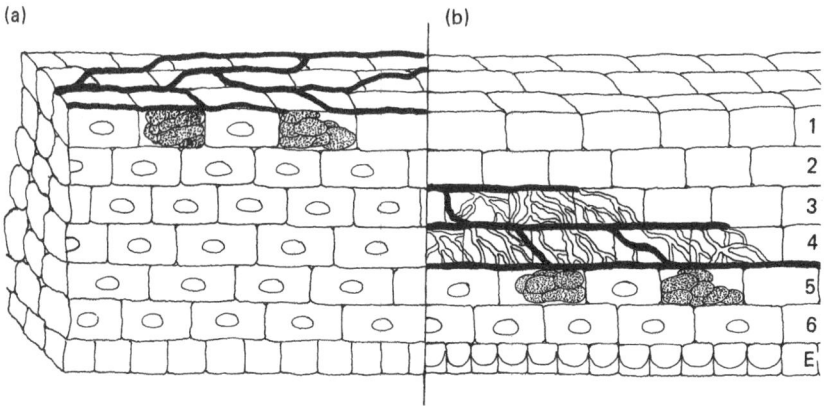

Fig. 14.2. The growth of *Phialophora graminicola* by dark runner hyphae (a) on the surface of young regions, and (b) within the cortex of older regions, of cereal roots. Nuclei in the cortex are indicated by circles; cortical cell layers are numbered 1–6, E = endodermis. The clusters of dark rounded cells are formed by *P. graminicola* in response to halted invasion of the cortex because of host-resistance.

a significant correlation between the depth at which *P. graminicola* grows within roots and the amount of RCD (Deacon, 1980). Moreover, the highest populations of *P. graminicola* on wheat occur in field plots showing most RCD (Deacon & Henry, 1981). Recently, Deacon & Lewis (1986) promoted RCD in sterile detached root pieces, by incubation on water agar, and demonstrated that this promoted subsequent invasion by *P. graminicola*.

In contrast to *P. graminicola*, the wheat take-all fungus, *G. graminis* var. *tritici (Ggt)*, is an aggressive pathogen that invades fully functional root cortices and thence the stele, where it causes vascular disruption. It should be noted that this 'conventional' view is based on studies with high inoculum levels of *Ggt* and, more importantly, with large inoculum 'units', whether these are artificially or naturally colonised pieces of crop residue. *Ggt* cannot, for example, infect roots from ascospores in normal, unsterile soil (Brooks, 1965). Moreover, if various grades (sizes) of organic matter containing *Ggt* are extracted from field soils, it is found that their infectivity decreases with decreasing size (Hornby, 1975).

How does *Ggt* infect from the smallest pieces of inoculum with low nutrient reserves? It seems that RCD may play a critical role. Using small pieces of agar inoculum (1.5–2.0 mm^3, Deacon & Henry (1978, 1980) found that *Ggt* caused more disease (vascular damage) if inoculated on to young (0-d) than on to older (15-d) regions of wheat roots, but more disease was caused on older than young regions of barley roots. Both RCD and endodermal development progress with age. It has been suggested that RCD can benefit the fungus because of reduced host resistance in the cortex and the release of nutrients to support infection, but this may be countered by increased resistance at the endodermis. RCD occurs so rapidly in wheat – e.g. in the epidermis of regions of root only five days old (Deacon & Henry, 1980) – that *Ggt* would gain no advantage from delayed inoculation but would, instead, meet increased endodermal resistance. The slower rate of RCD in barley suggests that *Ggt* might benefit from delayed inoculation by being better able to penetrate the cortex with sufficient inoculum potential to invade the stele. Data from Deacon & Henry (1980) support this argument. Differences in the resistance of cereals to *Ggt* may reside, in whole or in part, in differences in RCD (Deacon & Mitchell, 1985). This could have major implications for plant breeding.

Kirk (1984) inoculated young regions of barley roots with small inocula of *Ggt* supplemented with short lengths of excised root in the expectation that these would senesce and simulate RCD by providing nutrients for infection. Disease was significantly enhanced in the presence of root pieces, showing that an aggressive pathogen like *Ggt* can benefit from cell senescence if the inoculum potential is low. These findings may not have practical relevance when cereals are grown repeatedly on the same site, where high inoculum levels will occur; their chief significance may be for the early build-up of take-all on crops sown after a 1- or 2-year break from cereals. Inoculum can then be present in the soil in barely detectable amounts, and yet often take-all lesions are seen in the new crop later in the growing season (Hornby, 1975). It is likely that *Ggt* undergoes a prolonged 'feeding stage' on senescing cortices and eventually accumulates a nutritional base from which it can infect later-formed roots.

These comments suggest that *P. graminicola* might control take-all by competing for senescing root tissues (Deacon, 1981). To date this is the only tenable explanation according with the facts, summarised in Deacon (1981). For example, biocontrol is effective only against low inoculum levels of *Ggt*, which, itself, can interfere with the establishment of its biocontrol agents if their populations are lower than its own.

Phialophora zeicola; *maize root rot-stalk rot*

Several fungi closely related to *Ggt* and *P. graminicola* grow as dark runner hyphae on cereal and grass roots (Deacon, 1981; Walker, 1981). One such fungus, *P. zeicola*, is a common parasite of maize in South Africa, France and perhaps other countries (Deacon & Scott, 1983). *P. zeicola* normally grows as a weak parasite on or in the maize root cortex, though it may invade the stele of young root laterals and destroy them. When the crop is subjected to drought or other stresses *P. zeicola* aggressively invades the root system, causing extensive root rot, which may be followed by stalk rot. *P. zeicola* is one of several fungi involved in a root rot-stalk rot complex of maize, the different fungi apparently being favoured by different environmental conditions. There is now much evidence that the whole disease complex is stress-related (Dodd, 1980). During grain-filling, maize cobs act as powerful carbohydrate sinks. If drought or cloud-cover reduces photosynthesis at this stage, carbohydrate reserves are withdrawn from other tissues, especially the stalk. The resulting premature senescence enables weak parasites, including *P. zeicola,* to invade aggressively. *P. zeicola* apparently responds to the declining host resistance associated with stress. The relationships between stress conditions and RCD or other aspects of root function have received little study. Deacon *et al.* (1986) examined the effects of aeration on RCD and cortical gas space (aerenchyma) formation in solution- and soil-grown

Zea mays. RCD occurred in both aerated and non-aerated roots, though marginally faster in the non-aerated treatment, whereas aerenchyma developed only in non-aerated roots and bore no relationship to RCD

Fungi of the Phialophora-Gaeumannomyces *complex: general*

All fungi in the complex are specialised parasites of the Gramineae, and have poor competitive saprophytic abilities (CSA). Their survival and success in competition with saprophytes with a high CSA may lie in their characteristic ectotrophic infection habit (Deacon, 1980). As shown in Fig. 14.2, dark runner hyphae, which are lysis-resistant, usually lie between longitudinal files of cells, either on the root surface or (during and after RCD) in the intercellular spaces of the cortex; however, they always develop next to the outermost living layer of the cortex (Holden, 1976; Deacon, 1980). These hyphae are, thus, strategically positioned to act as bases from which files of host cells can be invaded by hyaline infection hyphae as the host cells begin to senesce, and in advance of the less-specialised saprophytes.

When attempts are made to isolate take-all fungi from infected roots, saprophytes grow on to the agar from the surface of all but very young roots, apparently suppressing outgrowth by fungi of the *Gaeumannomyces-Phialophora* complex. The latter grow primarily, if not exclusively, from the cut ends of root pieces – either from the stele *(Ggt)* or from the inner cortex (*P. graminicola* and *P. zeicola*). In nature, where freshly exposed root ends are not present, these fungi may become trapped in the roots, unable to grow out through the cortex to infect nearby healthy roots. Vojinovic (1973) sugested that this factor contributed to take-all decline in long sequences of cereals.

Microdochium bolleyi

M. bolleyi is naturally common on the roots and stem bases of cereals (Reinecke & Fokkema, 1981). At very high inoculum levels it can damage cereal root tips, but normally it causes little damage and is considered to be a 'minor pathogen' (Salt, 1979). Kirk & Deacon (1987a) concluded that *M. bolleyi* was affected by RCD in both wheat and grasses in the same way as was *P. graminicola*; in the former case analysis was based on isolations and distribution of microsclerotia in roots since *M. bolleyi* does not grow as dark runner hyphae. Henry & Deacon (1981) found that *M. bolleyi* does not enhance RCD in soil.

Most interest in *M. bolleyi* centres around its possible (but largely unproven) role in disease complexes and its potential to control more aggressive pathogens. It is reported to reduce damage to, or colonization of, cereal roots by *Pythium arrhenomanes* (J.M. Waller, quoted in Salt, 1979), *Ggt* and *P. graminicola* (Kirk & Deacon, 1987*b*) and infection of cereal stem bases by *Fusarium* spp. (Reinecke, Duben & Fehrmann, 1979) and the eyespot fungus (Reinecke & Fokkema, 1981). Its ability to respond to senescence may be important in these respects, as explained earlier for control of *Ggt* by *P. graminicola*. But could such a mechanism also inhibit stem base infections by specialised pathogens like the eyespot fungus, *Pseudocercosporella herpotrichoides*? The following points may be relevant.

In glasshouse experiments, Lewis & Deacon (1982) found that coleoptiles of wheat and barley senesce rapidly in the absence of pathogens: more rapidly in wheat than in barley (paralleling the difference in RCD between these cereals), and less rapidly in shaded than unshaded conditions. The eyespot pathogen has splash-dispersed spores that land on stem bases. It is known that wheat coleoptiles are much more susceptible to infection than are the underlying leaf sheaths (Higgins, 1984) and that differences between cultivars in eyespot-resistance are expressed at an early stage, during infection of the coleoptile (Bateman & Taylor, 1976). It is therefore conceivable that coleoptile senescence influences the infection process, and that *M. bolleyi* might compete for nutrients from the senescing cells. Clearly, more work is needed on coleoptile and stem base senescence in relation to the establishment of infection. Also, *M. bolleyi* merits further attention as a possible biocontrol agent of root and foot-rot fungi. Unlike *P. graminicola*, it is a prolific spore-former in culture, facilitating the production of inoculum, and it is a natural colonist of cereals whereas *P. graminicola* seems better adapted to grass swards (Deacon, 1981).

Common root rot: *Cochliobolus sativus*

Common root rot, caused principally by *C. sativus*, is a major disease of dryland wheat crops in North America and Australia. Atkinson, Neal & Larson (1975) identified sources of partial resistance to it in Canadian spring wheats and found that this resistance could be transferred by disomic substitution of chromosome 5B between cultivars. The resistant plants, whether commercial cultivars or 5B-substitution lines derived from them, support fewer bacteria in the rhizosphere than do susceptible plants, even in the absence of disease. However, whereas there is only a roughly 2-fold difference in total rhizosphere bacterial populations between susceptible and resistant plants, there can be up to 140-

fold differences in the numbers of cellulolytic, pectolytic and amylolytic bacteria. Assuming that RCD might provide substrates for these polymer-utilising bacteria, Deacon & Lewis (1982) compared the rates of RCD in resistant and susceptible breeding lines. Differences in RCD between individual cultivars and breeding lines were, with exceptions, not clear-cut, perhaps because growth conditions were different from those in which resistance is expressed. Nevertheless, combined results for three root rot-susceptible lines showed significantly more RCD than for three resistant lines ($P = 0.001$), suggesting that RCD may be implicated in susceptibility to *C. sativus*.

Rhizosphere bacteria

The Canadian work on spring wheat (Atkinson *et al.*, 1975; Larson & Neal, 1978) also showed that one breeding line, derived by substituting the chromosome pair 5D from cv. Rescue into cv. Cadet, supported naturally high numbers of nitrogen-fixing bacteria on its roots – more than did either parent cultivar. This breeding line had significantly less RCD than all others tested (Deacon & Lewis, 1982). Nitrogen fixation is known to depend on a supply of plant photosynthate, and this is more likely to be delivered through a living than a dead root cortex. The implications of these preliminary findings suggest that RCD is genetically determined (in this case by chromosome 5D among others) and that RCD might determine the activities of specific physiological groups of bacteria.

There is now increasing recognition that RCD might account for much of the organic carbon known to be released from roots (Newman, 1985). Van Vuurde, Kruyswyk & Schippers (1979) found that bacterial numbers in the rhizosphere increased markedly in regions of wheat roots (4–5 d-old); they suggested that this was a response to RCD, detected by nuclear staining in 0–3 d-old regions of root. L Morris & J.W. Deacon (unpublished) observed a similar relationship between RCD and the total number of rhizosphere bacteria (Fig. 14.3) on wheat roots. In these and many other respects the very early stages of RCD may be most significant, if only because the root epidermis senesces first, and in wheat seminal roots contains roughly 38% of all cells outside the endodermis, or 44% of all cells in the outer five cell layers which exhibit RCD (Kirk & Deacon, 1986).

With current renewed interest in plant growth-promoting rhizobacteria and bacterial antagonists of pathogens (e.g. Cook, 1985) it is surprising that little specific attention has been given to the sources of nutrients for these organisms in the root zone. RCD merits attention in this respect because it could determine success or failure of microbial inoculants to establish in the root zone. Indeed, inoculants might be selected for their abilities to exploit RCD.

VA mycorrhizal fungi

VA mycorrhizal fungi would be potentially disadvantaged by RCD because they are biotrophs. Hepper (1986) invoked possible differences in RCD to explain why clover (*Trifolium parviflorum*) roots became virtually immune to infection by *Glomus mosseae* in regions only 5 d-old, whereas leek roots remained susceptible even in 32 d-old regions. Unfortunately there are no data for rates of RCD in these plants. Macleod *et al.* (1986) found that inoculation with a VA mycorrhizal fungus can influence RCD. In phosphorus-deficient wheat plants inoculation with *Glomus fasciculatum* decreased RCD compared with that in uninoculated controls, but inoculation increased RCD in phosphorus-adequate plants.

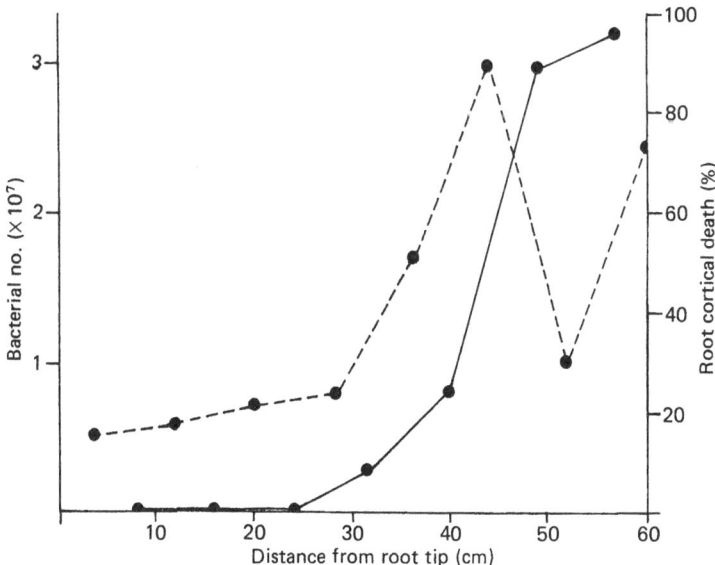

Fig. 14.3. Relationship between total number of rhizosphere bacteria growing on tryptic-soya agar (broken line) and percentage of cortical cells anucleate (solid line) in alternate 4-cm segments of first seminal root axes of wheat grown for 22 d in 1 m depths of soil in a glasshouse at 20°C. Means of 3 replicate roots; all showed a fall in bacterial numbers after the first peak.

These effects did not seem to operate through influences on phosphorus uptake or plant growth; they remain unexplained.

Environmental effects on RCD: the potential for manipulation

Genetic manipulation of RCD is a distinct possibility (Deacon & Lewis, 1982). Henry & Deacon (1981) found significant differences in rates of RCD between British commercial spring wheat cultivars which, like the marked and seemingly consistent difference in rate of RCD between wheat and barley, might be exploited in disease management. In all these respects, however, we need to know if RCD is beneficial or detrimental to whole-plant function or yield, and this has not been investigated.

Irradiation affects RCD and coleoptile senescence (Lewis & Deacon, 1982): shading plants to a quarter or a third of 'control' irradiation significantly reduced RCD on first seminal roots of wheat and barley. Shading drastically reduced total root production suggesting that the first-formed roots (on which RCD was assessed) experienced less

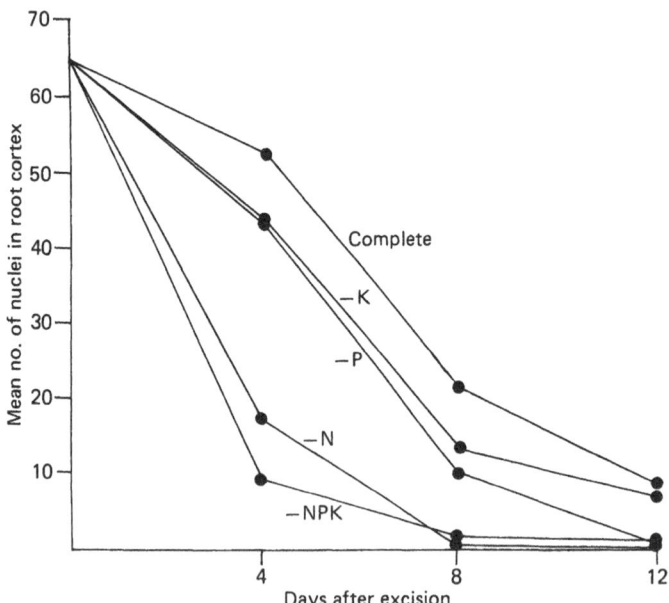

Fig. 14.4. Mean numbers of nuclei counted in microscope fields along 2-cm lengths of sterile excised wheat root incubated on agar containing dextrose (1%) and Heller's mineral salts solution with or without nitrogen, potassium and phosphorus. (I.M.M. Gillespie, unpublished.)

competition for available photosynthate than first-formed roots of unshaded plants. This suggestion presupposes that cortices die through carbohydrate starvation. Results of I.M.M. Gillespie (unpublished) support this; they show that RCD can be significantly delayed in sterile detached root pieces by supplying glucose. Surprisingly, infection of leaves by powdery mildew (*Erisyphe graminis*), which was expected to have the same effect as shading, did not affect RCD in either barley (Lewis & Deacon, 1982) or wheat (Deacon & Mitchell, 1985); the reason for this is unknown.

Macleod *et al.* (1986) found that soil phosphorus supply affected RCD, although the relationship was complex. In comparison with wheat seedlings adequately supplied with P, moderate P deficiency increased RCD, severe P deficiency had no effect, and supra-optimal P decreased RCD, all these assessments being made on first seminal roots. As with shading, interpretation of these results must take account of assimilate partitioning in the whole plant. This complication can be avoided, however, by using excised pieces of root on agar or in nutrient solution. In this way, I.M.M. Gillespie (unpublished) found that exogenous N, P and K absorbed through the cortex affect the rate of RCD, and that N was the most important of these mineral nutrients for maintaining cortical viability (Fig. 14.4).

P. graminicola grows abundantly on roots of turf and ley grasses in many parts of the world (Deacon, 1981) but is found in relatively low amounts on cereal roots unless the cereal crop follows a grass ley. The population can be maintained to a degree on cereal roots but apparently does not build up on them. In a replicated field trial, Deacon & Henry (1981) found that first to third wheat crops had significantly more RCD if they followed one or two years of grass than if following swedes and potatoes. Although the reasons for this are not known, and *P. graminicola* itself cannot be implicated in this difference, an earlier onset and faster rate of RCD would be significant in maintaining the population of *P. graminicola*. Crop rotation — one of the oldest forms of disease management — may yet prove to operate in unexpected ways.

References

Atkinson, T.G., Neal, J.L. & Larson, R.I. (1975). Genetic control of the rhizosphere microflora of wheat. In *Biology and Control of Soil-borne Plant Pathogens*, ed. G.W. Bruehl, pp. 116–22. St Paul: American Phytopathological Society.

Bateman, G.L. & Taylor, G.S. (1976). Seedling infection of two wheat cultivars by *Pseudocercosporella herpotrichoides*. *Transactions of the British Mycological Society*, **67**, 95–101.

Beckel, D.K.B. (1956). Cortical disintegration in roots of *Bouteloua gracilis* (H.B.K.) Lag. *New Phytologist*, **55**, 183–90.

Brooks, D.H. (1965). Root infection by ascospores of *Ophiobolus graminis* as a factor in epidemiology of the take-all disease. *Transactions of the British Mycological Society*, **48**, 237–48.

Cook, R.J. (1985). Biological control of plant pathogens: theory to application. *Phytopathology*, **75**, 25–9.

Deacon, J.W. (1973). *Phialophora radicicola* and *Gaeumannomyces graminis* on roots of grasses and cereals. *Transactions of the British Mycological Society*, **61**, 471–85.

Deacon, J.W. & Henry, C.M. (1978). Studies on virulence of the take-all fungus, *Gaeumannomyces graminis*, with reference to methodology. *Annals of Applied Biology*, **89**, 401–9.

Deacon, J.W. (1980). Ectotrophic growth by *Phialophora radicicola* var. *graminicola* and other parasites of cereal and grass roots. *Transactions of the British Mycological Society*, **75**, 158–60.

Deacon, J.W. & Henry, C.M. (1980). Age of wheat and barley roots and infection by *Gaeumannomyces graminis* var. *tritici*. *Soil Biology and Biochemistry*, **12**, 113–18.

Deacon, J.W. (1981). Ecological relationships with other fungi: competitors and hyperparasites. In *Biology and Control of Take-all*, ed. M.J.C. Asher & P.J. Shipton, pp. 75–101. London: Academic Press.

Deacon, J.W. & Henry, C.M. (1981). Death of the root cortex of winter wheat in field conditions; effects of break crops and possible implications for the take-all fungus and its biological control agent, *Phialophora radicicola* var. *graminicola*. *Journal of Agricultural Science, Cambridge*, **96**, 579–85.

Deacon, J.W. & Lewis, S.J. (1982). Natural senescence of the root cortex of spring wheat in relation to susceptibility to common root rot (*Cochliobolus sativus*) and growth of a free-living nitrogen-fixing bacterium. *Plant and Soil*, **66**, 13–20.

Deacon, J.W. & Scott, D.B. (1983). *Phialophora zeicola* sp. nov., and its role in the root-rot stalk-rot complex of maize. *Transactions of the British Mycological Society*, **81**, 247–62.

Deacon, J.W. & Mitchell, R.T. (1985). Comparison of rates of natural senescence of the root cortex of wheat (with and without mildew infection), barley, oats and rye. *Plant and Soil*, **84**, 129–31.

Deacon, J.W. & Lewis, S.J. (1986). Invasion of pieces of sterile wheat root by *Gaeumannomyces graminis* and *Phialophora graminicola*. *Soil Biology and Biochemistry*, **18**, 167–72.

Deacon, J.W., Drew, M.C. & Darling, A. (1986). Progressive cortical senescence and lysigenous gas space (aerenchyma) formation distinguished by nuclear staining in adventitious roots of *Zea mays*. *Annals of Botany*, **58**, 719–27.

Dodd, J.L. (1980). The role of plant stresses in development of corn stalk rots. *Plant Disease*, **64**, 533–7.

Henry, C.M. & Deacon, J.W. (1981). Natural (non-pathogenic) death of the cortex of wheat and barley seminal roots, as evidenced by nuclear staining with acridine orange. *Plant and Soil*, **60**, 255–74.

Hepper, C.M. (1985). Influence of age of roots on the pattern of vesicular-arbuscular mycorrhizal infection in leek and clover. *New Phytologist*, **101**, 685–93.

Higgins, S. (1984). Factors affecting spore germination and initial growth of *Pseudocercosporella herpotrichoides* on wheat. *Transactions of the British Mycological Society*, **82**, 443–8.

Holden, J. (1975). Use of nuclear staining to assess rates of cell death in cortices of cereal roots. *Soil Biology and Biochemistry*, **7**, 333–4.

Holden, J. (1976). Infection of wheat seminal roots by varieties of *Phialophora radicicola* and *Gaeumannomyces graminis*. *Soil Biology and Biochemistry*, **8**, 109–19.

Hornby, D. (1975). Inoculum of the take-all fungus: nature, measurement, distribution and survival. *EPPO Bulletin*, **5**, 319–33.

Kirk, J.J. (1984). Ability of *Gaeumannomyces graminis* to benefit from senescence of the cereal root cortex during infection. *Transactions of the British Mycological Society*, **82**, 107–11.

Kirk, J.J. & Deacon, J.W. (1986). Early senescence of the root cortex of agricultural grasses, and of wheat following root amputation or infection by the take-all fungus. *New Phytologist*, **104**, 63–75.

Kirk, J.J. & Deacon, J.W. (1987a). Invasion of naturally senescing root cortices of cereal and grass seedlings by *Microdochium bolleyi*. *Plant and Soil*, **98**, 239–46.

Kirk, J.J. & Deacon, J.W. (1987b). Control of the take-all fungus by *Microdochium bolleyi*, and interactions involving *M. bolleyi*, *Phialophora graminicola* and *Periconia macrospinosa* on cereal roots. *Plant and Soil*, **98**, 231–7.

Larson, R.I. & Neal, J.L. (1978). Selective colonization of the rhizosphere of wheat by nitrogen-fixing bacteria. *Ecological Bulletin (Stockholm)*, **26**, 331–42.

Lewis, S.J. & Deacon, J.W. (1982). Effects of shading and powdery mildew infection on senescence of the root cortex and coleoptile of wheat and barley seedlings, and implications for root- and foot-rot fungi. *Plant and Soil*, **69**, 401–11.

Macleod, W.J., Robson, A.D. & Abbott, L.K. (1986). The effect of phosphate supply and inoculation with vesicular-arbuscular (VA) mycorrhizal fungi on the death of the root cortex of agricultural plants. *New Phytologist*, **103**, 349–57.

Newman, E.I. (1985). The rhizosphere: carbon sources and microbial populations. In *Ecological Interactions in Soil*, ed. A.H. Fitter, pp. 107–21. Oxford: Blackwell.

Reinecke, P., Duben, J. & Fehrmann, H. (1979). Antagonism between fungi of the foot rot complex of cereals. In *Soil-borne Plant Pathogens*, ed. B. Schippers & W Gams, pp. 327–36. London: Academic Press.

Reinecke, P. & Fokkema, N.J. (1981). An evaluation of methods of screening fungi from the haulm base of cereals for antagonism to *Pseudocercosporella herpotrichoides* in wheat. *Transactions of the British Mycological Society*, **77**, 343–50.

Salt, G.A. (1979). The increasing interest in 'minor pathogens'. In *Soil-borne Plant Pathogens*, ed. B. Schippers & W. Gams, pp. 289–312. London: Academic Press.

Tommerup, I.C. (1984). Development of infection by a vesicular-arbuscular mycorrhizal fungus in *Brassica napus* L. and *Trifolium subterraneum* L. *New Phytologist*, **98**, 487–95.

Van Vuurde, J.W.L., Kruyswyk, C.J. & Schippers, B. (1979). Bacterial colonization of wheat roots in a root-soil model system. In *Soil-borne Plant Pathogens*, ed. B. Schippers & W. Gams, pp. 229–34. London: Academic Press.

Vojinovic, Z. (1973). The influence of micro-organisms following *Ophiobolus graminis* Sacc. on its further pathogenicity. *EPPO Bulletin*, **9**, 91–101.

Walker, J. (1981). Taxonomy of take-all fungi and related genera and species. In *Biology and Control of Take-all*, ed. M.J.C. Asher & P.J. Shipton, pp. 15–74. London: Academic Press.

Wong, P.T.W. (1981). Biological control by cross-protection. In *Biology and Control of Take-all*, ed. M.J.C. Asher & P.J. Shipton, pp. 417–31. London: Academic Press.

15

The role of the saprophytic phase in Dutch elm disease

JOAN F. WEBBER, C. M. BRASIER and
A. G. MITCHELL
Forest Research Station, Alice Holt Lodge, Farnham, Surrey GU10 4LH UK,

Ophiostoma (Ceratocystis) ulmi (Buism.) Nannfeldt, the causal agent of Dutch elm disease, exists as two distinct sub-groups – the highly pathogenic aggressive strain and the weakly pathogenic non-aggressive strain (Gibbs & Brasier, 1973; Brasier, 1979, 1982). Transmission of *O. ulmi* is primarily in the form of spores which are carried by several species of scolytid elm bark beetle (Lanier & Peacock, 1981). Following their emergence from pupal chambers in dead elm bark, the beetles fly to healthy elms and feed on the succulent phloem tissue in young twig crotches. The wounds or feeding grooves created by this activity serve as entry points for *O. ulmi* which gains access to the xylem vessels and thereby establishes infection. Even with a moderately susceptible elm, such as *Ulmus procera*, the pathogen succeeds in entering the xylem in only about 2–5% of all feeding grooves (Webber & Brasier, 1984). Once infection is initiated, the combination of an aggressive pathogen and susceptible elm species ensures that the outcome is severe wilting and usually death of the tree (Gibbs *et al.*, 1975).

In many necrotrophic pathogens, infection and subsequent sporulation on host tissue complete the disease cycle, but in Dutch elm disease successful completion of the cycle depends upon the pathogen being reunited with a new generation of vector beetles. For this to occur the scolytid beetles must breed in the bark of diseased or dying elms, specifically in the phloem or inner bark tissue which is also colonised by *O. ulmi*. The combined process of beetle and pathogen colonisation of elm bark is known as the saprophytic phase (Lea, 1977; Gibbs & Smith, 1978) and effectively lasts from the time beetles enter the bark to breed to the time their progeny emerge. This period may be as short as two months for beetle broods able to complete their development during a warm summer, but the overwintering of scolytid larvae also regularly produces a sapro-phytic phase of 6–10 months, extending from the late summer of one year to early summer in the following year. The result therefore, can be a very lengthy period of intimate interaction between host, vector and pathogen. In this chapter some of the more important aspects of the behaviour of *O. ulmi* in elm bark are described and their influence on the saprophytic phase of Dutch elm disease considered.

Initiation of the saprophytic phase

The fungal cycle of Dutch elm disease involves two clearly separated phases, a pathogenic phase involving the invasion and spread of *O. ulmi* in xylem vessels, and a saprophytic phase already briefly described. In contrast to those of many other pathogens, the propagules of *O. ulmi* that initiate the saprophytic phase are not necessarily derived from the preceeding pathogenic phase. Instead, two distinct sources of inoculum are available to contribute to bark colonisation in a tree dying of Dutch elm disease; these consist of (a) the pathogen present in the xylem and (b) the spores of *O. ulmi* introduced into the bark by beetles entering to breed.

Evidence for this comes from a series of experiments in which *O. ulmi* labelled with an easily recognised nuclear gene marker was introduced into the xylem of previously healthy elms (Webber & Brasier, 1984). Two of these experiments involved inoculating elms with isolates of the aggres-sive strain of the pathogen, one with an isolate of the non-aggressive strain. All the isolates rapidly became established in the xylem and the bark was then invaded by breeding beetles. Subsequent isolation of *O ulmi* from beetle breeding galleries revealed that both 'labelled' and unlabelled' forms of the fungus were colonising the bark, indicating that both xylem-derived and beetle-introduced inoculum contributed to the saprophytic phase. The new generation of beetles emerging from these trees also carried both types of inoculum (Table 15.1). Clearly, in each case the bark was thoroughly colonised by *O. ulmi*, as few beetles emerged without having acquired at least some spores of the elm pathogen on their body surfaces. However, there were very significant differences in the relative contribution of xylem-derived (pathogenic phase) *O. ulmi* to the total amount carried by emerging beetles when all experiments were com-pared.

A number of factors may be responsible for these differences, including the ease with which *O. ulmi* can escape from xylem vessels into the overlying bark, the quantity of *O. ulmi* brought into the bark by vector

Table 15.1. *Source of* Ophiostoma ulmi *inoculum on newly emerged* Scolytus scolytus *beetles*

Year	Strain introduced into xylem	% of emerging beetles carrying spores derived from each inoculum source			
		Labelled (xylem-derived) pathogen only	Both labelled and unlabelled pathogen	Unlabelled (beetle-introduced) pathogen only	No inoculum detected
1981[a]	Aggressive strain	10.4	31.5	56.2	2.1
1982[a]	Aggressive strain	35.4	45.8	8.3	10.4
1984[a]	Non-aggressive strain	0	68.6	30.0	1.4

[a]A total of 50–70 beetles was sampled each year. Chi^2 analysis indicated a significant difference ($P < 0.01$) in the relative composition of inoculum carried by beetles when the data for all three years were compared. (From Webber & Brasier, 1984 and A. G. Mitchell, unpublished)

beetles, and the outcome of intraspecific competition between *O. ulmi* individuals. Even small differences in the growth rates of individuals, by influencing competitive fitness, can play a part in resolving the extent to which one particular individual contributes to the saprophytic phase.

The escape of *O. ulmi* from the xylem is also likely to depend on hyphal extension rates; if the pathogen has to grow into adjacent bark tissue by penetrating the pits of intact xylem vessels (MacDonald, 1970; Krause & Wilson, 1972; Scheffer & Elgersma, 1982) the process of invading the bark may be slowed significantly. However, escape can be aided by the action of breeding beetles which frequently score the underlying wood during gallery excavation, and in so doing break open xylem vessels, permitting rapid release of the elm pathogen (Webber & Brasier, 1984).

If beetles enter the bark of branches or even whole trees killed by agents other than *O. ulmi*, then the inoculum introduced by beetles will be the sole contributor to the pathogenic phase. Conversely, if the search for breeding sites by vector beetles is prolonged such that much of the inoculum carried on the beetles is lost (Webber & Brasier, 1984), the *O. ulmi* derived from the pathogenic phase will be mainly responsible for the success of the saprophytic phase. In the short term, the existence of the two distinct souces of *O. ulmi* increases the chance of the pathogen being united with a new brood of vector beetles. In the long term, however, the extent to which *O. ulmi* from the pathogenic phase is recyled into the saprophytic phase is probably critical for the maintenance of pathogenic fitness in the aggressive strain, since it is during pathogenesis in the xylem that the most stringent host selection in favour of strongly pathogenic genotypes will operate.

Population structure during the saprophytic phase

O. ulmi is a heterothallic, outbreeding ascomycete (Buisman, 1932). Within either the aggressive or non-aggressive strains sexual reproduction is promoted between genotypes of greatest dissimilarity (Brasier, 1984), a mechanism which generally ensures a genetically heterogenous population of individuals. This genetic diversity is reflected in the quality of inoculum acquired and transmitted by vector beetles. Isolations made from samples of only 50 *Scolytus scolytus* beetles emerging from a small log, usually yield several different *O. ulmi* genotypes, and even the spore inoculum on an individual beetle can consist of a mixture of genotypes (J.F. Webber, unpublished). Hence, when the vector beetles invade the bark of dead and dying elms to make their breeding galleries, the phloem tissue is 'inoculated' with a genetic assortment of isolates and these are further augmented by any additional genotypes released from the xylem of the same tree.

In the early stages of colonisation, *O. ulmi* is largely confined to the points of beetle entry. Within a few days oval brown lesions marking the areas of colonisation are usually present around the maternal galleries (Fig. 15.1*a*). Initially the bark appears to have a strong resistance to microbial invasion and for a brief period it is certainly not the decaying substratum implied by the term saprophytic phase. Only *O. ulmi* and a few other fungi with the ability to invade and kill living elm tissue are able to establish successfully (Webber, 1979; Webber & Hedger, 1986).

It is not known whether the enzyme and toxin systems that contribute to the success of *O. ulmi* as a vascular wilt pathogen in the xylem vessels of elms (Scheffer & Elgersma, 1981; Takai, Richards & Stevenson, 1983; Brasier, 1986*b*) also make it an effective coloniser of living phloem tissue, but this would seem likely. The non-aggressive strain, which is only a

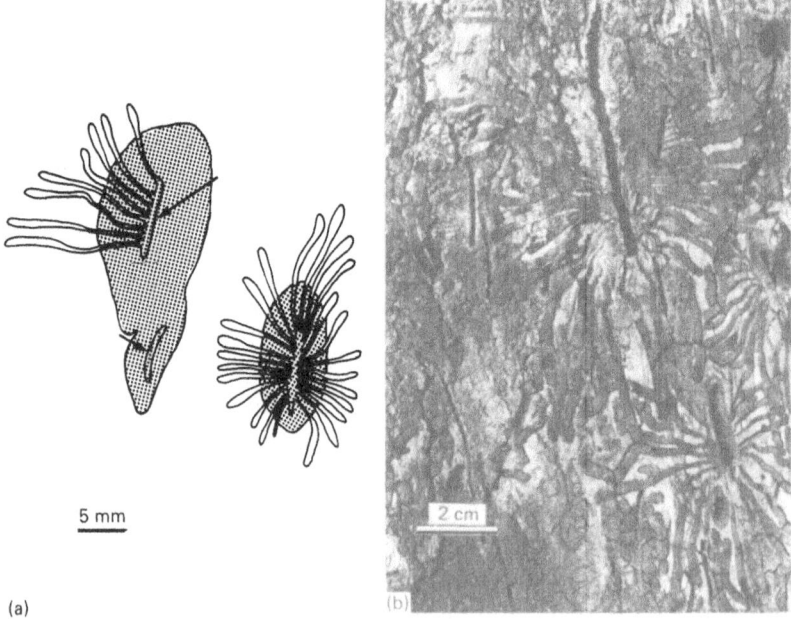

(a)   (b)

Fig. 15.1. Breeding galleries of scolytid beetles. (a) *O. ulmi* lesions (stippled) developing around *S. scolytus* maternal galleries. Colonisation has completely suppressed normal larval development associated with one maternal gallery, and partially suppressed it along one side of another (arrowed); (b) breeding galleries of *S. scolytus* in inner elm bark fully colonised by *O. ulmi*.

relatively weak vascular wilt pathogen, is also a poor coloniser of elm bark compared with the more highly pathogenic aggressive strain (Webber & Hedger, 1986). Whatever the basis of this ability, the early stage of inner bark colonisation is probably another period, albeit transient, in which a degree of host selection operates on the *O. ulmi* population.

A finely balanced relationship may also exist at this time between colonising *O. ulmi* and the developing larvae of the scolytid beetles. For the successful development of the 1st, 2nd and 3rd larval instars, it appears essential that they feed on phloem tissue free from any fungal colonisation (Fig. 15.1*a*). If *O. ulmi* lesions extend beyond the fringe of the feeding larvae, gallery excavation is usually aborted, presumably because the larvae have been killed (Webber, 1979). However, once larvae reach a later stage in their development (4–5th instar) normal growth generally continues even if feeding occurs in bark colonised by *O. ulmi* (Fig. 15.1*b*).

Over a period of 3-4 weeks the phloem tissue gradually dies and the saprophytic phase can be said to have truly begun. *O. ulmi* lesions around the galleries progressively extend and eventually coalesce resulting in an inner bark layer colonised by a complex mosaic of genotypes (Fig. 15.2).

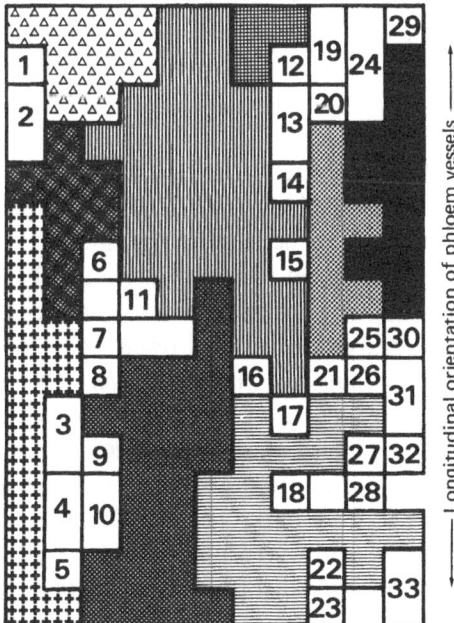

Fig. 15.2. Mosaic arrangement of *O. ulmi* in inner elm bark. Each shaded or numbered portion represents a distinct genotype; 39 different genotypes were present. Unshaded areas show the absence of *O. ulmi*. The smallest squares correspond to 1 cm^2.

In the mosaic some genotypes dominate, stretching over large areas of up to 20–50 cm^2 of phloem tissue. Each individual area tends to be elongated, since longitudinal colonisation is facilitated and lateral spread hindered by the vertical orientation of the phloem cells. The territory ultimately occupied by any one genotype is also three-dimensional as all the inner bark tissues, and eventually even the dead phelloderm arcs of the outer bark, are colonised by the pathogen (Lea, 1977).

Fruiting during the saprophytic phase

As a result of the activities of the egg-laying female beetles and later of the feeding larvae, the galleries of the scolytids gradually extend through much of the inner bark (Fig. 15.2). These spaces created by the beetles then become filled with the mycelium and fruiting structures of *O. ulmi*. Initially only mycelium bearing the *Sporothrix* stage conidia can be observed, but synnemata with sticky spore masses quickly follow and during the long overwintering saprophytic phase both these spore types can be readily observed soon after the beetles enter the bark to breed (Lea & Brasier, 1983). Many of these asexually produced spores are probably disseminated widely within the bark by the abundant mite population (Brasier, 1978) thus contributing to secondary foci of colonisation. With the onset of winter (November/December), however, both these fruiting structures, but particularly the synnemata, tend to decline and be replaced by perithecia.

Lea & Brasier (1983) suggested that perithecia and the ascospores they produce have two main functions; firstly, to act as an overwintering stage and, secondly, to act as a further source of genotype diversity during the saprophytic phase. Subsequent research has shown that some at least of the ascospores contribute to the later stages of bark colonisation. Novel, recombinant genotypes produced as a result of hybridisation between genetically-labelled pathogenic phase *O. ulmi* (Webber & Brasier, 1984) and vector-introduced *O. ulmi* can be isolated from elm bark following the initial flush of perithecia (see Table 15.2). Such recombinant genotypes can also contribute to the inoculum acquired by beetles before their emergence from bark (Table 15.2).

The role of perithecia in overwintering is less certain. Although in winter the onset of perithecial formation is strongly correlated with a fall in temperatures (Lea & Brasier, 1983), perithecia are also produced during the much shorter saprophytic phases which occur when scolytid beetles complete an entire breeding cycle during hot summer months (J.F. Webber, unpubl.). This suggests that the production of *O. ulmi* fruiting

Table 15.2. *Frequency of recombinant genotypes contributing to the saprophytic phase*

	Total no. of *Ophiostoma ulmi* individuals examined	No. of unchanged individuals originating from the xylem	No. of recombinant[b] individuals
Elm bark[a]			
January	75	10	0
February	200	62	2
March	150	19	11
April	108	6	1
Vector beetles[a]			
May–June	62	47	9

[a]*O. ulmi* individuals were isolated from a 700-cm² sample of bark each month, and from 56 *S. scolytus* beetles as they emerged between May–June.
[b]Recombinant genotypes resulted from xylem-derived × beetle-introduced *O. ulmi* matings.

structures tends to follow a largely predetermined, ontogenetic sequence of *Sporothrix* stage, synnemata and then perithecia. The timing of this sequence may also be governed by the availability of suitable cavities or crevices for fructification. Thus, beetle galleries contain the first flush of fruiting structures (Fig. 15.3*a*), while subsequent splitting apart of inner and outer bark layers, which tends to occur after galleries are first established, promotes a slightly later sequence of sporulation (Fig. 15.3*b*). Pupal chambers excavated by larvae a few weeks before emergence then provide a final opportunity for *O. ulmi* to fruit (Fig. 15.3*c*), and the spores produced here are largely responsible for contaminating young adult beetles prior to emergence (Webber & Brasier, 1984).

Intraspecific mycelial interactions between *O. ulmi* genotypes

Critical delineation of the mosaic of *O. ulmi* individuals which develops during elm bark colonisation (Fig. 15.2) is possible through experimental procedures which rely on the use of the fungus' vegetative compatibility (*v-c*) system. Such systems now appear to be widespread throughout the Ascomycotina and Basidiomycotina and permit the recognition of one mycelial genotype by another (Brasier, 1984; Rayner *et al.*, 1984).

Vegetative compatibility or incompatibility in *O. ulmi* is under polygenic control (Brasier, 1984) and its occurrence can be readily demonstrated in culture when two isolates are paired. If they are of different genotypes at the *v-c* loci, a reaction zone of proliferating hyphae and synnemata is formed and few, if any, viable hyphal fusions are formed between the adjacent mycelia. This is termed an incompatible reaction and may vary in appearance according to the degree of genetic difference between the paired isolates (Fig. 15.4*a, b, c*). In contrast, if the opposing isolates are compatible – by virtue of sharing the same *v-c* genes or having identical genotypes – their hyphae are able to fuse successfully and no reaction zone is formed (Fig. 15.4*d*; Brasier, 1984). Since the majority of mycelia involved in elm bark colonisation are of different genotypes it is inevitable that frequent and extensive incompatible reactions will occur during the saprophytic phase. These will not only delineate the area containing an individual genotype, but may act as a temporary barrier, defending the area occupied by an individual from invasion by adjacent mycelia (see Todd & Rayner, 1980; Brasier, 1984; Rayner *et al.*, 1984).

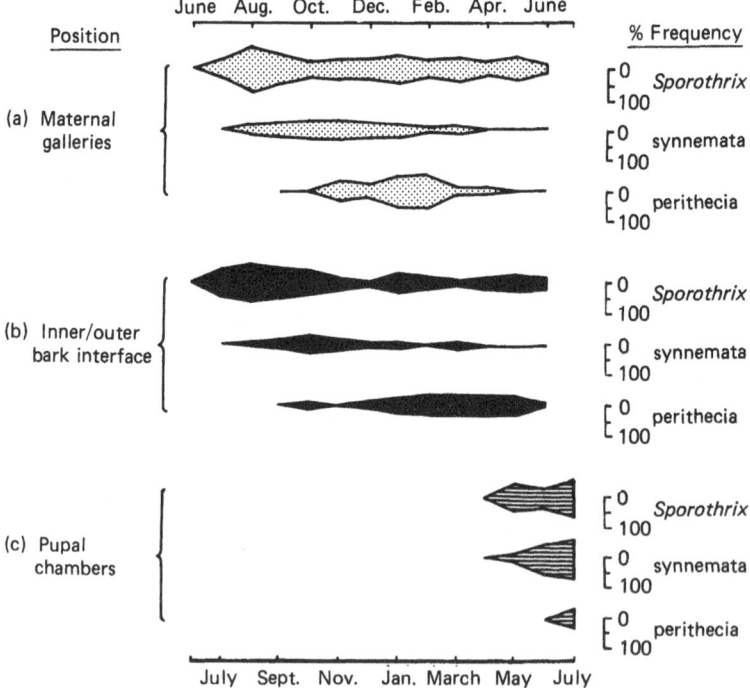

Fig. 15.3. Sequence of *O. ulmi* fruiting in (a) maternal galleries, (b) interface between inner and outer bark and (c) pupal chambers. % scale refers only to half the height of each plot.

Reaction zones are commonly breached; some genotypes appear strongly invasive and expand into territory already occupied by another individual. The arrangement of individuals in a mosaic may therefore alter considerably during the saprophytic phase. Brasier (1984) has termed this invasive process the penetration effect, and the extent to which one isolate penetrates another can be gauged from the broad bands of synnemata that are formed within a penetrated individual (Fig. 15.4*a, b*). Apart

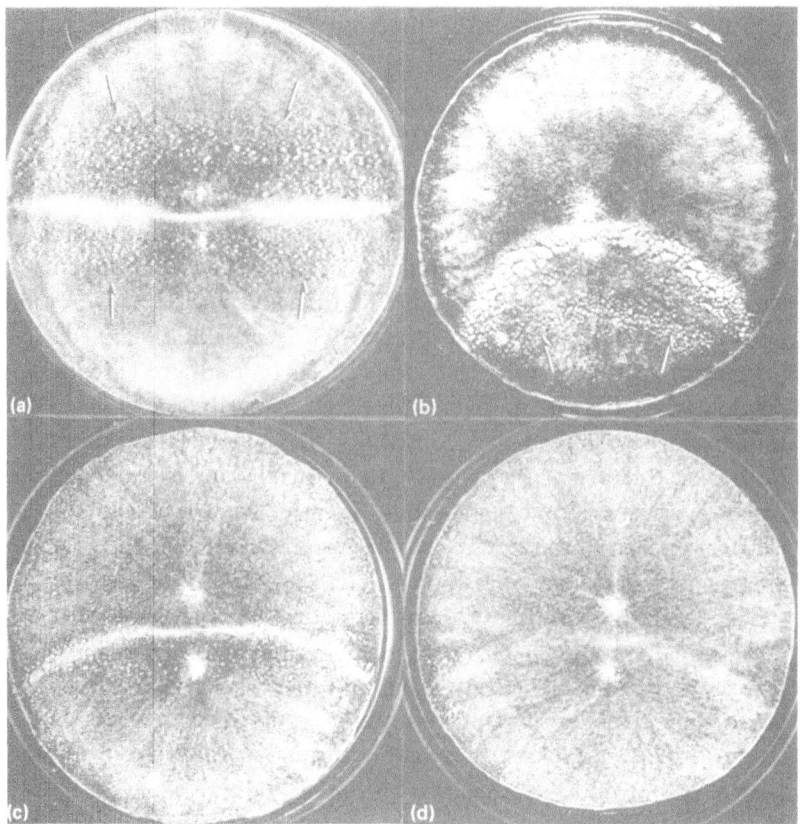

Fig. 15.4. Vegetative incompatibility reactions between paired *O. ulmi* individuals growing on elm sapwood agar. The inoculation points on each plate are indicated by O. (a) Complete incompatibility producing a wide reaction zone. Penetration of one individual into another is bidirectional; the extent of penetration is indicated by the lines of synnemata (arrowed). (b) Undirectional penetration across a wide reaction zone. The extent of penetration by the upper isolate into the lower is indicated by the lines of synnemata (arrowed). (c) Partial incompatibility producing a narrow reaction zone. (d) Compatible reaction showing no discernible effect.

from the intrinsic penetrating ability of certain genotypes, the depth of penetration which occurs between mycelia is dependent on the degree of incompatibility and the area of substratum occupied by the interacting individuals. Thus, penetration is most pronounced when interacting isolates have no v-c genes in common (Brasier, 1984) and where the penetrator operates from a base of at least the same area as the penetrated individual. Strong penetrating ability is probably an essential prerequisite if *O. ulmi* individuals generated from the overwintering flush of perithecia are to colonise elm bark after the mosaic has become established.

Prevention of cytoplasmic infection

By preventing viable hyphal fusions between genetically dissimilar mycelia, vegetative incompatibility has one further role, it restricts the transmission of cytoplasmic infection from one individual to another (Day, 1970; Caten, 1972). The importance of this particular function has recently been highlighted in *O. ulmi* by the discovery of virus-like infectious agents termed disease or d-factors (Brasier, 1983).

D-factors can exert very deleterious effects on both growth and reproduction when present in an isolate of *O. ulmi* (Brasier, 1983, 1986c). Initial evidence suggests they consist of dsRNA segments (Rogers, Buck & Brasier, 1986) and they appear to be a regular cellular component within aggressive strain populations. Transfer from diseased to healthy isolates is achieved during the cytoplasmic exchange that follows hyphal anastomoses (Brasier, 1983; Rogers *et al.*, 1986).

The greatest opportunity for d-factor transfer in nature must occur during the saprophytic phase when many individuals come into close contact. By using genetically labelled isolates, it has proved possible to show that healthy *O. ulmi* introduced into elm bark has on reisolation acquired d-factors, presumably through transfer from d-infected individuals introduced into bark by breeding scolytids (Brasier, 1986a,c). However, although transfer of d-factors is unrestricted between vegetatively compatible individuals, it is considerably restricted between incompatible isolates. In over 70 laboratory pairings between completely vegetatively incompatible healthy and d-infected individuals, d-factor transmission occurred in less than 4% of all the individuals tested (Brasier, 1984). During the saprophytic phase, therefore, vegetative incompatibility is likely to be a crucial factor in restricting d-factor transfer within *O. ulmi* mycelial mosaics.

D-infected isolates are also likely to be at a competitive disadvantage when colonising elm bark. They are slower growing and also tend to be strongly penetrated and replaced by healthy isolates (Brasier, 1984). Moreover, the fertility of crosses between d-infected and healthy isolates is reduced. Sexual reproduction between individuals that carry d-factors, however, has one significant feature. Although d-factors are readily transmitted to conidia produced from d-infected mycelia, they are rarely transmitted to ascospores produced through mating (Brasier, 1983, 1986c; Rogers *et al.*, 1986). Thus, in addition to vegetative incompatibility, sexual reproduction also influences d-factor spread in *O. ulmi* populations, by generating new genotypes which are free from dsRNA. (The role of dsRNA in other pathogenic fungi is described in chapter 17).

Interspecific competition during the saprophytic phase

A diverse range of fungal species can be isolated from bark of elms affected by Dutch elm disease; over fifty have been recorded in bark of *U. glabra* (Webber, 1979; Brayford, 1983) and many of these are also frequent colonisers in *U. procera* (M.E. Smith, unpublished). As with *O. ulmi,* some of these fungi are introduced into the bark by scolytid beetles. Others are ubiquitous colonisers of woody substrata, and a few already exist within the dead outer bark layers of healthy elms and invade the inner bark tissue as host resistance declines. Of this varied mycoflora, only a small number of species has the capacity to act as competitors of *O. ulmi* during the saprophytic phase (Webber & Hedger, 1986).

The competitors of *O. ulmi* can be categorised into two types (Gibbs & Smith, 1978). Primary antagonists, which as early colonisers of elm bark are able to exclude the elm pathogen by utilising space and available nutrients, and secondary antagonists, which can replace *O. ulmi* established in elm bark through mycoparasitism and antibiotic production (Table 15.3). Generally, the properties which make a fungus an effective early coloniser or replacer are mutually exclusive. Fungi with the greatest potential as primary antagonists have some pathogenic ability and hence are able to invade and establish in living or senescing phloem tissue. Those able to replace *O. ulmi* are poor at colonising live phloem and usually only become abundant weeks or months after the bark has been thoroughly invaded by primary colonisers (Webber, 1979).

In ecological terms, the secondary antagonists probably only have a relatively small impact on *O. ulmi* during the cycle of Dutch elm disease. There is little evidence to suggest that either *Trichoderma* spp. or *Gliocladium roseum* significantly reduce the number of spores acquired by vector beetles, although they may substantially replace the pathogen in old elm bark after the vector beetles have emerged (Webber, 1979). In contrast, primary antagonists are likely to be much more formidable as competitors since at least three species, *Nectria coccinea, Botryosphaeria stevensii* and *Phomopsis oblonga* were found to be more effective as bark colonisers than the aggressive strain of *O. ulmi* (Table 15.3). The extent of their impact is also likely to be governed by their abundance and the frequency with which they become established in elm bark before the pathogen. *B. stevensii* and *N. coccinea* tend to be somewhat sporadic colonisers and rarely become established within large areas of inner bark before beetle breeding is initiated. They must, therefore, be of limited effect as competitors of *O. ulmi* during the saprophytic phase.

Probably only *P. oblonga* is likely to have a major effect as a competitor as a consequence of its abundance in the dead outer bark layers of healthy elms where it exists in a latent mycelial form (Brayford, 1983; Webber & Gibbs, 1984). Following the onset of Dutch elm disease in such trees *P. oblonga* quickly invades the senescing inner bark from its base in the outer bark, often colonising large areas of phloem tissue before the bark is used as beetle breeding material. Any *O. ulmi* introduced by beetles into such bark typically fails to establish, and the new generation of beetles which eventually emerge do not carry any *O. ulmi* spores (Webber, 1979). However, colonisation of elm bark by *P. oblonga* not only excludes *O. ulmi*, but is usually highly detrimental to developing beetle broods (Webber, 1981, 1982). Thus, although *P. oblonga* can be a significant competitor of *O. ulmi* during the saprophytic phase, its main impact on Dutch elm disease is as a competitor of the scolytid beetles seeking bark for breeding.

Conclusions

The saprophytic phase in Dutch elm disease is considerably more than a period in which the pathogen colonises elm bark, ensuring the dissemination of its spores on the next generation of vector beetles. It is a period of significant selection and competition. Most selection, operating through the living phloem tissue, is probably imposed on the fungus during initial bark colonisation. The subsequent period of colonisation, geared to the capture and recapture of the highly nutritive inner bark tissue, is marked by waves of vegetative growth and cycles of fruiting and involves not only antagonistic interactions with bark saprophytes but intense competition between different *O. ulmi* individuals. It also represents a period of regeneration, providing an opportunity for *O. ulmi* to produce, via perithecial formation and the penetration effect, a reser-

Table 15.3. *Fungal competitors of* Ophiostoma ulmi *during the saprophytic phase*

	% occurrence in elm bark samples taken from 27 *Ulmus glabra* trees	Ability[a] to colonise fresh *U. glabra* bark expressed as lesion size (cm^2)	Ability[a] to replace *O. ulmi* in elm bark
Botryosphaeria stevensii	17	6.5 (range 2–21)	−
Fusarium solani	15	2.9 (range 2–4)	+
Gliocladium roseum	39	1.1 (range 0–3)	+ + + +
Nectria coccinea	20	7.6 (range 2–19)	−
Phomopsis oblonga	66	7.3 (range 1–16)	
Trichoderma viride	27	1.2 (range 0–3)	+ + + +
T. polysporum	1	2.2 (range 0–3)	+ + +
O. ulmi	73		
aggressive strain		5.5 (range 1–11)	
non-aggressive strain		3.0 (range 1–7)	

[a]Estimated from 12 replicate inoculation points.
−, No replacement; +, + + +, partial or total replacement of *O. ulmi* in < 10, and < 75% of all lesions respectively; + + + +, total replacement of *O. ulmi* in all lesions.
(From Webber & Hedger, 1986)

voir of genetic diversity to meet the conflicting demands of the complete disease cycle of the pathogen. Loss of d–factors or cytoplasmic rejuvenation is also achieved via sexual recombination during this time. Finally, the saprophytic phase is the means by which the gene pool from the pathogenic phase in the xylem feeds back into the main or saprophytic gene pool. The pathogenic phase *O. ulmi* is thereby able to contribute either unchanged or sexually recombined to the next cycle of the disease, a process which is probably essential for the maintenance of pathogenic fitness in the fungus.

References

Brasier, C.M. (1978). Mites and reproduction in *Ceratocystis ulmi* and other fungi. *Transactions of the British Mycological Society*, **70**, 81–9.

Brasier, C.M. (1979). Dual origin of recent Dutch elm disease outbreaks in Europe. *Nature* **281**, 78–80.

Brasier, C.M. (1982). Occurrence of three sub-groups within *Ceratocystis ulmi*. In: *Proceedings of the 1981 Dutch Elm Disease Symposium and Workshop*, ed. E.S. Kondo, Y. Hiratsuka & W.B.G. Denyer, pp. 298–321. Winnipeg, Manitoba: Department of Natural Resources.

Brasier, C.M. (1983). A cytoplasmically transmitted disease of *Ceratocystis ulmi*. *Nature*, **305**, 22–3.

Brasier, C.M. (1984). Inter-mycelial recognition systems in *Ceratocystis ulmi*: their physiological properties and ecological importance. In: *The Ecology and Physiology of Fungal Mycelium*, ed. D.H. Jennings & A.D.M. Rayner, pp. 451—97. Cambridge: University Press.

Brasier, C.M. (1986*a*). The population biology of Dutch elm disease: its principal features and some implications for other host-pathogen systems. In: *Advances in Plant Pathology*, vol. 5, ed. D.S. Ingram & P. Williams, pp. 53−118. London: Academic Press.

Brasier, C.M. (1986*b*). Some genetical aspects of necrotrophy with special reference to *Ophiostoma ulmi*. In: *Genetics of Plant Pathogenesis*, ed. P.R. Day & G.J. Jellis, pp. 297−310. Oxford: Blackwell Scientific Publications.

Brasier, C.M. (1986*c*). The d-factor in *Ophiostoma ulmi*: its biological properties and implications for Dutch elm disease. In: *Fungal Virology*, ed. K.W. Buck, pp. 177−208. Florida: CRS Press.

Brayford, D. (1983). *Phomopsis* as a biological control agent of Dutch elm disease. *Ph.D. Thesis, University College of Wales, Aberystwyth*.

Buisman, C. (1932). Over het voorkomen van *Ceratostomella ulmi* (Schwarz) Buisman in de natur. *Tijdschrift over Plantenziekten*, **38**, 203–4.

Caten, C.E. (1972). Vegetative incompatibility and cytoplasmic infection in fungi. *Journal of General Microbiology*, **72**, 221–9.

Day, P.R. (1970). The significance of genetic mechanisms in soil fungi. In: *Root Diseases and Soil-Borne Pathogens*, ed. T.A. Toussoun, R.V. Bega & P.E. Nelson, pp. 62–74. Berkeley: University of California Press.

Gibbs, J.N. & Brasier, C.M. (1973). Correlation between cultural characteristics and pathogenicity in *Ceratocystis ulmi* from Britain, Europe and America. *Nature*, **241**, 381–3.

Gibbs, J.N., Brasier, C.M., McNabb, H.S. & Heybroek, H.M. (1975). Further studies on pathogenicity in *Ceratocystis ulmi*. *European Journal of Forest Pathology*, 5, 161–74.

Gibbs, J.N. & Smith, M.E. (1978). Antagonism during the saprophytic phase of the life cycles of two pathogens of woody hosts – *Heterobasidion annosum* and *Ceratocystis ulmi*. *Annals of Applied Biology*, 89, 125–8.

Krause, C.R. & Wilson, C.L. (1972). Fine structure of *Ceratocystis ulmi* in elm wood. *Phytopathology*, 62, 1253–6.

Lanier, G.N. & Peacock, J.W. (1981). Vectors of the pathogen. In: *Compendium of Elm Diseases*, ed. R.J. Stipes & R.J. Campana, pp. 14–16. American Phytopathological Society.

Lea, J. (1977). A comparison of the saprophytic and parasitic stages of *Ceratocystis ulmi*. *Ph.D. Thesis, University of London*.

Lea, J. & Brasier, C.M. (1983). A fruiting succession in *Ceratocystis ulmi. Transactions of the British Mycological Society*, 80, 381–7.

MacDonald, W.L. (1970). Electron microscopy of elm infected with *Ceratocystis ulmi* (Buism.) C. Moreau. *Ph.D. Thesis, Iowa State University, Ames, Iowa.*

Rayner, A.D.M., Coates, D., Ainsworth, A.M., Adams, T.J.H., Williams, E.N.D. & Todd, N.K. (1984). The biological consequences of the individualistic mycelium. In: *The Ecology and Physiology of Fungal Mycelium*, ed. D.H. Jennings & A.D.M. Rayner, pp. 509–40. Cambridge: University Press.

Rogers, H.J., Buck, K.W. & Brasier, C.M. (1986). Transmission of double stranded RNA and a disease factor in *Ophiostoma ulmi. Plant Pathology*, 35, 277–87.

Scheffer, R.J. & Elgersma, D.M. (1981). Detection of a phytotoxic glycopeptide produced by *Ophiostoma ulmi* in elm by enzyme-linked immunospecific assay (ELISA). *Physiological Plant Pathology*, 18, 27–32.

Scheffer, R.J. & Elgersma, D.M. (1982). A scanning electron microscope study of cell wall degradation in elm wood by aggressive and non-aggressive isolates of *Ophiostoma ulmi. European Journal of Forest Pathology*, 12, 25–8.

Takai, S., Richards, W.C. & Stevenson, K.J. (1983). Evidence for involvement of ceratoulmin, the *Ceratocystis ulmi* toxin, in the development of Dutch elm disease. *Physiological Plant Pathology*, 23, 275–80.

Todd, N.K. & Rayner, A.D.M. (1980). Fungal individualism. *Science Progress*, 66, 331–54.

Webber, J.F. (1979). Interactions between elm bark saprophytes and the Dutch elm disease pathogen *Ceratocystis ulmi. Ph.D. Thesis, University College of Wales, Aberystwyth.*

Webber, J.F. (1981). A natural biological control of Dutch elm disease. *Nature*, 292, 449–51.

Webber, J.F. (1982). Natural biological control of Dutch elm disease by *Phomopsis oblonga*. In: *The Proceedings of the 1981 Dutch Elm Disease Symposium and Workshop*, ed. E.S. Kondo, Y. Hiratsuka & W.B.G. Denyer, pp. 24–35. Winnipeg: Manitoba, Department of Natural Resources.

Webber, J.F. & Brasier, C.M. (1984). The transmission of Dutch elm disease: a study of the processes involved. In: *Invertebrate-microbial Interactions*, ed. J. Anderson, A.D.M. Rayner & D. Walton, pp. 271–306. Cambridge: University Press.

Webber, J.F. & Gibbs, J.N. (1984). Colonisation of elm bark by *Phomopsis oblonga. Transactions of the British Mycological Society*, 82, 352–4.

Webber, J.F. & Hedger, J.N. (1986). Comparison of interactions between *Ceratocystis ulmi* and elm bark saprobes *in vitro* and *in vivo. Transactions of the British Mycological Society*, 85, 93–101.

16
Sporulation of foliar pathogens

Y. COHEN[1] and J. ROTEM[2]

[1]Department of Life Sciences, Bar Ilan University, Ramat Gan 52100, Israel and
[2]Department of Plant Pathology, Agricultural Research Organisation, The Volcani Center, Bet Dagan 50250, Israel

Introduction

Most studies of sporulation have been made on fungi growing *in vitro* (Cochrane, 1958; Hawker, 1966; Smith & Galbraith, 1971; Turian, 1974). As such they have failed to take account of (a) the influence of the host on reproduction of its parasite, (b) the effects of environmental factors which differ greatly from those in nature and (c) the density of sporulation in the field. Also, they have dealt exclusively with facultative parasites (Rotem *et al.*, 1978). There have been relatively few investigations of sporulation *in vivo* despite its importance in the development of disease in the field. Nelson & Tung (1973) stated that '...No one portion of the disease cycle exerts a greater influence on epidemic increase of many diseases than the production of inoculum...' Thus, from the first attempts to simulate epidemics (Waggoner & Horsfall, 1969), the number of spores produced has been a basic ingredient of calculations. In many pathosystems the number of spores produced has been found to reflect the pathogenicity of the pathogen and the resistance mechanisms of the host (see review by Johnson & Taylor, 1976). The range of conditions for sporulation is often narrower than that required for infection, e.g. in *Phytophthora infestans* on potato (Rotem & Sari, 1983); other examples follow.

From an epidemiological standpoint, the more restricted the conditions for the development of a given process, the more vulnerable it is to environmental hazards and the more critical it is in the life cycle of a pathogen (Rotem & Sari, 1983). The critical role of sporulation is further evidenced by cases in which increased sporulation compensated for environmental limitations in the field, or resistance phenomena in the host, as discussed in the following sections.

Direct effects of environmental factors

Temperature

Temperature optima are not absolute and may vary with changes in other conditions. Hildebrand & Sutton (1984) found that the optimum temperature for sporulation of *Peronospora destructor* on onions increased from 10 to 18°C when the onset of high humidity and darkness was delayed from 10 pm to 3 am. In many instances higher temperatures are optimal for sporulation during shorter periods of leaf wetness while lower temperatures are optimal in longer periods of leaf wetness, e.g. for *Phyllosticta maydis* on corn (Caster *et al.*, 1977), *Pyricularia oryzae* on rice (Kato & Kozaka, 1974), *P. infestans* on potatoes (J. Rotem, unpublished), and *Peronospora pisi* on peas (R. Neubauer, unpublished). In *Rhynchosporium secalis* infection of barley, the optimum temperature and leaf surface wetness conditions were affected by the antagonist *Pseudomonas fluorescens* growing in sporulation sites and causing lysis of spores. Through the presence of this antagonist maximum sporulation was achieved at a lower temperature during prolonged periods of wetness, while a relatively higher temperature was associated with maximum sporulation in short periods of wetness (Rotem *et al.*, 1976). The optimum temperature varied also in relation to the stage of lesion development.

In *Pseudoperonospora cubensis* infections of cucumbers the optimum temperature for sporulation increased from 15°C on yellow lesions to 20°C on lesions turning necrotic (Cohen & Rotem, 1969). The rate of sporulation of *Colletotrichum lagenarium* on cucumbers in the early stage of lesion development was greatest at 24°C, but total production of spores throughout the sporulation period was greatest at 16°C (Thompson & Jenkins, 1985). Similarly, it was found that a high temperature induced maximum daily sporulation early in lesion development, while a lower temperature maximised total spore production by downy mildew of cucumbers (Cohen & Rotem, 1971). *Helminthosporium turcicum* on corn (Levy & Cohen, 1980), *Pyricularia oryzae* on rice (Kato & Kozaka, 1974), and *Phyllosticta maydis* on corn (Castor *et al.*, 1977). The phenomenon probably also occurs in other pathosystems. Although relatively high temperatures may promote sporulation by a direct effect on the pathogen, prolonged exposure to higher temperatures may result in a faster depletion of nutrients from the host which thus decreases the amount of spores produced (Cohen, 1976; Cohen, 1981b).

Differences in temperature response have sometimes been observed in strains of the same pathogen. For example, race T of *Helminthosporium maydis* on corn produced more spores at relatively high temperatures than race O (Warren, 1975). However, race T consists of cool (16–20°C) and warm (28-31°C) environment populations that change from 'cool' to 'warm' and *vice versa* when exposed for six sporulation cycles to cool or warm environments (Hill & Nelson, 1976). The effects of temperature on sporulation are discussed further below (p. 318).

Relative humidity (RH) and free leaf moisture

Powdery mildews sporulate in RHs below saturation but among them individual species may prefer high RH, e.g. *Erysiphe graminis* on barley (Ward & Manners, 1974), low RH, e.g. *Sphaerotheca fuliginea* on squash (Reuveni & Rotem, 1974), or be intermediate in response to RH, e.g. *E. cichoracearum* on tobacco (Cole, 1971). Free leaf moisture inhibits sporulation of at least some powdery mildew species (Yarwood, 1939).

Surprisingly, there is little information regarding humidity requirements for sporulation of rusts, the exception is *Uromyces phaseoli* on beans, which was reported to sporulate best under intermittent short periods of free moisture on the leaf or at high RH (Yarwood, 1961). By contrast, Imhoff *et al.* (1982) concluded that RH has little effect on sporulation of this pathogen, although more spores were produced in low rather than high RH. Many downy mildews sporulate best at RHs near saturation but are inhibited by free leaf moisture, e.g. *Peronospora trifoliorum* on alfalfa (Fried & Stuteville, 1977), *P. tabacina* on tobacco (Pinckard, 1942) and *Pseudoperonospora cubensis* on cucumbers (Cohen, 1981).

Sporulation in the field is regularly associated with dewfall but its precise regulation is not clear, because there are no tools to measure or to control RH at the leaf surface, and it is possible that at RHs near saturation minute drops of free water may form on leaves under conditions of fluctuating temperatures. To illustrate this complexity, when *Helminthosporium turcicum* on corn was exposed to moist conditions for 36h at 20°C more spores were produced in a moisture-saturated atmosphere (about 100% RH) than under free leaf moisture, but the difference disappeared when the periods were shorter (Levy & Cohen, 1980). The inhibiting effect of free leaf moisture on sporulation depends on the size or density of the covering drops (Rotem *et al.*, 1978) and appears to be associated with restricted aeration (Inaba & Hino, 1980).

Despite more prolific sporulation under high RH than in free water in some pathosystems, the amount of spores produced in most pathosystems increases significantly with increased duration of free moisture, e.g. in *H. turcicum* on corn (Levy & Cohen, 1980). The minimal duration of leaf surface wetness needed to induce sporulation ranges from 2 to3h for *Phytophthora colocasiae* on taro (Trujillo, 1965), *Mycosphaerella musicola* on banana (Kranz, 1968) and *Sclerospora sorghi* on sorghum (Cohen & Sherman, 1977), to about 24h for *Alternaria porri* f.sp. *solani* on potatoes (Bashi & Rotem, 1975a). In many localities such long periods of wetness are extremely rare, but some pathogens are able to complete the required period of wetness by utilisation of short wet periods over several nights, e.g. *A. porri* f.sp. *solani* on potatoes and *Stemphylium botryosum* f.sp. *lycopersici* on tomatoes (Bashi & Rotem, 1975a), *Rhynchosporium secalis* on barley (Rotem *et al.*, 1976) and *Helminthosporium maydis* (Nelson & Tung, 1973) and *H. turcicum* on corn (Levy & Cohen, 1980). The species which can sporulate under such conditions are those whose conidiophores are formed in one night, withstand dryness on the following day and produce spores during the subsequent night (Bashi & Rotem, 1975b). When sporulation of these pathogens requires induction by light and/or dryness, the interrupted wetting period regime is associated with more prolific sporulation, as in the case of *A. porri* f.sp. *solani* on potatoes (Bashi & Rotem, 1976) and *H. turcicum* on corn (Levy & Cohen, 1980). The evolution and ecological importance of these phenomena are discussed elsewhere (Rotem *et al.*, 1978).

The relationship between the amount of spores produced and the duration of leaf surface wetness is usually simpler under controlled conditions than in the field. For instance, the number of sporangia produced by *P. infestans* in dew chambers was directly correlated with the duration of leaf surface wetness (J. Rotem & E. Bashi, unpublished). In the field it was correlated with the average duration of dew in various seasons, but not with the duration of dew periods on specific nights (Bashi *et al.*, 1982). However, the conditions governing sporulation of *H. maydis* on corn in the laboratory did not correlate with those in the field, where sporulation did not always increase as it did in the laboratory with increasing periods of high humidity, particularly in the range 90% RH and above (Wallin & Loonan, 1977). The relation of sporulation in controlled conditions to sporulation in the field is discussed elsewhere at more length (Rotem, 1988a,b).

In many cases a longer leaf surface wetness period is needed for sporulation than for infection. For instance, 3h and 7h of free moisture were needed for minimum infection (I) and sporulation (S), respectively, in *Phytophthora infestans* on potatoes (Rotem *et al.*, 1971; E. Bashi, unpublished); 2 h(I) and 3 h(S) in *P. cactorum* on strawberry fruit (Grove *et al.*, 1985a, b); 2 h(I) and 6–9h(S) in *Pseudoperonospora cubensis* on cucumbers (Cohen & Rotem, 1969; Cohen *et al.*, 1971; Cohen 1981); 4 h(I) and 12h(S) in *H. maydis* on corn (Nelson & Tung, 1972, 1973); 3h(I) and 12h(S) in *Peronospora pisi* on peas (Olofsson, 1966); 12 h(I) and 16 h(S) in *Stemphylium* f.sp. *lycopersici* in tomatoes and 8 h(I) and 16 h(S) in *Alternaria* f.sp. *solani* in potatoes (Bashi & Rotem, 1974, 1975*a*).

Light

Continuous illumination with visible light is inhibitory to spore differentiation, but not to sporophore formation, in many fungal species both in culture (Turian, 1974) and on host plant tissue. On infected tissues, blue light was more inhibitory than green light, while red light was least effective (Cruickshank, 1963; Zimmer & McKeen, 1969; Inaba & Kajiwara, 1975; Cohen, 1976; Cohen & Eyal, 1977; Brook, 1979; Levy & Cohen, 1981; Hildebrand & Sutton, 1984). A decrease in irradiance has been associated with a corresponding decrease of inhibition, for example in *A. porri* f.sp. *solani* and *S. botryosum* f.sp. *lycopersici* (Bashi & Rotem, 1975*c*), but pathogens differ in their sensitivity. Sporulation of several, such as *Rhynchosporium secalis* on barley, some rusts (J. Rotem, unpublished) and *Erysiphe cichoracearum* on tobacco (Cole, 1971), was not inhibited by continuous illumination. However, with sensitive species, spore production in culture as well as on host tissue was not affected by light at relatively low temperatures (Lukens, 1966; Bashi & Rotem, 1975*b*). Sporangial formation by *P. infestans* on potato leaves kept under continuous blue light of 25 μE m^{-2} s^{-1} at 10°C and 20°C was inhibited by 18 and 92% respectively, when compared with the dark controls (Cohen *et al.*, 1975), and that of *A. porri* f.sp. *solani* on illuminated tomato leaves (200–250 ft-c) at 14°C and 25°C was reduced by about 56 and 98%, respectively, compared with the dark controls (Luckens, 1966). Exposure to incandescent light of about 40 μE m^{-2} s^{-1} for 10 min h^{-1}, in *P. tabacina* on tobacco, *Pseudoperonospora cubensis* on cucumbers and *H. turcicum* on corn, reduced conidial yield by about 70–90% (Cohen *et al.*, 1978). A night temperature of 20°C is common in warm localities, or in heated greenhouses where the use of intermittent nocturnal illumination may be useful for the control of fungal diseases of foliage.

With downy mildews, a preceding dry, dark period of about 4 h increases spore yield remarkably under light conditions (Cruickshank, 1963; Cohen, 1976; Cohen & Eyal, 1977; Hildebrand & Sutton, 1984). Such recovery, observed only in leaves treated with a preceding dark period at 20°C, but not at 10°C (Cohen, 1976), supports the hypothesis that in the dark enzymic decay occurs of an antisporulant synthesised in the light. No such effect of a preceding dark treatment enhancing sporulation was observed in *P. infestans* on potato (Cohen *et al.*, 1975).

Effects of host and nutrition on sporulation

Sporulation, more than other events in a pathogen's life cycle, distinguishes the biotrophic (obligate) or necrotrophic (facultative) character of fungal species. Sporulation of biotrophs is associated with conditions favourable for the host. In *Bremia lactucae* on lettuce, any environmental factor which increased stress on the host, or decreased the stability of the host-parasite relationship, tended to shorten the duration of sporulation (Michelmore, 1981). In rusts, downy mildews and powdery mildews, the pathogen sporulated best on living, green or chlorotic rather than necrotic tissue (Cohen & Rotem, 1969, 1971). By contrast, greatest sporulation of many necrotrophic parasites is confined to necrotic lesions (Becker, 1963; Scharen, 1964; Kranz, 1968; Bashi & Rotem, 1975*b*). In *S. botryosum* f.sp. *lycopersici* on tomato, external factors that speeded up necrosis and death of the infected leaves increased the number of spores produced (Bashi & Rotem, 1975*b*). *Septoria nodorum* formed pycnidia when infected wheat was killed by a herbicide (Obst, 1971).

In epidemics caused by necrotrophs the highest spore production and dispersal may occur on dead leaves, as in the case of *A. porri* f.sp. *solani* on a completely wilted field of tomatoes (Rotem, 1964). In epidemics caused by biotrophs, such as *Pseudoperonospora cubensis* on cucumbers, the highest rate of spore production and dispersal may occur when lesions are yellow and the lowest rate may occur when most of the leaves become necrotic (Cohen & Rotem, 1971). In some cases the behaviour of necrotrophic and biotrophic pathogens is less extreme. Thus, *Phakospora pachyrhizi*, often regarded as a biotroph, will sporulate on necrotic tissue of soybeans (Melching *et al.*, 1979). *Phytophthora infestans* on potatoes sporulates on young rather than old and necrotic lesions and exemplifies those species which fill an intermediate position between biotrophic and necrotrophic fungi (Rotem *et al.*, 1978).

Photosynthetic activity of the host is the main factor affecting sporulation of biotrophic fungi. Stakman & Levine (1919) suggested that development of *Puccinia graminis* on wheat, as assessed by production of uredospores, is conditioned by host photosynthesis. Yarwood (1937) came to similar conclusions in relation to downy mildews. Later studies showed that in rusts, downy mildews and powdery mildews sporulation at night is conditioned by photosynthesis during the previous day and is inhibited by chemical inhibitors of photosynthesis despite otherwise favourable conditions (Cohen & Rotem, 1970; Inaba & Kajiwara 1971, 1975; Reuveni *et al.*, 1971; Inaba & Morinoka, 1983*b*). Similar effects are apparent from patterns of photosynthate accumulation and depletion in diseased plants. Amounts of photosynthates increase in the infected tissue until the onset of sporulation and decrease thereafter, e.g. with *Uromyces phaseoli* on beans (Inman, 1962), *Plasmodiophora brassiccae* in cabbage (Keen & Williams, 1969), and *Pseudoperonospora cubensis* on cucumbers (Perl *et al.*, 1972). In the latter system, photosynthates accumulated in lesions in the form of polysaccharides. During the period of sporulation these materials were converted into simple sugars needed for spore formation (Perl *et al.*, 1972; Inaba & Kajiwara, 1975). This process proceeds in darkness under wet as well as under dry conditions. Other downy mildews showing enhanced sporulation in darkness include *Peronospora tabacina* on tobacco (Cruickshank, 1963; Uozomi & Kröber, 1967). *P. destructor* on onions (Hildebrand & Sutton, 1984) and probably other species of this group. [For additional literature see Michelmore (1981) and Populer (1981).]

Populer (1981) discussed the epidemiological consequences of sporulation in dry dark conditions. His calculations showed that around 50° latitude in the summer, with nights extending from 2000 to 0400 hours, spores of *P. tabacina* are expected to be fully developed at 0700 hours; in the autumn, with nights from 1800 to 0600 hours, they will be fully developed at 0600 hours. Hildebrand & Sutton (1984) showed that in *P. destructor* on onions the optimum temperatures for sporulation vary in accordance with the onset of dry dark conditions, as mentioned previously. It was assumed that the depletion of photosynthates in darkness is less at the lower than at the higher temperature and that a temperature-dependant enzyme system is an additional regulator of sporulation. The effect of darkness on sporulation *via* changes in host metabolites is known so far only in downy mildews.

Host-mediated nutritional effects on sporulation of facultative parasites *in vivo* are almost unknown, or difficult to interpret. For example, sporulation of *P. infestans* on potatoes increased in plants supplied with a standard Hoagland fertiliser and decreased in plants supplied with double-strength Hoagland's solution (Rotem & Sari, 1983). In 1957 Horsfall & Dimond proposed that low sugar content in host plants predisposes them to diseases caused by some pathogens. The preference of necrotrophic parasites for sporulation on necrotic tissue may be associated with the low sugar content of these tissues but other, unknown factors are probably involved. In *Alternaria porri* f.sp. *solani* on potatoes and *Stemphylium botryosum* f.sp. *lycopersici* on tomatoes, artificially increasing the glucose concentrations inhibited sporulation either when glucose was applied during the entire 48h of the sporulation process or during the last 24 h of this process. However, in *A. porri* f.sp. *solani* additional glucose increased spore yields when applied to leaves in the first 24-h period, during which conidiophores rather than spores were being formed. In *Helminthosporium turcicum* on corn, sucrose applied to leaves during the formation of conidiophores increased subsequent sporulation, while sucrose applied during the formation of spores had no effect (Levy & Cohen, 1981). It appears, therefore, that the conidiophore-producing stage of sporulation resembles a vegetative rather than a reproductive process, and is facilitated by sugars. The inhibiting or indifferent effect of sugars applied during the second stage of the process derives from an excessive stimulation of mycelial development (Bashi & Rotem, 1975).

Quality of spores

Conditions affecting the host plant, or the sporulating pathogen directly, may influence not only the number of spores produced but also their quality. Information on this subject is available only for some biotrophic pathogens. In *Erysiphe graminis* on wheat the most infectious conidia were produced under conditions of 20°C and 100% RII (Ward & Manners, 1974). In *E. graminis* f.sp. *hordei* on barley the most infectious spores were produced in light regimes optimal for host photosynthesis (Aust, 1975). Conidia of *Sphaerotheca fuliginea* on cucumbers were produced in greater numbers in warm and dry than in relatively cooler and more humid conditions. However, the infectivity of the less-numerous spores produced in the cooler and humid conditions was much higher and compensated for the reduced numbers (Bashi & Aust, 1980).

The infectivity of spores also depends on the duration of their previous exposure to moisture. Fried & Stuteville (1977) demonstrated that incubation of *Peronospora trifoliorum*-infected lucerne plants for 10h under moist conditions was adequate to achieve morphologically mature spores. However, an additional 2h in moist conditions was required for those spores to reach their full physiological maturity. Bashi *et al.* (1982) studied diurnal changes in the infectivity of sporangia of *Phytophthora infestans* produced on potatoes during the previous night in the field. Detached sporangia became less infectious during the day, while attached sporangia became more infective towards the end of the day. Retention of germinability or infectivity in attached as compared with detached and dispersed spores is apparently typical of most fungi but has been recognised only in those that produce sensitive spores, e.g. *Phytophthora palmivora* on papaya (Hunter & Kunimoto, 1974), *Peronospora destructor* on onions (Yarwood, 1943) and *P. tabacina* on tobacco (Rotem *et al.*, 1985).

Induction processes

Some fungi sporulate *in vitro* only after exposure of their previously formed conidiophores to short wavelength, preferably ultraviolet (UV) radiation (Cochrane, 1958; Hawker, 1966; Smith & Galbraith, 1971; Turian, 1974). As discussed elsewhere, such induction phenomena also occur *in vivo*, but are less frequent (Rotem *et al.*, 1978). Pathogens that require induction for sporulation *in vitro* but sporulate without induction *in vivo* include *Phytophthora parasitica* on papaya (Aragaki & Hine, 1963), *S. botryosum* f.sp. *lycopersici* on tomatoes (Bashi & Rotem, 1975c) and *Botrytis cinerea* on geranium (Hyre, 1972). In species such as *Alternaria porri* f.sp. *solani* on potatoes, which need induction by sunlight or UV light to sporulate *in vitro*, weak fluorescent light, almost deficient of UV, will induce sporulation *in vivo*. In the latter case dryness partially substituted for, or promoted, the stimulatory effect of light (Bashi & Rotem, 1975b, 1976). Sporulation of *Helminthosporium turcicum* on corn was not induced by light, but at 15°C was triggered by drying of the conidiophore-bearing leaves (Levy & Cohen, 1980).

Honda *et al.* (1977) reported that, with pathogens triggered to sporulate by light, e.g. *Botrytis cinerea* in cucumbers and tomatoes, partial reduction of disease in greenhouses was accomplished by covering the greenhouses with UV-absorbing vinyl films, although they did not include measurements of spore production on hosts. Rotem *et al.* (1978) proposed that the decreased demand for light-induction of sporulation that can occur *in vivo* derives from the presence of sporogenic materials in the host tissue; the effect of radiation on sporulation of the same parasites in culture may be a substitute for natural processes proceeding in pathogenesis. More studies are needed to confirm these hypotheses.

Some workers claim that the confinement of sporulation of most fungi to darkness is evidence that darkness itself is inductive (e.g. Raffray & Sequeria, 1971). However, most fungi that sporulate in darkness under relatively high temperatures will also produce spores in light when the temperature is lower.

Formation of sexual spores

In many downy mildews the production of sexual spores is conditioned by availability of heterothallic strains (Michelmore, 1981; Shaw *et al.*, 1985), and the absence of sexual spores in specific geographical areas may be associated with absence of such strains. Thus, while *Phytho-phthora infestans* reproduces sexually in potato and tomato in Mexico, where the mating types A1 and A2 are present, the apparent absence of sexual reproduction in other countries was attributed to absence of the mating type A2 (Shaw *et al.*, 1985). Recently, however, this hypothesis has been questioned because A2 mating type was found in Switzerland (Hohl & Iselin, 1984), in North Wales (Shaw *et al.*, 1985), and in Israel (Fry, personal communication). Although heterothallism is very wide-spread, many other pathogens, or their specific strains, can produce sexual spores when cultured alone, e.g. certain isolates of *Bremia lactucae* on lettuce (Tommerup *et al.*, 1974).

In may pathogens, sexual spores and, or, their sporocarps appear at the end of the vegetative season and serve as a mean of perennation between seasons or crops, e.g. cleistocarps of *Erysiphe graminis* on barley, apothecia of *Venturia inaequalis* on apples and teliospores of many rusts. In other cases the sexual spores also participate in spreading the pathogen during the vegetative season, as in the case of *Mycosphaerella musicola* on bananas. In the latter, production of ascospores and of pycnidia is stimulated by rain and by dew, respectively, and the developmental flexibility assists the pathogen to spread in a variety of environmental conditions. Nevertheless pycnospores contribute much more inoculum than do ascospores (Stover, 1970).

For many biotrophic pathogens the biotic conditions that favour formation of sexual spores contrast with conditions that favour asexual sporulation. Oospores of *Peronospora viciae* on peas are formed when conditions are adverse for asexual sporulation (Pegg & Mence, 1970). In some downy mildews, e.g. *Bremia lactucae*, sexual reproduction is associ-ated with a regression of asexual spore formation (Tommerup, 1981). *P. manshurica* on soybean similarly produces conidia during the first few days after infection. The production of oospores starts several days later and suppresses the production of conidia (Inaba & Morinoka, 1983*a*). In some of these examples sexual spores are formed on senescent hosts which are partly depleted of nutrients. Thus, oospores of *P. parasitica* in crucifers are stimulated to form by factors favouring leaf senescence and a deficiency of nutrients (McMeekin, 1960). Oospores of *Sclerospora sorghi* on sorghum are formed in shredded, necrotic leaves (Cohen & Sherman, 1977). Teliospores of *Puccinia recondita* f.sp. *tritici* on wheat are stimulated to form when the host is also infected by *Septoria nodorum*

(Van der Wal *et al.*, 1970). A mycoparasite, *Aphanocladium album*, introduced into uredinia induced production of teliospores in several rusts (Biali *et al.*, 1972). Among factors listed by Waters (1928) the formation of teliospores in rusts is induced by darkness, slow deprivation of water and starvation. These examples suggest that sexual sporulation is induced by stress on the pathogen. However, many pathogens, including the biotrophs, produce their sexual spores in the absence of stress. Oospore formation in *Bremia lactucae* in lettuce is induced by conditions favourable for host development. At least in this pathosystem, senescence of the host induces formation of oospores rather than formation of oospores speeding up host senescence and necrosis (Michelmore, 1981). Oospore production in *Peronospora manshurica* on soybean is promoted by light regimes favouring photosynthesis of the host (Inaba & Morinaka, 1983b) thus resembling the mechanism that facilitates asexual sporulation in biotrophic fungi.

The role of sporulation in epidemics

Association between the amount of inoculum and the intensity of an epidemic have been observed in many diseases (Rotem, 1978). In some pathosystems, but especially in rusts, the amounts of inoculum present have been used for quantification and/or prediction of epidemics (e.g. Burleigh *et al.*, 1969; Eversmeyer & Burleigh, 1970). Data compiled from various sources enabled Schafer & Roelfs (1985) to calculate that spring wheat in the USA produces 2.3×10^4 uredospores of *Puccinia graminis* per uredinium per season. Stakman & Harrar (1957) calculated that one acre of rusted wheat in the USA may produce 5×10^{13} spores. High levels of sporulation in nature may overcome infection barriers, such as unfavourable environmental conditions and host resistance, and result in epidemics. In Indiana, occurrence of high numbers of spores of *Helminthosporium maydis* on corn explained the more severe epidemic under the relatively unfavourable weather conditions in 1970, than under the more favourable weather, but with lower spore load, in 1971 (Shaner *et al.*, 1972). Outbreaks of *Pseudoperonospora cubensis* on cucumbers in Israel have been associated more with a proximity to a source of inoculum than with the range of temperatures prevailing in various seasons (Cohen & Rotem, 1971). The case in which the amount of spores produced compensates for the relative resistance of the host is demonstrated in epidemics of *Stemphylium botryosum* f.sp. *lycopersici* on tomato. In this pathosystem, although ageing tomato plants are nearly resistant to infection, epidemics occur only on ageing plants in the field. This is because there is an abundance of inoculum in these fields, and relatively little in fields with plants that are susceptible and young (Rotem & Bashi, 1977). According to Zadoks & Schein (1979) the four factors that condition epidemic development are: (1) the length of the latent (incubation) period, namely the time between infection and sporulation, (2) the number of spores produced per lesion, (3) the infectious period, i.e. the duration of the sporulation period, and (4) the effectiveness of inoculum, i.e. the proportion of spores that initiate new infections. The first three factors are directly associated with sporulation. In general, the shorter the latent period, the higher the number of spores produced and the longer the infectious period, the more explosive an epidemic would be. These phenomena are obviously influenced by environmental and host conditions. In practice, relative efficiency in one of these factors is associated with a relative lack of efficiency in another (Rotem, 1978; Aust *et al.,* 1980). For example, the short latent period of about four days in *P. infestans* on potatoes, and the prolific sporulation of this pathogen compensate for a short infectious period of five to eight days (J. Rotem, unpublished: Rotem & Cohen, 1974), the low number of spores produced by *Alternaria porri* f.sp. *solani* on potatoes, 1×10^3 cm^{-2} lesion d^{-1} (Bashi & Rotem, 1975a), is compensated by an infectious period that lasts for several weeks and proceeds also in dead plants (Rotem, 1964). The latent period of *Eutypa armeniacae* on apricot trees lasts for 3-4 years, but its infectious period also lasts for several years (Carter, 1960).

The relative importance of various components of sporulation is not equal. For example, *Helminthosporium turcicum* and *H. maydis* on corn produce a total of 56×10^3 and 151×10^3 spores cm^{-2} infected tissue, respectively. Both pathogens have similar infectious periods. The latent periods of *H. turcicum* and *H. maydis* are 6.4 and 3.4d respectively. Due to a more prolific sporulation, and particularly to a shorter latent period, *H. maydis* is able to cuase more severe epidemics than *H. turcicum* (Y. Levy & M. Ballas, pers. comm.).

The number of spores produced is a density-dependent phenomenon. Teng & Close (1978) studied sporulation of *P. hordei* on barley as affected by ten uredinium densities. The size of uredosori, number of uredospores per sorus, latent period and the infectious period decreased with increasing density of uredosori on leaves.

Density-dependent sporulation phenomena were also observed in *Erysiphe graminis* f.sp. *tritici* on wheat (Rouse *et al.,* (1984) and *Uromyces phaseoli* on beans (Imhoff *et al.,* 1982). In wheat, an exponential decrease in cumulative spore production per colony occurred as number of colonies per leaf increased, with a maximum at 25 colonies per leaf. In beans, the average sporulation per day was inversely proportional to density of uredinia over all RHs tested (51–87%).

A long infectious period may represent a survival mechanism that secures continuation of the pathogen over periods temporarily adverse to infection (Zadoks, 1978). Thus, a sporulation period of up to 65d for *P. recondita* f.sp. *triticina* on wheat constitutes such a survival mechanism (Mehta & Zadoks, 1970). Depending on temperature, lesions of *Pseudoperonospora cubensis* on cucumbers sporulate for 2 to 3 weeks (Cohen & Rotem, 1971). The sporulation period of *P. infestans* on potato leaves lasts for only 5 to 8 d which hardly secures survival; the ability to survive spells of extremely hot and dry weather in Israel is secured by stem lesions that sporulate for about 20d (Rotem & Cohen, 1974). *Septoria passerinii* on barley straw produces conidia over a period of several months (Lutey & Feser, 1960). *A. porri* f.sp. *solani* on potato and tomato sporulates for several months on debris left on the soil surface in semi-arid conditions (Rotem, 1968). *H. turcicum* perennates in infected tissues and sporulates on infected corn leaves after being buried in soil for a year (Levy, 1985). *Cercospora zeaemaydis*, which also perennates in debris, sporulates on leaf tissues either exposed to air on the soil surface, or buried in the soil (Payne & Waldron, 1983). *Eutypa armeniacae* on apricots, with its year-long latent period and an even longer infectious period (Carter, 1960), is a further example of a species using such a survival mechanism. A unique sporulation-conditioned survival mechanism exists in *Phyllosticta maydis* on corn; in addition to pycnidia that release spores regularly, other pycnidia release spores slowly over a prolonged period (Castor *et al.*, 1977).

Sporulation on resistant host genotypes

Sporulation is often the most sensitive indicator of race-specific (vertical) or race-non-specific (horizontal) resistance (Johnson & Taylor, 1976). Among recent studies, Craig & Frederickson (1983) distinguished three pathotypes of *Sclerospora sorghi* on the basis of their sporulation on four inbred lines of sorghum. Resistance of peanut genotypes to *Cercospora arachidicola* was recognised by the pathogen's ability to sporulate on them (Gobina *et al.*, 1983). Also in peanut, about 22 to 84 and 1000 uredospores per pustule were formed, respectively, in cultivars resistant and susceptible to rust, *Puccinia arachidis* (Subrahmanyam *et al.*, 1983).

Sporulation is never an isolated component of resistance and there are no cultivars that allow regular lesion development in which the fungus fails to sporulate. In most cases, resistance is characterised by inhibited sporulation as well as inhibited infection and colonisation. For example, in *Cucumis meo* resistance to *Pseudoperonospora cubensis* and to *Sphaerotheca fuliginea* race 2 is characterised by the appearance of small lesions with few spores (Cohen *et al.*, 1984; Cohen & Cohen, 1986). In tobacco, resistance to powdery mildew does not result from a failure of the pathogen to infect but from inhibition of both mycelial growth and sporulation (Cohen, 1982). Resistance to *Puccinia graminis* on lines of *Avena sterilis* and *A. sativum* cv. Fulghum was manifest as a reduction in the number of spores produced and not by differences in infection type or in latent period (Sztejnberg & Wahl, 1976). On the other hand, both the number of spores produced and the length of the latent period determined the resistance of grapes to *Uncinula necator* (Doster & Schnathorst, 1985).

Reduced sporulation of a given pathogen is often observed in plants artificially sensitised by a prior inoculation with another pathogen. This induced resistance was reported to be systemic in several pathosystems (Cohen & Kuć, 1981; Salt & Kuć, 1985; see Chapter 20).

References

Aragaki, M. & Hine, R.B. (1963). Effect of radiation on sporangial production of *Phytopthora parasitica* on artificial media and detached papaya fruit. *Phytopathology*, **53**, 854–6.

Aust, H.J. (1975). Wirkung präinokulativer Anzuchtbedingungen auf die Keimung, Infektion, Inkubationszeit und Sporulation des Gerstenmehltaues (*Erysiphe graminis* DC. f.sp. *hordei* Marchal). *Phytopathologische Zeitschrift*, **82**, 326–32.

Aust, H.J., Bashi, E. & Rotem, J. (1980). Flexibility of plant pathogens in exploiting ecological and biotic conditions in the development of epidemics. In: *Comparative Epidemiology*, ed.J. Palti & J. Krans, pp. 46–56. Pudoc, Wageningen: The Netherlands.

Bashi, E. & Aust, H.J. (1980). Quality of spores produced in cucumber powdery mildew compensates for their quality. *Zeitschrift fur Pflanzenkrankheiten und Planzenshutz*, **87**, 594–9.

Bashi, E., Ben-Joseph, Y. & Rotem, J. (1982). Inoculum potential of *Phytophthora infestans* and the development of potato late blight epidemics. *Phytopathology*, **72**, 1043–7.

Bashi, E. & Rotem, J. (1974). Adaptation of four pathogens to semi-arid habitats as conditioned by penetration rate and germinating spore survival. *Phytopathology*, **64**, 1035–9.

Bashi, E. & Rotem, J. (1975a). Sporulation of *Stemphylium botryosum* f.sp. *lycopersici* in tomatoes and of *Alternaria porri* f.sp. *solani* in potatoes under alternating wet-dry regimes. *Phytopathology*, **65**, 532–5.

328 Sporulation of foliar pathogens

Bashi, E. & Rotem, J. (1975b). Host and biotic factors affecting sporulation of *Stemphylium botryosum* f.sp. *lycopersici* on tomatoes and of *Alternaria porri* f.sp. *solani* on potatoes. *Phytoparasitica*, 3, 27–38.

Bashi, E. & Rotem, J. (1975c). Effect of light on sporulation of *Alternaria porri* f.sp. *solani* and of *Stemphylium botryosum* f.sp. *lycopersici in vivo*. *Phytoparasitica*, 3, 63–7.

Bashi, E. & Rotem, J. (1976). Induction of sporulation of *Alternaria porri* f.sp. *solani* in vivo. *Physiological Plant Pathology* 8, 83–90.

Becker, G.L.F. (1963). Glume blotch of wheat caused by *Leptosphaeria nodorum* Müller *Stichting Nederlands Graan-Centrum, Technisch Bericht* No. 11.

Biali, M., Dinoor, A., Eshed, N. & Kenneth, R. (1972). *Aphanocladium album*, a fungus inducing teliospore production in rusts. *Annals of Applied Biology*, 72, 37–42.

Bonde, M.R., Peterson, G.L. & Duck, N.B. (1985). Effects of temperature on sporulation, conidial germination, and infection of maize by *Peronospora sorghi* from different geographical areas. *Phytopathology*, 75, 122–6.

Brook, P.J. (1979). Effect of light on sporulation of *Plasmopara viticola*. *New Zealand Journal of Botany*, 17, 135–8.

Burleigh, J.R., Romig, R.W. & Roelfs, A.P. (1969). Characterization of wheat rust epidemics by numbers of uredia and numbers of urediospores. *Phytopathology*, 59, 1229–37.

Carter, M.V. (1960). Further studies on *Eutypa armeniacae* Hansf. & Carter. *Australian Journal of Agricultural Research*, 11, 498–504.

Castor, L.L., Ayers, J.E. & Nelson, R.R. (1977). Controlled-environment studies of the epidemiology of yellow leaf blight of corn. *Phytopathology*, 67, 85–90.

Cochrane, V.W.(1958). *Physiology of Fungi*. Wiley: New York, 524pp.

Cohen, Y. (1976). Interacting effects of light and temperature on sporulation of *Peronospora tabacina* on tobacco leaves. *Australian Journal of Biological Sciences*, 29, 281–9.

Cohen, Y. (1981a). The processes of infection of downy mildews on leaf surfaces. In:*Microbial Ecology of the Phylloplane*, ed. J. P. Blackman, pp. 115–33, Academic Press: London.

Cohen, Y. (1981b). Downy mildew of Cucurbits. In: *The Downy Mildews*, ed. D.M. Spencer, pp. 341–54, Academic Press: London.

Cohen, Y. (1982). Cultivar resistance and species immunity in *Nicotiana* spp. against tobacco powdery mildew. *Colloq. INRA*, 11, 143–55.

Cohen, Y. & Eyal, H. (1977). Growth and differentiation of sporangia and sporangiophores of *Pseudoperonospora cubensis* on cucumber cotyledons under various combinations of light and temperature. *Physiological Plant Pathology*, 10, 93–103.

Cohen, S. & Cohen, Y. (1986). Nature and genetics of resistance to powdery mildew race 2 in *Cucumis melo* PI 124111. *Phytopathology* 76, 1165–7.

Cohen, Y., Levy, Y. & Eyal, H. (1978). Sporogenesis of some fungal plant pathogens under intermittent light conditions. *Canadian Journal of Botany*, 56, 2538–43.

Cohen, Y., Eyal, H., Cohen, A. & Thomas, C.E. (1984). Evaluating downy mildew resistance in *Cucumis melo* L. *Cucurbits Genetics Cooperative Report*, 7, 38–40.

Cohen, Y., Eyal, M. & Sadon, T. (1975). Light-induced inhibition of sporangial formation of *Phytophthora infestans* on potato leaves. *Canadian Journal of Botany*, 53, 2680–3.

Cohen, Y. & Kúc, J. (1981). Evaluation of systemic resistance to blue mould induced in tobacco by prior stem inoculation. *Phytopathology*, 71, 783–7.

Cohen Y., Perl M. & Rotem J. (1971). The effect of darkness and moisture on sporulation of *Pseudoperonospora cubensis* in cucumbers. *Phytopathology*, 61, 594–5.

Cohen, Y., Reuveni, M. & Eyal, H. (1979). The systemic antifungal activity of Ridomil against *Phytophthora infestans* on tomato plants. *Phytopathology*, 69, 645–9.

Cohen, Y. & Rotem, J. (1969). The effects of lesion development, air temperature and duration of moist periods on sporulation of *Pseudoperonospora cubensis* in cucumbers. *Israel Journal of Botany*, **18**, 135–40.

Cohen, Y. & Rotem, J. (1970). The relationship of sporulation to photosynthesis in some obligatory and facultative parasites. *Phytopathology*, **60**, 1600–4.

Cohen, Y. & Rotem, J. (1971). Rate of lesion development in relation to sporulating potential of *Pseudoperonospora cubensis* in cucumbers. *Phytopathology*, **61**, 265–8.

Cohen, Y. & Sherman, Y. (1977). The role of airborne conidia in the epiphytotics of *Sclerospora sorghi* on sweet corn. *Phytopathology*, **67**, 515–21.

Cole, J.S. (1971). Sporulation of powdery mildews, particularly *Erysiphe cichoracearum* DC. on tobacco. In: *Ecology of Leaf Surface Microorganisms*, ed. T.H. Preece & C.H. Dickinson, pp. 323–337, Academic Press: London.

Craig, J. & Frederikson, R.A. (1983). Differential sporulation of pathotypes of *Peronosclerospora sorghi* on inoculated sorghum. *Plant Disease*, **67**, 278–9.

Cruickshank, I.A.M. (1963). Environment and sporulation in phytopathogenic fungi. IV. The effect of light on the formation of conidia of *Peronospora tabacina* Adam. *Australian Journal of Biological Science*, **16**, 88–98.

Doster, M.A. & Schnathorst, W.C. (1985). Comparative susceptibility of various grapevine cultivars to the powdery mildew fungus *Uncinula necator*. *American Journal of Viticulture*, **36**, 2, 101–4.

Eversmeyer, M.G. & Burleigh, J.R. (1970). A method of predicting epidemic development of wheat leaf rust. *Phytopathology*, **60**, 805–11.

Fried, P.M. & Stuteville, D.L. (1977). *Peronospora trifoliorum* sporangium development and effects of humidity and light on discharge and germination. *Phytopathology*, **67**, 890–4.

Gobina, S.M., Melouk, H.A. & Banks, D.J. (1983). Sporulation of *Cercospora arachidicola* as a criterion for screening peanut genotypes for leaf spot resistance. *Phytopathology*, **73**, 556–8.

Grove, G.G., Madden, L.V. & Ellis, M.A. (1985a). Influence of temperature and wetness duration on sporulation of *Phytophthora cactorum* on infected strawberry fruit. *Phytopathology*, **75**, 700–3.

Grove, G.G., Madden, L.V., Ellis, M.A. & Schmitthenner, A.F. (1985b). Influence of temperature and wetness duration on infection of immature strawberry fruit by *Phytophthora cactorum*. *Phytopathology*, **75**, 165–9.

Hawker, L.E. (1966). Environmental influences on reproduction. In: *The Fungi*, ed. G.C. Ainsworth, A.S. Sussman, pp. 435–469, Academic Press: London.

Hildebrand, P.D. & Sutton J.C. (1984). Interactive effects of the dark period, humid period, temperature, and light on sporulation of *Peronospora destructor*. *Phytopathology*, **74**, 1444–9.

Hill, J.P. & Nelson, R.R. (1976). Ecological races of *Helminthosporium maydis* race T. *Phytopathology*, **66**, 873–6.

Hohl, H.R. & Iselin, K. (1984). Strains of *Phytophthora infestans* from Switzerland with A2 mating type behaviour. *Transactions of the British Mycological Society*, **83**, 529–30.

Honda, Y., Toki, T. & Yunoki, T. (1977). Control of grey mould of greenhouse cucumber and tomato by inhibiting sporulation. *Plant Disease Reporter*, **61**, 1041–4.

Horsfall, J.G. & Dimond, A.E. (1957). Interactions of sugar, growth substances and disease susceptibility. *Zeitschrift für Pflanzenkrankheiten und Pflanzenschutz*, **64**, 415–21.

Hunter, J.E. & Kunimoto, R.K. (1974). Dispersal of *Phytophthora palmivora* sporangia by wind-blown rain. *Phytopathology*, **64**, 202–6.

Hyre, R.A. (1972). Effect of temperature and light on colonization and sporulation of the *Botrytis* pathogen on geranium. *Plant Disease Reporter*, **56**, 126–30.

Imhoff, H.W., Leonard, K.J. & Main, C.E. (1982). Patterns of bean rust lesion size increase and spore production. *Phytopathology*, **72**, 441–6.

Inaba, T. & Hino, T. (1980). Influence of water and temperature on oogonium and oospore formation of downy mildew fungus, *Peronospora manshurica*, in soybean lesions. *Annals of the Phytopathological Society of Japan*, **46**, 480–6.

Inaba, T. & Kajiwara, T. (1971). Effect of light on lesion development in cucumber downy mildew. *Annals of the Phytopathological Society of Japan*, **37**, 340–7.

Inaba, T. & Kajiwara, T. (1975). Physiological studies of cucumber downy mildew disease. *Bulletin of the National Institute of Agricultural Science*: Series C, **29**, 65–139. (Japanese with English summary.)

Inaba, T. & Morinaka, T. (1983*a*). The relationship between conidium and oospore production in soybean leaves infected with *Peronospora manshurica*. *Annals of the Phytopathological Society of Japan*, **49**, 554–7.

Inaba, T. & Morinaka, T. (1983*b*). Effects of environmental factors on oospore production of soybean downy mildew fungus, *Peronospora manshurica*. *Bulletin of the National Institute of Agricultural Science*: Series C, 38, 121–30. (In Japanese with English summary.)

Inman, R.E. (1962). Disease development, disease intensity, and carbohydrate levels in rusted bean plants. *Phytopathology*, **52**, 1207–11.

Johnson, R. & Taylor, A.J. (1976). Spore yield of pathogens in investigations of the race specifity of host resistance. *Annual Review of Phytopathology*, **14**, 97–119

Kato, H. & Kozaka, T. (1974). Effect of temperature on lesion enlargement and sporulation of *Pyricularia oryzae* in rice leaves. *Phytopathology*, **64**, 828–30.

Keen, N.T. & Williams, P.H. (1969). Synthesis and degradation of starch and lipids following infection of cabbage by *Plasmodiophora brassicae*. *Phytopathology*, **59**, 778–85.

Kranz, J. (1968). Zur konidienbildung und-verbreitung bei *Mycosphaerella musicola* Leach. *Zeitschrift für Pflanzenkrankheiten und Pflanzenschutz*, **75**, 327–38.

Levy, Y. (1984). The overwintering of *Exserohilum turcicum* in Israel. *Phytoparasitica*, **12**, 177–82.

Levy, Y. & Cohen, Y. (1980). Sporulation of *Helminthosporium turcicum* on sweet corn: Effects of temperature and dew period. *Canadian Journal of Plant Pathology*, **2**, 65–9.

Levy, Y. & Cohen, Y. (1981). Sporulation of *Helminthosporium turcicum* on sweet corn: effects of light and sugars. *Physiological Plant Pathology*, **18**, 17–25.

Lukens, R.J. (1966). Interference of low temperature with the control of tomato early blight through use of nocturnal illumination. *Phytopathology* **56**, 1430–1.

Lutey, R.W. & Feser, K.D. (1960). The role of infested straw in the epiphytology of Septoria leaf blotch of barley. *Phytopathology*, **50**, 910–13.

McMeekin, D. (1960). The role of the oospores of *Peronospora parasitica* in downy mildew of crucifers. *Phytopathology*, **50**, 93–7.

Mehta, Y.R. & Zadoks, J.C. (1970). Uredospore production and sporulation period of *Puccinia recondita* f.sp. *triticina* on primary leaves of wheat. *Netherlands Journal of Plant Pathology*, **76**, 267–76.

Melching, J.S., Bromfield, K.R. & Kingsolver, C.H. (1979). Infection, colonization, and urediospore production on Wayne soybean by four cultures of *Phakopsora pachyrhizi*, the cause of soybean rust. *Phytopathology*, **69**, 1262–5.

Michelmore, R.W. (1981). Sexual and asexual sporulation in the downy mildews. In: *The Downy Mildews*, ed. D.M. Spencer, pp. 165–82. Academic Press: London.

Nelson, R.R. & Tung, G. (1972). Effect of dew temperature on infection of a male-sterile corn hybrid by race T of *Helminthosporium maydis*. *Plant Disease Reporter*, **56**, 767–9.

Nelson, R.R. & Tung, G. (1973). Influence of some climatic factors on sporulation by an isolate of race T of *Helminthosporium maydis* on a susceptible male-sterile corn hybrid. *Plant Disease Reporter*, **57**, 304–7.

Obst, A. (1971). Infektionsquellen für *Septoria nodorum*. *NachrBl. dt. Pflschutzdieust.*, Stuttg. **23**, 177–8.

Olofsson, J. (1966). Downy mildew of peas in Western Europe. *Plant Disease Reporter*, **50**, 257–61.

Payne, G.A. & Waldron, J.K. (1983). Overwintering and spore release of *Cercospora zeae-maydis* in corn debris in North Carolina. *Plant Disease*, **67**, 87–9.

Pegg, G.F. & Mence, M.J. (1970). The biology of *Peronospora viciae* on pea: laboratory experiments on the effects of temperature, relative humidity and light on the production, germination and infectivity of sporangia. *Annals of Applied Biology*, **66**, 417–28.

Perl, M., Cohen, Y. & Rotem, J. (1972). The effect of humidity during darkness on transfer of assimilates from cucumber leaves to sporangia of *Pseudoperonospora cubensis*. *Physiological Plant Pathology*, **2**, 113–22.

Pinckard, J.A. (1942). The mechanism of spore dispersal in *Peronospora tabacina* and certain other downy mildew fungi. *Phytopathology*, **32**, 505–11.

Populer, C. (1981). Epidemiology of downy mildews. In: *The Downy Mildews*, ed. D.M. Spencer, pp. 57–105. Academic Press: London.

Raffray, J.B. & Sequeira, L. (1971). Dark induction of sporulation in *Bremia lactucae*. *Canadian Journal of Botany*, **49**, 237–9.

Reuveni, R., Cohen, Y. & Rotem, J. (1971). Sporulation of *Erysiphe cichoracearum* as influenced by conditions favouring photosynthesis in the host. *Israel Journal of Botany*, **20**, 78–83.

Reuveni, R. & Rotem, J. (1974). Effect of humidity on epidemiological patterns of the powdery mildew (*Sphaerotheca fuliginea*) on squash. *Phytoparasitica*, **2**, 25–33.

Ritchie, D.F. (1983). Mycelial growth, peach fruit-rotting capability, and sporulation of strains of *Monilinia fructicola* resistant to dichloran, iprodione, procymidone, and vinclozolin. *Phytopathology*, **73**, 44–7.

Rotem, J. (1964). The effect of weather on dispersal of Alternaria spores in a semi-arid region of Israel. *Phytopathology*, **54**, 628–39.

Rotem, J. (1968). Thermoxerophitic properties of *Alternaria porri* f.sp. *solani*. *Phytopathology*, **58**, 1284–7.

Rotem, J. (1978). Climatic and weather influences on epidemics. In: *Plant Disease, Advanced Treatise*, ed. J.G. Horsfall & E.B. Cowling, vol. 2, pp. 317–337. Academic Press, New York.

Rotem, J. (1988a). Quantitative assessment of inoculum production, dispersal, deposition, survival and infectiousness. In: *Experimental Techniques in Plant Disease Epidemiology*, J. Kranz & J. Rotem, Springer-Verlag, pp. 69–83.

Rotem, J. (1988b). Techniques of controlled-condition experiments to studies of epidemiology in the field. In: *Experimental Techniques in Plant Disease Epidemiology*, ed. J. Kranz & J. Rotem, Springer-Verlag, pp. 21–33.

Rotem, J. & Bashi, E. (1977). A review of the present status of the *Stemphylium* complex in tomato foliage. *Phytoparasitica*, **5**, 45–58.

Rotem, J. & Cohen, Y. (1974). Epidemiological patterns of *Phytophthora infestans* under semi-arid conditions. *Phytopathology*, **64**, 711–14.

Rotem, J., Cohen, Y. & Bashi, E. (1978). Host and environmental influences on sporulation *in vivo*. *Annual Review of Phytopathology*, **16**, 83–101.

Rotem, J., Cohen, Y. & Putter, J. (1971). Relativity of limiting and optimum inoculum loads, wetting durations, and temperatures for infection by *Phytophthora infestans*. *Phytopathology*, **61**, 275–8.

Rotem, J., Clare, B.G. & Carter, M.V. (1976). Effects of temperature, leaf wetness, leaf bacteria and leaf and bacterial diffusates on production and lysis of *Rhynchosporium secalis* spores. *Physiological Plant Pathology*, **8**, 297–305.

Rotem, J. & Sari, A. (1983). Fertilization and age-conditioned predisposition of potatoes to sporulation of and infection by *Phytophthora infestans*. *Zeitschrift für Pflanzenkrankheiten und Pflanzenschutz*, **90**, 83–8.

Rotem, J., Wooding, B. & Aylor, D.E. (1985). The role of solar radiation, especially ultraviolet, in the mortality of fungal spores. *Phytopathology*, **75**, 510–14.

Rouse, D.I., MacKenzie, D.R. & Nelson, R.R. (1984). Density dependant sporulation of *Erysiphe graminis* f.sp. *tritici*. *Phytopathology* **74**, 1176–80.

Salt, S.D. & Kuć, J. (1985). Elicitation of disease resistance in plants by expression of latent genetic information. *American Chemical Society Symposium*, Series 276, 48–68.

Schafer, J.F. & Roelfs, A.P.(1985). Estimated relation between numbers of urediospores of *Puccinia graminis* f.sp. *tritici* and rates of occurrence of virulence. *Phytopathology*, **85**, 749–50.

Scharen, A.L. (1964). Environmental influences on development of glume blotch in wheat. *Phytopathology*, **54**, 300–3.

Shaner, G.E., Peart, R.M., Newman, J.E., Stirm, W.L. & Loewer, O.L. (1972). A plant disease display model: an evaluation of the computer simulator Epimay for southern corn leaf blight in Indiana. *Purdue University Agricultural Experiment Station*, RB-890. 15 pp.

Shaw, D.S., Fyee, A.M. & Hibberd, P.J. (1985). A2 mating type of *Phytophthora infestans* from imported Egyptian potatoes. *Plant Pathology*, **34**, 552–6.

Smith, J.E. & Galbraith, J.C. (1971). Biochemical and physiological aspects of differentiation in the fungi. In: *Advances of Microbial Physiology*, ed. A.H. Rose & J.F. Wilkinson, vol. 5, 45–134. Academic Press: London.

Stakman, E.C. & Harrar, J.G. (1957). *Principles of Plant Pathology*. Ronald Press: New York.

Stakman, E.C. & Levine, M.N. (1919). Effect of certain ecological factors on the morphology of the urediniospores of *Puccinia graminis*. *Journal of Agricultural Research*, **16**, 43–77.

Stover, R.H. (1970). Leaf spot of bananas caused by *Mycosphaerella musicola*: role of conidia in epidemiology. *Phytopathology* **60**, 856–60.

Subrahmanyam, P., McDonald, D. & Subba Rao, P.V. (1983). Influence of host genotype on uredospore production and germinability in *Puccinia arachidis*. *Phytopathology*, **73**, 726–9.

Szkolnik, M. (1983). Unique vapour activity by CGA-64251 (Vangard) in the control of powdery mildews roomwide in the greenhouse. *Plant Disease* **67**, 360–6.

Sztejnberg, A. & Wahl, I. (1976). Mechanisms and stability of slow stem rusting resistance in *Avena sterilis*. *Phytopathology*, **66**, 74–80.

Teng, P.S. & Close, R.C. (1978). Effect of temperature and uredinium density on urediniospore production, latent period, and infectious period of *Puccinia hordei* Otth. *New Zealand Journal of Agricultural Research*, **21**, 287–96.

Thompson, D.C. & Jenkins, S.F. (1985). Effect of temperature, moisture, and cucumber cultivar resistance on lesion size increase and conidial production by *Colletotrichum lagenarium*. *Phytopathology*, **75**, 828–32.

Tomerlin, J.R., Eversmeyer, M.G., Kramer, C.L. & Browder, L.E. (1983). Temperature and host effects on latent and infectious periods and on urediniospore production of *Puccinia recondita* f.sp. *tritici. Phytopathology*, **73**, 414–19.

Tommerup, I.C., Ingram, D.S. & Sargent, J.A. (1974). Oospores of *Bremia lactucae. Transactions of the British Mycological Society*, **62**, 145–50.

Trujillo, E.E. (1965). The effects of humidity and temperature on *Phytopthora* blight of taro. *Phytopathology*, **55**, 183–8.

Turian, G. (1974). Sporogenesis in fungi. *Annual Review of Phytopathology*, **12**, 129–37.

Uozomi, T. & Kröber, H. (1967). Der Einfluss des Lichtes auf die Konidienbildung von *Peronospora tabacina* Adam an Tabakblattern. *Phytopathologishe Zeitschrift* **59**, 372–84.

Van Der Wal, A.F., Shearer, B.L. & Zadoks, J.C. (1970). Interaction between *Puccinia recondita* f.sp. *triticina* and *Septoria nodorum* on wheat, and its effects on yield. *Netherlands Journal of Plant Pathology* **76**, 261–3.

Waggoner, P.E. & Horsfall, J.G. (1969). EPIDEM: A simulator of plant disease written for a computer. *Connecticut Agricultural Experiment Station Bulletin*, **689**, 1–80.

Wallin, J.R. & Loonan, D.V. (1977). Temperature and humidity associated with sporulation of *Helminthosporium maydis* race T. *Phytopathology*, **67**, 1370–2.

Ward, S.V. & Manners, J.G. (1974). Environmental effects on quantity and viability of conidia produced by *Erysiphe graminis. Transactions of the British Mycological Society*, **62**, 119–28.

Warren, H.L. (1975). Temperature effects on lesion development and sporulation after infection by races O and T of *Bipolaris maydis Phytopathology*, **65**, 623–6.

Waters, C.W. (1928). The control of teliospore and urediniospore formation by experimental methods. *Phytopathology*, **18**, 157–213.

Wicks, T. & Lee, T.C. (1982). Effect of fungicide volatiles on sporangial production of *Plasmopara viticola. Plant Disease* **66**, 945–6.

Yarwood, C.E. (1937). The relation of light to the diurnal cycle of sporulation of certain downy mildews. *Journal of Agricultural Research*, **54**, 365–73.

Yarwood, C.E. (1939). Control of powdery mildews with a water spray. *Phytopathology*, **29**, 288–90.

Yarwood, C.E. (1943). Onion downy mildew. *Hilgardia*, **14**, 595–691.

Yarwood, C.E. (1961). Uredospore production by *Uromyces phaseoli. Phytopathology*, **51**, 22–7.

Zadoks, J.C. (1978). Methodology of epidemiological research. *Plant Disease* II. ed. J.G. Horsfall & E.B. Cowling, pp. 63–96. Academic Press: New York.

Zadoks, J.C. & Schein, R.D. (1979). *Epidemiology and Plant Disease Management*. University Press: Oxford.

17

Double-stranded RNA viruses of pathogenic fungi: virulence and plant protection

YIGAL KOLTIN, ALIZA FINKLER and
BAT-SHEBA BEN-ZVI
Department of Microbiology, Faculty of Life Sciences, Tel-Aviv University, Ramat Aviv, Israel

Introduction

Two decades have passed since viruses were first isolated from a fungus (Kleinschmidt & Ellis, 1967). Viruses are now known to occur in many fungi including some 40 different species that are plant pathogens (Day & Dodds, 1979). The genetics, molecular biology and physical properties of these viruses have been described in detail by Lemke (1979) and more recently by Buck (1986). The information to date suggests that only in a very few cases do the carriers of these viruses display a unique phenotype that can be related to the viruses (Bevan, Herring & Mitchell, 1973; Wood & Bozarth, 1973; Castanho, Butler & Shepherd, 1978; Hollings, 1978; Ghabrial & Mernaugh, 1983). Rarely do phenotypes conferred by the viruses contribute a selective advantage to the carrier. The complexity of their organisation and maintenance (Bruenn & Brennan, 1980; Buck *et al.*, 1981*a*, 1981*c*; Welsh & Leibowitz, 1982; Field *et al.*, 1983) suggests that some selective forces must operate to maintain them.

A segmented genome of double-stranded RNA typifies the fungal viruses (Bozarth, 1979; Bozarth *et al.*, 1981). The segments are separately encapsidated, each with its unique information but with an antigenically identical coat, and an RNA-dependent RNA polymerase is associated with the viral coats. The general features of the system are thus similar to those of a major virus with a number of satellite viruses. No proviral state has yet been detected (Bruenn, 1980; Ben-Zvi *et al.*, 1984). Transmission of the viruses is through cytoplasmic exchange between compatible mating types of one species. Artificial infection methods have been devised very recently (Stanway & Buck, 1984).

Genetic instability of various characters in fungi, at a rate exceeding that attributable to gene mutations, has been reported for many years. Some of the unstable characters are related to sporulation, morphology, synthesis of secondary metabolites and virulence (Jinks, 1959; Grindle, 1964). As the factors related to genetic instability in prokaryotes and eukaryotes are resolved, and the role of plasmids and transposable elements is recognised (Shapiro, 1983), one overriding fact that emerges in consideration of fungal instability is that DNA plasmids are not abundant. In fungi the extent of instability related to transpositions is as yet unclear (Esser *et al.*, 1983; Roeder & Fink, 1983) but the abundance of dsRNA viruses suggests these factors are a much more significant source of instability. Genetic drift in RNA viruses has been documented both in molecular terms and in epidemiological outbreaks (Ortino *et al.*, 1980; Sugiyama, Bishop & Roy, 1981; Holland *et al.*, 1982). It seems that in the absence of an RNA repair mechanism, and lacking a DNA proviral state, such viruses show a high frequency of genetic alterations and a strong genetic drift which may manifest itself as instability in the host.

Phenotype expression in the fungal host resulting from viral infection includes the secretion of a toxin (Puhalla, 1968: Day & Dodds, 1979), degeneration of fungal cells (Castanho, Butler & Shepherd, 1978; Hollings, 1978: Ghabrial & Mernaugh, 1983) and, in one instance, the attenuation of virulence (Van Alfen, 1982). The latter findings are the first in which a positive selection is indicated and an adaptive advantage to the fungal host is conferred by the dsRNA virus. They suggest an explanation for the rapid establishment of equilibrium between the host and pathogen that was recorded for the epidemic of the chestnut blight caused by the pathogen *Endothia parasitica* (Grente & Sauret, 1969; Van Alfen *et al.*, 1975).

Although dsRNA viruses have been detected in many plant pathogens, in only a few has there been any examination of the relationship between virulence and the virus. Results have been often confusing. In some, the complexity of the system resulting from the occurrence of multiple viral infections, such as in the causative agent of the take-all disease, precluded a clear analysis (Rawlinson *et al.*, 1973). In others, incomplete characterisation of the viruses and their genome organisation made the analysis premature. For example, both diseased and normal strains of *Helminthosporium victoriae*, the causal agent of Victoria blight of oats, contain virus-like particles (Ghabrial, Sanderlin & Calvert, 1979), as do toxin and nontoxin producing strains of *Periconia circinata* (Dunkle, 1974). In addition, the multicomponent problem, i.e. the involvement of the fungal pathogen, viruses of the fungal pathogen and the host plant, all possibly interacting genetically, has led to premature conclusions in various studies. Also, since the definition of virulence may be subjective unless well defined criteria are used, interpretation of the data can be very confusing.

Generalisations cannot be made about the relationship of fungal dsRNA viruses and virulence and the range of effects expressed. The adaptation by attenuation of virulence in *E. parasitica,* described above which led to the restoration of host-pathogen equilibrium and permitted new planting of the chestnut in Italy, is only one form of expression. In addition, the viruses may act also as inducers of virulence, in a way similar to that of bacterial plasmids (Elwell & Shipley, 1980). Other forms of expression not recognised as relevant to the regulation of virulence may be applied to combat various plant pathogens. To illustrate these possibilities, studies of the regulation of virulence by a dsRNA virus, and the control of plant pathogens by a virus-encoded product, this chapter describes relationships between two pathogenic fungi *Rhizoctonia solani* and *Ustilago maydis,* and their viruses.

Rhizoctonia solani
Early studies and the occurrence of hypovirulent strains

Rhizoctonia solani causes damping off in some 130 different plant species (Baker, 1970; Castanho & Butler, 1978) and is found in most agricultural soils. Using debilitated strains assumed to be hypovirulent, Castanho, Butler & Shepherd (1978) reported the occurrence of attenuation, similar to that in *E. parasitica.* A survey of a natural population in Israel to determine the degree of virulence, defined by the effect on 11 host species, revealed that the population consisted of a mixture of virulent strains causing damage of varying degree of severity to one or more hosts and of avirulent strains that caused no discernible damage to any host (Ichielevich-Auster *et al.,* 1985*b*). Of 107 isolates, 32 were avirulent and confirmed as *R. solani* by compatibility tests with anastomosis group tester strains. These strains are referred to as hypovirulent strains.

Double-stranded RNA was detected in mycelia of the hypovirulent strains but was not found in any of virulent strains. Virus particles were not found in either the virulent or the hypovirulent strains. Debilitated growth was transmitted to a virulent strain via cytoplasmic exchange along with the dsRNA found in the hypovirulent strain. These early studies led to the conclusion that the properties of the hypovirulent strains of *R. solani* were similar to those of hypovirulent strains of *E. parasitica.*

However, the debilitated strains persisted in the soil for less than one month and the precise relation between debilitated growth and dsRNA, or any other cytoplasmic factor, was unclear. The reason was that, as shown in later studies, growth rate and hypovirulence are independent factors. Furthermore, if hypovirulence represents a state of adaptation with a positive selection value it should be anticipated that hypovirulent strains would not be debilitated. Also, one of the strains originally claimed to be virulent and devoid of dsRNA (Castanho, Butler & Shepherd, 1978) was re-examined recently and found to contain dsRNA (A. Finkler and Y. Koltin, unpublished).

A later study (Zanzinger, Bandy & Tavantzis, 1984) reported the occurrence of dsRNA in both virulent and hypovirulent strains but virulence was defined by the use of one host, thus nonstringent criteria were adopted that precluded precise characterisation of the level of virulence of each of the strains used. A DNA plasmid was also reported recently in *R. solani* with the implication that this plasmid is related to virulence (Hashiba *et al.*, 1984). However, the latter study also failed to define stringent criteria to distinguish between virulent and hypovirulent strains. In both studies the cytoplasmic transmission of virulence or attenuation of virulence was not tested and the effect of curing strains of the cytoplasmic factors on the degree of virulence of the fungal host was not examined.

Virulence and the dsRNA viruses

To approach the question of whether phenotypic expression of the dsRNA viruses affects virulence in *R. solani*, highly virulent and avirulent strains were examined for the presence of viruses. The strains examined were isolated from various locations representing a range of ecological niches and seven anastomosis groups. The relation of growth rate to the degree of virulence was also examined, since it is possible that slow-growing strains may be less virulent if their virulence is only effective over a short range, for example, if it depends on enzymatic degradation rather than toxic effect. The range in growth rate of all the strains was found to be similar. None of the strains was debilitated and those identified as avirulent caused no discernible damage such as damping off or growth inhibition to the 11 different species used as indicators of virulence. The virulent strains caused severe damage to all 11 hosts.

Nine strains were used to examine the relation of the dsRNA to virulence (Finkler *et al.*, 1985). Only the virulent strains contained viruses; neither viruses nor dsRNA were recovered from the avirulent strains.

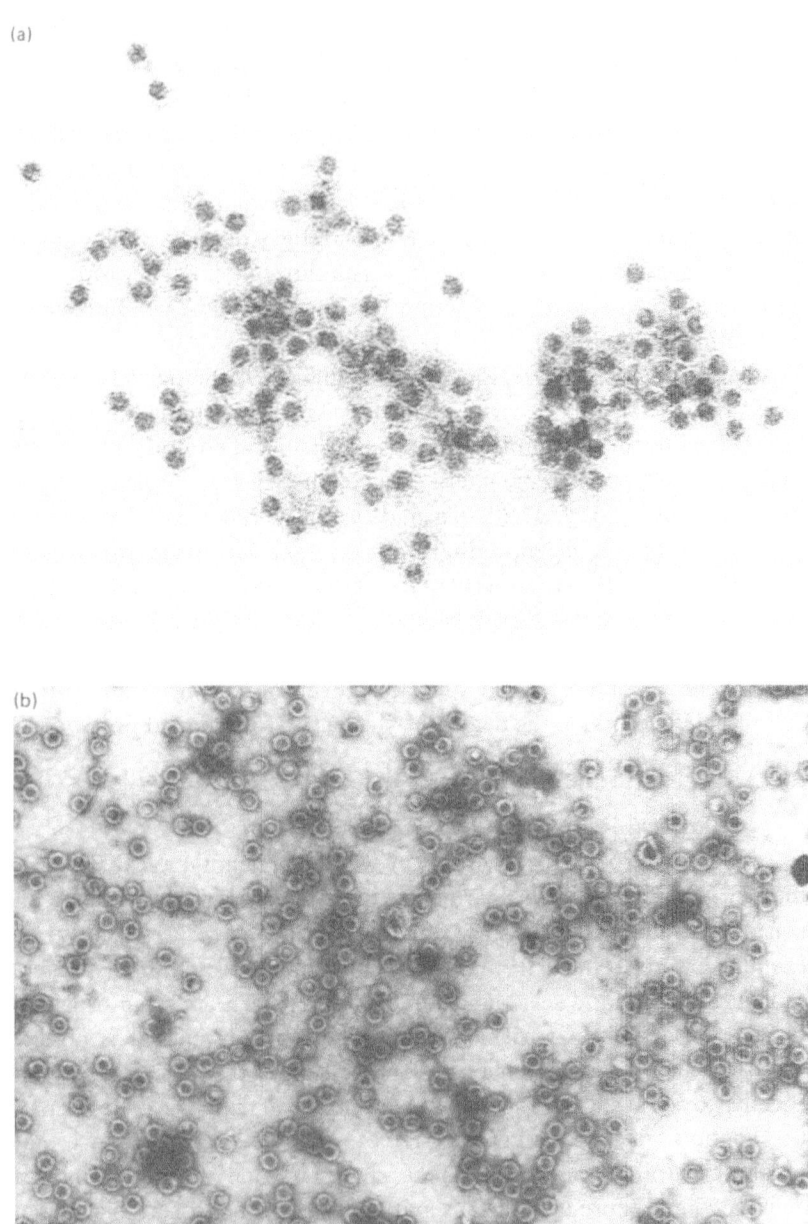

Fig. 17.1. Electron micrograph of negatively stained virus particles: (a) 30–35 nm in diameter, purified on sucrose density gradients, from a virulent strain of *Rhizoctonia solani*, (b) 41nm in diameter recovered from a killer strain of *Ustilago maydis*.

The viruses were 30–35nm in diameter (Fig. 17.1a) and contained two segments of dsRNA with molecular weights of 1.45 and 1.32 × 10^6 (Fig. 17.2). The virus particles from different strains seemed to be interrelated. Their coat protein was of a similar size, 55K, and viruses from different strains were immunoprecipitated by antibodies derived against one of the isolates. Furthermore, as in all other dsRNA viruses, an RNA polymerase acting as a transcriptase was coat-associated. *In vitro* transcription reactions have been defined for the *R. solani* virus and for a number of other fungal viruses (Buck, 1979; Bruenn *et al.*, 1980; Buck *et al.*, 1981*b*; Ben-Zvi *et al.*, 1984).

Viral transcripts were obtained *in vitro* using virus particles supplemented with the four ribonucleotide triphosphates. The RNA-directed

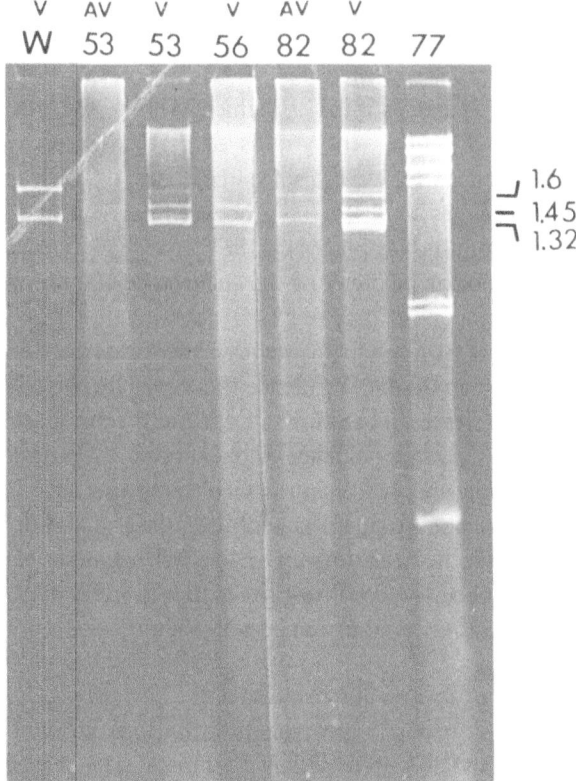

Fig. 17.2. Polyacrylamide gel electrophoresis of unique segments of dsRNA isolated from an *Ustilago maydis* killer strain (right lane) and from virulent strains of *R. solani* (other lanes). The dsRNA was stained with ethidium bromide and viewed with UV. All extracted strains contained virus particles.

RNA polymerase activity, which is currently a routine assay to monitor the viruses in purification procedures, was tested in hypovirulent strains but no traces were detected. The transcripts obtained were used in various studies with the fungal viruses for the resolution of viral-encoded proteins and also as a probe to resolve the relatedness of the viruses, since such probes avoid artefacts introduced by impurities of ribosomal RNA adhered to the dsRNA. Using such probes under stringent conditions ($65°C$, $0.6M$ NaCl, $60mM$ trisodium citrate) cross hybridisation between these probes and the viral dsRNA of various isolates revealed that all the viruses isolated from the virulent strains are interrelated. In addition, no variation of coat proteins from the viruses or in the products of *in vitro* translated viral transcript products could be resolved by fingerprinting using *Staphylococcus aureus* V8 protease (A. Finkler, B-S. Ben-Zvi & Y. Koltin, unpublished).

Virus curing and virulence

Since only virulent strains contained viruses, the question that emerged was how will curing the strains of dsRNA viruses affect the expression of virulence? However, since there are no practical, direct experimental approaches to distinguish between carriers and virus-free descendants, the emphasis was placed on the recovery of avirulent subcultures and the examination of these for the occurrence of a specific pattern of dsRNA.

Among subcultures of hyphal tips from virulent strains, about 1% of the new cultures were avirulent. Virulence was never recovered in subcultures from avirulent strains. Therefore, a clear unidirectional relationship was established and the high incidence of phenotypic alteration suggested that a loss of a cytoplasmic factor in the virulent strains led to a reduced level of virulence. In one avirulent subculture a loss of the dsRNA was noticed, while in another the loss of a specific segment of the viral information was noticed. To confirm further that dsRNA may be associated with virulence, additional evidence was sought.

Conversion of hypovirulent strains

The conversion of a hypovirulent strain to a virulent strain by the introduction of dsRNA viruses was attempted. The major difficulty was that infective methods are available for only one fungal virus and methods have not yet been devised for *R. solani*. Therefore, the natural mode of transmission of the fungal viruses was adopted, that is via cytoplasmic exchange in heterokaryons. The drawback of such a method is the lack of control over the cytoplasmic transmission of various cytoplasmic elements. Techniques might be refined if restriction polymorphism could be found in the mitochondrial DNA of the interacting strains. However, with the need to use genetically identifiable strains it is difficult to meet all the required conditions and so our experiments were based on cytoplasmic exchange using strains that could be identified and selected in defined conditions.

To test for the cytoplasmic nature of the agents associated with virulence, the classical heterokaryon transfer experiment was performed (Jinks, 1969). A virulent strain and an avirulent strain of the same compatibility group (Parmeter, Sherwood & Platt, 1969) were selected on the basis of their differential sensitivity to the fungicides Benomyl and Bay NTN, 19701 (N-[(4-chlorophenyl)methyl]-N-cyclopentyl-N'-phenylurea, Mobay Chemicals, USA). Differential sensitivity permits the identification of each of the parental strains and the dissociation of the heterokaryon. In addition to the differential resistance to two fungicides, the strains differed in a number of other morphological characteristics, such as a distinguishable difference in growth rate and the synthesis of melanin in culture. The virulent strain contained dsRNA viruses (Finkler et al., 1985) whereas the avirulent strain contained neither the viruses nor dsRNA.

Two methods were used for sampling cells after conjugation between the two strains. In the first, strains were confronted while growing on nonselective medium and were allowed to interact 24 or 48 h. Mycelium was taken from the line of confrontation and plated on medium that selected for the avirulent strain. In the second, the hypovirulent strain was inoculated on nonselective medium and allowed to grow in optimal conditions until two-thirds of the plate area was colonised. The virulent strain was then inoculated in the centre of the plate, thus establishing immediate contact with the mycelium of the avirulent strain. Hyphal tips were sampled from the periphery of the avirulent mycelium after 24 and 48h, grown in nonselective medium and then tested for resistance to the fungicides, for morphological characteristics and for virulence. The advantage of this method is that the area from which the hyphal tips were isolated was at a distance from the growing points of the virulent strains and thus the probability of recovering only the avirulent strain was very high.

Of 25 colonies isolated in both methods, two virulent colonies were identified which possessed all the markers typical of the hypovirulent strain. These cultures were resistant to Bay NTN, sensitive to Benomyl, did not secrete melanin and displayed a growth rate typical of the avirulent strain. Both were among those isolated after 48h of interaction. Both strains contained dsRNA, which was absent from the parental avirulent strain, and the pattern of segmentation was typical of the dsRNA pattern characterised in viruses recovered from the virulent strain that served as the donor.

These results, together with the earlier indications from curing experiments and the occurrence of the viruses in virulent strains and their absence in avirulent strains, provide good reason to suggest that in *Rhizoctonia solani* the dsRNA viruses are associated with the regulation of virulence. However, contrary to the situation in *Endothia parasitica*, the dsRNA viruses induce virulence and do not suppress virulence.

Viral information and virulence

To determine whether the content of viral information extends beyond the capacity to regulate the synthesis of structural proteins, efforts have been made to reveal its full coding potential. Characterisation of the proteins encoded by the viral dsRNA indicates a coding capacity extending two to three times beyond that of the dsRNA segments, if one assumes no overlapping genes. The viral coat protein is the major translation product, but clearly other proteins are produced. Their reaction to polyclonal antiviral antibodies suggests they are nonstructural. The nature of these proteins is unknown and neither enzymatic activities thought to be related to virulence, nor their indirect involvement in the induction of virulence, have yet been examined.

One of the major difficulties in the elucidation of the role of the viral encoded proteins is the absence of sufficient biological characterisation of virulent and hypovirulent strains. The pectolytic enzymes of a number of virulent and hypovirulent strains used in our studies were examined as markers for virulence. The role of pectolytic enzymes in the pathogenesis of *R. solani* has been well established (Weinhold & Motta, 1973). Four enzymes, known as endopolygalacturonase I (endo PG-I) and II, pectinesterase (PE) and endopectinolyase (endo PL), were purified to homogeneity from a number of virulent strains (Marcus *et al.*, 1986). Endo PL was found in all virulent strains examined but was not detected in the five hypovirulent strains used in this study. It is not yet known whether the 'converted' hypovirulent strains derived from the heterokaryon transfer experiment also display endo PL activity but this enzyme is clearly a good marker to follow since the potential for its production may be introduced directly to the fungus by the viruses or induced in the fungus by the viruses.

Hypovirulence and biological control

It became clear from a series of experiments including field trials that hypovirulent strains can be used for biological control of root rot caused by *R. solani* (Ichielevich *et al.*, 1985*a*). Seedlings infected by the hypovirulent strain were protected against superinfection by virulent strains of *R. solani* (Fig. 17.3). In a series of experiments with one of the hypovirulent strains the damage was reduced by 75–94% in various experiments with different crop plants including cotton, radish, wheat, lettuce and carrots. The protection was effective against a challenge by virulent strains and only occurred in the presence of the hypovirulent strain. However, the protection by one hypovirulent strain was effective in a number of very different ecological niches differing in climatic conditions and soil type. The only limitation was the specificity displayed since the protection was effective against *Rhizoctonia* but not against other soil-borne pathogens. Effective protection was displayed using a hypovirulent strain of one anastomosis group and individually against a number of virulent strains of different anastomosis groups. It is unclear whether the hypovirulent strain induces a resistance mechanism in the host plant or creates a barrier against superinfection merely by efficient superficial invasion of the plant tissue, i.e. exploitation of a limited resource. Nonetheless, the protective mechanism operates irrespective of the incompatibility system and there is no need for the selection of hypovirulent strains from the different anastomosis groups. The hypovirulent strains

Fig. 17.3. Protection of cotton seedlings by hypovirulent isolate of *Rhizoctonia solani* (521). From left to right, uninfected control, infected with virulent strain 82, infected with hypovirulent strain 521, infected with strain 521 and challenged after 3 d with the virulent strain 82.

in addition to, or as a part of, their protective effect appear to promote host growth; thus, in all experiments the protected plants clearly displayed vigorous growth that was expressed in higher yields than in controls.

It is unclear why hypovirulent strains are common in nature in spite of the fact that the mere acquisition of the viral information can convert a hypovirulent strain to a virulent strain. The hypovirulent strains studied were not limited to a unique or rare anastomosis group and, unless vegetative incompatibility occurs within anastomosis groups, they are widely accessible to cytoplasmic exchange and viral transmission. Information available from a better-characterised system, the viruses of *Saccharomyces cerevisiae*, suggests that many nuclear genes are involved in the maintenance of such viruses (see review by Tipper & Bostian, 1984). In that system as many as 30 genes play some part in the maintenance and replication of the fungal viruses. In the viruses of the corn smut pathogen, *Ustilago maydis*, another well characterised system, exchange of viral information is dependent on a nuclear gene that confers resistance to a viral-encoded system (Finkler, Peery & Koltin, 1984). Furthermore, other cytoplasmic factors may impose constraints on the exchange of viral information. Mechanisms of this nature may play a role in the persistence of hypovirulent strains of *R. solani*.

Ustilago maydis (Corn Smut)

Virus-related interstrain inhibition

Ustilago maydis represents a class of fungal pathogens in which the expression of a virus harboured by the pathogen inhibits growth of sensitive strains of the fungus. This phenotypic expression has stimulated consideration of the use of this phenomenon for plant protection.

In the course of studies on the sexual interaction among haploid lines of *U. maydis* growing on solid medium the occurrence of interstrain inhibition was reported (Puhalla, 1968) and referred to as the killer phenomenon. We now know that the factors associated with this phenomenon are polypeptides secreted by some strains of *U. maydis*. Such strains constitute c. 1% or less of the natural isolates examined. The polypeptides are highly specific and affect other members of the Ustilaginales including the plant pathogens of graminaceous plants *U. tritici*, *U. hordei* and *U. nuda*. The phenotype associated with the secretion of these polypeptides is related to a group of dsRNA viruses found in *U. maydis* (Fig. 17.1b) (Wood & Bozarth, 1973; Kiltin & Day, 1976). Almost all strains of *U. maydis* contain dsRNA viruses but very few secrete the toxic polypeptides. The gentic organisation of these viruses is complex: a major viral component provides the genetic information coding the coat protein and also the replicative machinery that provides for the maintenance of a number of satellite viruses (Koltin, Levine & Perry, 1980). Some of the satellite viruses contain the information that encodes the inhibitory polypeptides. A detailed account of the *Ustilago* killer system was recently given by Koltin (see Buck, 1986).

Use of viral information for plant protection

The broad spectrum of the inhibitory phenomenon among the Ustilaginales merits some consideration for future plant protection. Resistance to the *U. maydis* virus-encoded toxin was not detected in six other members of the Ustilaginales that were tested. The species barriers, and the lack of common hosts in this order, suggest that many species have not been evolving under the selective pressure of similar killer systems. (A killer system has not been reported in other Ustilaginales.) Therefore, the toxin may be effective against these species. The use of the toxic polypeptides as fungicides is impractical but the introduction of the viral sequences that encode these polypeptides as cDNA, cloned in a plant expression vector, appears as a feasible approach. The dsRNA encoding the toxin is comprised of c. 1000 base pairs (Field *et al.*, 1983). If plant cells of wheat, barley, oats and other hosts of the various Ustilaginales can process the information and express the virus-encoded toxin, the plant tissues should be protected against infection by their respective smut fungus. It is proposed that the information derived from studies of viruses of *U. maydis* could be applied to programmes for the control of infections of other members of the Ustilaginales in other Graminaceous hosts.

The feasibility of this nonconventional approach in plant protection is difficult to assess at this stage in the development of plant molecular genetics and genetic engineering. Notwithstanding the lack of the proper vectors it is clear, however, that certain prerequisites must be met for the expression of the viral information by the plant. Specific processing by the plant tissue will be required and, as yet, it is unclear whether the processing signals recognised by the fungus will be also recognised by the plant.

Conclusions

Although they are abundant in many different species and within each species, dsRNA viruses of fungi have been largely unexplored beyond the morphological description of many and physical characterisation of a few. Elements with such an adaptive potential should have drawn much attention, as did the bacterial plasmids with their drug resistance factors and virulence factors but, possibly because of their indirect impact via their fungal host cell, viruses have attracted much less attention than the human pathogens. At present, information relating to the ecological implications of fungal viruses is scarce and even those cases that have been reported here require further elucidation. However, agents affecting at least four forms of phenotypic expression, debilitation of growth, supression of virulence, induction of virulence and secretion of species-specific inhibitory polypeptides, have been reported. Of these phenotypes, clearly the aspects related to virulence are the most complicated to study. Yet with heterokaryosis and parasexuality so prominent among plant pathogens, it is imperative that we understand better the role of cytoplasmic factors as adaptive agents which can play a role in each of the above-mentioned phenotypes. Molecular and genetic studies of viruses found in plant pathogens could resolve their specific role in virulence and explain the persistence of various forms *in vivo*.

References

Baker, K.F. (1970). In: *Rhizoctonia solani: Biology and pathology*, ed. J.R. Parmeter, pp. 125–48. University of California Press.

Ben-Zvi, B., Koltin, Y., Mevarech, M. & Tamarkin, A. (1984). RNA polymerase activity in virions from *Ustilago*. *Molecular and Cellular Biology*, **4**, 188–94.

Bevan, E.A., Herring, J.A. & Mitchell, D.J. (1973). Preliminary characterization of two species of dsRNA in yeast and their relationship to the killer character. *Nature*, **245**, 81–6.

Bozarth, R.F. (1979). The physico-chemical properties of mycoviruses. In: *Viruses and Plasmids in Fungi*, ed. P.A. Lemke, pp. 44–91. New York: Marcell Dekker Inc.

Bozarth, R.F., Koltin, Y., Weissman, M.B., Parker, A.C., Dalton, R.K. & Steinlauf, R. (1981). The molecular weight and packaging of dsRNAs in the mycovirus from *Ustilago maydis* killer strains. *Virology*, **57**, 492–502.

Bruenn, J & Brennan, V. (1980). Yeast viral double-stranded RNAs have heterogenous 3′-termini. *Cell*, **19**, 928–33.

Bruenn, J.A. (1980). Virus-like particles of yeast. *Annual Review of Microbiology*, **34**, 49–68.

Bruenn, J., Bobek, L., Brennan, V. & Held, W. (1980). Yeast viral RNA polymerase is a transcriptase. *Nucleic Acids Research*, **8**, 2985–97.

Buck, K.W. (1979). Replication of double-stranded RNA mycoviruses. In: *Viruses and Plasmids in Fungi*, ed. P.A. Lemke, pp. 93–160. New York: Marcel Dekker Inc.

Buck, K.W., Almond, M.R., McFadden, J.J.P., Romanos, M.A. & Rawlinson, C.J. (1981a). Properties of thirteen viruses and virus variants obtained from eight isolates of wheat take-all fungus. *Gaeumannomyces graminis* var. *tritici*. *Journal of General Virology*, **53**, 235–45.

Buck, K.W., Romanos, M.A., McFadden, J.J.P. & Rawlinson, C.J. (1981b). *In vitro* transcription of double-stranded RNA by virus-associated RNA polymerase of viruses from *Gaeumannomyces graminis*. *Journal of General Virology*, **57**, 157–68.

Buck, K.W. (1986). *Fungal Virology*. Florida: CRC Press Inc.

Castanho, B. & Butler, E.E. (1978). Rhizoctonia decline: a degenerative disease of *Rhizoctonia solani*. *Phytopathology*, **68**, 1505–10.

Castanho, B., Butler, E.E. & Shepherd, R.J. (1978). The association of double-stranded RNA with Rhizoctonia decline. *Phytopathology*, **65**, 1515–19.

Day, P.R. & Dodds, J.A. (1979). Viruses of plant pathogenic fungi. In: *Viruses and Plasmids in Fungi*, ed. P.A. Lemke, pp. 202–38. New York: Marcel Dekker Inc.

Dunkle, L.D. (1974). Double-stranded RNA mycovirus in *Periconia circinata*. *Physiological Plant Pathology*, **4**, 107–16.

Elwell, L.P. & Shipley, P.L. (1980). Plasmid-mediated factors associated with virulence of bacteria to animals. *Annual Review of Microbiology*, **34**, 465–96.

Esser, K., Kuck, U., Stahl, U. & Tudzynski, P. (1983). Cloning vectors of mitochondrial origins for eukaryotes: a new concept in genetic engineering. *Current Genetics*, **7**, 239–43.

Field, L.J., Bruenn, J.A., Chang, T.H., Pinchasi, O. & Koltin, Y. (1983). Two *Ustilago maydis* viral dsrNAs of different size code for the same product. *Nucleic Acids Research*, **11**, 2765–78.

Finkler, A., Peery, T. & Koltin, Y. (1984). Mitotic transmission of the *Ustilago* virus M dsRNA is dependent on the nuclear resistance genes. *Genetics*, **107**, s32.

Finkler, A., Koltin, Y., Barash, I., Sneh, B. & Pozniak, D. (1985). Isolation of a virus from virulent strains of *Rhizoctonia solani*. *Journal of General Virology*, **66**, 1221–32.

Ghabrial, S.A. & Mernaugh, R.L. (1983). Biology and transmission of *Helminthosporium victoriae* mycoviruses. In *Double Stranded RNA Viruses*, ed. R.W. Compans & D.H.L. Bishop, pp. 441–9. New York: Elsevier.

Ghabrial, S.A., Sanderlin, R.S. & Calvert, L.A. (1979). Morphology and virus-like particles content of *Helminthosporium victoriae* colonies regenerated from protoplasts of normal and diseased isolates. *Phytopathology*, **69**, 312–15.

Grente, J. & Sauret, S. (1969). L'hypovirulence exclusive phénomène original enpathologic végétable. *C.R. Academic Science Ser. D.*, **268**, 2347–50.

Grindle, M. (1964). Nucleo-cytoplasmic interactions in the 'red' cytoplasmic variant of *Aspergillus midulans*. *Heredity*, **19**, 75–95.

Hashiba, T., Homma, Y., Hyakumachi, M. & Matsuda, I. (1984). Isolation of a DNA plasmid in the fungus *Rhizoctonia solani*. *Journal of General Microbiology*, **130**, 2067–70.

Holland, J., Spindler, K., Horadyshi, E., Grabau, E., Nichol, S. & Vandepol, S. (1982). Rapid evolution of RNA genomes. *Science*, **215**, 1577–85.

Hollings, M. (1978). Mycoviruses: viruses that infect fungi. *Advances in Virus Research*, **22**, 3–53.

Ichielevich-Auster, M., Sneh, B., Koltin, Y. & Barash, I. (1985a). Suppression of damping off caused by *Rhizoctonia solani* spp. by non–pathogenic *R. solani*. *Phytopathology*, **75**, 1080–4.

Ichielevich-Auster, M., Sneh, B., Koltin, Y. & Barash, I. (1985b). Pathogenicity, host specificity and anastomosis groups of *Rhizoctonia* spp. isolated from soils in Israel. *Phytoparasitica*, **13**(2), 103–12.

Jinks, J.L. (1959). Lethal suppressive cytoplasm in aged clones of *Aspergillus glaucus*. *Journal of General Microbiology*, **21**, 397–409.

Jinks, J.L. (1969). *Extrachromosomal Inheritance*. Prentice–Hall.

Kleinschmidt, W.J. & Ellis, L.F. (1967). In *Ciba Foundation Symposium on Interferon*, ed. G.E. Wilstemholme & M. O'Connor, pp. 39–46. London: Churchill.

Koltin, Y. & Day, P.R. (1976). Inheritance of killer phenotypes and double stranded RNA in *Ustilago maydis*. *Proceedings of the National Academy of Sciences, U.S.A.*, **73**, 594–8.

Koltin, Y., Levine, R. & Peery, T. (1980). Assignment of functions to the segments of the dsRNA genome of the *Ustilago* viruses. *Molecular and General Genetics*, **178**, 173–8.

Lemke, P.A. (1979). *Viruses and Plasmids in Fungi*. New York: Marcel Dekker, Inc.

Marcus, L., Barash, I., Sneh, B., Koltin, Y. & Finkler, A. (1986). Characterization of pectolytic enzymes produced by virulent and hypovirulent isolates of *Rhizoctonia solani*. *Physiological and Molecular Plant Pathology*, **29**, 325–36.

Ortino, H., Najera, R., Lopez, C., Davila, M. & Domingo, E. (1980). Genetic variability of Hong-Kong (H3N2) Influenza viruses: spontanous mutations and their location in the viral genome. *Gene*, **11**, 319–31.

Parmeter, J.R., Sherwood, R.T. & Platt, W.D. (1969). Anastomosis grouping among isolates of *Thanatephorus cucumeris*. *Phytopathology*, **59**, 1270–8.

Puhalla, J.E. (1968). Compatibility reactions on solid medium and interstrain inhibition in *Ustilago maydis*. *Genetics*, **60**, 461–75.

Roeder, G.S. & Fink, G.R. (1983). Transposable elements in yeast. In: *Mobile Genetic Elements*, ed. J. Shapiro, pp. 299–328. London, New York: Academic Press Inc.

Shapiro, J. (1983). *Mobile Genetic Elements*. London, New York: Academic Press, Inc.

Stanway, C.A. & Buck, K.W. (1984). Infection of protoplasts of the wheat take-all fungus, *Gaeumannomyces graminis* var. *tritici*, with double-stranded RNA viruses. *Journal of General Virology*, **65**, 2061–5.

Sugiyama, K., Bishop, D.H.L. & Roy, P. (1981). Analysis of the genomes of Bluetongue viruses recovered in the United States. *Virology*, **114**, 210–17.

Tipper, D.J. & Bostian, K.A. (1984). Double-stranded RNA killer systems in yeast. *Microbiology Reviews*, **48**, 125–56.

Van Alfen, N.K. (1982). Biology and potential for disease control of hypovirulence of *Endothia parasitica*. *Annual Review Phytopathology*, **20**, 349–62.

Van Alfen, N.K., Jaynes, R.A., Anagnostakis, S.L. & Day, P.R. (1975). Chestnut blight: Biological control by transmissible hypovirulence in *Endothia parasitica*. *Science*, **189**, 890–1.

Weinhold, A.R. & Motta, J. (1973). Initial host responses in cotton infection by *Rhizoctonia solani*. *Phytopathology*, **63**, 157–62.

Welsh, J.D. & Leibowitz, M.S. (1982). Localization of genes for the double-stranded RNA killer virus of yeast. *Proceedings of the National Academy of Science, U.S.A.*, **79**, 786–9.

Wood, H.A. & Bozarth, R.F. (1973). Heterokaryon transfer of virus-like-particles associated with a cytoplasmically inherited determinant in *Ustilago maydis*. *Phytophathology*, **63**, 1019–21.

Zanzinger, D.H., Bandy, B.P. & Tavantzis, S.M. (1984). High frequency of finding double-stranded RNA in naturally occurring isolates of *Rhizoctonia solani*. *Journal of General Virology*, **65**, 1601–5.

18

Inoculum production and survival in fungi which form sclerotia

A. R. ENTWISTLE

Institute of Horticultural Research, Wellesbourne, Warwick CU35 9EF, UK

Fungi which form sclerotia are among the more destructive plant pathogens. The ability to survive in adverse conditions and remain viable for long periods makes sclerotial plant pathogens difficult to control. Not all sclerotia however are long lived nor do they always provide the pathogen's primary inoculum. This chapter reviews recent work on:

(1) the different forms of inocula and their importance,
(2) the dispersal of inoculum,
(3) changes in sclerotial populations and their spatial patterns in the field.

Details of sclerotial structure have been reviewed by Townsend & Willetts (1954), while sclerotial germination and survival have been reviewed by Coley-Smith & Cooke (1971).

Survival as sclerotia and hyphae

There are many reports of the occurrence of plant diseases caused by sclerotial fungi after long periods in the absence of hosts. For example, Croxall, Sidwell & Jenkins (1953) reported over 90% loss of salad onions by *Allium* white rot (*Sclerotium cepivorum*) following the ploughing of 20-year-old grassland or derelict orchards. Since *S. cepivorum* survives only as sclerotia in soil and infection is restricted to the genus *Allium*, this report indicates sclerotia of *S. cepivorum* can survive for up to 20 years. However, farm records are not usually so easily verified, e.g. when farms have changed ownership. There is also the possibility that apparently long survival may be due to the non-documented introduction of the fungus from other infested areas in periods when the host crop was not grown.

The long periods of survival of some sclerotial fungi obviously present problems for experimentation. These difficulties have been overcome where *Rhizoctonia tulipae* and *S. cepivorum* are concerned. In the late 1960s Coley-Smith cultured *S. cepivorum*, recently isolated from onion, on maize-meal and sand; known numbers of sclerotia were then sealed in bags of nylon fabric and buried 0–22.5 cm deep in the field. Fifteen years later, 88–100% sclerotia were still recoverable and 62–98% proved viable as demonstrated by their ability to grow on agar medium (Coley-Smith, 1983). In a similar test, sclerotia of *R. tulipae*, produced on agar medium, remained viable for at least 10 years (Coley-Smith, Humphreys-Jones & Gladders, 1979).

Some fungi survive naturally as both sclerotia and hyphae, e.g. *Botrytis, Rhizoctonia* spp. and *Sclerotinia sclerotiorum*. Sclerotia may be short-lived in soil, as in the case of *Botrytis allii* where survival was less than six months (Maude, Bambridge & Presley, 1982) and in *Botrytis cinerea* where 90% sclerotia survived less than nine months (Thomas, Kotze & Matthee, 1983). The presence of hyphae within infected onion seed or possibly in plant residues in soil is the main mechanism of survival of *B. allii* (Tichelaar, 1967; Maude, Bambridge & Presley, 1982).

In *Rhizoctonia solani*, the relative importance of sclerotia and hyphae for survival shows considerable variation between the various anastomosis groups (AG). Sharp eyespot, *R. solani*, possibly AG-4, survived six months as sclerotia but only two months as hyphae within wheat straw (Pitt, 1964). In contrast, AG-3 survived as hyphae within dead plant tissue or free in soil but not apparently as sclerotia (Murray, 1981). The survival of *R. solani* can be for much longer periods if the fungus is protected within hard host tissue. Thus, AG-4 survived 9–10 months as hyphae within the shell of peanut pods (Bell & Sumner, 1984).

Differences in the longevity of *R. solani* sclerotia reflect differences in sclerotial structure (Naiki & Ui, 1978). AG-1 sclerotia have a three-layered structure, and survive the longest because they have a darkly pigmented outer mucilaginous layer. The sclerotia of AG-2 type-1 have no outer pigmented layer and survival was less than 10% after 150d incubation, that is, much less than the *c.* 50% survival of AG-1 sclerotia. Sclerotia of the remaining AG-2 type-2, -3, -4 and -5 consist of loosely grouped brown cells within which viable cells are distributed at random. Individual cells within these latter sclerotia are capable of growth and infection; there is, however, a progressive weakening of the sclerotia as cells die and sclerotial longevity is intermediate between that of sclerotia of AG-1 and AG-2 type-1.

Sclerotinia sclerotiorum survives as sclerotia on infected host tissue and also, to a lesser extent, as hyphae in colonised seed (Cook, Steadman & Boosalis, 1975). Host infection by *S. sclerotiorum* is mostly from ascospore inoculum, which causes petal blight or head rot (Newton & Sequeira, 1972; Williams & Stelfox, 1980; Patterson & Grogan, 1985). In rape and sunflower, however, there is also stem infection directly from mycelium emanating from sclerotia, and this results in wilt (Haung & Deuck, 1980; Williams & Stelfox, 1980). Sometimes *S. sclerotiorum* forms aberrant, tan-coloured sclerotia in lettuce (Newton & Sequeira, 1972; Garrabrandt, Johnston & Peterson, 1983) and annual sunflower (Haung, 1983). Such sclerotia fail to achieve dormancy and are readily infected by the mycoparasite *Coniothyrium minitans*; consequently, survival is minimal (Garrabrandt *et al.*, 1983; Huang, 1983).

Phymatotrichum omnivorum forms sclerotia on the roots of susceptible hosts and, less abundantly, in root systems of the non-host *Sorghum* (Rush & Gerik, 1984). The sclerotia always develop on hyphal strands and it is from the attached strand fragments that host infection is initiated (Alderman & Hine, 1982). Hyphal strands without attached sclerotia remain infective for many years, particularly when 30–60 cm deep in soil. Therefore, hyphal strands rather than sclerotia may represent the primary inoculum.

Effects of the origin of sclerotia on their germination and survival
Various methods have been used to produce sclerotia for experiments (Table 18.1). There is, however, the possibility that the behaviour of cultured and natural sclerotia may differ (Linderman & Gilbert, 1973; Merriman, 1976). Coley-Smith (1985) measured the germination of sclerotia of *S. cepivorum* isolate J412 placed under field-grown onions during summer. Sclerotia cultured on onions in non-sterile soil or obtained from naturally infected onions had a three to four times higher percentage germination than sclerotia cultured on perlite and maize meal. In contrast, sclerotia of isolate J11, characterised in previous experiments by a rapid response to germination stimulants, had a high percentage germination even when cultured on perlite and maize-meal (Esler & Coley-Smith, 1984).

Sclerotia of *S. sclerotiorum* from naturally infected beans or lettuce survived for shorter periods in soil than sclerotia cultured on sterile potato cubes (Merriman, 1976). Conversely, Cook *et al.* (1975) observed no differences in survival in soil, or on subsequent apothecial development and infection of beans, between sclerotia of *S. sclerotiorum* cultured on

Table 18.1. *Sources of sclerotia used in experiments*

Fungus	Source	Reference
Aerial plant parts		
Sclerotinia sclerotiorum	sunflower	Huang & Deuck (1980)
Botrytis allii	onion	Maude, Bambridge & Presley (1982)
Inoculated sterile plant tissue		
Rhizoctonia solani	barley grain	Ruppel (1985)
S. sclerotiorum	potato	Merriman (1976)
Inoculated host tissue in non-sterile soil		
Helicobasidium purpureum	sugar beet	Hull & Wilson (1946)
Sclerotium cepivorum	onion	Coley-Smith (1985)
Naturally infested soil		
S. cepivorum	—	Merriman, Samson & Schippers (1981)
Macrophomina phaseolina	—	Wyllie *et al.* (1984)
Artificially infested soil		
R. solani	potato dextrose broth	Naiki & Ui (1978)
Solid and liquid culture		
Rhizoctonia tulipae	malt extract agar	Coley-Smith, Humphreys-Jones & Gladders (1979)
Verticillium dahliae	potato dextrose agar	Filonow & Lockwood (1983)

potato dextrose agar (PDA) and those collected from bean residues. Punja, Jenkins & Grogan (1984) compared the behaviour of *Sclerotium rolfsii* sclerotia cultured on non-sterile oat seed, on sterile oat seed, or on PDA, with sclerotia from naturally infected carrots. When tested in non-sterile soil, eruptive sclerotial germination (germination by a plug of mycelium) was 70–74% when sclerotia were from cultured sources and 16–21% when sclerotia originated from the natural source. When treated with sodium hypochlorite solution, however, germination of PDA-cultured sclerotia was considerably less than that of other sclerotia; this difference was attributed to there being less melanisation of rind cells and less cortical development in sclerotia produced on PDA. Other evidence of differences between the rinds of natural and cultured sclerotia was found in *Sclerotinia sclerotiorum*. When naturally produced sclerotia were buried in soil the rind cells collapsed and became perforated whereas sclerotia cultured on sterile potato pieces remained intact when buried (Merriman, 1976). Rinds of *Sclerotium cepivorum* sclerotia also became pitted in soil, but in this instance there was no difference between sclerotia cultured on perlite/maize-meal medium and sclerotia cultured on onions (New, Coley-Smith & Georgy, 1984).

Differences in behaviour between cultured and natural sclerotia may be the result of differences in their size (Henis & Ben-Yephet, 1970) or nutritional status (Willetts & Wong, 1980; Filonow & Lockwood, 1983). Thus the numbers and size of *R. solani* sclerotia were markedly affected by the C : N ratio of the substrate (McCoy & Kraft, 1984). Large sclerotia germinated spontaneously, and the resulting hyphae readily invaded bean hypocotyls, whereas small sclerotia were dependent on external nutrients for germination (van Bruggen & Arneson, 1985). The germination of cultured sclerotia of both *S. cepivorum* and *Macrophomina phaseolina* was also dependent on an external source of nutrients but only when they had been leached (Filonow & Lockwood, 1983) indicating the possibility that, in natural conditions, sclerotial germination may become nutrient-dependent with age. In *M. phaseolina* a higher sclerotial nutrient status may have contributed to germination being faster in sclerotia from culture than in sclerotia from roots or soil (Wyllie *et al.*, 1984).

Although the use of naturally produced sclerotia in experiments is desirable, it may result in unacceptably low and erratic responses (Punja *et al.*, 1984). In these circumstances the use of cultured sclerotia is justified, provided there is evidence that their behaviour is generally similar to that of naturally produced sclerotia. A practical difference between cultured and natural sclerotia is that as naturally produced sclerotia mature in non-sterile conditions, bacteria or fungi may become incorporated within the sclerotial medulla, cortex or rind. The presence of such organisms reduces sclerotial longevity thus contributing to natural biological control (Merriman, 1976; Huang, 1977; Utkhede & Rahe, 1980; Trutmann, Keane & Merriman, 1982).

A second cycle of sclerotia formation may occur in some fungi. Such secondary sclerotia have been reported in *Sclerotium rolfsii* (Punja *et al.*, 1984), *Sclerotinia trifoliorum* and *S. sclerotiorum* (Williams & Western, 1965; Cook *et al.*, 1975), and *S. minor* (Adams, 1975) and may prolong the survival of these fungi. However, in *Sclerotium cepivorum* secondary sclerotia are formed only infrequently (Entwistle & Munasinghe, 1981; Merriman, Samson & Schippers, 1981; New *et al.*, 1984) and are probably of minor importance to the survival.

S. cepivorum is also able to form abnormally large sclerotia which have the typical layered structure but enclose other sclerotia of normal size and structure within the medulla (Georgy & Coley-Smith, 1982). These large sclerotia are not found frequently and their importance to survival is not known. The facility to develop stroma, sclerotia and microconidia is common in the Sclerotiniaceae (Kimbrough, 1970), e.g. *Stromatinia (Sclerotinia) gladioli* to which Whetzel (1945) considers *S. cepivorum* to be taxonomically related. Black stroma-like tissue of *S. cepivorum* has been recorded on onion leaf bases in winter (A.R. Entwistle, unpublished data) and the growth of hyphae from such tissues during warm periods in winter indicates that they could initiate hyphal spread between plants.

Microconidia have been reported in *B. cinerea* (Groves & Drayton, 1939), *Sclerotinia minor* (Willetts & Wong, 1980), *S. sclerotiorum* (Haung & Hoes, 1976), *Sclerotium cepivorum* (New *et al.*, 1984) and *S. gladioli* (Drayton, 1934). In *S. cepivorum* the function of microconidia is uncertain. In other fungi, microconidia are more likely to function as spermatia than propagules. Thus, in *Sclerotinia trifoliorum*, microconidia derived from strains of the fungus forming large ascospores fertilised the sclerotia from small-spored strains leading to the formation of apothecia and ascospores (Uhm & Fujii, 1983).

Dispersal of inoculum

Sclerotial fungi are often soil-borne and this restricts their dispersal, for example in comparison with air-borne fungi. However, several factors aid dispersal.

Livestock carry *Sclerotinia trifoliorum* sclerotia on their hooves (Scott, 1984), while sclerotia of *Sclerotium cepivorum* and *Aspergillus leporis* (*A.*

flavus group) survive ingestion by sheep and rabbits respectively (Mikhail *et al.*, 1974; Wicklow, 1985). *Macrophomina phaseolina* is an example of a fungus dispersed by insects; sclerotia are ingested by *Cylindrocopturus adspersus*, the sunflower stem boring weevil, and infection is transmitted to healthy plants during oviposition (Yang, Rogers & Luciani, 1983).

Irrigation is responsible for the dispersal of sclerotia of *Sclerotinia minor* and *Sclerotium cepivorum* (Adams, 1975, 1979), *Sclerotium oryzae* (Webster *et al.*, 1976) and *S. sclerotiorum* (Schwartz & Steadman, 1978). The spread of *Allium* white rot in Egypt by flood and irrigation water is a demonstration of effective water dispersal of a soil-borne disease. *S. cepivorum* was known in only one field in 1929 but now it affects land along most of the Nile Valley (Georgy, 1983).

The infected organic refuse remaining after crops have been harvested, graded and cleaned is potentially a source of inoculum. This danger can be reduced if the infected plant material is composted. Thus, sclerotia and tissue-protected hyphae of *Helicobasidium purpureum* are killed by composting at 47–56°C, while those of *R. solani, S. rolfsii* and *Verticillium dahliae* are killed at 60–70°C (Hull & Wilson, 1946; Yuen & Raabe, 1984).

Sclerotial fungi are also dispersed in seed colonised by hyphae and for some fungi this may be the major method of dispersal, e.g. *Botrytis allii* (Maude *et al.*, 1982). The unharvested pods of groundnut infected by *R. solani* provide a source of inoculum for the infection of a variety of future crops (Bell & Sumner, 1984). Clean groundnut seed may also be contaminated by *Calonectria crotalariae*-(stat. conid. *Cylindrocladium crotalariae*)-infected pods during processing (Johnson, 1985). Seed can become contaminated by sclerotia, for example with *S. trifoliorum* in clover (Scott, 1984) and *A. flavus* in maize (Wicklow *et al.*, 1984). Sclerotia of *S. sclerotiorum* have also been found with rape seed but they do not survive the high temperatures during processing (Dueck *et al.*, 1981).

Changes in soil sclerotial populations

The changes in soil sclerotial populations that occur in different cropping situations provide information for the management of sclerotial diseases and evidence for the presence of naturally occurring biological control. Monocropping with susceptible hosts may result in rising disease incidence, e.g. of *Sclerotinia minor* in lettuce, as described below (Dillard & Grogan, 1985); rotation with non-susceptible hosts is clearly advisable in such circumstances.

The size of soil sclerotial populations and disease incidence are not

Table 18.2. *Effect of lettuce crops on soil sclerotial populations of* Sclerotinia minor *(Dillard & Grogan, 1985)*

Crop details		Number of sclerotia (per 100 cm^3 soil)
Commercial crop		
ploughed	7.6.82	(not stated)
Experimental crop 1		
planted	7.7.82	5.8
harvested	14.9.82	13.5
ploughed	22.9.82	15
Experimental crop 2		
planted	15.3.83	11
harvested	8.6.83	8

always obviously related to cropping history. Thus, Hull & Wilson (1946) observed no increase in violet root rot (*H. purpureum*) in sugar beet after 6 years of susceptible crops. Other examples of a lack of correlation between cropping history and disease incidence are found with *S. sclerotiorum* in beans, rape and lettuce (Schwartz & Steadman, 1978; Morrall & Deuck, 1982; Patterson & Grogan, 1985); each is possibly explained by the long-distance spread of infection by ascospores, but the lack of correlation in *V. dahliae* in potatoes (Smith & Rowe, 1984) is more difficult to explain.

Naturally occurring biological control reduced sclerotial survival in *Sclerotinia minor*. The planting of an experimental crop of lettuces immediately after the removal of an earlier commercial lettuce crop infected by *S. minor*, was accompanied by an increase in *S. minor* sclerotial populations as residues from the previous crop were colonised (Table 18.2) (Dillard & Grogan, 1985). After the experimental lettuce crop had been harvested there was a further increase in sclerotial populations, again due to the colonisation of lettuce tissue remaining in the soil. Sclerotial populations then declined due to natural biological control. Ultimately, there was little difference between initial sclerotial populations and those found after the one commercial crop and two additional experimental lettuce crops, even though sclerotial populations had more than doubled in the intervening period.

Naturally occurring biological control of *S. minor* was also reported by Adams (1975). In naturally infested fields, *S. minor* sclerotial populations declined to 30% of their original value over a two-year period (Adams, 1975). Later, a similar reduction in sclerotial populations was associated with the presence of the sclerotial parasite *Sporidesmium sclerotivorum* (Adams & Ayers, 1981).

Sporidesmium sclerotivorum is an obligate parasite which uses sclerotia of several species as a source of protected carbon (Adams, Ayers & Marois, 1985). The parasite obtains energy from soluble extracellular monosaccharides within the sclerotium. These monosaccharides are derived by the hydrolysis of β-1,3-glucans in the extracellular matrix of the sclerotium cortex and medulla (Bullock, Ashford & Willetts, 1980). *S.sclerotivorum* then forms large numbers of macroconidia on the surface of the host sclerotium and on hyphae which grow through the soil from one sclerotium to another (Adams *et al.*, 1985). The successful invasion and colonisation of *S. sclerotivorum* depends on the population size of both host and parasite (Adams, Marois & Ayers, 1984). In natural conditions, therefore, the spread of *S. sclerotivorum* is likely to be affected by the clustering of inoculum (Marois & Adams, 1985).

Coniothyrium minitans has been reported as a parasite of *Sclerotinia trifoliorum* sclerotia in natural conditions in the UK (Tribe, 1957). The same fungus parasitises sclerotia of *S. sclerotiorum* in Canada (Huang, 1977) and Australia (Trutmann, Keane & Merriman, 1980). In contrast to *Sporidesmium sclerotivorum, Coniothyrium* has only a limited ability to grow in soil and close contact with sclerotia is needed for parasitism. The opportunities for contact, thus parasitism, may be increased by the transport of *Coniothyrium* spores by soil Collembola (Trutmann *et al.*, 1980).

The degree of invasion by sclerotial parasites is influenced by host sclerotial vigour. Solarisation, the process of using sunlight to heat soil under transparent plastic sheets in the field, is intended to kill soil-borne pathogens. Temperatures may not be sufficiently high to be immediately lethal, but they may weaken sclerotia. Such sclerotia may become susceptible to colonisation by antagonistic bacteria, as in *S. rolfsii* (Lifshitz *et al.*, 1983), or take longer to germinate, as in *V. dahliae* and *R. solani* (Pullman, DeVay & Garber, 1981). The stress resulting from sublethal heating during solarisation or composting may have weakened the sclerotia of *S. cepivorum, S. minor* and *S. rolfsii* and thus been responsible for their reduced viability (Porter & Merriman, 1983; Yuen & Raabe, 1984). Weakened sclerotia from whatever cause would not be expected to recover and the inoculum efficiency of the fungus would be reduced (Pullman *et al.*, 1981; Black & Beute, 1984). Sclerotia weakened, e.g. by drying or by exposure to metham sodium, are readily invaded by *Trichoderma harzianum* (Henis & Papavizas, 1983). The presence of *Trichoderma* as a coloniser rather than as a pathogen may also weaken sclerotia and thus reduce their capacity for survival (Henis *et al.*, 1983).

The presence of microorganisms naturally resident in soil may have been responsible for the restricted occurrence of *Allium* white rot in Burnaby, Canada. Out of ten fields in which *S. cepivorum* sclerotia were detected, only one field had plants with symptoms of white rot. Sclerotia from the white rot-free fields had low viability and were associated with bacteria which proved antagonistic *in vitro* and, when subsequently added to field plots, reduced white rot incidence (Utkhede & Rahe, 1980; Linderman *et al.*, 1983).

Another form of natural biological control is the capacity of a soil to become suppressive to fungi. Thus, soils in which potatoes had been grown became suppressive to *R. solani* due to the presence of *Verticillium biguttatum* which parasitises *R. solani* sclerotia (Velvis & Jager, 1983; Jager & Velvis, 1984).

The use of disease-resistant host genotypes can result in a reduction in pathogen populations in soil. Davis *et al.* (1985) monocropped with potatoes a field with a history of *Verticillium* wilt. After five years, microsclerotial populations were 81–199 colony-forming units (cfu) g^{-1} soil when *V. dahliae*-resistant genotypes had been planted but 228-315 cfu g^{-1} soil with susceptible genotypes.

Spatial patterns resulting from the dispersal of sclerotial fungi

Soil-borne plant pathogens characteristically appear in nonrandom spatial patterns (Nicot, Rouse & Yandell, 1984). Violet root rot, *Helicobasidium purpureum* (Stirrup, 1939), cotton root rot, *Phymatotrichum omnivorum* (Scofield, 1919) *and Allium* white rot (Crowe *et al.*, 1980; Entwistle, 1985) are often restricted to a few areas of a field. Other diseases occur in numerous, widely scattered areas, e.g. lettuce leaf drop, *Sclerotinia minor* (Marois & Adams, 1985) and *Rhizoctonia* patch in cereals (Murray, 1981; MacNish, 1985).

These spatial patterns reflect the presence of soil-borne inoculum and will in turn determine the position of new inoculum as diseased plants are incorporated into soil. Thus, 80% of *R. solani* sclerotia occurred within 0–10 cm of infected sugar beet roots (Naiki & Ui, 1977).

Maps of spatial patterns of inoculum have been produced by recording the sclerotial contents in soil samples, and are complementary to maps of spatial patterns of disease (Smith & Rowe, 1984; Punja *et al.*, 1985). Techniques for recovering sclerotia from soil are very laborious and restrict the amount of data available for mathematical analysis. Maps of diseased plants or of sclerotial populations are readily understood. They are not easy to analyse statistically but, nevertheless, such analyses may provide clues to underlying biological mechanisms. For example, a reduced variance : mean ratio for the frequency distribution of soil sclerotial populations of *Sclerotinia minor* was interpreted by Dillard & Grogan (1985) as resulting from the reduced clustering of inoculum following decay of most of the sclerotia in each cluster. It was regarded as evidence of biological control.

Observed horizontal patterns of populations of *Cylindrocladium crotalariae* microsclerotia in peanut and peanut/soybean fields were in close agreement with frequencies expected from the negative binomial distribution, indicating that inoculum was clustered (Taylor, Griffin & Garren, 1981). The clusters probably originated from roots and adjacent peanut pods containing microsclerotia; microsclerotia were thought to be logarithmically distributed within the cluster. The presence of 21–92 sclerotia of *S. minor* in pieces of decaying lettuce tissue in soil was also accepted as evidence for clustering (Imolehin & Grogan, 1980). A similar situation may exist with *Allium* white rot. Sclerotia form on the outside of surviving plants and within the tissue of dead plants and become incorporated into soil during post-harvest cultivations. The subsequent decay of onion tissue would leave a clump of several hundred sclerotia. This clustering might provide an explanation for the large variability observed in post-harvest soil sclerotial populations of *S. cepivorum* (Crowe *et al.*, 1980).

Grogan, Sall & Punja (1980) and Gilligan (1985) have introduced the concepts of *competence volume* and *pathozone* to define the volume of soil within which sclerotia, or other inoculum, are able to infect host roots. For example, sclerotia of *S. minor* are able to infect roots in the upper 5 cm of soil when they are within a distance of 1 cm from a root. The estimated competence volume for *S. minor* sclerotia would therefore be 19 cm^3 for a seedling lettuce with a root radius of 0.1 cm and 63 cm^3 for a mature plant with a root radius of 1.5 cm. The competence volume could be used as the basis for selecting the size and shape of soil samples suitable for the investigation of inoculum density-disease incidence relationships (Grogan *et al.*, 1980; Punja *et al.*, 1985).

There have been several studies of relationships between spatial patterns of disease and of inoculum density. Mueller *et al.* (1983) found that hyphal strands of *Phymatotrichum omnivorum* were present in patches of cotton root rot but not in surrounding disease-free areas. The situation is less clear in *Allium* white rot. No consistent differences were found between soil sclerotial populations of *S. cepivorum* in patches where there was more than 50% white rot and parts of the same field where plants were healthy; sometimes no sclerotia were detected in areas of severe disease (Entwistle, 1986). In contrast, Amein, El-Shabrawy & Abd Elrazik (1982) reported that at least 1.2×10^4 sclerotia kg^{-1} soil were present in naturally infected garlic fields in Egypt. These high populations were rather exceptional, however, and 20–500 sclerotia kg^{-1} soil are more usual (Adams, 1979; Utkhede & Rahe, 1979).

A failure to detect sclerotia or, alternatively, a failure to detect differences in sclerotial populations between diseased and non-diseased areas of the field may reflect inadequacies of technique. Sampling may fail to intercept clustered inoculum, or a lack of resources may prevent the size of sample for analysis being sufficiently large to distinguish between low and high sclerotial populations, the former being inherently more variable (Griffin & Tomimatsu, 1983). The extraction technique itself may also be insufficiently sensitive to detect spatial patterns (Punja *et al.*, 1985).

Alternatively, lack of correlation between the size of pathogen populations and the incidence of disease may be due to population changes occurring between the times of soil sampling and recording of disease. Sclerotia which have initiated infection may no longer be present at sampling, or sclerotia which were not present at sampling may subsequently form on colonised plant debris. Thus, while sclerotia of *Sclerotia minor* can survive in soil for only a few months, Imolehin & Grogan (1980) found that sclerotial numbers can increase rapidly on freshly buried lettuce debris. Lack of correlation between sclerotial populations and disease incidence at a particular site can also occur where infection is initiated from external sources. For example, ascospores of *Sclerotinia sclerotiorum* originating from neighbouring fields represent the primary inoculum for lettuce crops in California (Patterson & Grogan, 1985) and rape in Canada (Williams & Stelfox, 1980).

References

Adams, P.B. (1975). Factors affecting survival of *Sclerotinia sclerotiorum* in soil. *Plant Disease Reporter*, **59**, 599–603.

Adams, P.B. (1979). A rapid method for quantitative isolation of sclerotia of *Sclerotinia minor* and *Sclerotium cepivorum* from soil. *Plant Disease Reporter*, **63**, 349–51.

Adams, P.B. & Ayers, W.A. (1981). *Sporidesmium sclerotivorum*: distribution and function in natural biological control of sclerotial fungi. *Phytopathology*, **71**, 90–3.

Adams, P.B., Ayers, W.A. & Marois, J.J. (1985). Energy efficiency of the mycoparasite *Sporidesmium sclerotivorum in vitro* and in soil. *Soil Biology and Biochemistry*, **17**, 155–8.

Adams, P.B., Marois, J.J. & Ayers, W.A. (1984). Population dynamics of the mycoparasite

Sporidesmium sclerotivorum, and its host *Sclerotinia minor*, in soil. *Soil Biology and Biochemistry*, **16**, 627–33.

Alderman, S.C. & Hine, R.B. (1982). Vertical distribution in soil and induction of disease by strands of *Phymatotrichum omnivorum*. *Phytopathology*, **72**, 409–12.

Amein, A.M., El-Shabrawy, A.M. & Abd Elrazik, A.A. (1982). Density of *Sclerotium cepivorum* Berk. sclerotia in soil in relation to severity of white rot of garlic. *Assiut Journal of Agricultural Science* (In press).

Bell, D.K. & Sumner, D.R. (1984). Unharvested peanut pods as a potential source of inoculum of soil-borne plant pathogens. *Plant Disease*, **68**, 1039–42.

Black, M.C. & Beute, M.K. (1984). Relationships among inoculum density, microsclerotium size, and inoculum efficiency of *Cylindrocladium crotalariae* causing root rot on peanuts. *Phytopathology*, **74**, 1128–32.

Bruggen, van A.H.C. & Arneson, P.A. (1985). A quantifiable type of inoculum of *Rhizoctonia solani*. *Plant Disease*, **69**, 966–9.

Bullock, S., Ashford, A.E. & Willetts, H.J. (1980). The structure and histochemistry of sclerotia of *Sclerotinia minor* Jagger. II. Histochemistry of extracellular substances and cytoplasmic reserves. *Protoplasma*, **104**, 333–51.

Coley-Smith, J.R. (1983). Integrated control of *Sclerotium cepivorum*. *Proceedings of the Second International Workshop on Allium White Rot*, June 22–4, 1983, ed. P.B. Adams & A.R. Entwistle, pp. 87–98. Beltsville, U.S.A.: Soilborne Diseases Laboratory, U.S.D.A.

Coley-Smith, J.R. (1985). Methods for the production and use of sclerotia of *Sclerotium cepivorum* in field germination studies. *Plant Pathology*, **34**, 380–4.

Coley-Smith, J.R. & Cooke, R.C. (1971). Survival and germination of fungal sclerotia. *Annual Review of Phytopathology*, **9**, 65–92.

Coley-Smith, J.R., Humphreys-Jones, D.R. & Gladders, P. (1979). Long-term survival of sclerotia of *Rhizoctonia tuliparum*. *Plant Pathology*, **28**, 128–30.

Cook, G.E., Steadman, J.R. & Boosalis, M.G. (1975). Survival of *Whetzelinia sclerotiorum* and initial infection of dry edible beans in Western Nebraska. *Phytopathology*, **65**, 250–5.

Crowe, F.J., Hall, D.H., Greathead, A.S. & Baghott, K.G. (1980). Inoculum density of *Sclerotium cepivorum* and the incidence of white rot of onion and garlic. *Phytopathology*, **70**, 64–9.

Croxall, H.E., Sidwell, R.W. & Jenkins, J.E.E. (1953). White rot (*Sclerotium cepivorum*) of onions in Worcestershire, with special reference to control by seed treatment with calomel. *Annals of Applied Biology*, **40**, 166–75.

Davis, J.R., Pavek, J.J., Corsini, D.L. & Sorensen, L.H. (1985). Stability of *Verticillium* resistance of potato clones and changes in soil-borne populations with potato monoculture. In: *Ecology and Management of Soilborne Plant Pathogens*, ed. C.A. Parker, A.D. Rovira, K.J. Moore, P.T.W. Wong & J.F. Kollmorgen, pp. 165–6. St. Paul, U.S.A.: The American Phytopathological Society.

Dillard, H.R. & Grogan, R.G. (1985). Influence of green manure crops and lettuce on sclerotial populations of *Sclerotinia minor*. *Plant Disease*, **69**, 579–82.

Drayton, F.L. (1934). The sexual mechanism in *Sclerotinia gladioli*. *Mycologia*, **26**, 46–72.

Dueck, J., Morrall, R.A.A., Klassen, A.J. & Vose, J. (1981). Heat inactivation of sclerotia of *Sclerotinia sclerotiorum*. *Canadian Journal of Plant Pathology*, **3**, 73–5.

Entwistle, A.R. (1985). White rot in Cambridgeshire and Lincolnshire. *Report of the National Vegetable Research Station, Wellesbourne for 1984*, p. 68.

Entwistle, A.R. (1986). The relationship between soil sclerotial populations of *Sclerotium cepivorum* and the incidence of *Allium* white rot. *Proceedings of the Third International Workshop on Allium White Rot*, September 17–19, 1986, ed. A.R. Entwistle: (in press). Wellesbourne, U.K.: Institute of Horticultural Research.

Entwistle, A.R. & Munasinghe, H.L. (1981). Formation of secondary sclerotia by *Sclerotium cepivorum*. *Transactions of the British Mycological Society*, **77**, 432–4.

Esler, G. & Coley-Smith, J.R. (1984). Resistance of *Sclerotium cepivorum* in *Allium* and other genera. *Plant Pathology*, **33**, 199–204.

Filonow, A.B. & Lockwood, J.L. (1983). Loss of nutrient-independence for germination by fungal propagules incubated on soils or on a model system imposing diffusive stress. *Soil Biology and Biochemistry*, **15**, 567–73.

Garrabrandt, L.E., Johnston, S.A. & Peterson, J.L. (1983). Tan sclerotia of *Sclerotinia sclerotiorum* from lettuce. *Mycologia*, **75**, 451–6.

Georgy, N.I. (1983). *Allium* white rot: geographic distribution and economic importance (Egypt). *Proceedings of the Second International Workshop on Allium White Rot*, June 22–4, 1983, ed. P.B. Adams & A.R. Entwistle, pp. 16–17. Beltsville, U.S.A.: Soilborne Diseases Laboratory, U.S.D.A.

Georgy, N.I. & Coley-Smith, J.R. (1982). Variation in morphology of *Sclerotium cepivorum* sclerotia. *Transactions of the British Mycological Society*, **79**, 534–6.

Gilligan, C.A. (1985). Probability models for host infection by soil-borne fungi. *Phytopathology*, **75**, 61–7.

Griffin, G.J. & Tomimatsu, G.S. (1983). Root infection pattern, infection density-disease incidence relationships of *Cylindrocladium crotalariae* on peanut in field soil. *Canadian Journal of Plant Pathology*, **5**, 81–8.

Grogan, R.G., Sall, M.A. & Punja, Z.K. (1980). Concepts for modelling root infection by soil-borne fungi. *Phytopathology*, **70**, 361–3.

Groves, J.W. & Drayton, F.L. (1939). The perfect stage of *Botrytis cinerea*. *Mycologia*, **31**, 485–9.

Henis, Y., Adams, P.B., Lewis, J.A. & Papavizas, G.C. (1983). Penetration of sclerotia of *Sclerotium rolfsii* by *Trichoderma* spp. *Phytopathology*, **73**, 1043–6.

Henis, Y. & Ben-Yephet, Y. (1970). Effect of propagule size of *Rhizoctonia solani* on saprophytic growth, infectivity, and virulence on bean seedlings. *Phytopathology*, **60**, 1351–6.

Henis, Y. & Papavizas, G.C. (1983). Factors affecting germinability and susceptibility to attack of sclerotia of *Sclerotium rolfsii* by *Trichoderma harzianum* in field soil. *Phytopathology*, **73**, 1469–74.

Huang, H.C. (1977). Importance of *Coniothyrium minitans* in survival of sclerotia of *Sclerotinia sclerotiorum* in wilted sunflower. *Canadian Journal of Biology*, **55**, 289–95.

Huang, H.C. (1983). Pathogenicity and survival of the tan-sclerotial strain of *Sclerotinia sclerotiorum*. *Canadian Journal of Plant Pathology*, **5**, 245–7.

Haung, H.C. & Deuck, J. (1980). Wilt of sunflower from infection by mycelial-germinating sclerotia of *Sclerotinia sclerotiorum*. *Canadian Journal of Plant Pathology*, **2**, 47–52.

Huang, H.C. & Hoes, J.A. (1976). Penetration and infection of *Sclerotinia sclerotiorum* by *Coniothyrium minitans*. *Canadian Journal of Botany*, **54**, 406–10.

Hull, R. & Wilson, A.R. (1946). Distribution of violet root rot (*Helicobasidium purpureum* Pat.) of sugar beet and preliminary experiments on factors affecting the disease. *Annals of Applied Biology* **33**, 420–33.

Imolehin, E.D. & Grogan, R.G. (1980). Factors affecting survival of sclerotia, and effects of inoculum density, relative position, and distance of sclerotia from the host on infection of lettuce by *Sclerotinia minor*. *Phytopathology*, **70**, 1162–7.

Jager, G. & Velvis, H. (1984). Biological control of *Rhizoctonia solani* on potatoes by antagonists. 2. Sprout protection against soil-borne *R. solani* through seed inoculation with *Verticillium biguttatum*. *Netherlands Journal of Plant Pathology*, **90**, 29–33.

Johnson, G.I. (1985). Occurrence of *Cylindrocladium crotalariae* on peanut (*Arachis hypogaea*) seed. *Plant Disease*, **69**, 434–6.

Kimbrough, J.W. (1970). Current trends in the classification of Discomycetes. *The Botanical Review*, **36**, 91–161.

Lifshitz, R., Tabachnik, M., Katan, J. & Chet, I. (1983). The effect of sublethal heating on sclerotia of *Sclerotium rolfsii*. *Canadian Journal of Microbiology*, **29**, 1607–10.

Linderman, R.G. & Gilbert, R.G. (1973). Behaviour of sclerotia of *Sclerotium rolfsii* produced in soil or in culture regarding germination stimulation by volatiles, fungistasis and sodium hypochlorite treatment. *Phytopathology*, **63**, 500–4.

Linderman, R.G., Moore, L.W., Baker, K.F. & Cooksey, D.A. (1983). Strategies for detecting and characterising systems for biological control of soil-borne plant pathogens. *Plant Disease*, **67**, 1058–64.

MacNish, G.C. (1985). Mapping rhizoctonia patch in consecutive cereal crops in Western Australia. *Plant Pathology*, **34**, 165–74.

Marois, J.J. & Adams, P.B. (1985). Frequency distribution analyses of lettuce drop caused by *Sclerotinia minor*. *Phytopathology*, **75**, 957–61.

Maude, R.B., Bambridge, J.M. & Presley, A.H. (1982). The persistence of *Botrytis allii* in field soil. *Plant Pathology*, **31**, 247–52.

McCoy, R.J. & Kraft, J.M. (1984). Comparison of techniques and inoculum sources in evaluating peas (*Pisum sativum*) for resistance to stem rot caused by *Rhizoctonia solani*. *Plant Disease*, **68**, 53–5.

Merriman, P.R. (1976). Survival of sclerotia of *Sclerotinia sclerotiorum* in soil. *Soil Biology and Biochemistry*, **8**, 385–9.

Merriman, P.R., Samson, I.M. & Schippers, B. (1981). Stimulation of germination of sclerotia of *Sclerotium cepivorum* at different depths in soil by artificial onion oil. *Netherlands Journal of Plant Pathology*, **87**, 45–53.

Mikhail, S., Stewart, D.M., El Haggagy, M. Kh. & Wilkinson, R.E. (1974). The role of grazing animals in the dissemination of the onion white rot pathogen in Egypt. *Food and Agriculture Organisation Plant Protection Bulletin*, **22**, 37–41.

Morrall, R.A.A. & Deuck, J. (1982). Epidemiology of *Sclerotinia* stem rot on rapeseed in Saskatchewan. *Canadian Journal of Plant Pathology*, **4**, 161–8.

Mueller, J.P., Hine, R.B., Pennington, D.A. & Ingle, S.J. (1983). Relationship of soil cations to the distribution of *Phymatotrichum omnivorum*. *Phytopathology*, **73**, 1365–8.

Murray, D.I.L. (1981). *Rhizoctonia solani* causing barley stunt disorder. *Transactions of the British Mycological Society*, **76**, 383–95.

Naiki, T. & Ui, T. (1977). Population and distribution of sclerotia of *Rhizoctonia solani* Kuhn. in sugar beet field soil. *Soil Biology and Biochemistry*, **9**, 377–81.

Naiki, T. & Ui, T. (1978). Ecological and morphological characteristics of the sclerotia of *Rhizoctonia solani* Kuhn. produced in soil. *Soil Biology and Biochemistry*, **10**, 471–8.

New, C.M., Coley-Smith, J.R. & Georgy, N.I. (1984). Scanning electron microscopy of sclerotial germination in *Sclerotium cepivorum*. *Transactions of the British Mycological Society*, **83**, 690–3.

Newton, H.C. & Sequeira, L. (1972). Ascospores as the primary infective propagule of *Sclerotinia sclerotiorum* in Wisconsin. *Plant Disease Reporter*, **56**, 798–802.

Nicot, P.C., Rouse, D.I. & Yandell, B.S. (1984). Comparison of statistical methods for studying spatial patterns of soil-borne plant pathogens in the field. *Phytopathology*, **74**, 1399–1402.

Patterson, C.L. & Grogan, R.G. (1985). Differences in epidemiology and control of lettuce drop caused by *Sclerotinia minor* and *S. sclerotiorum*. *Plant Disease*, **69**, 766–70.

Pitt, D. (1964). Studies on sharp eyespot disease of cereals II. Viability of sclerotia: persistence of the causal fungus, *Rhizoctonia solani* Kuhn. *Annals of Applied Biology*, **54**, 231–40.

Porter, I.J. & Merriman, P.R. (1983). Effects of solarisation of soil on nematode and fungal pathogens at two sites in Victoria. *Soil Biology and Biochemistry*, 15, 39–44.

Pullman, G.S., DeVay, J.E. & Garber, R.H. (1981). Soil solarisation and thermal death: a logarithmic relationship between time and temperature for four soil-borne pathogens. *Phytopathology*, 71, 959–64.

Punja, Z.K., Jenkins, S.F. & Grogan, R.G. (1984). Effect of volatile compounds, nutrients, and source of sclerotia on eruptive sclerotial germination of *Sclerotium rolfsii*. *Phytopathology*, 74, 1290–5.

Punja, Z.K., Smith, V.L., Campbell, C.L. & Jenkins, S.F. (1985). Sampling and extraction procedures to estimate numbers, spatial pattern and temporal distribution of sclerotia of *Sclerotium rolfsii* in soil. *Plant Disease*, 69, 469–74.

Ruppel, E.G. (1985). Susceptibility of rotation crops to a root rot isolate of *Rhizoctonia solani* from sugar beet and survival of the pathogen in crop residues. *Plant Disease*, 69, 871–3.

Rush, C.M. & Gerik, T.J. (1984). Interactions between *Phymatotrichum omnivorum* and *Sorghum bicolor*. *Plant Disease*, 68, 500–1.

Schwartz, H.F. & Steadman, J.R. (1978). Factors affecting sclerotium populations of, and apothecium production by, *Sclerotinia sclerotiorum*. *Phytopathology*, 68, 383–8.

Scofield, C.S. (1919). Cotton root rot spots. *Journal of Agricultural Research*, 28, 305–10.

Scott, S.W. (1984). Clover rot. *The Botanical Review*, 50, 491–504.

Smith, V.L. & Rowe, R.C. (1984). Characteristics and distribution of propagules of *Verticillium dahliae* in Ohio potato field soils and assessment of two assay methods. *Phytopathology*, 74, 553–6.

Stirrup, H.H. (1939). Violet root rot in sugar beet. *British Sugar Beet Review*, 13, 232–4.

Taylor, J.D., Griffin, G.J. & Garren, K.H. (1981). Inoculum pattern, inoculum density-disease incidence relationships, and population fluctuations of *Cylindrocladium crotalariae* microsclerotia in peanut field soils. *Phytopathology*, 71, 1297–302.

Thomas, A.C., Kotze, J.M. & Matthee, F.N. (1983). Development of a technique for the recovery of soil-borne sclerotia of *Botrytis cinerea*. *Phytopathology*, 73, 1374–6.

Tichelaar, G.M. (1967). Studies on the biology of *Botrytis allii* on *Allium cepa*. *Netherlands Journal of Plant Pathology*, 73, 157–60.

Townsend, B.B. & Willetts, H.J. (1954). The development of sclerotia of certain fungi. *Transactions of the British Mycological Society*, 37, 213–21.

Tribe, H.T. (1957). On the parasitism of *Sclerotinia trifoliorum* by *Coniothyrium minitans*. *Transactions of the British Mycological Society*, 40, 489–99.

Trutmann, P., Keane, P.J. & Merriman, P.R. (1980). Reduction of sclerotial inoculum of *Sclerotinia sclerotiorum* with *Coniothyrium minitans*. *Soil Biology and Biochemistry*, 12, 461–5.

Trutmann, P., Keane, P.J. & Merriman, P.R. (1982). Biological control of *Sclerotinia sclerotiorum* on aerial parts of plants by the hyperparasite *Coniothyrium minitans*. *Transactions of the British Mycological Society*, 78, 521–9.

Uhm, J.Y. & Fujii, H. (1983). Ascospore dimorphism in *Sclerotinia trifoliorum* and cultural characters of strains from different-sized spores. *Phytopathology*, 73, 565–9.

Utkhede, R.S. & Rahe, J.E. (1979). Wet-sieving flotation technique for isolation of sclerotia of *Sclerotium cepivorum* from muck soil. *Phytopathology*, 69, 295–7.

Utkhede, R.S. & Rahe, J.E. (1980). Biological control of onion white rot. *Soil Biology and Biochemistry*, 12, 101–4.

Velvis, H. & Jager, G. (1983). Biological control of *Rhizoctonia solani* on potatoes by antagonists. I. Preliminary experiments with *Verticillium biguttatum*, a sclerotium-inhibiting fungus. *Netherlands Journal of Plant Pathology*, 89, 113–23.

Webster, R.K., Bolstad, J., Wick, C.M. & Hall, D.H. (1976). Vertical distribution and survival of *Sclerotium oryzae* under various tillage methods. *Phytopathology*, 66, 97–101.

Whetzel, H.H. (1945). A synopsis of the genera and species of the Sclerotiniaceae, a family of stromatic inoperculate discomycetes. *Mycologia*, **37**, 648–714.

Wicklow, D.T. (1985). *Aspergillus leporis* sclerotia form on rabbit dung. *Mycologia*, **77**, 531–4.

Wicklow, D.T., Horn, B.W., Burg, W.R. & Cole, R.J. (1984). Sclerotium dispersal of *Aspergillus flavus* and *Eupenicillium ochrosalmoneun* from maize during harvest. *Transactions of the British Mycological Society*, **83**, 299–303.

Willetts, H.J. & Wong, J.A.L. (1980). The biology of *Sclerotinia sclerotiorum, S. trifoliorum* and *S. minor* with emphasis on specific nomenclature. *The Botanical Review*, **46**, 101–65.

Williams, J.R. & Stelfox, D. (1980). Influence of farming practices in Alberta on germination and apothecium production of sclerotia of *Sclerotinia sclerotiorum*. *Canadian Journal of Plant Pathology*, **2**, 169–72.

Williams, G.H. & Western, J.H. (1965). The biology of *Sclerotinia trifoliorum* Erikss. and others species of sclerotium-forming fungi II. The survival of sclerotia in soil. *Annals of Applied Biology*, **56**, 261–8.

Wyllie, T.D., Gangopadhyay, S., Teague, W.R. & Blanchar, R.W. (1984). Germination and production of *Macrophomina phaseolina* microsclerotia as affected by oxygen and carbon dioxide concentration. *Plant and Soil*, **81**, 195–201.

Yang, S.M., Rogers, C.E. & Luciani, N.D. (1983). Transmission of *Macrophomina phaseolina* in sunflower by *Cylindrocopturus adspersus*. *Phytopathology*, **73**, 1467–9.

Yeun, G.Y. & Raabe, R.D. (1984). Effects of small-scale aerobic composting on survival of some fungal plant pathogens. *Plant Disease*, **68**, 134–6.

19
Genetic analysis of interactions between microbes and plants

ALBERT H. ELLINGBOE

Departments of Plant Pathology and Genetics, University of Wisconsin, Madison, Wisconsin 53706, USA

Manipulation of host plant resistance has been a major means of controlling plant diseases caused by fungi. The development of resistant cultivars has usually led to selection of new strains of the pathogens. Plant breeders have studied the inheritance of resistance in an attempt to determine the means of transferring effective resistance into new cultivars. From these studies has come extensive information concerning the number of loci controlling resistance to a pathogen, allelic arrangements of genes, dominance relationships, mutability, etc. Numerous studies on inheritance of avirulence and virulence in pathogens, mutation frequencies, mechanisms of recombination, dominance (if diploid or dikaryotic), linkages, etc., have similarly given much information about the genetics of pathogens.

Mendelian genetics

Flor (1946, 1947) studied the inheritance of resistance to rust in flax and also made genetic analyses of the pathogen, *Melampsora lini*. The same isolates used to analyse the reaction of the plant were also used in the crosses of the pathogen. Flor's analyses of inheritance of interactions in both host and pathogen revealed the interdependence of the genotypes of both host and pathogen in determining the interaction between the two organisms. Flor's data showed the corresponding nature of host and pathogen genes and the specificity of the interactions. He concluded that a plant cannot be resistant to a pathogen strain unless the pathogen has the corresponding gene for avirulence. Conversely, a pathogen strain cannot be avirulent on a host line unless the host has a corresponding gene for resistance.

The number of host-pathogen combinations in which inheritance in both host and pathogen has been studied is relatively small. The pattern of interactions that has emerged from studies of inheritance in one of the two organisms, together with the variability present in the other, suggests that the gene-for-gene pattern of interactions is probably involved in many more combinations.

A conservative estimate of the proportion of the naturally-occurring genetic variability in host-parasite interactions suggests that at least 95% of the variability can be attributed to the gene-for-gene interactions as originally described by Flor (1956). However, not all genes involved in host-parasite interactions appear to follow this hypothesis. The genetics of interactions involving host-specific toxins produced by the pathogens suggest a different pattern (Ellingboe, 1976). Mutation studies with pathogens suggest that genes for which there is no naturally-occurring variability are intimately involved in the ability of the pathogen to grow through host tissue (pathogenicity genes) (Ellingboe & Gabriel, 1977). Whether there are host genes which correspond to these pathogenicity genes is unknown (Ellingboe, 1979, 1982b). These induced mutations suggest that there are pathogen genes controlling interactions with host plants that do not seem to follow a gene-for-gene pattern. Why the induction of mutations permits the identification of genes for which naturally-occurring variability is not available is not clear. Moreover, there are difficulties concerning the interpretation of such mutations. They may affect a basic life process that indirectly affects the pathogen's interaction with the host rather than affecting a gene that has a primary role in the pathogen's interaction with the host (Ellingboe & Gabriel, 1977). There are several ways to approach this problem of primary v. secondary effects. As more of the genes controlling interactions are cloned, and sequences and gene products are known, the ability to distinguish between the mutations of genes controlling different processes should become clearer.

Genetic control of infection

Fungal pathogens develop a variety of structures to gain entrance into plants. If spores constitute the inoculum, they may germinate, produce hyphae, and may produce appressoria. They may penetrate the host directly or enter through plant openings such as stomata. Subsequently the pathogens may penetrate into host cells or grow intercellularly with or without haustoria and may be restricted to particular cells or tissues. Extensive descriptions at the ultrastructural level of the infection process and the interface of host and pathogen have also been given by

Bracker & Littlefield (1973). See also Chapter 4. The differential sensitivity of the various stages in the ontogeny of the interactions to various environmental factors further suggests a complex process of infection.

Two types of question have been asked in studies of the genetics of infection. Do parasite-host genes that together condition incompatibility affect unique steps in the infection process (Ellingboe, 1978)? Can mutations be induced in the pathogen that affect unique stages in infection, growth in the host tissue, and expression of disease symptoms (Daniels *et al.*, 1984; Anderson & Mills, 1985)? Most research has centred around the first question.

The genetic control of the infection processes of *Erysiphe graminis* f.sp. *hordei* and f.sp. *tritici* on barley and wheat, respectively, have received much attention (Ellingboe, 1978). Given the appropriate sequence of environmental conditions, over 80% of the spores applied to a leaf will produce successful infections with a compatible pathogen-host genotype. Moreover, the population of spores applied to a leaf will go through the morphologically identifiable stages of infection with a high degree of synchrony. The two host-parasite systems investigated provided an opportunity to determine the effects of several different pathogen-host gene pairs that condition incompatibility on the ontogeny of interactions between host and parasite.

Chancellor was the recurrent parent in the development of the highly isogenic wheat series (Table 19.1). One pathogen strain (MS-1) was used with the genotype.

$$P1\ P2\ P3a\ P4 = \text{MS-1}$$

Host line $Pm1$ was resistant to MS-1 due to the $P1/Pm1$ interactions. Host line $Pm2$ was resistant to MS-1 based on the $P2/Pm2$ interaction, etc.

None of the five different parasite-host genotypes affected spore germination, formation of an elongating germ tube or the formation of appressoria. Three of the four parasite-host gene pairs, namely $P1/Pm1$, $P3a/Pm3a$ and $P4/Pm4$, affected the proportion of parasite units that produced primary haustoria. Parasite units that produced haustoria usually produced elongating secondary hyphae whereas parasite units that did not produce haustoria did not produce elongating secondary hyphae. The observations suggest that three parasite/host gene pairs act to prevent the establishment of haustoria. But if 100 spores were applied to a leaf, not all 100 parasite units were prevented from producing haustoria. The percentages that produced haustoria and elongating secondary hyphae

Table 19.1. *The five highly isogenic wheat lines used and their respective genotypes*

Genotype					Designation
$\dfrac{Pm1 \quad pm2 \quad pm3 \quad pm4}{Pm1 \quad pm2 \quad pm3 \quad pm4}$				=	$Pm1$
$\dfrac{pm1 \quad Pm2 \quad pm3 \quad pm4}{pm1 \quad Pm2 \quad pm3 \quad pm4}$				=	$Pm2$
$\dfrac{pm1 \quad pm2 \quad Pm3a \quad pm4}{pm1 \quad pm2 \quad Pm3a \quad pm4}$				=	$Pm3a$
$\dfrac{pm1 \quad pm2 \quad pm3 \quad Pm4}{pm1 \quad pm2 \quad pm3 \quad Pm4}$				=	$Pm4$
$\dfrac{pm1 \quad pm2 \quad pm3 \quad pm4}{pm1 \quad pm2 \quad pm3 \quad pm4}$				=	Chancellor (pm)

were approximately 20, 30, and 10% for plants with $Pm1$, $Pm3$ and $Pm4$, respectively. The fate of the parasite units that produced haustoria was also dependent on the parasite/host genotype. The $P2/Pm2$ genotype had its primary effect in the formation of secondary haustoria approximately 60 h after inoculation (Hayward, 1975).

The four parasite/host genotypes $P1/Pm1$, $P2/Pm2$, $P3a/Pm3a$ and $P4/Pm4$, each seemed to affect unique stages in the ontogeny of interaction. If the products of a P gene and/or Pm gene are important for a particular stage in the ontogeny of the interactions, then a mutation of a P or $R(Pm)$ gene may lead to a change in the intensity of expression of the interactions but should not lead to change of time of effects. The general argument is that if a particular P and/or R gene (in this example, Pm) is important for a particular step or set of steps in the ontogeny of interaction, then a mutation of that P or R gene would be expected to affect one or more of those steps, but not lead to a new pattern of times of effects. The argument has been tested, in part, by use of a series of four parasite strains and the host lines with $Pm4$ and $pm4$ (Martin & Ellingboe, 1976). One parasite strain gave infection type O on plants with $Pm4$ and infection type 4 on plants with $pm4$. A second strain gave infection-type 4 on plants with $Pm4$, but infection efficiency was reduced to 20%. Infection efficiency was 80% on plants with $pm4$. No other effect during the ontogeny of the interactions was noted with the second strain on plants with $Pm4$. A third strain gave infection-type 4 on plants with either $Pm4$ or $pm4$, but infection efficiency was reduced to 35% on plants with $Pm4$. No other effect during infection was noticed with the third strain on plants with $Pm4$. A fourth gave infection type 4 on plants with both $Pm4$ or $pm4$. The unique phenotype of the fourth strain was that its growth rate was reduced about 10% on plants with $Pm4$. The infection process took longer as did all subsequent development on the host plants with $Pm4$. Strain 4 clearly did not have a gene for slow growth since it had normal infection kinetics on plants with $pm4$. The slow growth was only on plants with $Pm4$.

The four parasite strains illustrated three basic phenotypes associated with plants containing $Pm4$. With strain one there is reduced infection efficiency (to 10%), collapse of most parasite units and discolouration of infected host cells at 21-22 h after inoculation, and a final infection type O on plants with $Pm4$. Strains two and three showed reduced infection efficiency but no other subsequent effects on plants with $Pm4$. Strain four had slow growth only on plants with $Pm4$. What then is the phenotype of $Pm4$? With strain one it is reduced infection efficiency, collapse of pathogen units that did successfully produce primary haustoria and discolouration of host cells in which the haustoria are located, giving a final infection type of O. With two strains, only primary infection efficiency is affected. The parasite units that do produce primary haustoria continue normal development and eventually produce large pustules. With the fourth strain, $Pm4$ is a gene for slow growth of the pathogen. Based on final infection type the four strains would have the genotypes $P4$, $p4$, $p4$, and $p4$. However, it is clear that strains 2, 3 and 4 are not identical.

How are the observations with these four pathogenic strains reconciled with the concept that each parasite/host gene pair affects a unique step or set of steps in the ontogeny of interactions? It must be concluded that it cannot. The phenotype of the interaction is dependent on the genotype of the particular pathogen strain and the genotype of the particular host line. A particular phenotype does not appear to be unique for a pathogen locus and/or a host locus.

There are several genetic interpretations to the data on the $P4/Pm4$ interactions. In the above analysis it was assumed that the four pathogen strains had different alleles at the $P4$ locus. There is also the possibility that the strains differ by genes that modify the expression of the $P4/Pm4$ phenotype. It is difficult to conceive how a gene that modifies the $P4/Pm4$ interaction can give such diverse phenotypes as are seen with strains one and four, and additionally several modifier genes would be necessary. The alternative interpretations could be tested by making a genetic analysis of the four strains discussed above, or by making a series of mutations in the pathogen for increased virulence and mapping the induced mutations. Such a genetic analysis should answer the questions as to the existence of a series of alleles at a P locus, the existence of modifiers of a P locus that can have a major effect on the phenotype observed, and the validity of the concept that different parasite-host gene pairs for incompatibility affect unique stages in the ontogeny of interactions.

Selection of genetic stocks

Numerous studies have been made of the effect of parasite-host genes on the ontogeny of the infection process. The general conclusion has been that each set of genes affects a unique stage or stages in the infection process from spore germination to sporulation of the pathogen in the host. Most of these studies have been made with either a single culture of the pathogen and two host lines or two strains of the pathogen and a single host line. The use of only one sample of a pathogen isolate that is avirulent on a particular host gives no opportunity to determine the within (avirulence) variability. Similarly the selection of only one strain of a resistant plant and one strain of a susceptible plant gives no opportunity to determine the within-treatment variability necessary for statistical analysis. Any difference between the two host lines, even if they are 'highly isogenic', is absolutely associated with resistance or susceptibility. One technique that has been used to estimate within-treatment variability has been to examine in several host lines, with and without the gene of interest, whether identical, similar or different phenomena observed are absolutely associated with the gene in question. Another way has been to test whether the correlated events segregate in absolute association in segregating populations. For example, the wheat and barley lines 'isogenic' for reaction to powdery mildew were developed by backcrossing for several generations followed by selfing heterozygous plants for several generations. Homozygous resistant and homozygous susceptible plants were selected in each of several generations (Yang, Moseman & Ellingboe, 1972). Regardless of the generation from which the homozygous plants were derived, the absolute association of resistance, as evaluated by final infection type, with the observations during the early stages of infection suggested that the effects on early stages of infection were due to the same gene that affected final infection type. These similarities were based on reactions of all the host lines to a single

strain of the pathogen. With a different strain of the pathogen the phenotype was different, but was still related to the alleles at a single locus in
the host plant. Analysis of the independent segregants showed that different phenotypes were associated with a single host locus, either one gene
or with very closely linked genes within which no recombination was
detected.

The analyses of one pathogen strain and two host lines, or two pathogen
strains and one host line, possess innumerable opportunities for associ-
ating differences between two host lines or between two pathogen strains
with resistance and susceptibility, virulence and avirulence. Whether the
differences correlated have any cause and effect relationship, or are common effects of a single cause, or merely chance associations in the one
sample of the biological materials investiaged is commonly unknown. The
tendency to draw extensive conclusions from such cursory analyses is
widespread in the literature.

Induced mutations of gene-for-gene interactions

The arguments used for developing a protocol for obtaining
mutants are based on several different assumptions (Ellingboe & Gabriel,
1977). The concept that resistance, conditioned by one parasite-host gene
pair, is due to the production of a gene product for the plant to be
resistant, has developed from genetic analyses of naturally-occurring
genetic variability in interactions. If the plant does not produce that
product, or the product of that cistron is altered, the plant would be
susceptible (Ellingboe, 1982a). As predicted by this argument, it is possible
to induce mutations to susceptibility at an R locus (McIntosh, 1977). It
would be difficult to induce a mutation at an R locus to resistance because
the new mutant needs a particular specificity.

The genes in the pathogens (generically called P genes) involved in the
naturally-occurring variability in interactions appear to have an active
function for the pathogen to be avirulent on a plant with a corresponding
R gene. As predicted by this argument, mutations to increased virulence
are relatively easy to produce (Gabriel Lisker & Ellingboe, 1982; Statler,
1985). Mutations to decreased virulence on a host line with a given R
gene would be difficult to produce because they would be mutations to
a particular specificity (corresponding to a given R gene), and have rarely
been recovered. As stated earlier, two classes of mutations are expected.
Some mutations would be expected to map to the P locus that corresponds
to the host R gene. Some mutations might map at loci that suppress the
allele for avirulence (Lawrence, Mayo & Shepherd, 1981). The two types

of mutations are easily distinguished in crosses, but unfortunately, mutations to increased virulence, or decreased virulence, have not been mapped (Statler, 1985).

Mutations to loss of pathogenicity

A large number of mutations to loss of virulence have been observed. It has been generally observed that pathogens lose pathogenicity after culture *in vitro*. Most of these reported changes have represented loss of pathogenicity on any member of a host species and are not changes in virulence solely for a particular host genotype. The frequency of changes to loss of pathogenicity varies from pathogen species to pathogen species. In organisms like *Kabatiella zeae* the frequency of occurrence is so high that it seems more probable that the change from being pathogenic to nonpathogenic is a physiological change rather than a gene mutation (Ellingboe, unpublished). Many induced auxotrophs lead to a loss in pathogenicity, and pathogenicity is sometimes restored to auxotrophic strains by supplementation of the requirement in the inoculum (Boone *et al.*, 1956). The temptation to conclude that parasitism and pathogenesis are the result of nutrients provided by the host satisfying the growth requirements of the pathogen has been very great. There are many organisms however, such as *Neurospora crassa*, that require only essential minerals and a simple source of energy to grow but which are not parasitic or pathogenic.

Loss of pathogenicity can be the result of mutations of many genes. Mutations in genes that control enzymes involved in secondary metabolism may lead to a loss in pathogenicity, e.g., mutations that lead to auxotrophy. The pertinent question, however, is whether each mutation has a primary or secondary effect on the pathogenicity.

Distinguishing types of mutants

Two approaches have been used to distinguish between mutations in genes that have a general 'housekeeping' function for the pathogen from mutations in the pathogen genes for a primary role in parasitism and pathogenesis. The first is simply to discard the mutants that are auxotrophic and assume that the remainder of mutants may have a primary effect on pathogenesis (Anderson & Mills, 1985; Daniels *et al.*, 1984). The second is to screen for conditional mutants, e.g. primarily temperature-sensitive (*ts*) mutants (Ellingboe & Gabriel, 1977).

Wild-type cultures of *Colletotrichum lindemuthianum* will grow on agar media at both 22 and 28°C and produce expanding lesions on plants at these temperatures. Wild-type cultures treated with nitrosoguanidine produced mutants that lacked the ability either to grow on an agar medium at 28°C or to produce expanding lesions at 28°C (Ellingboe & Gabriel, 1977). The assumption was that most of the high-temperature-sensitive mutations were missense mutations that led to a temperature-labile protein. The protein would be presumed functional at 22°C, but not at 28°C. Genes in which *ts* mutations were produced would be considered crucial to growth and/or production of lesions. Growth and/or the production of expanding lesions would occur at the lower temperature but not at the higher, inhibiting temperature. Three classes of mutants were recovered in *C. lindemuthianum* Class I mutants grew at 22°C on agar media but not, or only very slowly, at 28°C, even on a complete medium. Class I mutants produced expanding lesions at 22 but not at 28°C (the inoculated plants were incubated at 22°C in a mist chamber for 48 h prior to the shift to the higher temperature).

Class I mutants probably contained mutations in genes responsible for basic life processes. Their temperature-sensitivity on a complete medium suggested that they were not due to mutations leading to auxotrophy. The inability to produce expanding lesions in plants at 28°C was considered to be the result of the inability of the fungus to grow at 28°C.

Class II mutants grew at 22°C but not, or only slowly, at 28°C on complete media but produced expanding lesions in plants at both 22°C and 28°C. The plant in this case provided something that permitted the development of expanding lesions at 28°C, that was not available in the agar medium, i.e. permitted the wild-type phenotype in the plant at 28°C but the mutant phenotype on agar at 28°C.

Class III mutants grew at both 22 and 28°C on agar media but produced expanding lesions on plants at 22°C but not at 28°C. Class III mutants were temperature-sensitive only on the plant. These mutants are of intrinsic interest since the mutations appear to be in genes that are not essential for basic life processes but are important to the development of expanding lesions.

The interpretation of data with the *ts* mutants is that the genes produce products that are important specifically to the development of expanding lesions, and that the gene product is functional at 22 but not at 28°C.

The procedures used to produce the temperature-sensitive mutations also led to the production of auxotrophic mutants (Poplawsky & Ellingboe, unpublished). Some of the temperature-sensitive mutants also contained mutations leading to auxotrophy, i.e. mutants would not grow on a minimal medium. None of the auxotrophic mutations was a temperature-sensitive mutation. The mutations that led to auxotrophy would not permit growth on minimal medium at either 22 or 28°C. The possibility that the temperature-sensitive effects were due to auxotrophy, therefore, seems very low. Unfortunately, temperature-sensitive mutations have been produced in organisms in which to date it has been impossible to study the inheritance of the *ts* mutations. It would be of great interest to know the number of loci at which mutations can be produced to give the Class III mutants, the patterns of interactions between independently induced mutations, and whether there are corresponding host genes.

A different, and very unexpected, pattern of temperature-sensitive mutants was found in *Phyllosticta maydis*, a pathogen that grows *in vitro* and produces lesions on corn at 22 and 30°C (Gabriel, Ellingboe & Rossman, 1979). Mutants were recovered that grew on a supplemented agar medium at 22°C but not at 30°C. Nine mutants produced symptoms more slowly than wild type at 22°C. Of these nine, three produced more-rapidly expanding lesions at 30°C than the wild type, four were similar to the wild type at 30°C, and two produced more-slowly developing lesions at 30°C than the wild type. No mutants of the Class I-type recovered in *C. lindemuthianum* were recovered in *P. maydis*. Seven mutants were recovered that were temperature-sensitive on agar but not temperature-sensitive on the host (Class II). Four mutants were not temperature-sensitive on agar media but were temperature-sensitive in interactions with the host. All four mutants formed slowly developing lesions at 22°C and either normal or very rapidly developing lesions at 30°C. These four mutations were classified as Class III mutations because they were not temperature-sensitive on agar media but were temperature-sensitive on the host; however, they clearly had a different pattern compared to the mutations induced in *C. lindemuthianum*. Temperature-sensitive mutations affecting host-parasite interactions induced in the obligate parasite *Erysiphe graminis* f.sp. *tritici* grew in the host at 22°C but not at 25°C. No temperature-sensitive mutations in *E. graminis* f.sp. *tritici* were obtained that had increased virulence on host lines with known *R* genes.

All temperature-sensitive mutations induced in *C. lindemuthianum*, *P. maydis* and *E. graminis* f.sp. *tritici* were identified in a total isolation procedure. The pathogen was treated with a mutagen and survivors were screened for their growth on agar media and, or, in the host at the normal temperature, and at an elevated temperature that was near the highest temperature at which the pathogen would still produce disease in the plant. The screening for mutants with the first two pathogens involved screening for growth on agar media. The test for temperature-sensitivity with the plants involved inoculation and incubation at normal temperatures during the time required for the pathogen to gain entrance into the host. Thus, the test for temperature-sensitive mutants would detect mutations of genes that are crucial for successful development of disease but, since the shift to the higher temperature occurred after infection, the procedure would not be likely to detect mutations that affected spore germination, growth of germ tubes, formation of appressoria, or any processes crucial to infection but not morphologically identifiable as a particular infection stage. When the screening for mutants included the infection process, as with *E. graminis* f.sp. *tritici*, the yield of mutants was high (3.6%) (Gabriel, Lisker & Ellingboe, 1982). This yield is considered a conservative estimate because a mixed infection of wild type and mutant would be scored as non-mutant.

The physiological changes associated with spore germination and the infection processes have been studied extensively. The cause and effect of the extensive biochemical and physiological changes that occur during spore germination and germ tube differentiation have been the subject of much conjecture. Constitutive mutations of genes of an obligate parasite whose products were crucial to spore germination and infection processes would be assumed to be lethal and, therefore, difficult to work with. Conditional lethals, such as temperature-sensitivity, of genes whose products are crucial to germination and infection should be easily amenable to experimental analysis.

Perception of the system and molecular approaches

Research on the process of infection by *Rhizobium* illustrates how induced mutations and the arguments of recombinant DNA technology may be used to examine interactions between a microorganism and a plant (Long *et al.*, 1982; Long, Buikena & Ausubel, 1982) and shows that prior assumptions have a definite effect on results obtained. Whereas plant pathologists have typically used naturally-occurring genetic variability to study the phenomena of resistance v. susceptibility and virulence v. avirulence, research on *Rhizobium* was begun with a basic assumption that strains of *Rhizobium* that have the ability to infect, nodulate, and fix nitrogen possess positive capabilities that are lacking in a strain unable to infect, nodulate and fix nitrogen on a plant that permits such processes. This is analogous to what pathologists call 'basic compatibility'. Thus, if a number of genes exist whose functions are necessary for infection, nodulation, and nitrogen fixation, then mutations in those genes should lead to loss or modification of one or more of the processes observed. These arguments are very similar to the arguments used in the selection of temperature-sensitive mutations. Mutations resulting in the inability to infect, nodulate and fix nitrogen were induced by nitrosoguanidine or Tn5 mutagenesis (Beringer, Johnston & Wills, 1977; Long *et al.*, 1982). A considerable clustering of mutations was observed (Kondorosi, Banfalvi & Kondorosi, 1984). It may be asked whether all mutations to nodulate map to a single, or several DNA restriction fragments. Two procedures have been used to clone these genes. One is the use of Tn5 as a molecular tag. The more common procedure has been to prepare a library of the wild type strains and genetically to transform the mutants with the members of this library. The DNA fragment that converts the mutant into the wild type phenotype is considered to possess the wild type allele of the mutant gene (Long, Buikema & Ausubel, 1982). The combination of transposon tagging and complementation of mutants has been a very effective means to identify and clone genes that have an active function for nodulation and nitrogen fixation.

The basic supposition with *Rhizobium* is that there are genes whose products are necessary for successful nodulation and nitrogen fixation (Long *et al.*, 1982). The results support this assumption, but reports of host genes that give resistance to nodulation provide evidence that there may be naturally-occurring variability in host species and their *Rhizobium* symbionts that is consistent with the gene-for-gene hypotheses (Vest, 1970, 1972). The genes identified in *Medicago sativa* and *Glycine max* indicate that resistance is usually dominant and that the resistance is specific for particular strains. Strains of *R. japonicum* that can nodulate a soybean with *Rj*4, for example, can also nodulate a soybean with *rj*4. The patterns of epistasis are also consistent with the gene-for-gene pattern. Assuming that the naturally-occurring genetic variability of *Rhizobium* and host plant follows the gene-for-gene pattern, it is therefore logical further to assume that there is an active function on the part of the *Rhizobium* gene(s) that conditions a lack of nodulation on a plant with a corresponding host gene, such as *Rj*4 in soybean. A mutation in a *Rhizobium* gene that had an active function to be unable to nodulate on a plant of a given genotype would never have been identified by a screen for mutations that detected a loss of ability to nodulate and fix nitrogen. The perception of the biological phenomenon (nodulation and nitrogen fixation) determined the kinds of mutations sought. Since induced mutations do not follow a pattern consistent with the naturally-occurring genetic variability, it is doubtful that they will tell us much about the genes controlling this variability. It does not mean that they have no value for studying interactions between microorganism and plant. The difference in genetic patterns is a clear reminder that there are several kinds of gene actions that appear to be crucial for compatible and/or incompatible relationships (Ellingboe, 1976).

The approach used with the *Rhizobia* has also been used with the bacterial pathogens *Pseudomonas syringae* and *Xanthomonas campestris* (Anderson & Mills, 1985; Daniels *et al.*, 1984). Mutants that lack, or have altered, pathogenicity have been induced with Tn5 or chemical mutagenesis. The Tn5-induced mutations have lead to auxotrophy as well as altered or lost pathogenicity (path$^-$). The basic procedure with both symbionts and pathogens has been to discard auxotrophic mutants that affect interactions with the plant and retain non-auxotrophic mutants. The assumption has been that non-auxotrophic mutations are associated in some way with pathogenesis, an argument similar to that used with the nodulation-deficient *Rhizobia*. It is obvious that the mutations in the path$^-$ mutants affect the ability of the bacteria to infect, grow and produce disease in the host plants. The genes in which the mutations have been produced can be cloned by transposon tagging or by complementation by genetic tranformation of the mutant with a library of the wild type culture. If the mutation led to a loss of pathogenicity, complementation of the mutation with a library of the wild type parent would be expected.

Predicted types of path$^-$ mutants

An obvious question is whether the mutations described as being path$^-$ have a primary or secondary role in host-parasite interaction. Even though auxotrophs may affect pathogenicity, the conclusion that the wild$^-$ type allele plays a role in pathogenicity is obviously weak. The path$^-$ mutants are expected to be of at least two types. Some will be the physiological equivalent of auxotrophs but will not lead to a specific nutritional requirement. Others will probably be in genes that play a primary role in the success of the pathogen. How might these classes of mutations be distinguished? The ability of members of a library of the wild type parent to transform a path$^-$ mutant to a path$^+$ phenotype will probably not distinguish between the different types of mutants. It is possible that mutants might be distinguished by investigating whether any member of a library of DNA from *E. coli* (or other distantly related non-phyto-pathogenic bacteria) might restore the path$^+$ phenotype. If the path$^-$ mutant could be converted to path$^+$ through genetic transformation by a clone of *E. coli* DNA, then the path$^-$ mutation is probably in an essential 'housekeeping' gene that affects pathogenicity in a secondary manner. If only clones from the wild type parent, or closely related species, can restore the path$^+$ phenotype of a mutant, then the path$^-$ mutation is most likely in a gene whose function is of primary importance to parasitism and pathogenesis. The use of a closely related plant pathogen as a source of clones for restoring the path$^+$ phenotype will probably have the disadvantage of confounding influences of *P* genes. This argument is based on the accumulating data which suggest that the difference between a 'host' and 'nonhost', 'pathogen' and 'nonpathogen' are the numbers of pathogen/host gene pairs that specify incompatibility, $-$ i.e. the probability of incompatibility $-$ if a microbe and plant are selected at random and tested for compatibility.

Molecular genetics of fungal pathogens

The arguments used for studying nodulation competence in *Rhizobia* or pathogenicity in phytopathogenic bacteria should also be applicable in pathogenic fungi. However, the ability genetically to transform pathogenic fungi is only beginning to be available, and not yet routine. On the assumption that genetic transformation will become routine in plant pathogenic fungi, it should be possible to study genes involved in gene-for-gene relationships, basic compatibility, and other characteristics associated or correlated with the ability to be a successful pathogen, irrespective of that organism's ability to go through the sexual cycle. A sexually-reproducing fungal pathogen is an advantage for experimental analysis, but not a requirement if routine genetic transformation is available.

The development of genetic transformation systems in fungi should permit a much more detailed analysis of the genes controlling a particular phenomenon. For example, are the differences observed between four isolates of *Erysiphe graminis* f.sp. *tritici* on host plants with *Pm*4 due to alleles at one locus (Martin & Ellingboe, 1976)? An analysis of the inheritance of the differences was not possible because the isolates could not be crossed. If the arguments used to clone avirulence genes for bacteria (Staskawicz, Dahlbeck & Keen, 1984; Gabriel, 1985) were applied to *E. graminis*, the avirulence allele that gives a high level of avirulence could be used to clone the avirulence locus. That clone could be used to clone the allele(s) from the other three isolates, and all four clones, could be used used genetically to transform an isolate virulent on wheat plants with *Pm*4. The phenotypes of the four transformants could thus be compared. If differences between the four isolates of *E. graminis* were due to different alleles at the *P* locus, then all four transformants should have different phenotypes. If the differences between the four isolates were due to different modifier genes, then the differences between the transformants would be negligible. It should then be possible to clone the modifier genes particularly since any modifiers identified to date have led to increased virulence. If changes in messenger RNA or peptides had been correlated with a particular effect in the infection process, it would be possible to determine whether the molecular changes were associated with the *P* gene product or the modifiers.

Why is the cloning of a *P* gene and the identification of its product considered so important? There are the obvious reasons for basic knowledge of the host-parasite interactions, particularly identifying which molecules are of primary importance in determining the interactions crucial to particular phenotypes. There are also some potential practical benefits. If the genetic models of host-parasite interactions are correct (Ellingboe, 1976, 1982*a*), the *P* gene product should be useful in isolating the products of the corresponding *R* gene products. If there is an affinity between the products of the *P* and *R* genes, then the product of one should permit the isolation of the product of the other. The isolation of the gene product will then permit the cloning of the gene by several different strategies.

Fungi should provide a rich source of material to study interactions of a pathogen with a host because, in part, they undergo extensive morphological changes during infection, namely growth and sporulation in host tissue, etc. Most of the morphologically identifiable stages of a fungus have been implicated as being important in determining the type of interaction with a host plant. What is the significance of the correlation of morphological, biochemical, physiological changes, etc., with the phenomena of resistance v. susceptibility, and virulence v. avirulence, etc? The means by which to tell whether these are chance associations or have real cause-and-effect relationships are now available. A major approach is to ask specific questions about the kinds of mutants expected, the numbers of genes in which mutations will affect a phenomenon, the kinds of revertants expected, the complementation patterns of mutations in heterokaryons and in genetic transformation, etc. The use of mutations to reach an understanding of host-parasite interactions has received very little emphasis in plant pathology, in sharp contrast to the emphasis in comparative physiology and biochemistry. The ability to select for specific types of mutants will undoubtedly become very important if the arguments of molecular genetics and the recombinant DNA technology are to be fully utilised in attempts to study host-parasite interactions. There are several reasons to believe that the ability genetically to transform fungi will develop rapidly (Rodriquez, Turgeon & Yoder, 1986), as for example did the cloning of genes controlling race specificity in bacterial pathogens (Staskawicz, Dahlbeck & Keen, 1984; Gabriel, 1985). The identification of transposable elements in fungi that can be used as molecular tags of genes of special interest will also become available in the next few years. These studies will undoubtedly have a profound effect on our perception of host-parasite interactions.

References

Anderson, D.M. & Mills, D.I. (1985). The use of transposon mutagenesis in the isolation of nutritional and virulence mutations in two pathovars of *Pseudomonas syringae*. *Phytopathology*, **75**, 104–8.

Beringer, J.E., Johnston, A.W.B. & Wills, B. (1977). The isolation of conditional ineffective mutants of *Rhizobium leguminosarum*. *Journal of General Microbiology*, **98**, 339–43.

Boone, D.M., Stouffer, J.F., Stakman, M.A. & Keitt, G.W. (1956). *Venturia inaequalis* (Cke.) Wint. VII. Induction of mutants for studies on genetics, nutrition, and pathogenicity. *American Journal of Botany*, **43**, 199–204.

Bracker, C.E. & Littlefield, L.J. (1973). Structural concepts of host-pathogen interfaces. In: *Fungal Pathogenicity and the Plant's Response*, ed. R.J.W. Byrde & C.V. Cutting, pp. 159–318. London: Academic Press.

Daniels, M.J., Barber, C.E., Turner, P.C., Cleary, W.G. & Sawczyc, M.K. (1984). Isolation of mutants of *Xanthomonas campestris* pv. *campestris* showing altered pathogenicity. *Journal of General Microbiology*, **13**, 2447–55.

Daniels, M.J., Barber, C.E., Turner, P.C., Sawczyc, M.K., Byrde, R.J.W. & Fielding, A.H. (1984). Cloning of genes involved in pathogencity of *Xanthomonas campestris* pv. *campestris* using the broad host range cosmid pLAFR. *Embo*, **3**, 3312–28.

Ellingboe, A.H. (1976). Genetics of host-parasite interactions. In: *Encyclopedia of Plant Physiology, New Series Volume 4 Physiological Plant Pathology*, ed. R. Heitefuss & P.H. Williams, pp. 761–78. Heidelberg: Springer–Verlag.

Ellingboe, A.H. (1978). A genetic analysis of host-parasite interactions. In: *In The Powdery Mildews*, ed. D.M. Spencer, pp. 159–81. London: Academic Press.

Ellingboe, S.H. (1979). Inheritance of specificity: the gene-for-gene hypothesis. In: *Recognition and Specificity in Plant Host-Parasite Interactions*, ed. J.M. Daly & I. Uritani, pp. 3–15. Tokyo: Japanese Scientific Societies Press; and Baltimore: University Park Press.

Ellingboe, A.H. (1982a). Genetical aspects of active defence. In: *Active Defence Mechanisms in Plants*, ed. R.K.S. Wood, pp. 179–92. London: Plenum Publishing Corporation.

Ellingboe, A.H. (1982b). Host resistance and host-parasite interactions: a perspective. In: *Phytopathogenic Prokaryotes*, Vol. 2, ed. M.S. Mount & G. Lacy, pp. 103–17. New York: Academic Press.

Ellingboe, A.H. & Gabriel, D.W. (1977). Induced conditional mutants for studying host/ pathogen interactions. In: *Use of Induced Mutations for Improving Disease Resistance in Crop Plants*, ed. A. Micke, pp. 35–46. Vienna: International Atomic Energy Agency.

Flor, H.H. (1946). Genetics of pathogenicity in *Melampsora lini*. *Journal of Agricultural Research*, **73**, 335–57.

Flor, H.H. (1947). Inheritance of reactions to rust in flax. *Journal of Agricultural Research*, **74**, 241–62.

Flor, H.H. (1956). The complementary genic systems in flax and flax rust. *Advances in Genetics*, **8**, 29–54.

Gabriel, D.W. (1985). Molecular cloning of specific avirulence genes from *Xanthomonas malvacearum*. In: *Advances in the Molecular Genetics of the Bacteria-Plant Interaction*, ed. A.A. Szalay & R.P. Legocki, pp. 202–3. Ithaca, NY: Media Services, Cornell University.

Gabriel, D.W., Ellingboe, A.H. & Rossman, E.C. (1979). Mutations affecting avirulence in *Phyllosticta maydis*. *Canadian Journal of Botany*, **57**, 2639–43.

Gabriel, D.W., Lisker, N. & Ellingboe, A.H. (1982). The induction and analysis of two classes of mutations affecting pathogenicity in an obligate parasite. *Phytopathology*, **72**, 1026–8.

Hayward, M.J. (1975). Genetic control of the development of haustoria of *Erysiphe graminis* f.sp. *tritici* on wheat. *Ph.D. Dissertation*, Michigan State University.

Kondorosi, E., Banfalvi, Z. & Kondorosi, A. (1984). Physical and genetic analysis of a symbiotic region of *Rhizobium meliloti*: identification of nodulation genes. *Molecular and General Genetics*, **193**, 445–52.

Lawrence, G.J., Mayo, G.M.E. & Shepherd, K.W. (1981). Interactions between genes controlling pathogenicity in flax rust. *Phytopathology*, **71**, 12–19.

Long, S.R., Buikema, W.J. & Ausubel, F.M. (1982). Cloning of *Rhizobium meliloti* nodulation genes by direct complementation of nod-mutants. *Nature*, **298**, 485–8.

Long, S.R., Meade, H.M., Brown, S.E. & Ausubel, F.M. (1982). Transposon-induced symbiotic mutants of *Rhizobium meliloti*. In: *Genetic Engineering in the Plant Sciences*, ed. N.J. Panopoulos, pp. 129–43, Praeger.

Martin, T.J. & Ellingboe, A.H. (1976). Differences between compatible parasite/host genotypes involving the *Pm4* locus of wheat and the corresponding genes in *Erysiphe graminis* f.sp. *tritici*. *Phytopathology*, **66**, 1435–8.

McIntosh, R.A. (1977). Nature of induced mutations affecting disease reaction in wheat. In: *Induced Mutations against Plant Diseases*, ed. A. Micke, pp. 115–18. Vienna: International Atomic Energy Agency.

Rodriguez, R.J., Turgeon, G. & Yoder, O.C. (1986). Transformation and molecular analysis of the filamentous plant pathogen *Colletotrichum lindemuthianum (Glomerella cingulata)*. *Journal of Cellular Biochemistry* (Supplement 10C), p. 26.

Staskawicz, B.J., Dahlbeck, D. & Keen, N.T. (1984). Cloned avirulence gene of *Pseudomonas syringae* pv. *glycinea* determines race-specific incompatibility on *Glycine max* L. Merr. *Proceedings of the National Academy of Sciences, U.S.A.*, **81**, 6024–8.

Statler, G.D. (1985). Mutations affecting virulence in *Puccinia recondita*. *Phytopathology*, **75**, 565–7.

Vest, G. (1970). *Rj3* – a gene conditioning ineffective nodulation in soybean. *Crop Science*, **10**, 34–5.

Vest, G. (1972). *Rj4* – a gene conditioning ineffective nodulation in soybean. *Crop Science*, **12**, 692–3.

Yang, S.L., Moseman, J.G. & Ellingboe, A.H. (1972). The formation of elongating secondary hyphae of *Erysiphe graminis* and the segregation of *Ml* genes in barley. *Phytopathology*, **62**, 1219–23.

20

Immunisation against disease: the plant fights back

RALPH A. DEAN and JOSEPH A. KUĆ

Department of Plant Pathology, University of Kentucky, Lexington, Ky 40546, USA

'It remains incredible that any such discovery should be made, and when it has been made, it appears incredible that it should so long escaped men's research. All which affords good reason for the hope that a vast mass of investigation yet remains.' Francis Bacon, 1620.

The emphasis of the preceding chapters has been on the development of the fungal pathogen in plant tissues. This chapter is concerned with the strategies and mechanisms developed by plants to combat invading organisms. The emphasis is on induced systemic resistance ('immunisation') against the development of pathogens which is manifested in part by a sensitisation of the tissue to respond rapidly in their presence. Substantial evidence exists that susceptible plants often contain the genetic information for the expression of resistance. Resistance through immunisation is accomplished without adding genes to the plant's genome. Some reference will be made to bacteria and viruses in view of the nonspecific nature of this phenomenon in plants.

Strategies of defence

Plants have evolved over many millenia several effective mechanisms for resistance to disease. They include pre-existing chemical and physical barriers, general metabolic responses to injury and disease as well as more specific responses against particular groups of pathogens. These general and specific metabolic responses are often activated by infection. They are expressed, however, without preconditioning and may be considered as constitutive sensitisation.

Induced sensitisation

Induced senitisation requires an initial stimulus which causes subtle, persistent systemic biochemical changes in the plant. Upon subsequent infection these changes result in a rapid and intense expression of disease resistance. This strategy of defence in plants, in outcome but not mechanism, is comparable to the inducible antibody system in animals. Although specific immunoglobulins have not been found in plants, plants have inducible defence mechanisms which are effective against diseases caused by fungi, bacteria, and viruses (Kuć, 1985a,b; Kuć & Rush, 1985). An important metabolic advantage of inducible defence mechanisms is that a major diversion of energy and metabolites away from the main stream of metabolism occurs only when and where needed. In addition, the phytotoxicity often associated with the accumulation of defence compounds is avoided in the absence of infection.

Disease outcome

The longer the pathogen remains undetected, by avoiding or suppressing host recognition, the greater its chance of success. Once the plant detects that it is under attack, it activates metabolic processes to counter development of the pathogen. The pathogen may be able to suppress or inactivate these processes. The timing and magnitude of induced responses however are generally critical to the outcome of disease. Induced protection may make genetically susceptible plants resistant to disease by enhancing recognition and/or promoting a more efficient, faster and more intense host defence response.

The phenomenon of immunisation

A large number of different agents induce protection which has been documented since the beginning of the twentieth century (Chester, 1933). These include synthetic chemicals, pathogens, metabolic products of hosts or infectious agents, races of pathogens that are nonpathogenic to a host cultivar and nonpathogens of a host.

Abiotic induction

Numerous examples of natural and synthetic chemicals have been reported systemically to enhance host resistance, e.g. aluminum tris(o-ethylphosphite) (Aliette) (Bompeix et al., 1980; Fenn & Coffey, 1984, 1985; Guest, 1984), 3-(allyloxy)-1,2-benzisothiazole-1,1-dioxide (probenazole) (Shimura et al., 1981), 2,2-dichloro-3,3-dimethylcyclopropane carboxylic acid (DDCC) (Cartwright, Langcake & Ride, 1980), D-phenyl-alanine, D-alanine, α-aminoisobutyric acid (Kuć et al., 1959; Kuć & Williams, 1962; MacLennan, Kuć & Williams, 1963), polyacrylic acid, salicylic acid and acetylsalicylic acid (Gianinazzi & Kassanis, 1974;

Kassanis & White, 1975; White, 1979; Mills & Wood, 1984). Although these compounds appear to have little direct effect on pathogens *per se*, when coupled with the host's natural defence mechanisms, they may provide the competitive edge required to reduce disease.

Biotic induction

Immunisation using biotic agents has been extensively studied. Green beans were systemically immunised against disease caused by cultivar-pathogenic races of *Colletotrichum lindemuthianum* by prior infection with either cultivar-nonpathogenic races (Rahe *et al.*, 1969; Elliston, Kuć & Williams, 1971; Skipp & Deverall, 1973), cultivar-pathogenic races attenuated by heat in host tissue prior to symptom appearance (Rahe & Kuć, 1970), or nonpathogens of bean. The anthracnose pathogen of cucumber, *Colletotrichum lagenarium*, was equally effective as non-pathogenic races as an inducer of systemic protection against all races of bean anthracnose. Protection was induced by *C. lagenarium* in cultivars resistant to one or more races of *C. lindemuthianum* as well as in cultivars susceptible to all reported races of the fungus and which accordingly had been referred to as 'lacking genetic resistance' to the pathogen (Elliston, Kuć & Williams, 1976*a,b*). These results suggest that the same mechanisms may be induced in cultivars reported as 'possessing' or 'lacking' resistance genes (Elliston *et al.*, 1977). It also is apparent that cultivars susceptible to all races of *C. lindemuthianum* do not lack genes for resistance mechanisms against the pathogen.

Kuć, Shockley & Kearney (1975) showed that cucumber plants could be systemically protected against disease caused by *Colletotrichum lagenarium* by prior inoculation of the cotyledons or the first true leaf with the same fungus. Subsequently cucumbers have been systemically protected against fungal, bacterial, and viral diseases by prior localised infection with either fungi, bacteria or viruses (Hammerschmidt, Acres & Kuć, 1976; Jenns & Kuć, 1977; Caruso & Kuć, 1979; Staub & Kuć, 1980; Bergstrom, Johnson & Kuć, 1982; Gessler & Kuć, 1982*a*; Basham & Cohen, 1983). Non-specific protection induced by infection with *C. lagenarium* or tobacco necrosis virus (TNV) was effective against at least 13 pathogens (Table 20.1), including obligatory and facultative parasitic fungi, local lesion and systemic viruses, wilt fungi and bacteria. Similarly, protection was induced by and was also effective against root pathogens. Other curcurbits, including watermelon and muskmelon have been systemically protected against *C. lagenarium* (Caruso & Kuć, 1977*a*).

Systemic protection in tobacco has also been induced against a wide

Table 20.1. *The biological spectrum of induced systemic protection in cucumber*

Pathogen	Disease
Fungi	
Colletotrichum lagenarium	Anthracnose, local lesions
Cladosporium cucumerinum	Scab, local lesions
Mycosphaerella melonis	Gummy stem blight, unrestricted lesions
Fusarium oxysporum f.sp. *cucumerinum*	Fusarium wilt
Verticillium albo-atrum	Verticillium wilt
Fusarium solani f.sp. *cucurbitae*	Fusarium root rot
Sphaerotheca fuliginea	Powdery mildew
Pseudoperonspora cubensis	Downy mildew
Phytophthora infestans	Late blight of potato, local necrosis
Bacteria	
Pseudomonas lachrymans	Angular leaf spot, local lesions
Erwinia tracheiphila	Bacterial wilt
Viruses	
Tobacco necrosis virus	Local lesions
Cucumber mosaic virus	Local lesions and systemic mosaic

variety of diseases (Kuć & Tuzun, 1983). Necrotic lesions caused by tobacco mosaic virus (TMV) enhanced resistance in the upper leaves to disease caused by the virus (Ross, 1961, 1966), *Phytophthora parasitica* var. *nicotianae, P. tabacina* and *Pseudomonas tabaci* and reduced reproduction of the aphid *Myzus persicae* (McIntyre & Dodds, 1979; McIntyre, Dodds & Hare, 1981). Infiltration of heat-killed *P. tabaci* (Lovrekovich & Farkas, 1965) and *Pseudomonas solanacearum* (Sequeira, Gaard & DeZoeten, 1977) into tobacco leaves induced resistance against the same bacteria used for infiltration. Tobacco plants were also protected by the nematode *Pratylenchus penetrans* against *P. parasitica* var. *nicotiana* (McIntyre & Miller, 1978).

Cruickshank & Mandryk (1960) were the first to report immunisation of tobacco foliage against blue mould (*P. tabacina*) by stem injection with the fungus, which also involved dwarfing and premature senescence. It was recently discovered that injection external to the xylem not only alleviated stunting but also promoted growth and development. Immunised tobacco plants, in both glasshouse and field experiments, were approximately 40% taller, had a 40% increase in dry weight, 30% increase in fresh weight, and 4–6 more leaves than control plants (Tuzun, Nesmith & Kuć, 1984). These plants flowered approximately 2-3 weeks earlier than control plants (Tuzun & Kuć, 1985a).

Effectiveness of immunisation

Systemic protection does not confer absolute immunity against infection, but reduces the severity of the disease and delays symptom development. Lesion number, lesion size and extent of sporulation of fungal pathogens are all decreased. The diseased area may be reduced by more than 90%.

When cucumbers were given a 'booster' inoculation 3–6 weeks after the initial inoculation, immunisation induced by *C. lagenarium* lasted through flowering and fruiting (Kuć & Richmond, 1977). Protection could not be induced once plants had set fruit. Tobacco plants were immunised for the growing season by stem injection with sporangia of *P. tabacina.* However, to prevent systemic blue mould development, this technique was only effective when the plants were above 20 cm in height.

Removal of the inducer leaf from immunised cucumber plants did not reduce the level of immunisation of pre-existing expanded leaves. However, leaves which subsequently emerged from the apical bud were progressively less protected than their predecessors (Dean & Kuć, 1986a). Similar results were reported by Ross (1966) with tobacco (local lesion host) immunised against TMV by prior infection with TMV. In contrast, new leaves which emerged from scions excised from tobacco plants immunised by stem-injection with *P. tabacina* were highly protected (Tuzun & Kuć, 1985b). Plants regenerated via tissue culture from leaves of immunised plants showed a significant reduction in blue mould compared to plants regenerated from leaves of non-immunised parents. Young regenerants only showed reduced sporulation. As plants aged, both lesion development and sporulation were reduced (Tuzun & Kuć, unpublished data). Other investigators, however, did not reach the same conclusion, although a significant reduction in sporulation in one experiment was reported (Lucas, Dolan & Coffey, 1985).

Protection of cucumber and watermelon is effective in the glasshouse, and in the field (Caruso & Kuć, 1977b). In one trial the total lesion area of *C. lagenarium* on protected cucumber was less than 2% of the lesion area on unprotected control plants. Similarly, only 1 of 66 protected, challenged plants died, whereas 47 of 69 unprotected, challenged water-melons died. In extensive field trials in Kentucky and Puerto Rico, stem injection of tobacco with sporangia of *P. tabacina* was at least as effective in controlling blue mould as the best fungicide, metalaxyl. Plants were protected 95–99%, based on the necrotic area and degree of sporulation, leading to a yield increase of 10–25% in cured tobacco (Turzun *et al.*, unpublished data).

Induced resistance against bacteria and viruses appears to be expressed as suppression of disease symptoms or pathogen multiplication or both (Caruso & Kuć, 1979; Doss & Hevisi, 1981; Jenns, Caruso & Kuć, 1979).

Dynamics of induction

The diversity of biotic agents capable of inducing protection precludes a common factor of pathogen origin as the immunity signal. More probably, inducers exert a common effect which results in synthesis or release of immunity signals which sensitise the host to respond rapidly when infected or injured. These may be produced by the stressed (infected) dying cells and exported from the leaf.

Induction stimuli

The common effect of inducing agents is slow progressive host stress (or injury) and cell death; however, necrosis *per se* is not effective. Stem necrosis in tobacco caused by a range of Peronosporales other than *P. tabacina* did not protect the foliage against blue mould. However, stems injected with β-ionone, a degradation product of β-carotene found in stems of tobacco injected with *P. tabacina*, caused necrosis, induced resistance to blue mould, and mimicked the morphological changes associated with stem-injection of the fungus (Salt, Tuzun & Kuć, 1986). Injury caused by various chemical and physical agents which resulted in rapid cucumber cell death did not protect the plant from *C. lagenarium*. Recently, we have extracted a thermostable, low molecular weight fraction from spinach leaves which causes progressive damage and confers protection against *C. lagenarium* in cucumbers. It would appear unlikely that the compounds and plant extracts which cause restricted damage and induce systemic protection are the translocated signals, since protected leaves above the sites of induction generally appear undamaged.

Systemic alarm signals

Biological signals are well documented in living organisms and are an essential component for the survival and performance of complex multicellular beings. Recently, several signals associated with plant disease resistance have been identified (see Chapter 1). Plant cell wall fragments containing galacturonic acid residues may be released during infection by the action of pathogen hydrolases or pathogen-induced plant hydrolases. These wall fragments elicit localised resistance mechanisms, in particular, phytoalexin accumulation (Hahn, Darvill & Albersheim, 1981; Bruce & West, 1982; Davies *et al.,* 1983; Nothnagel *et al.,* 1983; Walker-Simmons, Hadwiger & Ryan, 1983). Fungal glucans also elicit localised phytoalexin accumulation (Keen, 1982; Sharp, McNeil & Albersheim, 1984). These elicitors may be directly responsible for eliciting resistance or, perhaps due to phytotoxicity, cause plant hydrolases to release other active components.

Galacturonate-containing oligosaccharides from tomato cell walls elicit the accumulation of the putative phytoalexin casbene in castor beans (Bruce & West, 1982) and glyceollin in soybeans (Albersheim *et al.,* 1983). Such fragments have also been postulated to be the systemic alarm signal for the proteinase inhibitor-inducing factor (PIIF) in tomato and potato (Ryan *et al.,* 1985). The accumulation of proteinase inhibitors arising from damage or infection is suggested as at least part of the induced resistance mechanism against insects and microorganisms (Ryan, 1978, 1984). However, evidence for their role in insect resistance is contradictory and their role in resistance to microbial plant pathogens has not been demonstrated (Weiel & Hapner, 1976; Cleveland & Black, 1983).

The minimum structural moiety that exhibits PIIF activity has recently been reported as 4-0-α-D galacturonosyl-D-galacturonic acid (Ryan *et al.,* 1985). In contrast, the soybean phytoalexin glyceollin was elicited by fragments of about 12 galacturonic acid residues (Albersheim *et al.,* 1983). However, it is intriguing that similar signals may regulate the localised accumulation of phytoalexins and induce systemic increases in proteinase inhibitors. The systemic movement of these fragments has not been demonstrated; it may be that the localised reactions produce secondary messengers which are the true alarm signals.

Extracting a non-phytotoxic fraction with signal activity from infected or protected cucumber and tobacco tissues has been only partially successful (Garas & Kuć, 1984). The inducing substance may be irreversibly bound, modified, or rapidly degraded. This is not incompatible with the concept that the signal is ionic. Sustained stimuli, such as injury or disease, cause systemic electrical disturbances in plants, including cucumber (Van Sambeck & Pickard, 1976). Davies & Schuster (1981*a,b*) suggested that changes in membrane potentials and ion fluxes may play a role in the rapid formation of polyribosomes in distal pea tissues in response to wounding.

Latent period

Following the inducing inoculation of cucumber or tobacco there is a lag period before systemic protection is detectable. Systemic protection of cucumber was first observed about 72 h after initial infection with *C. lagenarium* (Kuć & Richmond, 1977; Dean & Kuć, 1986a,b) or *P. lachrymans* (Caruso & Kuć, 1979). The lag was about 48 h with TNV (Jenns & Kuć, 1980). In all cases, the development of protection coincided with the appearance of symptoms. Protection increased to a maximum over the next 72 h. There was a direct correlation between the extent of symptoms on the inducer leaf and the maximum level of protection. While a single *C. lagenarium* lesion induced some protection, the response was saturated by approximately ten lesions.

In tobacco there is also a lag period. Protection against *P. tabacina* was not evident until about 9 days after stem injection with the fungus and gradually increased until about 21 days after induction (Tuzun & Kuć, 1985b).

Time of signal production

The latent period is probably due to lack of signal production rather than lack of its movement. Systemic protection was graft-transmissible in both cucumbers and tobacco (Jenns & Kuć, 1979; Tuzun & Kuć, 1985b). When scions from uninfected cucumber plants were grafted onto infected protected rootstocks, protection was conferred shortly after the formation of the graft union, reaching a maximum in 48-96 h. When the inoculated inducer leaf was removed from cucumber plants before symptoms appeared, plants were not systemically protected regardless of when challenged (Dean & Kuć, 1986a).

The signal produced in cucumbers is not genus, species, or cultivar specific. Infected protected cucumber rootstocks conferred protection to cucumber, watermelon, and muskmelon scions. Protection could not be transmitted from a protected rootstock lacking an infected leaf but still bearing a protected leaf (Dean & Kuć, 1986b). The inoculated leaf is, therefore, the only source of signal and the signal cannot be produced or remobilised from systemically protected tissues.

In both cucumber and tobacco systems, protection eventually reaches a maximum level. Graft transmission of protection still occurred in cucumber after the rootstock had become maximally protected (Dean & Kuć, 1986a,b). It appears unlikely, therefore, that protection reaches a maximum because the inducer leaf stops producing a signal and more likely that the number of signal receptor sites is finite and/or the plant is maximally 'sensitised'.

Signal movement

The signal appears to move out of the inducer tissue to other parts of the plant where it conditions protection; however, it is possible, though unlikely, that signal activity involves movement of disease-stimulating factors away from uninoculated tissues into infected tissues. However, uninfected areas on infected cucumber leaves are protected and girdling the petiole of the inducer leaf or the challenge leaf of cucumber plants prior to induction inhibits the transmission of protection. Placing infected plants or leaves beside unprotected plants in a sealed chamber to detect a possible volatile substance did not induce protection (Guedes, Richmond & Kuć, 1980). Girdling the outer phloem of tobacco stems above the site of *P. tabacina* stem injection also prevented the transmission of protection to the foliage (Tuzun & Kuć, 1985*b*). These results suggest that the signal is not volatile and does not move outside the plant. Although it was reported that gassing plants with ethylene induced protection of muskmelon to *C. lagenarium* (Esquerré-Tugayé *et al.,* 1979), similar results were not obtained with ethylene, ethrel, kinetin or zeatin in the cucumber *C. lagenarium* system (Jenns, 1979). Signal activity moves acropetally and basipetally in the cucumber plant, but protection is greatest above the inducer leaf.

It is apparent that the movement of the signal and its processing in the receiving tissues are not the limiting factors in the development of protection. When the inducer leaf was removed from a cucumber plant at any time after the leaf was inoculated, the time of challenge (either immediately after excision or 5–6 d later) did not affect the extent of systemic protection. Protection required an approximately 3-d lag period between induction and challenge regardless of whether the plants were challenged immediately after the inducer leaf was excised or 5–6 days later (Dean & Kuć, 1986*a*). The signal was not identified but it seems that the limiting factor in the development of protection was the rate of signal production and not its amount.

Mechanisms of induced resistance

The non-specificity of induced resistance suggests that several defence mechanisms may be involved. These include lignification and the production of stress proteins, antiviral factors, ethylene, plant hydrolases, hydroxyproline-rich glycoproteins and phytoalexins.

Stress proteins

A general response of plants to stress is the accumulation of low molecular weight proteins. A group of proteins, the b-proteins or pathogenesis-related (PR) proteins, have been detected in leaves of tobacco responding hypersensitively to TMV (Van Loon & Van Kammen, 1970). These proteins, of which there are predominantly four with molecular weights ranging from 10 000 to 64 000 daltons, are also induced by polyanions and acetylsalicylic acid or its derivatives (Gianinazzi & Kassanis, 1974; Gianinazzi *et al.*, 1977; Rohloff & Leach, 1977; Antoniw & White, 1980; Antoniw *et al.*, 1980). Similar proteins have been detected in beans, tomatoes and cucumbers (Tas & Peters, 1977; Coutts, 1978; Henriquez & Sanger, 1981*a,b*), and accumulate in response to viral, bacterial and fungal infection (Ahl *et al.*, 1980, 1981; Andebrahan *et al.*, 1980; Gianinazzi *et al.*, 1981; Gessler & Kuć, 1982*b*). Their appearance is associated with induced resistance, especially in tobacco (Gianinazzi, 1984). However, there are several reports to the contrary. Fraser (1981) detected the same proteins in leaves of uninfected but flowering and senescing tobacco, but found no correlation between their concentration and TMV resistance (Fraser, 1982). Moreover, the presence of b-proteins, though at low levels, did not prevent the delocalisation and systemic spread of TMV in tobacco plants transferred from 20°C to 30°C (Van Loon, 1975).

The presence of the N gene for hypersensitivity to TMV is not essential for the production of these proteins in *Nicotiana* species. The tobacco variety, Judy's Pride, which lacks the N gene, produces b-proteins in response to local lesions formed by the fungus *Thielaviopsis basicola* (Gianinazzi, 1982). Though a direct role as part of a resistance mechanism has not been reported for the b-proteins, they may have a regulatory role. One of the roles for stress-induced proteins may be to act as phytointerferons or antiviral factors (Chessin, 1983; Pierpoint, 1983).

In animals, interferons are acid-stable proteins of approximately 20000 daltons which may or may not be glycosylated or serologically related. They are usually species-specific and are restricted to the cells which produce them. Cells produce them in response to viral or certain bacterial infections (Lengyel, 1982).

A number of laboratories have recently identified proteinaceous molecules from plants which induce resistance and show many properties of phytointerferons. Sela and colleagues extracted an antiviral factor (AVF) from potato virus Y (PVY)- and TMV-infected plants which when mixed with viral inoculum reduced symptoms in a number of hosts. AVF was also extracted from uninfected systemic leaves of plants infected with TMV (Sela & Applebaum, 1962). An AVF from local lesion tobacco cultivars infected with TMV has been partially characterised and suggested to be a phosphoglycoprotein with a molecular weight of approximately 22 000 daltons. This AVF was only produced in infected tobacco cultivars with the N gene for resistance to TMV and was not produced by mechanical wounding. Its synthesis was inhibited by actinomycin-D, but its effect *in vivo* of inhibiting virus multiplication, as determined by infectivity assay and ELISA, was not inhibited by the antibiotic (Antignus, Sela & Harpaz, 1977; Mozes *et al.*, 1978; Sela, 1981).

Sela postulated that the N gene determines an enzyme which processes an inactive AVF to produce an active AVF. The increased enzyme activity is coupled with cyclic nucleotide-dependent protein phosphorylation (Sela, 1981). Sela has reported the release of c-AMP in TMV-infected *N. glutinosa* (Rosenberg, Pines & Sela, 1982). However, only polyinosinic-polycytidylic acid (poly I:C) was an absolute requirement to induce the antiviral state in tobacco leaves and callus (Gat-Edelbaum, Altman & Sela, 1983). More recently, human interferon (Orchansky, Rubinstein & Sela, 1982) and synthetic 2′,5′-linked oligoadenylates (2-5A) (Devash *et al.*, 1984) have been shown to be potent inhibitors of TMV multiplication, strengthening the case for a phytointerferon. A nucleotide fraction from TMV-infected leaves has been reported to have antiviral activity, but the fraction from AVF or interferon-treated leaves did not. An enzyme preparation from tobacco callus treated with AVF or interferon and poly I : C, in the presence of ATP, synthesised a product which inhibited virus multiplication (Reichman *et al.*, 1983). The authors concluded that AVF and interferon induced double-stranded RNA-dependent synthesis of oligoadenylates from ATP in plants, producing nucleotides with antiviral activity. Interestingly, poly I : C in the absence of AVF or interferon is a very effective inducer of 'synthetase' activity. It has been suggested from these data that these inducers produce a 'sensitised state' which requires double-stranded RNA (viral or synthetic) to turn on the antiviral machinery. Poly I : C has been shown to induce resistance to virus infections in plants (Stein & Loebenstein, 1970). Nevertheless, 2-5A or its binding protein in radiobinding assays have not been detected (Cayley *et al.*, 1982). Rosenberg *et al.* (1985) suggested that the plant oligoadenylates do not compete with 2-5A in such assays. It has been reported that the $2',5'$ cordycepin trimer analogue is a more effective inhibitor of TMV replication than 2-5A in tobacco tissues and protoplasts. These plant oligoadenylates inhibited protein synthesis *in vitro*. However, the mechanism appeared to be different from the 2-5A-dependent endonuclease activation in animals (Devash *et al.*, 1985). Cayley *et al.* (1982) did not detect a double-stranded RNA-dependent protein kinase, another essential component of interferon action in animals, in tobacco treated with poly I:C, interferon and TMV or interferon followed by TMV. The b-proteins were not induced by interferon or poly I:C.

Loebenstein & Gera (1981) isolated a factor (IVR) from TMV- infected protoplasts prepared from tobacco, cv. Samsun NN, which inhibited viral replication. IVR, isolated from the liquid media and from intact infected protoplasts, was destroyed by heating to 60°C for 10 min or treatment with trypsin and chymotrypsin. It was separated into two acid-stable components of 26 000 and 57 000 daltons. Uninfected Samsun NN protoplasts, or infected protoplasts prepared from the systemically infectable host Samsun, did not produce IVR. However, IVR inhibited TMV multiplication in protoplasts from both cultivars even when applied 18 h after inoculation. Virus replication in leaf discs and in intact leaves was also inhibited by IVR (Gera & Loebenstein, 1983). In addition, IVR inhibited replication of TMV, cucumber mosaic virus (CMV) and potato virus X (PVX) in leaf discs from several hosts, which suggested IVR was neither virus- nor host-specific. IVR did not appear to affect TMV directly, since mixing IVR with purified TMV for 1 h did not affect the infectivity of the virus recovered by ultracentrifugation.

It is plausible that IVR and AVF are identical. Both are isolated by similar methods and have similar physical and biological properties. An antiserum specific to IVR, which neutralised IVR's biological activity, reacted weakly with AVF, but not with α or δ interferon (Loebenstein & Stein, 1985). The research with IVR and AVF is potentially extremely important, but it awaits verification from independent laboratories. It has recently been demonstrated that recombinant interferon does not inhibit replication of alfalfa mosaic virus in cowpea (Huisman *et al.*, 1985), tobacco and alfalfa protoplasts, or in tobacco leaf discs (Loesh-Fries *et al.*, 1985).

Extracts from uninfected tissues from many plant species have been reported to induce systemic resistance. A non-phytotoxic component from the roots of *Boerhaavia diffusa* had a broad range of antiviral activity, including activity against TMV in *N. glutinosa*. Lesion numbers produced by four viruses were reduced in leaves distant from those treated with extract 24 h after treatment. The reduction in lesion number was reversed by actinomycin-D or exposure of treated infected plants to temperatures of 35°C or above (Verma & Awasthi, 1979). It was suggested that the active material, possibly a glycoprotein of 16 000–20 000 daltons (Verman, Awasthi & Saxena, 1979), caused the synthesis of a proteinaceous antiviral agent (AVA) (Verma & Awasthi, 1980). AVA activity has been extracted from a number of plant species treated with extracts from leaves of *Datura metel* and roots of *B. diffusa* (Verma & Awasthi, 1980; Mukerjee, Awasthi & Verma, 1981). The mechanism of action of AVA is unreported.

Stress ethylene

Little is known about the significance of stress ethylene. Some information has appeared on its possible regulation of inducible defence mechanisms in plants (Carlton, Peterson & Tolbert, 1961; Chalutz & Stahmann, 1969; Pegg, 1976). Prolonged exposure of melon plants to ethylene was reported to enhance resistance against *C. lagenarium* (Esquerré-Tugayé *et al.*, 1979). However these results could not be reproduced with the cucumber-*C. lagenarium* system (Jenns, 1979).

Plant hydrolases

Synthesis of the plant hydrolases, chitinase and β-1,3-glucanase, is strongly induced by ethylene in several plant species (Abeles, Forrence & Habig, 1971; Boller, 1985b). These enzymes, proteins of approximately 30 000 daltons, are capable of degrading isolated cell walls of some fungal pathogens. Although evidence of their role in disease resistance is limited (Wargo, 1975; Young & Pegg, 1982), host glycan hydrolases have been proposed as an *in vivo* disease resistance mechanism (Abeles *et al.*, 1970; Pegg & Vessey, 1973; Pegg, 1977).

Infection of young cucumber plants with a number of pathogens caused systemic induction of chitinase (Boller, 1985a; Metraux & Boller, 1985). Toppan & Rody (1982) reported that ethylene treatment or infection of melon seedlings with *C. lagenarium* induced chitinase. Rabenantoandro, Auriol & Touzé (1976) had previously concluded that enhanced β-1,3-glucanase activity in infected melons was produced by the pathogen. Recently, using immature pea-pods, Mauch, Hadwiger & Boller (1984) reported that ethylene, fungal infection, and phytoalexin elicitors were independent stimuli of chitinase and β-1,3-glucanase. The two enzymes were induced even when endogenous ethylene was inhibited by amino-ethoxyvinyl-glycine (AVG). Thus, ethylene production appears more the result of biotically-induced systemic resistance than its cause.

The correlation between induced enzyme activity and resistance is unclear. Netzer, Kritzmann & Chet (1979) found that β-1,3-glucanase was increased more in a resistant muskmelon variety infected with *Fusarium oxysporum* than in a susceptible variety. Pegg & Young (1981) did not find this relationship in tomato infected with *Verticillium albo-atrum*. In pea-pods, there was no difference in the time-course or the level of induction of chitinase and β-1,3-glucanase following infection with a pathogenic or nonpathogenic strain of *Fusarium solani* (Mauch *et al.*, 1984).

Hydroxyproline-rich glycoprotein (HPRG)

Heat shock has been reported to induce localised resistance in cucumber seedlings to *C. cucumerinum* (Stermer & Hammerschmidt, 1984). Plants challenged 12 h after treatment showed a 50% reduction in disease after four days which persisted for at least two days. Ethylene and 1-aminocyclopropane-1-carboxylic acid (ACC) production doubled by 12 h but peroxidase activity did not change until 12 h later. Hydroxyproline, a major component of the cell wall glycoprotein extensin, showed some increase at this time, reaching a maximum 50% above controls at 72 h. It was suggested that heat shock increased ethylene production which in turn regulated peroxidase activity. Moreover, the crosslinking of extensin to phenolic residues might result from peroxidase activity and strengthen plant cell walls against degradation by pathogens (Stermer & Hammerschmidt, 1985).

The evidence for a role of HPRG in disease resistance is growing. The mRNAs for HPRG have been reported to accumulate earlier in green beans infected with cultivar nonpathogenic races of *C. lindemuthianum* than with pathogenic races. In the compatible interaction, the mRNAs were delayed until lesion formation. Interestingly the mRNAs were observed to accumulate some distance from the site of infection (Showalter *et al.*, 1985).

Esquerré-Tugayé *et al.* (1979) reported a close correlation between the amount of HPRG in cell walls of muskmelon and resistance to *C. lagenarium*. Endogenous ethylene production was reported to parallel HPRG accumulation in cell walls following infection (Toppan, Rody & Esquerré-Tugayé, 1982). Exogenous ethylene enriched the HPRG of cell walls from stems and decreased fungal growth as measured by the glucosamine content of infected tissues, as reported by Pegg (1976) for *Verticillium* in tomato. Conversely, free hydroxyproline supplied to the plants reduced the HPRG content of the cell walls and increased the growth of the fungus. Ethylene inhibitors, L-canaline and AVG lowered levels of ^{14}C-hydroxyproline deposition in cell walls of infected tissues. Conversely, treatment of healthy tissues with ACC stimulated ethylene production and ^{14}C-hydroxyproline deposition. However, the accumulation of HPRG in cell walls of infected tissues mediated by ethylene has been questioned (Mauch *et al.*, 1984). AVG significantly inhibited ethylene production in infected tissues, but only reduced HPRG content by approximately 25%. Exogenous ethylene at 500 ppm only increased HPRG in cell walls by a factor of two in seven days, compared to a 10-fold stimulation in the amount of infections.

Compounds, proposed to be phosphoglycopeptides, have been isolated from *C. lagenarium* which induce the production of ethylene (Toppan & Esquerré-Tugayé, 1984) and HPRG (Esquerré-Tugayé *et al.*, 1985) in muskmelon hypocotyls. Materials released from muskmelon cell walls by autoclaving have also been reported to be effective elicitors (Rody, Toppan & Esquerré-Togayé, 1985). The elicitor fraction from *C. lagenarium* also has been reported to stimulate chitinase activity systemically in muskmelon leaves (Esquerré-Tugayé *et al.*, 1985). Another role for HPRG, although not associated with ethylene, is as bacterial agglutins or lectins. Plant lectins have been proposed to function in binding bacteria to plant surfaces and may have some function in induced resistance of tobacco against pathogenic pseudomonads (Goodman, 1980; Sequeira, 1983; Pueppke, 1984).

Phytoalexins and inhibitins

Phytoalexins, low molecular weight antimicrobial substances, accumulate around sites of infection in many plant families (see Chapter 9).

The isoflavonoid phytoalexin, phaseollin, has been shown to accumulate rapidly in green bean tissues infected with cultivar-nonpathogenic races of *C. lindemuthianum*. In the compatible interaction, accumulation was delayed and, although phaseollin eventually reached high levels, it was deemed to be too late to restrict the development of the fungus (Bailey & Deverall, 1971; Rahe, 1973*a,b*). Phytoalexins were not detected systemically in hypocotyls of green beans following localised infection with *C. lagenarium* or cultivar-nonpathogenic races of *C. lindemuthianum*. However, on challenge with pathogenic races of the fungus, phaseollin accumulated rapidly in systemically protected tissues, containing the infection (Elliston *et al.*, 1977). The rapid accumulation of phytoalexins in induced, systematically protected tissues after infection strengthens the case for their involvement as part of a non-specific defence mechanism. We have recently demonstrated (Reuveni & Kuć, unpublished data), that dipping Ky 14 tobacco leaves in acetone for one second increased their susceptibility to *P. tabacina.* Cruickshank, Perrin & Mandryk (1977) identified fungitoxic leaf surface compounds as α and β isomers of 4,8,13-duvatriene-1,3-diols (DVT), but did not establish a role for the compounds in disease resistance. The quantity of these substances extracted from systemically protected leaves from plants stem-injected with *P. tabacina* by a one second dip in acetone, was approximately twice as great as that from leaves of control plants. Dipped protected leaves, however, were still more resistant to blue mould than dipped control leaves. There was little correlation between the extracted quantities of DVT and varietal resistance (Tuzun & Kuć, unpublished data). It would seem unlikely, therefore, that DVT represents the only defence mechanism of induced resistance in tobacco against blue mould.

Quiesone (3-isobutyroxy-β-ionone) has been extracted from tobacco plants infected with *P. tabacina* (Leppik, Holloman & Bottomley, 1972). It has been shown to be a very potent inhibitor of spore germination in this fungus ($ED_{50} = 0.0001$ ppm). However, penetration of *P. tabacina* into immunised leaves does not appear to be markedly reduced (Cohen & Kuć, 1981; Tuzun & Kuć, 1985b).

Lignification

Strong evidence exists that lignification is an effective mechanism of disease resistance (Vance, Kirk & Sherwood, 1980; Aist, 1983; Ride, 1983). Lignification may form a structural barrier as well as a chemical barrier (lignin precursors and their free radicals are toxic). Rapid localised lignification appears to be primarily responsible for inhibiting pathogen growth in immunised tissues, at least for fungal pathogens.

The penetration of *C. lagenarium* from appressoria into the epidermis of protected leaves was reduced by approximately 90% (Richmond, Kúc & Elliston, 1979; Hammerschmidt & Kúc, 1982). Spore germination and appressorial formation were unaffected. The direct penetration of *C. cucumerinum* was unaffected, but early hyphal development was restricted. In the case of *C. cucumerinum,* penetration occurred within 18 h, whereas with *C. lagenarium,* penetration from appressoria required 36–60 h (Hammerschmidt & Kúc, 1982). If the epidermis was bypassed (by infiltrating the spores of *C. lagenarium* into the leaf or by wounding), protection was diminished and if the epidermis was completely removed, protection of some cultivars was barely observable (Richmond *et al.,* 1979;

Jenns & Kúc, 1980). Classical phytoalexins have not been detected in infected cucumber leaves (Deverall, 1977). Histological and histochemical studies indicated that the invaded epidermal cells and a few adjacent cells of challenged protected tissues were rapidly lignified; in control tissues the response was weak and diffuse (Hammerschmidt & Kúc, 1982). Coniferyl alcohol has been isolated from challenged leaves and was highly fungitoxic *in vitro* and *in vivo* to both *C. lagenarium* and *C. cucumerinum* (0.1–0.3 μg) (Hammerschmidt & Kúc (1982). The invading organism may also have been directly lignified. Hammerschmidt & Kúc (1982) reported that fungal mycelia of *C. lagenarium* and *C. cucumerinum* were lignified in the presence of coniferyl alcohol, hydrogen peroxide and peroxidase prepared from immunised leaves.

^{14}C-*trans*-cinnamic acid and ^{14}C-phenylalanine were more rapidly incorporated into the lignin of cell walls in immunised leaves as compared to control leaves before challenge. After challenge with *C. lagenarium* or wounding, the rate of label incorporation increased further (Dean & Kúc, 1985, and unpublished data). Touzé & Rossingnol (1977) also reported increased hypocotyl growth following infection with *C. lagenarium*. The lignin content, determined as thioglycolignin, in unchallenged immunised cucumber cell walls also appeared to be greater than in control cell walls (Dean & Kúc, 1985, and unpublished data). It is possible that metabolism in general is enhanced through immunisation, thus allowing the plant to respond more rapidly to stress or infection.

Peroxidase

A necessary step in the polymerisation of lignin is the generation of free radicals from cinnamic alcohols catalysed by certain (probably cell wall associated) isoenzymes of peroxidase or phenoloxidase. A threefold increase in systemic peroxidase activity in cucumber (marked by enhanced activity of several anionic isoenzymes) was evident approximately four days after the inducer inoculation with *C. lagenarium*, roughly coinciding with the first signs of disease and systemic protection. Peroxidase activity was apparently of host origin, since induction with a number of pathogens, including TNV, caused similar increases. Increasing the number of lesions on the inducer leaf stimulated increases in both peroxidase activity and systemic protection. Physical injury resulted in only a localised increase in peroxidase activity. The activity of other enzymes in the phenylpropanoid pathway (including phenylalanine ammonia lyase, *p*-coumaryl CoA ligase and coumaryl reductase) appeared to be unchanged in these protected unchallenged tissues (Hammerschmidt *et al.*, 1982). Increases in peroxidase have not been detected in immunised tobacco leaves (Nadolny & Sequira, 1980; Stein & Loebenstein, 1976).

Enhanced recognition and/or expression of resistance mechanisms

The containment of diverse fungal, bacterial and viral pathogens, as well as other organisms, by immunisation may require the participation of a particular combination of multicomponent resistance mechanisms. Immunistion may be effective because of the induction of new mechanisms for resistance and, or, the amplification of a pre-existing system.

The induction of antiviral factors may be an example of a new defence mechanism. Conversely, the more rapid and intense lignification response may be seen as amplification of a pre-existing mechanism. Enhanced lignification may be the major mechanism responsible for the observed reduction in penetration of *C. lagenarium*. Immunisation in cucumbers appears to add an increment of resistance which is independent of the underlying cultivar resistance. In two cultivars, approximately a 1000-fold increase in inoculum concentration of *C. lagenarium* was required to produce a similar level of disease in systemically protected leaves compared with control leaves (Dean & Kúc, 1986c).

Not all defence mechanisms may be sensitised by immunisation. In some cucumbers, cultivar resistance to *C. lagenarium* is associated with rapid host cell death following penetration. The frequency of necrotisation appeared to be relatively unchanged in systemically protected leaves (Richmond *et al.*, 1979). In the green bean-*C. lindemuthianum* interaction, however, immunisation resulted in rapid host cell death in response to inoculation with cultivar-pathogenic races of the fungus (Elliston *et al.*, 1971).

Although many of the mechanisms associated with induced resistance may be non-specific and function in many plant species, others may be specifically dependent on the host-pathogen combination. For example, once *C. lagenarium* was established in a protected cucumber leaf, the lesion, although smaller, continued to expand at the same rate as a lesion in a control leaf (Dean & Kuć, 1986c). In contrast, it appeared that there was little or no reduction of penetration by *P. tabacina*, and lesions in immunised tobacco remained small and confined (Cohen & Kuć, 1981; Tuzun & Kuć, 1985a,b).

Induced resistance may be not only determined by the enhanced rate of expression of disease-resistance mechanisms following infection or physiological stress. Enhanced recognition by the plant may also be important. The cucumber cultivar SMR-58 is susceptible to *C. lagenarium* but resistant to *C. cucumerinum*. It may be that *C. cucumerinum* is sensitive to a mechanism to which *C. lagenarium* is resistant, but in SMR-58 both induced resistance against *C. lagenarium* and cultivar resistance to *C. cucumerinum* were associated with rapid lignification of epidermal cells (Hijwegen, 1963; Hammerschmidt & Kuć, 1982; Hammerschmidt, Lamport & Muldoon, 1984). The difference, therefore, between susceptibility and resistance is probably not due to the lack of genetic information for effective defence mechanisms, but rather to the ability and rapidity of the host's recognition of the pathogen and/or expression of the defence response.

Conclusion

Immunisation may provide a further dimension for the control of plant disease. It has been demonstrated that many plant species can be effectively protected against disease caused by a variety of pathogens by both abiotic and biotic agents. Immunisation can be systemic and persist in the field throughout the growing season. In contrast to race-specific varietal resistance, immunisation is non-specific and may be highly durable. It may contribute to the survival of the species under pathogen pressure.

References

Abeles, F.B., Bosshart, R.P., Forrence, L.E. & Habig, W.H. (1970). Preparation and purification of glucanase and chitinase from bean leaves. *Plant Physiology*, **47**, 129–34.

Ahl, P., Benjama, A., Samson, R. & Gianinazzi, S. (1980). New host proteins induced by bacterial infection together with resistance to secondary infection in tobacco. *Phytoalexins et phenomênes d'elicitation 18 eme colloque societe française de phytopathologie*. Abstract No. 22.

Ahl, P., Benjama, A., Samson, R. & Gianinazzi, S. (1981). Induction chez le tabac par *Pseudomonas syringae* de nouvelles proteines (proteines 'b') associees au developpement d'une resistante non specifique à une deuxieme infection. *Phytopathologische Zeitschrift*, **102**, 201–12.

Aist, J.R. (1983). Structural responses as resistance mechanisms. In: *The Dynamics of Host Defence*. ed. J.A. Bailey & B.J. Deverall, p. 33-70. Sydney: Academic Press.

Albersheim, P., Darvill, A.G., McNeil, M., Valent, B., Sharp, J.K., Nothnagel, E.A., Davis, K.R., Yamasaki, N., Gollin, D., York, W., Dudman, W.P., Darvill, J. & Dell, A. (1983). Oligosaccharins: naturally-occurring carbohydrates with biological regulatory functions. In: *Structure and Function of Plant Genomes*, ed. O. Ciferri & L. Dure, p. 293. NATO Advanced Science Institute Series, vol. 63. New York: Plenum.

Andebrahan, T., Coutts R.H.A., Wagih, E.E. & Wood, R.K.S. (1980). Induced resistance and changes in the soluble protein fraction of cucumber leaves infected with *Colletotrichum lagenarium* or tobacco necrosis virus. *Phytopathologische Zeitschrift*, **98**, 47–52.

Antignus, Y., Sela, I. & Harpaz, I. (1977). Further studies on the biology of an antiviral factor (AVF) from virus-infected plants and its association with the N-gene of *Nicotiana* species. *Journal of General Virology*, **35**, 107–16.

Antoniw, J.F., Ritter, C.E., Pierpoint, W.S. & Van Loon, L.C. (1980). Comparison of three pathogen-related proteins from plants of two cultivars of tobacco-infected with TMV. *Journal of General Virology*, **47**, 79–87.

Antoniw, J.F. & White, R.F. (1980). The effect of aspirin and polyacrylic acid on soluble proteins and resistance to virus infection in five cultivars of tobacco. *Phytopathologische Zeitschrift*, **98**, 333–41.

Bailey, J.A. & Deverall, B.J. (1971). Formation and activity of phaseollin in the interaction between bean hypocotyls (*Phaseolus vulgaris*) and physiological races of *Colletotrichum lindemuthianum*. *Physiological Plant Pathology*, **1**, 435–49.

Balazs, E., Sziraki, I. & Kiraly, Z. (1977). The role of cytokinins in the systemic acquired resistance of tobacco hypersensitive to tobacco mosaic virus. *Physiological Plant Pathology*, **11**, 29–37.

Basham, B. & Cohen, Y. (1983). Tobacco necrosis virus induces systemic resistance in cucumbers against *Sphaerotheca fuliginea*. *Physiological Plant Pathology*, **23**, 137–44.

Bergstrom, G.C., Johnson, M.C. & Kuć, J. (1982). Effects of local infection of cucumber by *Colletotrichum lagenarium*, *Pseudomonas lachrymans* or tobacco necrosis virus on systemic resistance to cucumber mosaic virus. *Phytopathology*, **72**, 922–6.

Boller, T. (1985a). Local and systemic induction of hydrolases which can attack fungal cell walls: a reaction of plants to pathogen stress. *1st International Congress of Plant Molecular Biology*, p. 69. Savannah: University of Georgia.

Boller, T. (1985b). Induction of hydrolases as a defence reaction against pathogens. In: *Cellular and Molecular Biology of Plant Stress*, ed. J.L. Key & T. Kosuge, pp. 247–62. U.C.L.A. symposia on Molecular and Cellular Biology, New Series, Vol. 22. New York: Alan R. Liss Inc.

Bompeix, G., Ravise, A., Raynal, G., Fettouche, F. & Durrand, M.C. (1980). Modalities de l'obtention des necroses bloquantes sur feuilles detachees de Tomate par l'action du tris-O-ethyl phosphonate d'aluminium (phosethyl d'aluminium), hypotheses sur son mode d'action *in vivo*. *Annales Phytopathologie*, **12**, 337–51.

Bruce, R. & West, C. (1982). Elicitation of casbene synthetase activity in castor bean, the role of pectic fragments of the plant cell wall in elicitation by fungal endopolygalacturonase. *Plant Physiology*, **69**, 1181–8.

Carlton, B.C., Peterson, C.E. & Tolbert, N.E. (1961). Effects of ethylene and oxygen on production of a bitter compound by carrot roots. *Plant Physiology*, **36**, 550.

Cartwright, D.W., Langcake, P. & Ride, J.P. (1980). Phytoalexin production in rice and its enhancement by a dichlorocyclopropane fungicide. *Physiological Plant Pathology*, **17**, 259–67.

Caruso, F.L. & Kuć, J. (1977a). Protection of watermelon and muskmelon against *Colletotrichum lagenarium* by *Colletotrichum lagenarium*. *Phytopathology*, **67**, 1285–9.

Caruso, F.L. & Kuć, J. (1977b). Field protection of cucumber against *Colletotrichum lagenarium* by *C. lagenarium*. *Phytopathology*, **67**, 1290–2.

Caruso, F.L. & Kuć, J. (1979). Induced resistance of cucumber to anthracnose and angular leaf spot by *Pseudomonas lachrymans* and *Colletotrichum lagenarium*. *Physiological Plant Pathology*, **14**, 191–201.

Cayley, P.J., White, R.F., Antoniw, J.F., Walesby, N.J. & Kerr, I.M. (1982). Distribution of the ppp(A2p')A-binding protein and interferon-related enzymes in animals, plants and lower organisms. *Biochemical & Biophysical Research Communications*, **108**, 1243–50.

Chalutz, E. & Stahmann, M.A. (1969). Induction of pisatin by ethylene. *Phytopathology*, **59**, 1972–3.

Chessin, M. (1983). Is there a plant interferon? *The Botanical Gazette*, **49**, 1–28.

Chester, K.S. (1933). The problem of acquired physiological immunity in plants. *Quarterly Review of Biology*, **8**, 129–54, 275–324.

Clevelanc, T. & Black, L. (1983). Partial purification of proteinase inhibitors from tomato plants infected with *Phytophthora infestans*. *Phytopathology*, **73**, 664–70.

Cohen, Y. & Kuć, J. (1981). Evaluation of systemic resistance to blue mould induced in tobacco leaves by prior stem inoculation with *Peronospora hyoscyami tabacina*. *Phytopathology*, **71**, 783–7.

Coutts, R.H.A. (1978). Alteration in the soluble protein patterns of tobacco and cowpea leaves following inoculation with tobacco necrosis virus. *Plant Science Letters*, **12**, 189–97.

Cruickshank, I.A.M. & Mandryk, M. (1960). The effect of stem infestation of tobacco with *Peronospora tabacina* Adam on foliage reaction to blue mould. *Journal of the Australian Institute of Agricultural Science*, **26**, 369–72.

Cruickshank, I.A.M., Perrin, D.R. & Mandryk, M. (1977). Fungitoxicity of duvatrienediols associated with the cuticular wax of tobacco leaves. *Phytopathologische Zeitschrift*, **90**, 133–46.

Davies, E. & Schuster, A. (1981*a*). Intercellular communication in plants: evidence for a rapidly generated, bidirectionally transmitted wound signal. *Proceedings of the National Academy of Science, U S A*, **78**, 2422–6.

Davies, E. & Schuster, A. (1981*b*). Polyribosome formation in response to wounding. *Plant Physiology*, **65** (Supplement), 33.

Davies, K.R., Nothnagel, E.A., McNeil, M., Darvell, A.G. & Albersheim, P. (1983). Endopolygalacturonic acid lyase isolated from *Erwinia carotovora* elicits phytoalexins in soybean by releasing an oligosaccharide elicitor from the plant cell walls. *Plant Physiology*, **72** (Supplement), 161.

Dean, R.A. & Kuć, J. (1985). Immunization in cucumber: is the immunized plant sensitized to lignify rapidly after wounding? *Newsletter of the Phytochemical Society of North America*, **25**(2). 25th Annual meeting.

Dean, R.A. & Kuć, J. (1986*a*). Induced systemic protection in cucumber: time of production and movement of the 'signal'. *Phytopathology*, **76**, 966–70.

Dean, R.A. & Kuć, J. (1986*b*). Induced systemic protection in cucumbers: the source of the signal. *Physiological Plant Pathology*, **28**, 227–33.

Dean, R.A. & Kuć, J. (1986*c*). Induced systemic protection in cucumber: effects of inoculum density on symptom development caused by *Colletotrichum lagenarium* in previously infected and uninfected plants. *Phytopathology*, **76**, 186–9.

Devash, Y., Gera, A., Willis, D.H., Reichman, M., Pfleiderer, W., Charabala, R., Sela, I. & Suhadolnik, R.J. (1984). 5'-Dephosphorylated 2',5'-adenylate trimer and its analogs. Inhibition of tobacco mosaic virus replication in TMV-infected leaf discs, protoplasts and intact tobacco plants. *Journal of Biological Chemistry*, **259**, 3482–6.

Devash, Y., Reichman, M., Sela, I., Reichenbach, N.L. & Suhadolnik, R.J. (1985). Plant oligoadenylates: enzymatic synthesis, isolation and biological activities. *Biochem-*

istry, **24**(3), 593–9.

Deverall, B.J. (1977). In: *Defence Mechanisms of Plants*, Cambridge: Cambridge University Press.

Doss, M. & Hevesi, M. (1981). Systemic acquired resistance of cucumber to *Pseudomonas lachrymans* as expressed in suppression of symptoms, but not in multiplication of bacteria. *Acta Phytopathologia Academiae Scientiarum Hungaricae*, **16**(3–4), 269–72.

Elliston, J., Kuć, J. & Williams, E. (1971). Induced resistance to anthracnose at a distance from the site of the inducing interaction. *Phytopathology*, **61**, 1110–12.

Elliston, J., Kuć, J. & Williams, E. (1976a). Protection of bean against anthracnose by *Colletotrichum* species nonpathogenic on bean. *Phytopathologische Zeitschrift*, **86**, 117–26.

Elliston, J., Kuć, J. & Williams, E. (1976b). A comparative study on the development of compatible, incompatible and induced incompatible interactions between *Colletotrichum* species and *Phaseolus vulgaris*. *Phytopathologische Zeitschrift*, **87**, 289–303.

Elliston, J., Kuć, J., Williams, E. & Rahe, J. (1977). Relation of phytoalexin accumulation to local and systemic protection of bean against anthracnose. *Phytopathologische Zeitschrift*, **88**, 114–30.

Esquerré-Tugayé, M-T., Lafitte, C., Rody, D., Toppan, A. & Touze, A. (1979). Cell surfaces in plant-microorganism interactions. II. Evidence for the accumulation of hydroxyproline-rich glycoproteins in the cell wall of diseased plants as a defence mechanism. *Plant Physiology*, **64**, 320–6.

Esquerré-Tugayé, M-T., Mazau, D., Pelissier, B., Rody, D., Rumeau, D. & Toppan, A. (1985). Induction by elicitors and ethylene of proteins associated to the defence of plants. In: *Cellular and Molecular Biology of Plant Stress*, ed. J.L. Key, & T. Kosuge, pp. 459–73. U.C.L.A. Symposia on Molecular and Cellular Biology, New Series, Vol. 22. New York: Alan R. Liss, Inc.

Fenn, M.E. & Coffey, M.D. (1984). Studies on the *in vitro* and *in vivo* antifungal activity of fosetyl-Al and phosphorous acid. *Phytopathology*, **74**, 606–11.

Fenn, M.E. & Coffey, M.D. (1985). Further evidence for the direct mode of action of fosetyl-Al and phosphorous acid. *Phytopathology*, **75**, 1064–8.

Fraser, R.S.S. (1981). Evidence for the occurrence of the 'pathogenesis-related' proteins in leaves of healthy tobacco plants during flowering. *Physiological Plant Pathology*, **19**, 69–76.

Fraser, R.S.S. (1982). Are 'pathogenesis-related' proteins involved in acquired systemic resistance of tobacco plants to tobacco mosaic virus? *Journal of General Virology*, **58**, 305–13.

Garas, N. & Kuć, J. (1984). The extraction, assay and general properties of an inducing resistance factor from cucumber leaves induced with *Pseudomonas lachrymans*. *Phytopathology*, **74**, 873.

Gat-Edelbaum, O., Altman, A. & Sela, I. (1983). Polyinosinic-polycytidylic acid in association with cyclic nucleotides activates the antiviral factor (AVF) in plant tissues. *Journal of General Virology*, **64**, 211–14.

Gera, A. & Loebenstein, G. (1983). Further studies of an inhibitor of virus replication from tobacco mosaic virus-infected protoplasts of a local lesion-responding tobacco cultivar. *Phytopathology*, **73**, 111–15.

Gessler, C. & Kuć, J. (1982a). Induction of resistance to *Fusarium* wilt in cucumber by root and foliar pathogens. *Phytopathology*, **72**, 1439–41.

Gessler, C. & Kuć, J. (1982b). Appearance of a host protein in cucumber plants infected with viruses, bacteria and fungi. *Journal of Experimental Botany*, **33**(132), 58–66.

Gianinazzi, S. (1982). Antiviral agents and inducers of virus resistance: analogies with interferon. In: *Active Defence Mechanisms in Plants*, ed. R.K.S. Wood. New York: Plenum Publishing Co.

Gianinazzi, S. (1984). Genetic and molecular aspects of resistance induced by infection or chemicals. In: *Plant-Microbe Interactions: Molecular and Genetic Perspectives*, vol. 1, ed. T. Kosuge & E.W. Nester, pp. 321–42. New York: MacMillian.

Gianinazzi, S., Ahl, P., Cornu, A., Scalla, R. & Cassni, R. (1981). First report of host b-protein appearance in response to a fungal infection in tobacco. *Physiological Plant Pathology*, **16**, 337–42.

Gianinazzi, S. & Kassanis, B. (1974). Virus resistance induced in plants by polyacrylic acid. *Journal of General Virology*, **23**, 1–9.

Gianinazzi, S., Pratt, M., Shewry, P.R. & Miflin, B.J. (1977). Partial purification and preliminary characterization of soluble leaf proteins specific to virus infected plants. *Journal of General Virology*, **34**, 345–51.

Goodman, R. (1980). Defences triggered by previous invaders: bacteria. In: *Plant Diseases*, vol. 5, ed. J. Horsfall & E. Cowling, pp. 305–17. New York: Academic Press.

Guedes, M.E.M., Richmond, S. & Kuć, J. (1980). Induced systemic resistance to anthracnose in cucumber as influenced by the location of the inducer inoculation with *Colletotrichum lagenarium* and the onset of flowering and fruiting. *Physiological Plant Pathology*, **17**, 229–33.

Guest, D.I. (1984). Modification of defence responses in tobacco and Capsicum following treatment with fosetyl-A1 [Aluminium tris (O-ethyl phosphonate)]. *Physiological Plant Pathology*, **25**, 125–34.

Hahn, M., Darvill, A. & Albersheim, P. (1981). Host-pathogen interactions, XIX: The endogenous elicitor, a fragment of a plant cell wall polysaccharide that elicits phytoalexin accumulation in soybeans. *Plant Physiology*, **68**, 1161–9.

Hammerschmidt, R., Acres, S. & Kuć, J. (1976). Protection of cucumbers against *Colletotrichum lagenarium* and *Cladosporium cucumerinum*. *Phytopathology*, **66**, 790–3.

Hammerschmidt, R. & Kuć, J. (1982). Lignification as a mechanism for induced systemic resistance in cucumber. *Physiological Plant Pathology*, **20**, 61–71.

Hammerschmidt, R., Lamport, D.T.A. & Muldoon, E.P. (1984). Cell wall hydroxyproline enhancement and lignin deposition as an early event in the resistance of cucumber to *Cladosporium cucumerinum*. *Physiological Plant Pathology*, **24**, 43–9.

Hammerschmidt, R., Nuckles, E. & Kuć, J. (1982). Association of enhanced peroxidase activity with induced systemic resistance of cucumber to *Colletotrichum lagenarium*. *Physiological Plant Pathology*, **20**, 73–82.

Henriquez, A.C. & Sanger, H.L. (1981a). Gel electrophoresis of phenol-extractable leaf proteins from different viroid/host combinations. *Archives of Virology*, **74**, 167–80.

Henriquez, A.C. & Sanger, H.L. (1981b). Analysis of acid-extractable tomato leaf proteins after infection with a viroid, two viruses and a fungus, and partial purification of the 'pathogenesis-related' protein p 14. *Archives of Virology*, **74**, 181–96.

Hijwegen, T. (1963). Lignification, a possible mechanism of active resistance against pathogens. *Netherlands Journal of Plant Pathology*, **69**, 314–17.

Holowczak, J., Kuć, J. & Williams, E.B. (1962). Metabolism of DL- and L- phenylalanine in malus related to susceptibility and resistance to *Venturia inaequalis*. *Phytopathology*, **52**, 699–703.

Huisman, M.J., Broxterman, H.J.G., Schellekens, H. & Van Vloten-Doting, L. (1985). Human interferon does not protect cowpea plant cell protoplasts against infection with alfalfa mosaic virus. *Virology*, **143**, 622–5.

Jenns, A.E. (1979). Induction of systemic resistance in cucurbits to anthracnose and tobacco necrosis virus. *Ph.D. Thesis, University of Kentucky.*

Jenns, A.E., Caruso, F.L. & Kuć, J. (1979). Non–specific resistance to pathogens induced systemically by local infection of cucumber with tobacco necrosis virus, *Colletotrichum lagenarium* or *Pseudomonas lachrymans*. *Phytopathologia Mediterranea*, **18**, 129–34.

Jenns, A.E. & Kuć, J. (1977). Localized infection with tobacco necrosis virus protects cucumber against *Colletotrichum lagenarium*. *Physiological Plant Pathology*, **11**, 207–12.

Jenns, A.E. & Kuć, J. (1979). Graft transmission of systemic resistance of cucumber to anthracnose induced by *Colletotrichum lagenarium* and tobacco necrosis virus. *Phytopathology*, **69**, 753–6.

Jenns, A.E. & Kuć, J. (1980). Characteristics of anthracnose resistance induced by localized infection of cucumber with tobacco necrosis virus. *Physiological Plant Pathology*, **17**, 81–91.

Kassanis, B. & White, R. (1975). Polyacrylic acid-induced resistance to tobacco mosaic virus in tobacco cv Xanthi. *Annals of Applied Biology*, **79**, 215–20.

Keen, N. (1982). Phytoalexins – progress in regulation of their accumulation in gene-for-gene interactions. In: *Plant Infection*, ed. Y. Asada, W. Bushnell, S. Ouchi & C. Vance, pp. 281–99. Tokyo: Japan Scientific Press; Berlin, Heidelberg, New York: Springer-Verlag.

Kuć, J. (1985*a*). Expression of latent genetic information for disease resistance in plants. In: *Cellular and Molecular Biology of Plant Stress*, ed. J.L. Key & T. Kosuge, pp. 303–18, U.C.L.A. Symposium on Molecular and Cellular Biology, New Series, vol. 22. New York: Alan R. Liss, Inc.

Kuć, J. (1985*b*). Increasing crop productivity and value by increasing disease resistance by non-genetic techniques. *Forest Potentials, Productivity, and Value*, pp. 147–90. Weyerhaeuser Science Symposia.

Kuć, J., Barnes, E., Daftsios, A. & Williams, E.B. (1959). The effect of amino acids on susceptibility of apple varieties to scab. *Phytopathology*, **49**, 313–15.

Kuć, J., Shockley, G. & Kearney, K. (1975). Protection of cucumber against *Colletotrichum lagenarium* by *Colletotrichum lagenarium*. *Physiological Plant Pathology*, **7**, 195–9.

Kuć, J. & Preisig, C. (1984). Fungal regulation of disease resistance mechanisms in plants. *Mycologia*, **76**(5), 767–84.

Kuć, J. & Richmond, S. (1977). Aspects of the protection of cucumber against *Colletotrichum lagenarium* by *Colletotrichum lagenarium*. *Phytopathology*, **67**, 533–6.

Kuć, J. & Rush, J.S. (1985). Phytoalexins. *Archives of Biochemistry and Biophysics*, **236**, 455–72.

Kuć, J. & Tuzun, S. (1983). Immunization for disease resistance in tobacco. *Recent Advances in Tobacco Science*, **9**, 179–213.

Lengyel, P. (1982). Biochemistry of interferons and their actions. *Annual Review of Biochemistry*, **51**, 251–82.

Leppik, R., Holloman, D. & Bottomley, W. (1972). Quiesone: an inhibitor of germination of *Peronospora tabacina* conidia. *Phytochemistry*, **11**, 2055–63.

Loebenstein, G. & Gera, A. (1981). Inhibitor of virus replication released from tobacco mosaic virus-infected protoplasts of a local lesion-responding tobacco cultivar. *Virology*, **114**, 132–9.

Loebenstein, G. & Stein, A. (1985). Plant defence responses to viral infections. In: *Cellular and Molecular Biology of Plant Stress*, ed. J.L. Key & T. Kosuge, pp. 413–33, U.C.L.A. Symposia on Molecular and Cellular Biology, New Series, vol. 22. New York: Alan R. Liss, Inc.

Loesch-Fries, L.S., Halk, E.L., Nelson S.E. & Krahn, K.J. (1985). Human leukocyte interferon does not inhibit alfalfa mosaic virus in protoplasts or tobacco tissue. *Virology*, **143**, 626–9.

Lovrekovich, L. & Farkas, G.L. (1965). Induced reaction against wildfire disease in tobacco leaves treated with heat-killed bacteria. *Nature*, **205**, 823–4.

Lucas, J.A., Dolan, T.E. & Coffey, M.D. (1985). Nontransmissibility to regenerants from protected tobacco explants of induced resistance to *Peronospora hyoscyami*. *Phytopathology*, **75**, 1222–5.

MacLennan, D.H., Kúc, J. & Williams, E.B. (1963). Chemotherapy of the apple scab disease with butyric acid derivatives. *Phytopathology*, **53**, 1261–6.

Mauch, F., Hadwiger, L.A. & Boller, T. (1984). Ethylene: symptom, not signal for the induction of chitinase and β-1,3-glucanase in pea pods by pathogens and elicitors. *Plant Physiology*, **76**, 607–11.

McIntrye, J.L. & Dodds, J.A. (1979). Induction of localized and systemic protection against *Phytophthora parasitica* var. *nicotianae* by tobacco mosaic virus infection of tobacco hypersensitive to the virus. *Physiological Plant Pathology*, **15**, 321–30.

McIntyre, J.L., Dodds, A. & Hare, J.D. (1981). Effects of localized infections of *Nicotiana tabacum* by tobacco mosaic virus on systemic resistance against diverse pathogens and an insect. *Phytopathology*, **71**, 297–301.

McIntyre, J.L. & Miller, P.M. (1978). Protection of tobacco against *Phytophthora parasitica* var. *nicotianae* by cultivar-nonpathogenic races, cell-free sonicates and *Pratylenchus penetrans*. *Phytopathology*, **68**, 235–9.

Metraux, J.P. & Boller, T. (1985). Local and systemic activation of chitinase in cucumber plants after fungal, bacterial or viral infection. *Plant Physiology*, **77**(Supplement), 50.

Mills, P.R. & Wood, R.K.S. (1984). The effects of polyacrylic acid, acetyl-salicylic acid and salicylic acid on resistance of cucumber to *Colletotrichum lagenarium*. *Phytopathologische Zeitschrift*, **111**, 209–16.

Mozes, R., Antignus, Y., Sela, I. & Harpaz, I. (1978). The chemical nature of an antiviral factor (AVF) from virus-infected plants. *Journal of General Virology*, **38**, 241–9.

Mukerjee, K., Awasthi, L.P. & Verma, H.N. (1981). The inhibitory activity of an interfering agent, extracted from the leaves of host plants treated with Datura leaf extract, on plant virus infection. *Zeitschrift für Pflanzenkrankheiten und Pflanzenschutz*, **88**(4), 228–34.

Nadolny, L. & Sequeira, L. (1980). Increases in peroxidase are not directly involved in induced resistance in tobacco. *Physiological Plant Pathology*, **16**, 1–8.

Netzer, D., Kritzman, G. & Chet, I (1979). β-1,3-glucanase activity and quantity of fungus in relation to *Fusarium* wilt in resistant and susceptible near-isogenic lines of muskmelon. *Physiological Plant Pathology*, **14**, 47–55.

Nothnagel, E., McNeil, M., Albersheim, P. & Dell, A. (1983). Host-pathogen interactions XXII. A galacturonic acid oligosaccharide from plant cell walls elicits phytoalexins. *Plant Physiology*, **71**, 916–26.

Orchansky, P., Rubinstein, M. & Sela, I. (1982). Human interferons protect plants from virus infection. *Proceedings of the National Academy of Sciences, U.S.A.*, **79**, 2278–80.

Pegg, G.F. (1976). The response of ethylene-treated plants to infection by *Verticillium albo-atrum*. *Physiological Plant Pathology*, **9**, 215–26.

Pegg, G.F. (1977). Glucanohydrolases of higher plants: a possible defence mechanism against parasitic fungi, pp. 305–45. In: *Cell Wall Biochemistry Related to Specificity in Host-Plant Pathogen Interactions*, ed. B. Solheim & J. Raa Universitetsforlaget: Troms, Oslo, Bergen.

Pegg, G.F. & Vessey, J.C. (1973). Chitinase activity in *Lycopersicon esculentum* and its relationship to the *in vivo* lysis of *Verticillium albo atrum* mycelium. *Physiological Plant Pathology*, **3**, 207–22.

Pegg, G.F. & Young, D.H. (1981). Changes in glycosidase activity and their relationship to fungal colonization during infection of tomato by *Verticillium albo-atrum*. *Physiological Plant Pathology*, **19**, 371–82.

Pierpoint, W.S. (1983). Is there a phyto-interferon? *Trends in Biological Science*, **8**, 5–7.

Pueppke, S.G. (1984). Absorption of bacteria to surfaces. In: *Plant-Microbe Interactions. Molecular and Genetic Perspectives*, vol. 1, ed. T. Kosuge & E.W. Nester, pp. 215–61. New York: MacMillan.

Rabenantoandro, Y., Auriol, P. & Touze, A. (1976). Implication of β-(1,3)-glucanase in melon anthracnose. *Physiological Plant Pathology*, **8**, 313–24.

Rahe, J.E. & Kuć, J. (1970). Metabolic nature of the infection-limiting effect of heat on bean anthracnose. *Phytopathology*, **60**, 1005–9.

Rahe, J.E. (1973a). Phytoalexin in the nature of heat-induced protection against bean anthracnose. *Phytopathology*, **63**, 572–7.

Rahe, J.E. (1973b). Occurrence and levels of the phytoalexin phaseollin in relation to delimitation of sites of infection of *Phaseolus vulgaris* by *Colletotrichum lindemuthianum*. *Canadian Journal of Botany*, **51**, 2423–30.

Rahe, J.E., Kuć, J., Chuang, C. & Williams, E.B. (1969). Induced resistance in *Phaseolus vulgaris* to bean anthracnose. *Phytopathology*, **59**, 1641–5.

Reichman, M., Devash, Y., Suhadolnik, R.J. & Sela, I. (1983). Human leukocyte interferon and the antiviral factor (AVF) from virus-infected plants stimulate plant tissues to produce nucleotides with antiviral activity. *Virology*, **128**, 240–4.

Richmond, S., Kuć, J. & Elliston, J.E. (1979). Penetration of cucumber leaves by *Colletotrichum lagenarium* is reduced in plants systemically protected by previous infection with the pathogen. *Physiological Plant Pathology*, **14**, 329–38.

Ride, J.P. (1983). Cell walls and other structural barriers in defence. In: *Biochemical Plant Pathology*, ed. J.A. Callow, pp. 215–36. Chichester: John Wiley & Sons.

Rody, D., Toppan, A. & Esquerré-Tugayé, M-T. (1985). Cell surfaces in plant-microorganism interactions. V. Elicitors of fungal and of plant origin trigger the synthesis of ethylene and of cell wall hydroxyproline-rich glycoprotein in plants. *Plant Physiology*, **77**, 700–4.

Rohloff, H. & Leach, B. (1977). Soluble leaf proteins in virus-infected plants and acquired resistance. I. Investigation in *Nicotiana tabacum* cvs. 'Xanthi-n.c.' and 'Samsun'. *Phytopathologische Zeitschrift*, **89**, 306–16.

Rosenberg, N., Pines, M. & Sela, I. (1982). Adenosine 3' : 5'-cyclic monophosphate – its release in higher plant by an exogenous stimulus as detected by radioimmunoassay. *FEBS Letters*, **137**, 105–7.

Rosenberg, N., Reichman, M., Gera, A. & Sela, I. (1985). Antiviral activity of natural and recombinant human leukocyte interferons in tobacco protoplasts. *Virology*, **140**, 173–8.

Ross, A.F. (1961). Systemic acquired resistance induced by localized virus infections in plants. *Virology*, **14**, 340–58.

Ross, A.F. (1966). Systemic effects of local lesion formation. In: *Viruses of Plants*, ed. B.R. Beemstra & J. Dykstra, pp. 127–50. Amsterdam: North Holland Publishing Co.

Ryan, C.A. (1978). Proteinase inhibitors in plant leaves: a biochemical model for natural plant protection. *Trends in Biological Science*, **5**, 148–50.

Ryan, C.A. (1984). Systemic responses to wounding. In: *Plant-Microbe Interactions: Molecular and Genetic Perspectives*, vol. 1, ed. T. Kosuge & E.W. Nester, pp. 307–20. New York: MacMillan.

Ryan, C.A., Bishop, P.D., Walker-Simmons, M., Brown, W.E. & Graham, J.S. (1985). Pectic fragments regulate the expression of proteinase inhibitor genes in plants. In: *Cellular and Molecular Biology of Plant Stress*, ed. J.L. Key & T. Kosuge, pp. 319–34. U.C.L.A. Symposia on Molecular & Cellular Biology, New Series, vol. 22. New York: Alan R. Liss, Inc.

Salt, S.D., Tuzun, S. & Kuć, J. (1986). Effects of ionone-related compounds on the growth of tobacco and resistance to blue mould. Mimicry of effects of stem infection by *Peronospora tabacina* Adam. *Physiological Plant Pathology* **28**, 287–97.

Sela, I. (1981). Plant-virus interactions related to resistance and localization of viral infections. *Advances in Virus Research*, **26**, 201–37.

Sela, I. & Applebaum, S.W. (1962). Occurrence of antiviral factor in virus- infected plants. *Virology*, **17**, 543–8.

Sequeira, L. (1983). Mechanisms of induced resistance in plants. *Annual Review of Microbiology*, **37**, 51–79.

Sequeira, L., Gaard, G. & Dezoeten, G.A. (1977). Interaction of bacteria and host cell walls: its relation to mechanisms of induced resistance. *Physiological Plant Pathology*, **10**, 43–50.

Sharp, J.K., McNeil, M. & Albersheim, P. (1984). The primary structures of one elicitor-active and seven elicitor-inactive Hexa (β-D-glucopyranosyl)-D-glucitols isolated from the mycelial walls of *Phytophthora megasperma* f.sp. *glycinea*. *Journal of Biological Chemistry*. **259**, 11321–6.

Shimura, M., Iwata, M., Tashiro, N., Sekizawa, Y., Suzuki, Y., Mase, S. & Watanabe, T. (1981). Anti-conidial germination factors induced in the presence of probenazole in infected host leaves. I. Isolation and properties of four active substances. *Agricultural and Biological Chemistry*, **45**, 1431–5.

Showalter, A.M., Bell, J.N., Cramer, C.L., Bailey, J.A., Varner, J.E. & Lamb, C.J. (1985). Accumulation of hydroxyproline-rich glycoprotein mRNAs in response to fungal elicitor and infection. *Proceedings of the National Academy of Science, U.S.A.*, **82**, 6551–5.

Skipp, R. & Deverall, B.J. (1973). Studies on cross protection in the anthracnose disease of bean. *Physiological Plant Pathology*, **3**, 299–313.

Staub, T. & Kuć, J. (1980). Systemic protection of cucumber plants against disease caused by *Cladosporium cucumerinum* and *Colletotrichum lagenarium* by prior localized infection with either fungus. *Physiological Plant Pathology*, **17**, 389–93.

Stein, A. & Loebenstein, G. (1970). Induction of resistance to tobacco mosaic virus by poly I:poly C in plants. *Nature*, **226**, 363–4.

Stein, A. & Loebenstein, G. (1976). Peroxidase activity in tobacco plants with polyanion-induced interference to tobacco mosaic virus. *Phytopathology*, **66**, 1192–4.

Stermer, B.A. & Hammerschmidt, R. (1984). Heat shock induces resistance to *Cladosporium cucumerinum* and enhances peroxidase activity in cucumbers. *Physiological Plant Pathology*, **25**, 239–49.

Stermer, B.A. & Hammerschmidt, R. (1985). The induction of resistance by heat shock. In: *Cellular and Molecular Biology of Plant Stress*, ed. J.L. Key & T. Kosuge, pp. 291–302. U.C.L.A. Symposia on Molecular and Cellular Biology, New Series, vol. 22. New York: Alan R. Liss, Inc.

Tas, P.W.L. & Peters, D. (1977). The occurrence of a soluble protein (E_1) in cucumber cotyledons infected with plant viruses. *Netherlands Journal of Plant Pathology*. **83**, 5–12.

Toppan, A. & Esquerré-Tugayé, M-T. (1984). Cell surfaces in plant-microorganism interactions IV. Fungal glycopeptides which elicit the synthesis of ethylene in plants. *Plant Physiology*, **75**, 1133–8.

Toppan, A. & Rody, D. (1982). Activite chitinasique de plants de melon infectees par *Colletotrichum lagenarium* ou traitees par l'ethylene. *Agronomie*, **2**, 829–34.

Toppan, A., Rody, E. & Esquerré-Tugayé, M-T. (1982). Cell surfaces in plant-microorganism interactions III. *In vivo* effects of ethylene on hydroxyproline-rich glycoproteins in the cell wall of diseased plants. *Plant Physiology*, **70**, 82–6.

Touzé, A. & Rossingnol, M. (1977). Lignification and the onset of premunition in muskmelon plants. In: *Cell Wall Biochemistry Related to Specificity in Host-Plant Pathogen Interactions*, ed. B. Solheim & J. Raa, pp. 289–93. Universitetsforlaget Oslo, Norway.

Tuzun, S. & Kuć, J. (1985a). Movement of a factor in tobacco infected with *Peronospora tabacina* Adam. which systemically protects against blue mould. *Physiological Plant Pathology*, **26**, 321–30.

Tuzun, S. & Kuć, J. (1985b). Transfer of induced resistance in tobacco to blue mould (*Peronospora tabacina*Adam.) via callus. *Phytopathology*, **75**, 1304.

Tuzun, S., Nesmith, W. & Kuć, J. (1984). The effect of stem injections with *Peronospora tabacina* and metalaxyl treatment on growth of tobacco and protection against blue mould in the field. *Phytopathology*, **74**, 804.

Vance, C.P., Kirk, T.K. & Sherwood, R.T. (1980). Lignification as a mechanism of disease resistance. *Annual Review of Phytopathology*, **18**, 259–88.

Van Loon, L.C. (1975). Polyacrylamide disk electrophoresis of the soluble leaf proteins from *Nicotiana tabacum* var. 'Samsun' and 'Samsun NN'. III. Influence of temperature and virus strain on changes induced by tobacco mosaic virus. *Physiological Plant Pathology*, **6**, 289–300.

Van Loon, L.C. & Van Kammen, A. (1970). Polyacrylamide disc electrophoresis of the soluble leaf proteins from *Nicotiana tabacum* var. 'Samsun NN' II. Changes in protein constitution after infection with tobacco mosaic virus. *Virology*, **40**, 199–211.

Van Sambeck, J.W. & Pickard, B.G. (1976). Mediation of rapid electrical, metabolic, transpirational, and photosynthetic changes by factors released from wounds. I. Variation potentials and putative action potentials in intact plants. II. Mediation of the variation potential by Ricca's factor. III. Measurements of CO_2 and H_2O fluxes. *Canadian Journal of Botany*, **54**, 2642–71.

Verma, H.N. & Awasthi, L.P. (1979). Antiviral activity of *Boerhaavia diffusa* root extract and the physical properties of the virus inhibitor. *Canadian Journal of Botany*, **57**, 926–32.

Verma, H.N. & Awasthi, L.P. (1980). Occurrence of highly antiviral agent in plants treated with *Boerhaavia diffusa* inhibitor. *Canadian Journal of Botany*, **58**, 2141–4.

Verma, H.N., Awasthi, L.P. & Saxena, K.C. (1979). Isolation of the virus inhibitor from the root extract of *Boerhaavia diffusa* inducing systemic resistance in plants. *Canadian Journal of Botany*, **57**, 1214–17.

Walker-Simmons, M., Hadwiger, L. & Ryan, C. (1983). Chitosans and pectic polysaccharides both induce the accumulation of the antifungal phytoalexin pisatin in pea pods and anti-nutrient proteinase inhibitors in tomato leaves. *Biochemical and Biophysical Research Communications*, **110**, 194–9.

Wargo, P.M. (1974). Lysis of the cell wall of *Armillaria mellea* by enzymes from forest trees. *Physiological Plant Pathology*, **5**, 99–105.

Weiel, J. & Hapner, K. (1976). Barley proteinase inhibitors. A possible role in grasshopper control? *Phytochemistry*, **15**, 1885–7.

White, R.F. (1979). Acetylsalicylic acid (aspirin) induces resistance to tobacco mosaic virus in tobacco. *Virology*, **99**, 410–12.

Young, D.H. & Pegg, G.F. (1982). The action of tomato and *Verticillium albo-atrum* glycosidases on the hyphal wall of *V. albo-atrum*. *Physiological Plant Pathology*, **21**, 411–25.

Index